Bacterial Glycomics

Current Research, Technology and Applications

Edited by

Christopher W. Reid
Department of Science and Technology
Bryant University
Smithfield, RI
USA

Susan M. Twine
National Research Council Canada
Institute for Biological Sciences
Ottawa, ON
Canada

Anne N. Reid
Cumberland, RI
USA

Caister Academic Press

Caister Academic Press
Norfolk, UK

www.caister.com

British Library Cataloguing-in-Publication Data
A catalogue record for this book is available from the British Library

ISBN: 978-1-904455-95-0

Cover design adapted from Figure 2.2

Printed and bound in Great Britain

Contents

Contributors

Luke J. Alderwick
School of Biosciences
University of Birmingham
Edgbaston
Birmingham
UK

l.alderwick@bham.ac.uk

Sarah M. Batt
School of Biosciences
University of Birmingham
Edgbaston
Birmingham
UK

s.batt@bham.ac.uk

Gurdyal Singh Besra
School of Biosciences
University of Birmingham
Edgbaston
Birmingham
UK

g.besra@bham.ac.uk

Catherine Bougault
Institut de Biologie Structurale
UMR5075 (CEA/CNRS/UJF)
Grenoble
France

catherine.bougault@ibs.fr

Anthony J. Clarke
Department of Molecular and Cellular Biology
University of Guelph
Guelph, ON
Canada

aclarke@uoguelph.ca

Chelsea A. Clarke
Department of Molecular and Cellular Biology
University of Guelph
Guelph;
Department of Microbiology and Immunology
University of Western Ontario
London, ON
Canada

cclar3@uwo.ca

Stuart J. Cordwell
School of Molecular Bioscience
The University of Sydney
Sydney, NSW
Australia

stuart.cordwell@sydney.edu.au

Jon Cuccui
London School of Hygiene and Tropical Medicine
London
UK

jon.cuccui@lshtm.ac.uk

Leslie Cuthbertson
Department of Biochemistry and Biomedical Sciences and M.G. DeGroote Institute for Infectious Disease Research
McMaster University
Hamilton, ON
Canada

cuthber@mcmaster.ca

Miroslaw Cygler
Biotechnology Research Institute
National Research Council Canada
Department of Biochemistry
McGill University
Montreal, QC
Canada

mirek.cygler@cnrc-nrc.gc.ca

Danielle H. Dube
Department of Chemistry & Biochemistry
Bowdoin College
Brunswick, ME
USA

ddube@bowdoin.edu

Klaus Futterer
School of Biosciences
University of Birmingham
Edgbaston
Birmingham
UK

k.futterer@bham.ac.uk

Sabine Hediger
Laboratoire de Chimie Inorganique et Biologique
UMR-E3 (CEA/UJF), FRE3200 (CEA/CNRS)
Institut Nanosciences et Cryogénie, CEA
Grenoble
France

sabine.hediger@cea.fr

Ken F. Jarrell
Department of Microbiology and Immunology
Queen's University
Kingston, ON
Canada

jarrellk@queensu.ca

John F. Kelly
Institute for Biological Sciences
National Research Council Canada
Ottawa, ON
Canada

john.kelly@nrc-cnrc.gc.ca

Rebecca H. Langdon
London School of Hygiene and Tropical Medicine
London
UK

r.h.langdon@talk21.com

Susan M. Logan
Institute for Biological Sciences
National Research Council Canada
Ottawa, ON
Canada

susan.logan@nrc-cnrc.gc.ca

Allan Matte
Biotechnology Research Institute
National Research Council Canada
Montreal, QC
Canada

allan.matte@nrc-cnrc.gc.ca

Arun K. Mishra
School of Biosciences
University of Birmingham
Edgbaston
Birmingham
UK

a.k.mishra@bham.ac.uk

Madeleine G. Moule
London School of Hygiene and Tropical Medicine
London
UK

madeleine.moule@lshtm.ac.uk

Patrick J. Moynihan
Department of Molecular and Cellular Biology
University of Guelph
Guelph, ON
Canada

pmoyniha@uoguelph.ca

John M. Pfeffer
Department of Molecular and Cellular Biology
University of Guelph
Guelph, ON
Canada

jpfeffer@uoguelph.ca

Robert A. Pon
Institute for Biological Sciences
National Research Council Canada
Ottawa, ON
Canada

rob.pon@nrc-cnrc.gc.ca

Anne N. Reid
Cumberland, RI
USA

anne.n.reid@gmail.com

Christopher W. Reid
Department of Science and Technology
Bryant University
Smithfield, RI
USA

creid@bryant.edu

Ian C. Schoenhofen
Institute for Biological Sciences
National Research Council Canada
Ottawa, ON
Canada

ian.schoenhofen@nrc-cnrc.gc.ca

Nichollas E. Scott
School of Molecular Bioscience
The University of Sydney
Sydney, NSW
Australia

nichollas.scott@syndey.edu.au

Jean-Pierre Simorre
Institut de Biologie Structurale
UMR5075 (CEA/CNRS/UJF)
Grenoble
France

simorre@ibs.fr

Traian Sulea
Biotechnology Research Institute
National Research Council Canada
Montreal, QC
Canada

traian.sulea@cnrc-nrc.gc.ca

Susan M. Twine
Institute for Biological Sciences
National Research Council Canada
Ottawa, ON
Canada

susan.twine@nrc-cnrc.gc.ca

Chris Vandenende
Department of Molecular and Cellular Biology
University of Guelph
Guelph, ON
Canada

cvandene@uoguelph.ca

Warren Wakarchuk
Institute for Biological Sciences
National Research Council Canada
Ottawa, ON
Canada

warren.wakarchuk@nrc.gc.ca

Brendan W. Wren
London School of Hygiene and Tropical Medicine
London
UK

brendan.wren@lshtm.ac.uk

N. Martin Young
Institute for Biological Sciences
National Research Council Canada
Ottawa, ON
Canada

martin.young@nrc-cnrc.gc.ca

Preface

Carbohydrates constitute the most structurally diverse class of natural products, and can serve many functions in cells and organisms. Bacteria can synthesize a diverse array of glycan moieties, that are in many cases distinct from the carbohydrates elaborated by eukaryotes. Bacterial glycans may be found attached to proteins and lipids, or as loosely associated polysaccharides on the cell surface (Fig. 1). Cell-surface polysaccharides represent the predominant structures on all bacterial cells. These structures are often important players in the interactions between pathogens, their hosts and the environment. In many cases, the precise role of these glycan moieties in bacterial symbiosis, pathogenesis and cell–cell interactions are just beginning to be understood.

The glycome can be loosely defined as the entire complement of glycan structures within a cell. The study of the glycome has led to the emerging field of glycomics, which is the study of the significance of glycoconjugate assembly and expression in biological systems. Because glycans play such diverse roles in bacterial physiology, the study of bacterial glycomics is of relevance to carbohydrate experts, microbiologists, immunologists, as well as researchers in many other fields of life sciences. As a field, glycomics has lagged behind genomics and proteomics, owing to the inherent difficulties in the isolation of glycans, the subsequent analysis of their structure and function and the limited number of suitable analytical techniques available for their study (Fig. 2). The importance of glycomics research is perhaps best illustrated by examining potential applications of these findings. For instance, a comprehensive understanding of the bacterial glycome will undoubtedly shed light on mechanisms of bacterial pathogenesis, and could lead to the development of novel drugs, bioactive glycans and glycoconjugate vaccines. Bacterial glycomics is a rapidly growing field, whose progress will demand important improvements to existing analytical methodologies as well as the development of new and innovative approaches for glycan isolation, characterization and synthesis.

This book aims to provide the reader with an overview of our current understanding of bacterial glycomes. To this end, the book is divided into three sections. The first section includes overviews of several microbial glycoconjugates, and covers lipopolysaccharide, capsular polysaccharide, and lipoarabinomannan biosynthesis, cell wall metabolism, and glycosylation of bacterial and archaeal proteins. The second section provides an overview of analytical approaches used in the characterization of the bacterial glycome. The final section describes a few applications of bacterial glycomics, including metabolic oligosaccharide labelling and the synthesis of bioactive glycans and glycoconjugate vaccines.

Christopher W. Reid
Susan M. Twine
Anne N. Reid

Reference

Reid, C.W., Fulton, K.M., and Twine, S.M. (2010). Never take candy from a stranger: the role of the bacterial glycome in host–pathogen interactions. Future Microbiol. 5, 267–288.

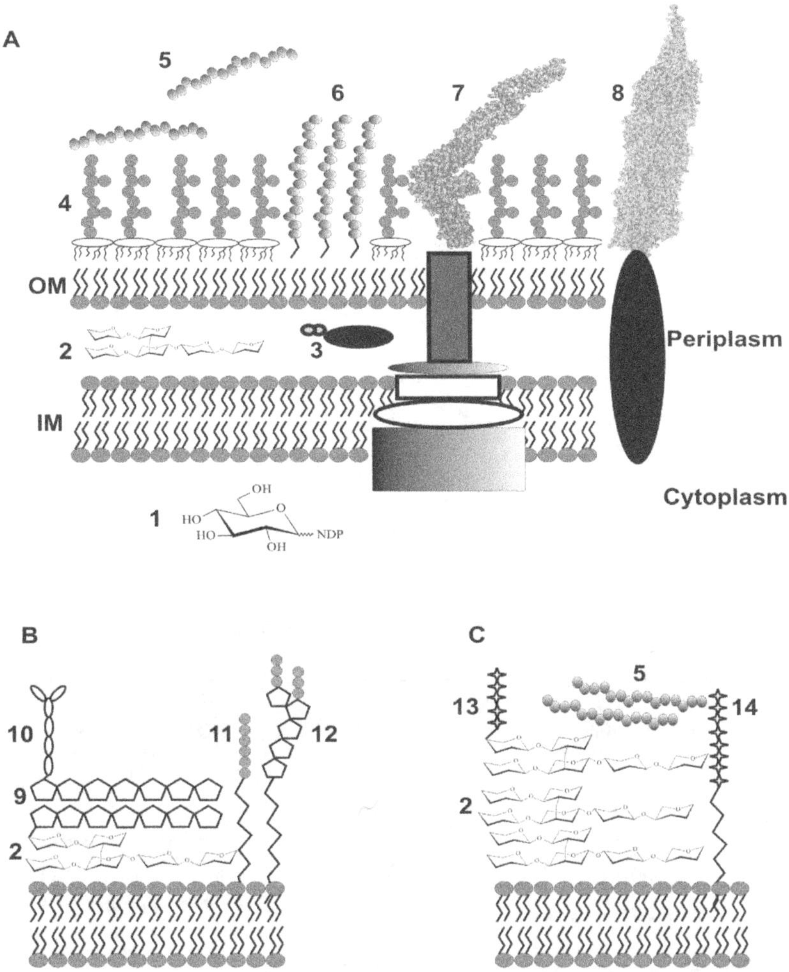

Figure 1 Schematic representation of microbial glycoconjugates. (A) The Gram-negative cell showing the various components that constitute the glycome: (1) nucleotide-linked sugar precursors; (2) peptidoglycan; (3) *N-linked* glycoproteins; (4) lipopolysaccharide/oligosaccharide; (5) extracellular polysaccharide; (6) capsular polysaccharide; (7) *O-linked* protein glycosylation, FliC, *Salmonella enterica*; (8) pilin, *Neisseria gonorrhoeae*. (B) The mycobacterial glycome: in addition to possessing nucleotide-linked sugar precursors and peptidoglycan, the cell surface contains: (9) arabinogalactan, (10) mycolic acids, (11) lipomannan and (12) lipoarabinomannan. (C) The Gram-positive cell produces many of the same glycoconjugates as the Gram-negative cell, such as capsular polysaccharide, extracellular polysaccharide and a thicker peptidoglycan layer but, in addition, produces the cell surface components (13) teichoic and (14) lipoteichoic acids. IM, inner membrane; OM, outer membrane (Reid *et al.*, 2010).

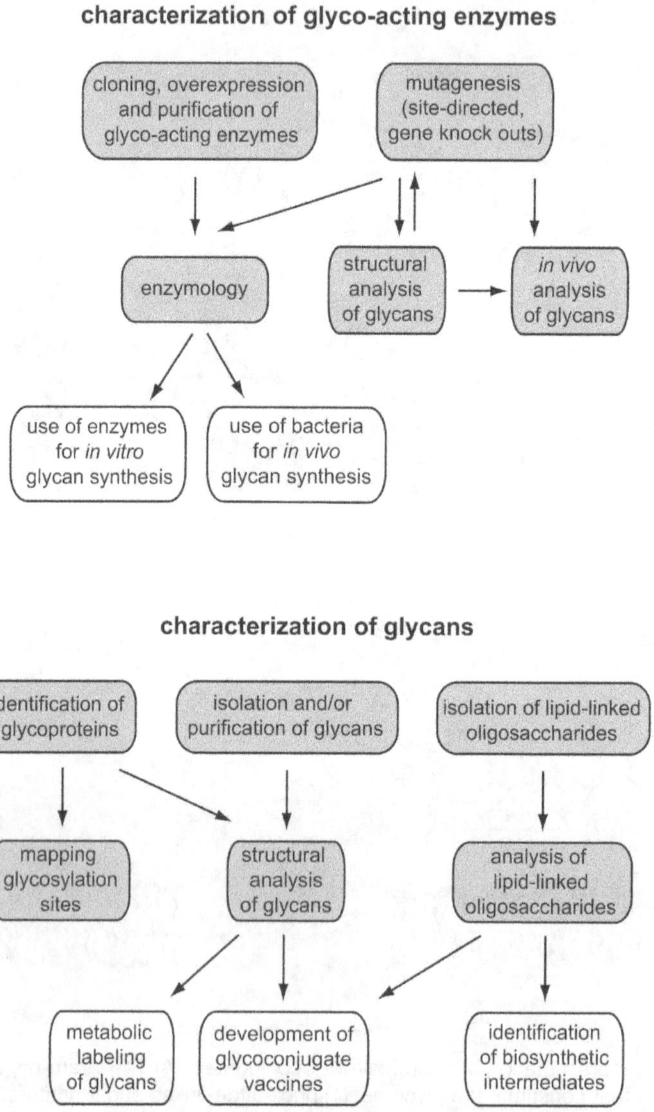

Figure 2 Glycomics research workflow. Possible steps towards the isolation and characterization of glycans and glyco-acting enzymes are presented (grey shading), as well as some practical applications that could potentially derive from such studies (white shading).

Lipopolysaccharide Biosynthesis

1

Leslie Cuthbertson

Abstract

Lipopolysaccharide (LPS) constitutes the major portion of the outer leaflet of the outer membrane and plays a major role in the physiology of Gram-negative bacteria. LPS can be divided into three structurally distinct regions: lipid A, core oligosaccharide and O-antigenic polysaccharide. Each of these regions, as well as regulated modifications, is important in the overall functions of the LPS molecule. Synthesis of lipid A and the core oligosaccharide occurs in the cytoplasm and is separate from that of the O-antigenic polysaccharide. These two portions of the LPS molecule are then ligated in the periplasm prior to transport to the outer membrane. This chapter will describe the structure and cytoplasmic synthesis of LPS, modifications to these structures regulated by environmental conditions or phage-encoded genes, and the transfer of LPS to its final destination at the cell surface.

The importance of LPS in bacterial physiology

Lipopolysaccharide (LPS) constitutes the majority of the outer leaflet of the outer membrane in Gram-negative bacteria (Nikaido, 2003). The LPS molecule can be divided into three structural regions: lipid A, core oligosaccharide (core-OS) and O-antigenic polysaccharide (O-PS). An LPS molecule consisting of lipid A and core-OS is termed rough or R-LPS. When substituted with O-PS the LPS molecule is termed smooth or S-LPS. Most laboratory strains of *Escherichia coli* express R-LPS owing to a mutation in the O-PS biosynthesis locus (Liu and Reeves, 1994). The minimum LPS molecule required for survival of *E. coli* K-12 under normal laboratory conditions consists of lipid A and the first two 3-deoxy-D-*manno*-oct-2-ulosonic acid (Kdo) residues of the core-OS (discussed below). This was once believed to be the minimum requirement for all Gram-negative bacteria but it is now clear that some bacteria (e.g. *Yersinia pestis*) are viable when expressing more minimal LPS structures (Tan and Darby, 2005) and at least two organisms, *Moraxella catarrhalis* and *Neisseria meningitidis*, are able to survive without LPS in the outer membrane (Steeghs *et al.*, 1998; Bos and Tommassen, 2005; Peng *et al.*, 2005). Why these organisms are able to survive without LPS is still a subject of debate.

Lipid A serves as the hydrophobic anchor for the LPS molecule and elicits production of immunomodulatory molecules from macrophages and endothelial cells and is responsible for the septic shock that often accompanies a Gram-negative infection (Trent *et al.*, 2006; Raetz *et al.*, 2007). Owing to this property of lipid A, the terms LPS and endotoxin are often used synonymously. The core-OS is distal to lipid A and serves as the attachment point for O-PS. Negatively charged residues in lipid A and the core-OS are required for stabilizing the outer membrane through interactions with divalent cations (Frirdich and Whitfield, 2005). In many bacteria, O-PS has been implicated in protection of the bacterium against serum killing by the complement system. It does so by shielding underlying outer membrane proteins from serum components and may provide an alternative site for the formation of the membrane attack complex at a distance from the cell membrane (Joiner, 1988; Reeves, 1995; Albertí *et al.*, 1996). LPS is not a static structure. Regulated modifications and phase variation play a role in LPS diversity and are important for some of its physiological functions.

Lipid A biosynthesis

Diversity of lipid A structures

Lipid A is responsible for anchoring the LPS molecule in the outer membrane. Fig. 1.1 shows the structure of the minimum unmodified lipid A molecule from *E. coli* K-12. It consists of a β-linked diglucosamine backbone, phosphorylated at the 1′ and 4′ positions, and six lauroyl and myristoyl residues. While lipid A is the most highly conserved portion of the LPS molecule, diversity is seen among Gram-negative bacteria and contributes to varying immune stimulating effects (Kabanov and Prokhorenko, 2010). The diverse lipid A structures of *Pseudomonas aeruginosa*, *Helicobacter pylori* and *Campylobacter jejuni* are shown in Fig. 1.1 for comparison with that of unmodified lipid A from *E. coli*. Differences exist in the number, length and linkage of the acyl chains and the number of Kdo and phosphate residues.

Lipid A biosynthesis: the Raetz pathway

The lipid A biosynthetic pathway consists of both conserved and variable components. The conserved or constitutive pathway for lipid A biosynthesis, also referred to as the Raetz pathway, occurs in the cytoplasm and inner face of the

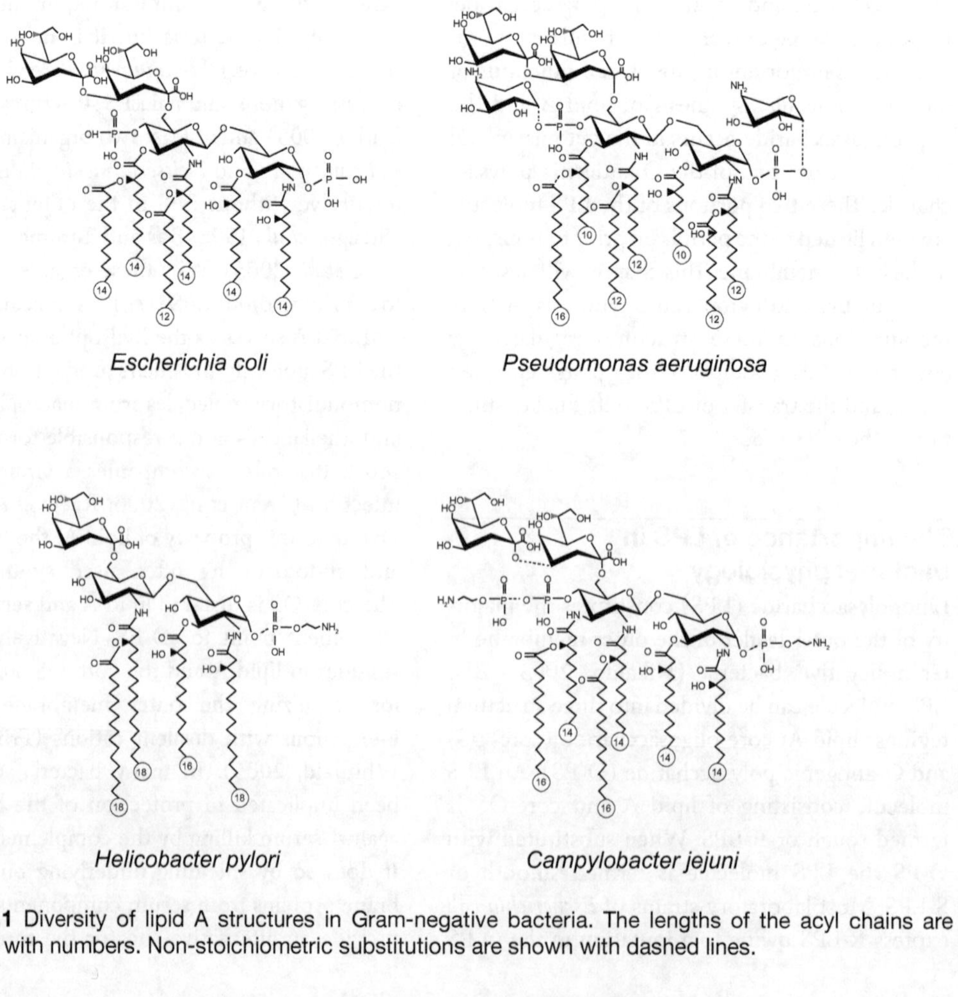

Figure 1.1 Diversity of lipid A structures in Gram-negative bacteria. The lengths of the acyl chains are indicated with numbers. Non-stoichiometric substitutions are shown with dashed lines.

inner membrane, while the variable components are typically regulated and occur extracytoplasmically.

Nine enzymes constitute the conserved biosynthetic pathway for lipid A synthesis in *E. coli* (Fig. 1.2) (Raetz and Whitfield, 2002; Trent *et al.*, 2006; Raetz *et al.*, 2007, 2009). The initial steps in lipid A biosynthesis are catalysed by the soluble enzymes LpxA, LpxC and LpxD. LpxA, the first enzyme in lipid A synthesis, transfers a fatty acyl chain to UDP-*N*-acetylglucosamine (GlcNAc) to create UDP-monoacyl-GlcNAc (Anderson and Raetz, 1987; Coleman and Raetz, 1988). The equilibrium constant for UDP-GlcNAc acylation is unfavourable (Anderson *et al.*, 1993), making the second step in lipid A synthesis, catalysed by LpxC, the first committed step in lipid A biosynthesis. LpxC is responsible for the de-*N*-acetylation of UDP-monoacyl-GlcNAc (Young *et al.*, 1995) to provide a free amino group for addition of a second acyl chain catalysed by LpxD to yield UDP-2,3-diacyl-glucosamine (GlcN) (Kelly *et al.*, 1993).

Two peripheral membrane proteins, LpxH and LpxB, catalyse the next two steps of lipid A biosynthesis. LpxH cleaves a portion of the UDP-2,3-diacyl-GlcN pool at the pyrophosphate bond, releasing UMP and producing lipid X (Babinski *et al.*, 2002a,b). The discovery of lipid X represents a crucial step in the elucidation of the lipid A biosynthetic pathway (Takayama *et al.*, 1983). LpxB then condenses one molecule of UDP-2,3-diacyl-GlcN with one molecule of lipid X to produce the β-1′,6-linked disaccharide (Crowell *et al.*, 1986). Many Gram-negative bacteria (e.g. *Caulobacter crescentus*) lack the *lpxH* gene found in *E. coli*. In these organisms, LpxI performs the same function as LpxH although the two enzymes are not related (Metzger and Raetz, 2010).

The remaining steps in lipid A biosynthesis are catalysed by the integral membrane proteins LpxK, WaaA (formerly KdtA), LpxL and LpxM. LpxK is responsible for the production of an important intermediate in lipid A biosynthesis, lipid-IV_A, through phosphorylation of the β-1′,6-linked disaccharide (Ray and Raetz, 1987; Garrett *et al.*, 1997). In *E. coli*, the first step in core-OS biosynthesis, the addition of two Kdo residues by WaaA, occurs next resulting in the formation of Kdo_2-lipid-IV_A (Clementz and Raetz, 1991). Lipid A biosynthesis is then completed by the

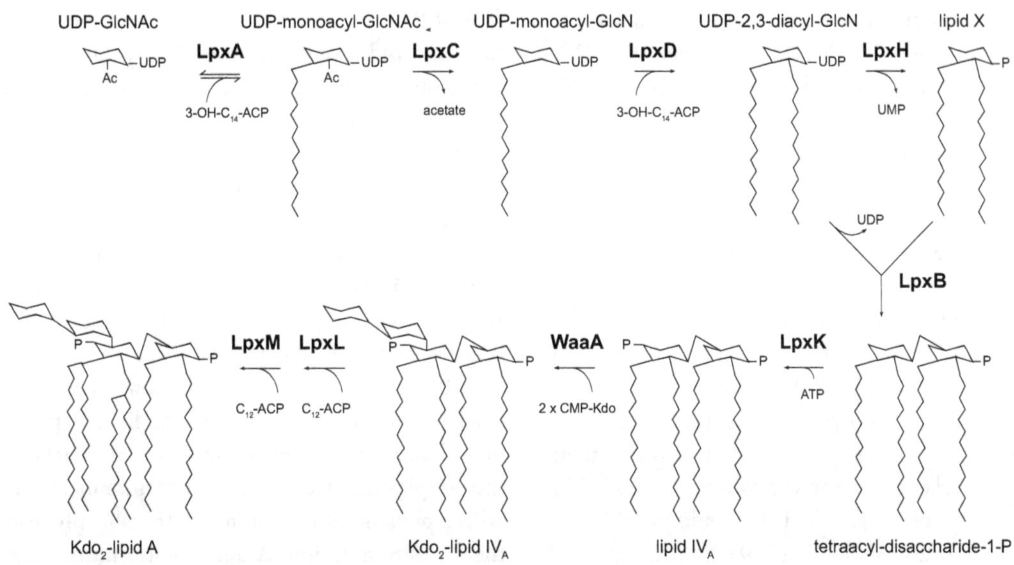

Figure 1.2 The conserved pathway for lipid A biosynthesis.

action of two acyltransferases LpxL and LpxM (Rick *et al.*, 1977, 1985; Brozek and Raetz, 1990).

In *Salmonella*, LpxO is responsible for the hydroxylation of lipid A (Gibbons *et al.*, 2000). This modification occurs in the cytosol and is found under normal laboratory conditions but is not present in *E. coli*. In *E. coli*, approximately 30% of the lipid A molecules contain a diphosphate at the 1 position (Zhou *et al.*, 1999). Addition of this phosphate residue occurs on the outer leaflet of the inner membrane and is catalysed by LpxT, which uses undecaprenol pyrophosphate (Und-PP) as the phosphate donor (Touze *et al.*, 2008).

Despite its conservation, species-specific differences in lipid A structure are observed. Many of these differences are a result of the substrate preferences of the acyltransferases LpxA, LpxD, LpxL and LpxM. For example, in *E. coli*, LpxA shows preference for C_{14} hydroxyacyl chains (Williamson *et al.*, 1991; Wyckoff *et al.*, 1998) whereas the LpxA homologues of other bacteria show preference for C_{10} (e.g. *P. aeruginosa*) (Williamson *et al.*, 1991; Wyckoff *et al.*, 1998) or C_{12} chains (e.g. *N. meningitidis*) (Odegaard *et al.*, 1997; Sweet *et al.*, 2004b).

Certain bacteria (e.g. *C. jejuni*, *Leptospira interrogans*) produce a lipid A species where one or more of the GlcN residues is replaced with 2,3-diamino-2,3-dideoxy-D-glucopyranose (GlcNAc3N) (Moran *et al.*, 1991a,b). Substitution occurs at the level of the sugar-nucleotide precursor where UDP-GlcNAc3N replaces UDP-GlcNAc as the LpxA substrate. UDP-GlcNAc3N is synthesized from UDP-GlcNAc by the action of two enzymes GnnA and GnnB (Sweet *et al.*, 2004a,b). It is thought that an increased number of amide-linked fatty acyl chains may improve stability of the outer membrane in harsh conditions (Trent *et al.*, 2006).

In many species including *Haemophilus influenzae* and *Bordetella pertussis* the Kdo transferase WaaA transfers only one Kdo residue to lipid-IV_A, as opposed to the two Kdo residues transferred in *E. coli*. In these species, the kinase KdkA then phosphorylates the same position on the Kdo residue where the second Kdo would be added in *E. coli* (White *et al.*, 1997, 1999). In other species, including *H. pylori*, the WaaA enzyme transfers two Kdo residues with one being subsequently removed by a Kdo hydrolase at the outer face of the inner membrane (Stead *et al.*, 2010).

Modifications to lipid A

Modifications to lipid A are involved in adaptive responses during infection and may provide resistance to cationic antimicrobial peptides or alter the recognition of LPS by the immune system (Trent *et al.*, 2006; Raetz *et al.*, 2007). LPS modifications are probably best characterized in *E. coli* and *Salmonella* and many of these modifications are regulated by the PhoP–PhoQ and PmrA–PmrB two component systems which respond to conditions probably experienced within the host (Gunn, 2008; Kato and Groisman, 2008; Prost and Miller, 2008). The PhoP–PhoQ system is activated at low concentrations of divalent cations and increasing concentrations of antimicrobial peptides while the PmrA–PmrB system responds to extracellular iron (Garcia Vescovi *et al.*, 1996; Wösten *et al.*, 2000). In *Salmonella*, these systems are integrated through PmrD, which is inactive in *E. coli* (Winfield and Groisman, 2004). In other organisms such as *Helicobacter* and *Francisella*, modifications to the conserved lipid A structure found in *E. coli* occur under normal laboratory conditions (Trent *et al.*, 2006; Raetz *et al.*, 2007).

In contrast to the conserved constitutive pathway for lipid A biosynthesis, which occurs primarily within the cytoplasm and at the inner face of the inner membrane, regulated modifications to lipid A and the core-OS typically occur at the outer face of the inner membrane or in the outer membrane. The localization of LPS modification enzymes has proven useful in studies of LPS export (discussed below; Ruiz *et al.*, 2009).

Addition of 4-amino-4-deoxy-L-arabinose (L-Ara4N) and phosphoethanolamine occurs at the periplasmic face of the inner membrane. ArnT is the enzyme responsible for L-Ara4N transfer to the 4′ phosphate of lipid A utilizing Und-P-L-Ara4N as the L-Ara4N donor (Gunn *et al.*, 1998; Trent *et al.*, 2001b). The lipid A phosphoethanolamine transferase EptA also acts at the periplasmic face of the inner membrane. EptA adds a phosphoethanolamine to the phosphate at the 1 position of lipid A but can also modify the 4′ phosphate under certain conditions (Lee *et al.*, 2004).

Addition of acyl chains to lipid A occurs within the outer membrane and is catalysed by PagP (Bishop *et al.*, 2000). In *Salmonella*, but not *E. coli* K-12, two lipases, LpxR and PagL, are responsible for the removal of the acyl chain at the 3′ and 3 positions of lipid A respectively (Trent *et al.*, 2001a; Reynolds *et al.*, 2006). Some bacteria possess a fifth cytoplasmic acyltransferase, LpxP, which is induced during cold shock (Carty *et al.*, 1999).

Core oligosaccharide synthesis

Diversity of core structures

The core-OS is located distal to lipid A and is more structurally diverse. The core-OS can be further subdivided into the inner and outer core-OS based on structural features of the two regions. The inner core-OS is typically more conserved and contains Kdo and L-*glycero*-D-*manno*-heptopyranose (Hep) while the outer core-OS contains hexose residues with the composition and linkages dependent on the specific core type (Holst and Brade, 1992; Holst, 1999, 2002, 2007). In *E. coli* five core types, K-12, R1, R2, R3 and R4, have been identified (Fig. 1.3). The R1 core type is the most prevalent while an R3 core is associated with verotoxigenic *E. coli* (Amor *et al.*, 2000). In *Salmonella enterica* and *Klebsiella pneumoniae* two and three core types have been identified respectively (Olsthoorn *et al.*, 1998; Holst, 2007). Both the inner and outer core-OS can be modified with non-stoichiometric substitutions including rhamnose (Rha), galactose (Gal), GlcN, GlcNAc, Kdo, phosphate and phosphoethanolamine.

Core oligosaccharide biosynthesis

Core-OS biosynthesis takes place at the cytoplasmic face of the inner membrane (Schnaitman and Klena, 1993; Heinrichs *et al.*, 1998c). The first step of inner core-OS biosynthesis in *E. coli* is the transfer of two Kdo residues to lipid-IV_A by WaaA and is required for completion of lipid A (Figs. 1.2 and 1.3A) (Clementz and Raetz, 1991). WaaC and WaaF act next, each transferring a Hep residue (Wilkinson *et al.*, 1972; Sirisena *et al.*, 1992). WaaP then phosphorylates HepI prior to addition of HepIII by WaaQ and phosphorylation of HepII by WaaY (Yethon *et al.*, 2000). Further non-stoichiometric additions to the inner core-OS are core-type specific. In *E. coli* K-12, two glycosyltransferases, WaaS and WaaZ, are responsible for modification of the outer Kdo residue with Rha and Kdo respectively (Frirdich *et al.*, 2003; Frirdich and Whitfield, 2005). Overexpression of WaaZ controls the amount of full-length core-OS synthesized, and therefore available for ligation with O-PS, which may play a role in virulence (Muller-Loennies *et al.*, 2003). Expression of WaaY and WaaZ is induced by superoxide stress and salicylate through the global regulators SoxS and MarA (Lee *et al.*, 2009). Truncation of the core oligosaccharide is also seen under conditions of PagP activation (Smith *et al.*, 2008). Increasingly, modifications to the core-OS are being found to be under regulatory control indicating that these modifications are just as important as those to the lipid A portion of LPS in Gram-negative physiology. EptB is regulated by Ca^{2+} and is responsible for the addition of phosphoethanolamine to the outer Kdo in the core-OS (Kanipes *et al.*, 2001; Reynolds *et al.*, 2005). Another phosphoethanolamine transferase CptA is responsible for the addition of phosphoethanolamine to a heptose residue within the core-OS and is regulated by the PmrA–PmrB two-component system, also involved in regulating lipid A modifications (Tamayo *et al.*, 2005). In R3 core types, WabB is responsible for addition of a GlcNAc residue to HepIII. WabB is located on a plasmid and its expression is thermoregulated (Kaniuk *et al.*, 2004; Yoon *et al.*, 2004). The added GlcNAc residue is then thought to be deacetylated by Shf, which is encoded in the same operon as WabB (Frirdich and Whitfield, 2005). The same GlcN modification is found in R1 core types but the enzymes responsible for modification have not been identified. Analysis of available *E. coli* genome sequences indicates that the loci of at least some R1 core types (identified by the presence of WaaV) encode WabB and Shf homologues (L. Cuthbertson, personal observation).

In *E. coli* K-12, five glycosyltransferases are required for outer core-OS biosynthesis (Fig. 1.3B). WaaG, WaaO and WaaR are

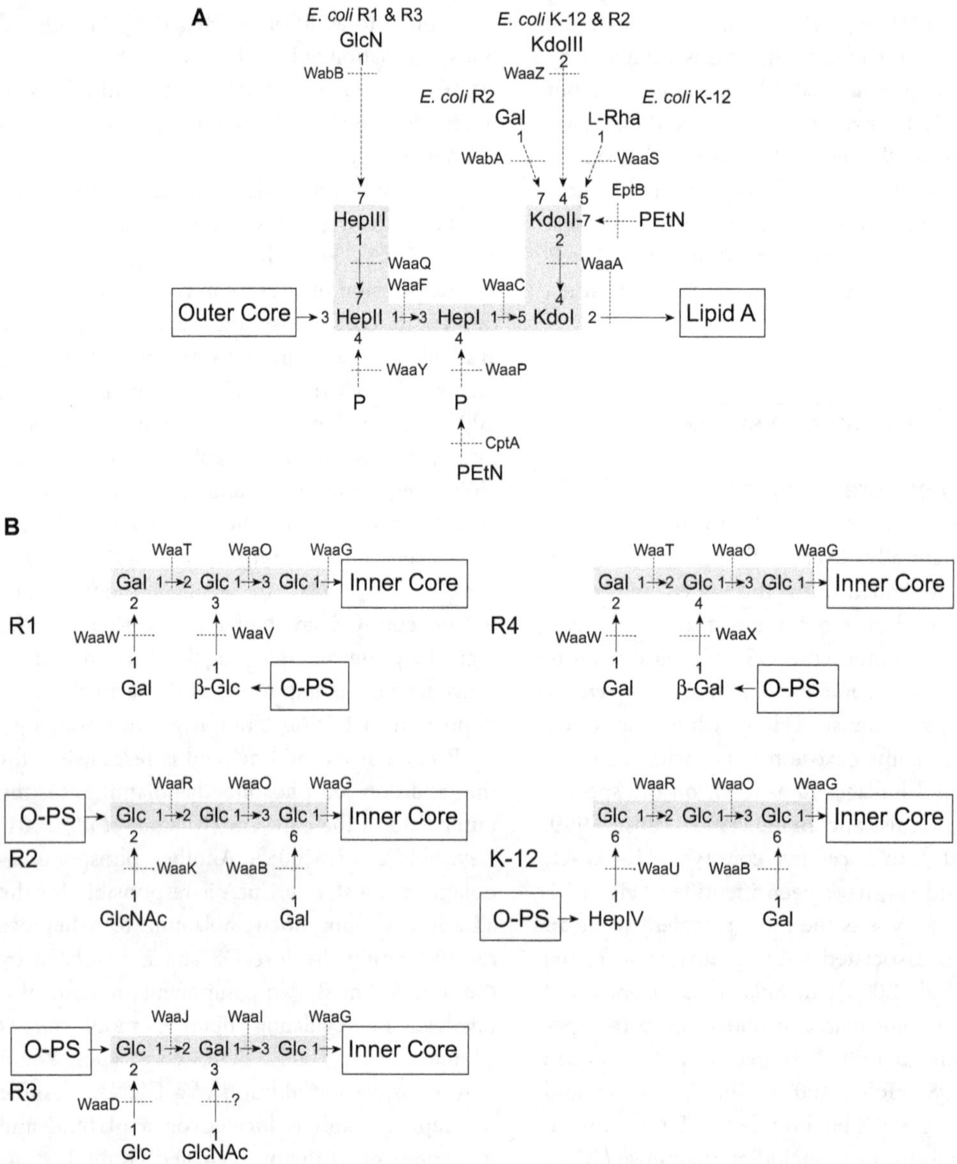

Figure 1.3 Structure of the five core-OS types in *E. coli.* (A) Structure of the inner core-OS. The conserved Kdo and Hep residues of the inner core-OS are highlighted in grey. (B) Structure of the outer core-OS. The outer core-OS backbone is highlighted in grey. Where known, the enzymes responsible for biosynthesis are shown. Non-stoichiometric substitutions are shown with dashed lines.

glucosyltransferases responsible for creation of the generalized outer core-OS backbone, containing three hexose residues (Holst and Brade, 1992). WaaG is found in all *E. coli* core types and is responsible for addition of the first residue of the outer core-OS (a Glc), providing a link between inner and outer core-OS biosynthesis (Yethon *et al.,* 2000). WaaO is found in all core types with the exception of R3 where WaaI functions instead (Shibayama *et al.,* 1998). WaaO transfers a Glc residue to the outer core-OS while WaaI transfers a Gal in the R3 core type. Addition of the last hexose to complete the generalized core-OS structure is more diverse. WaaR is responsible for the addition of Glc in K-12 and R2 core types, WaaT adds Gal in the R1 and R4 core types and WaaJ transfers Glc in the R3 core type (Leipold *et al.*, 2007a,b). WaaR, WaaT and WaaJ all belong to the same glycosyltransferase family but recognize different substrates and transfer different residues. Further additions to the outer core-OS backbone are even more core-type specific. All the glycosyltransferases responsible for these additions have been identified with the exception of an unidentified GlcNAc transferase in the R3 core types (Heinrichs *et al.*, 1998c). In *E. coli* K-12 two glycosyltransferases, WaaB and WaaU, further modify the outer core-OS backbone (Wollin *et al.*, 1983; Kadam *et al.*, 1985; Klena *et al.*, 1992). WaaB adds a Gal to the first glucose and WaaU is responsible for Hep addition to the terminal glucose. This Hep residue serves as the site of O-PS attachment in K-12 strains expressing O-PS.

Lipo-oligosaccharide biosynthesis

In a number of mucosal pathogens including *C. jejuni* and *H. influenzae*, the core-OS is further modified to produce lipo-oligosaccharide (LOS) (Raetz and Whitfield, 2002). Such pathogens do not produce a typical O-PS capped S-LPS. In many cases, the LOS resembles host antigens and is thought to allow the bacterium to evade the immune response through molecular mimicry (Guerry *et al.*, 2002). LOS diversity, similar to that of O-PS (below) is a result of differences in sugar composition and linkages as well as substitution with non-carbohydrate moieties. The phase variation seen in these structures also highlights the importance of antigenic diversity in survival of the bacterium in the host. In *C. jejuni*, at least 19 classes of LOS structures have been identified (Parker *et al.*, 2008), demonstrating the diversity of LOS in this species. LOS biosynthesis in *C. jejuni* 81-176 has been studied in some detail (Fig. 1.4) (Gilbert *et al.*, 2002; Guerry *et al.*, 2002; Kanipes *et al.*, 2004, 2006, 2008; Naito *et al.*, 2010). The initial steps in LOS assembly are identical to the synthesis of the inner core-OS and are similarly carried out by WaaC and WaaF. The remaining steps in LOS synthesis are carried out by additional glycosyltransferases.

Lipid A-core oligosaccharide export

Lipid A-core OS is exported across the inner membrane by the ABC transporter MsbA (Polissi and Georgopoulos, 1996). The requirement of the first two Kdo residues of the inner core for cell viability in *E. coli* is a result of the inefficient export of lipid A without these residues (Karow and Georgopoulos, 1993). Addition of the first two residues of the core-OS is required prior to completion of lipid A acylation in *E. coli* K-12 and efficient export of lipid A. This is the basis for the inclusion of two core-OS residues in the minimal LPS molecule in *E. coli*. Overexpression of MsbA, or the acyl transferases LpxL and LpxM, is able to suppress the requirement for completion of the lipid A-Kdo$_2$ molecule for efficient export (Karow and Georgopoulos, 1993; Meredith *et al.*, 2006; Klein *et al.*, 2009; Reynolds and Raetz, 2009). Initial reports of MsbA depletion in *E. coli* suggested that MsbA may also be involved in phospholipid export (Doerrler *et al.*, 2001, 2004). However, further work in an *msbA* mutant in *N. meningitidis*, which does not require LPS for viability (Steeghs *et al.*, 1998), indicates that MsbA is not involved

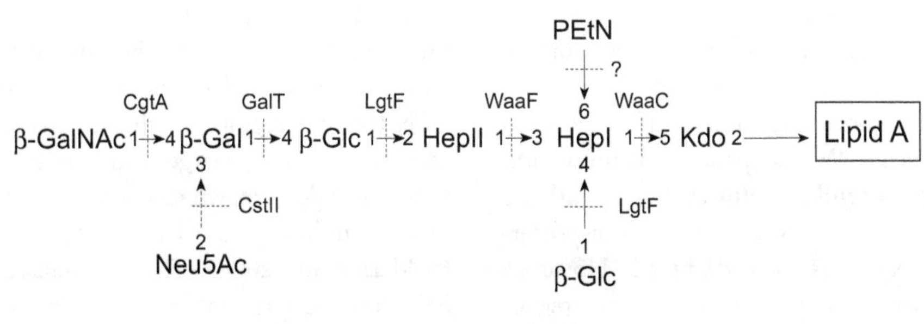

Figure 1.4 Structure of LOS in *C. jejuni* 81-176. The enzymes responsible for biosynthesis are shown.

in phospholipid transport. The accumulation of phospholipids observed in MsbA depleted cells in *E. coli* may have been a secondary effect of loss of LPS transport (Tefsen *et al.*, 2005a). Further investigations will be required to resolve these discrepancies and it is conceivable that the range of substrates for MsbA differs between *E. coli* and *N. meningitidis*. Recent work has demonstrated that MsbA is capable of transporting phospholipids *in vitro* but it is still unclear if this transport occurs *in vivo* (Eckford and Sharom, 2010).

O-antigenic polysaccharide synthesis

Diversity of O-antigenic polysaccharide structures and biosynthetic pathways

O-PS is the most distal portion of the LPS molecule and provides one of the first contacts with the host immune system. It is the most diverse portion of the LPS molecule with over 170 different O-PSs in *E. coli* alone (Stenutz *et al.*, 2006). Great diversity also exists in the O-PSs of other Gram-negative bacteria. More than 50 O-PS structures have been identified in *S. enterica*, which contribute to the almost 2500 serovars that have been identified (Popoff and Le Minor, 1997; Popoff *et al.*, 2001). In *P. aeruginosa* 20 O-PS serotypes have been described and close to 200 serogroups in *Vibrio cholerae* have been identified (Raymond *et al.*, 2002; Chatterjee and Chaudhuri, 2003). O-PS structures vary based on the composition, order and linkage of sugars in the repeat units. O-PS repeat-units contain between two and six residues and may be linear or branched. Activated sugar-nucleotide precursors serve as the donor molecules for O-PS synthesis. Many unusual sugars are found in O-PS structures, the biosynthesis of which is encoded within the O-PS gene cluster (Samuel and Reeves, 2003).

O-PS synthesis takes place through one of three pathways: the Wzy-dependent pathway, the ABC transporter-dependent pathway (also referred to as the Wzy-independent pathway) and the synthase-dependent pathway. O-PS synthesis by the Wzy-dependent and ABC transporter-dependent pathways is widespread. However, there is currently only one known example of O-PS synthesis by a synthase-dependent pathway. Similar pathways are utilized in the synthesis of other cell-surface polysaccharides including capsular polysaccharides (see Chapter 2) and polysaccharides destined for protein glycosylation (see Chapter 4) with ligation to a specific acceptor determining the type of polysaccharide synthesized. In addition to sugar-nucleotide biosynthesis genes, O-PS gene clusters encode genes for glycosyltransferases and O-PS processing, including genes for transport and polymerization. The type of polysaccharide processing genes present are indicative of the type of pathway used (i.e. Wzx/Wzy/Wzz for Wzy-dependent synthesis or Wzm/Wzt for ABC transporter-dependent synthesis) in polysaccharide synthesis and may provide an indication of the type of polysaccharide produced (i.e. Wzm/Wzt in O-PS synthesis or KpsM/KpsT in capsular polysaccharide synthesis) (Cuthbertson *et al.*, 2009, 2010). Despite differences in the location of repeat unit polymerization and the mechanism of export, the three pathways share common initiation and ligation reactions.

Initiation reactions

The first step in O-PS biosynthesis is the transfer of a glycose-1-P residue to the carrier lipid Und-P by a polyprenyl-phosphate glycosyl-1-phosphate transferase. There are currently two known classes of these enzymes represented by WbaP from *S. enterica* serovar Typhimurium and WecA from *E. coli* (Valvano, 2003).

WbaP is a member of the polyprenyl-P hexose-1-P transferase (PHPT) family and is responsible for the transfer of a Galactose-1-P to Und-P (Jiang *et al.*, 1991). PHPT family enzymes are involved in O-PS biosynthesis in different species as well as the synthesis of capsular polysaccharides and polysaccharides utilized in protein glycosylation (Hug and Feldman, 2010). WbaP contains five predicted transmembrane helices (Wang *et al.*, 1996) and is divided into three structural regions: an N-terminal region, which contains four transmembrane helices, a large periplasmic loop, and a C-terminal domain, which contains the last transmembrane helix and a large cytoplasmic region (Saldias *et al.*, 2008). The last transmembrane helix and C-terminal cytoplasmic domain of WbaP are sufficient for *in vivo* function (Wang *et al.*, 1996; Saldias *et al.*, 2008) and residues important in catalysis have been identified (Patel *et al.*, 2010). It has been suggested that the N-terminal domains of WbaP may play a role in the transport of O-PS repeat units (Wang and Reeves, 1994), release of Und-PP-linked substrate (Wang *et al.*, 1996) and regulation of polymerization (Saldias *et al.*, 2008).

WecA belongs to the family of polyprenyl-P *N*-acetylhexosamine-1-P transferases (PNPTs). In *E. coli*, *wecA* is encoded within the biosynthetic gene cluster for enterobacterial common antigen but is also involved in O-PS biosynthesis (Rick *et al.*, 1994). PNPTs are also involved in the biosynthesis of other bacterial polymers including peptidoglycan (Ikeda *et al.*, 1991) and teichoic acid (Soldo *et al.*, 2002). WecA is predicted to contain 11 transmembrane helices with residues important for catalysis located in the cytosol (Lehrer *et al.*, 2007).

Wzy-dependent O-antigenic polysaccharide synthesis

The O-PS of *S. enterica* serovar Typhimurium consists of a four-residue branched repeat unit and serves as the paradigm for O-PS synthesis by a Wzy-dependent pathway (Fig. 1.5). Individual O-PS repeat units are synthesized at the inner surface of the inner membrane by the sequential action of four glycosyltransferases WbaP, WbaN, WbaU and WbaV (Robbins *et al.*, 1966; Schnaitman and Klena, 1993; Raetz and Whitfield, 2002). Repeat units are transported to the periplasmic face of the inner membrane in a reaction involving the flippase, Wzx (Liu *et al.*, 1996). Individual repeat units are then polymerized at the periplasmic face of the inner membrane through addition of new repeat units to the reducing end of the growing chain through the action of the polymerase Wzy (Bray and Robbins, 1967; Robbins *et al.*, 1967; Mulford and Osborn, 1983; Collins and Hackett, 1991). Wzz is involved in regulating polysaccharide chain length (Batchelor *et al.*, 1991, 1992; Macpherson *et al.*, 1991). While the components involved in Wzy-dependent O-PS synthesis are known, their precise mechanisms of action are still unresolved.

Wzx homologues contain 12 membrane-spanning helices with both the N- and C-termini located in the cytoplasm (Mazur *et al.*, 2005; Cunneen and Reeves, 2008; Islam *et al.*, 2010; Marolda *et al.*, 2010). They belong to the multidrug/oligosaccharidyl-lipid/polysaccharide (MOP) flippase superfamily (Hvorup *et al.*, 2003). This superfamily consists of eight distantly related subfamilies of which Wzx belongs to the polysaccharide transport family (Paulsen *et al.*, 1997). Also included within the MOP superfamily are the oligosaccharidyl-lipid flippase family, which is involved in transporting similar dolichylpyrophosphate-linked intermediates across the ER membrane and the mouse virulence factor family, which includes MviN (MurJ). MviN is essential in *E. coli* and has been implicated in peptidoglycan biosynthesis, possibly through the transport of Und-PP linked peptidoglycan precursors across the cytoplasmic membrane (Ruiz, 2008; Inoue *et al.*, 2008). However, *Bacillus subtilis* encodes four putative MviN homologues all four of which could be knocked out without observed growth defects (Fay and Dworkin, 2009). More recently, FtsW has been shown to be responsible for the transport of peptidoglycan precursors (Mohammadi *et al.*, 2011) leaving the precise role of MviN in peptidoglycan biosynthesis unknown. The MOP superfamily also includes members of the multiantimicrobial extrusion (MATE) family. Although the mechanism of transport has not been established for Wzx homologues, they are proposed to energize substrate efflux through cation antiport similar to members of the MATE family (Hvorup *et al.*, 2003). Wzx homologues have a relaxed specificity towards their oligosaccharide substrates and are able to transport incomplete repeat units, as well as repeat units from different biosynthesis systems (Feldman *et al.*, 1999; Marolda *et al.*, 2004, 2006). However, Wzx homologues show specificity for the reducing terminal sugar of the repeat unit (Marolda *et al.*, 2004), either a hexose or an *N*-acetylhexosamine, either through direct recognition or possibly through interaction with the initiating transferase.

Wzy homologues are specific for their cognate polysaccharides (Schnaitman and Klena, 1993, Raetz and Whitfield, 2002). Topology analysis indicates that Wzy homologues have 12 (Daniels *et al.*, 1998; Mazur *et al.*, 2003) or 14 (Islam *et al.*, 2010) transmembrane segments. The difference

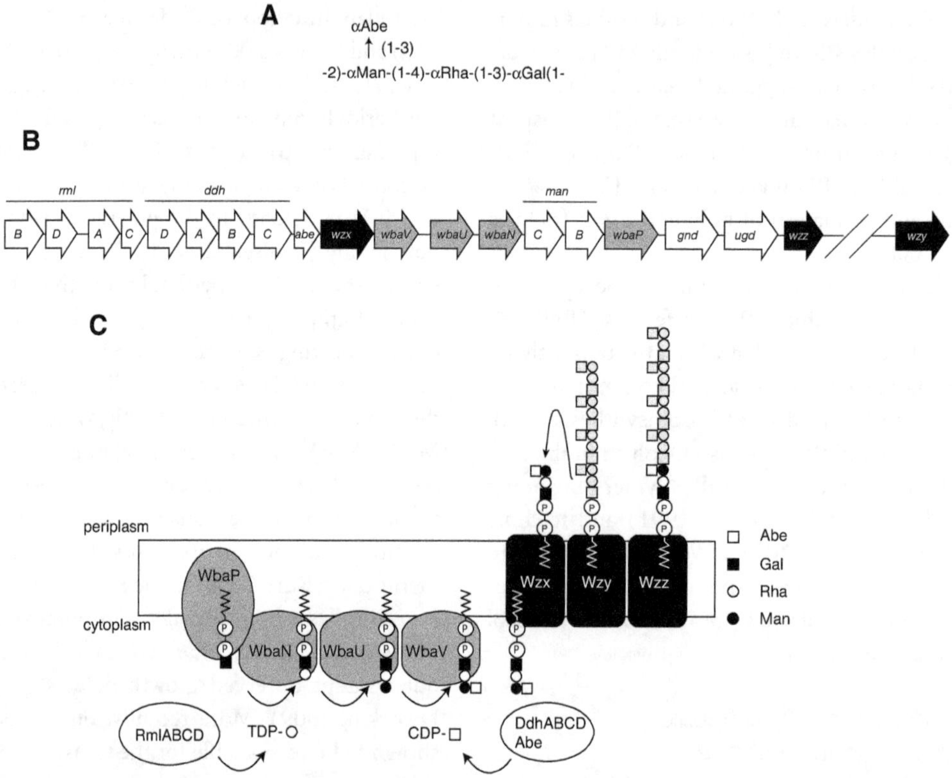

Figure 1.5 The Wzy-dependent pathway for O-PS biosynthesis in *S. enterica* serovar Typhimurium. (A) Structure of the O-PS repeat unit. (B) Organization of the O-PS genetic locus. (C) Model for O-PS biosynthesis. Abe, abequose; Gal, galactose; Man, mannose; Rha, rhamnose.

in the number of transmembrane helices has been proposed to reflect differing substrate specificities (Islam *et al.*, 2010). The catalytic mechanism of Wzy proteins is unknown although residues important for function have been identified (Kim *et al.*, 2010). They share a conserved domain (pfam04932) with WaaL proteins (discussed below) indicating that the two proteins may share a common mechanism of action.

Wzz homologues belong to the polysaccharide co-polymerase (PCP) family (formerly the cytoplasmic-membrane-periplasmic auxiliary (MPA) family) and are involved in regulating polysaccharide chain length (Paulsen *et al.*, 1997; Morona *et al.*, 2000). Members of the PCP family include homologues of Wzz, as well as Wzc and KpsE (Wzf), which are involved in capsular polysaccharide assembly (see Chapter 2). Regulation of the chain length of cell-surface polysaccharides is important for proper protective functions. Polymer chain-length must be sufficient to provide a protective barrier while still allowing access of adhesins and other virulence factors to host cells (Weiss *et al.*, 1986; Morona *et al.*, 2003; Morona and Van Den Bosch, 2003a,b). In at least some organisms, multiple Wzz homologues are encoded that confer different O-PS chain lengths (Murray *et al.*, 2003). Wzz homologues are not specific for polysaccharide structures and exchange of Wzz homologues between strains results in an O-PS chain length similar to that of the Wzz parent strain (Franco *et al.*, 1998).

PCP proteins share a similar topology and contain two membrane-spanning helices with a predicted coiled–coiled region in the periplasm (Morona *et al.*, 2000). Crystal structures of the periplasmic domains of three Wzz homologues have been reported (Tocilj *et al.*, 2008) and reveal a conserved monomeric structure consisting of an α/β base domain and an extended α-helical hairpin. The oligomeric structure of PCP proteins remains an area of debate. While Tocilj *et al.* (2008) reported a different oligomeric structure for each PCP protein investigated, other reports suggest that PCP proteins share a conserved quaternary structure (Larue *et al.*, 2009) and that the different oligomeric states observed by Tocilj *et al.* (2008) were a result of crystallization artefacts. Precise residues involved in Wzz function have not been identified and it has been suggested that the function is not determined by one particular region of the protein but rather that Wzz function is related to its ability to oligomerize (Papadopoulos and Morona, 2010).

Interactions between Wzx, Wzy, Wzz and WaaL are hypothesized to result in the preferred O-PS chain length, with the action of Wzz believed to be the regulating factor (Bastin *et al.*, 1993; Morona *et al.*, 1995). The involvement of WaaL in determining O-PS chain length is difficult to rationalize given that chain length is determined prior to ligation (Daniels *et al.*, 2002). Genetic evidence indicates that the Wzx, Wzy and Wzz proteins form a complex (Marolda *et al.*, 2006) and that the stoichiometry of different proteins within this complex is important for O-PS chain-length regulation (Carter *et al.*, 2009). Additional evidence indicates that domains within the initiating transferase are also important for proper chain-length control (Saldias *et al.*, 2008) leading to the suggestion that Wzz may control the flow of O-PS repeat units through the system thus regulating chain length (Larue *et al.*, 2009). Recent *in vitro* reconstitution of the O-PS pathway of *E. coli* O86 resulted in modal O-PS chain length in the presence of only Wzy and Wzz (Woodward *et al.*, 2010). This system will prove crucial to further investigating the functions of these proteins.

ATP-binding-cassette transporter-dependent O-antigenic polysaccharide synthesis

The O-PSs from a number of Gram-negative pathogens including *Bordetella bronchiseptica*, *Bordetella parapertussis*, *E. coli*, *K. pneumoniae*, *P. aeruginosa*, *V. cholerae* and *Yersinia enterocolitica* are synthesized by an ABC transporter-dependent pathway. While it was once thought that only linear homopolymers were synthesized by ABC transporter-dependent pathways, it is now becoming clear that both heteropolymers and even branched polymers can be synthesized by ABC transporter-dependent pathways (Vinogradov *et al.*, 2002; Izquierdo *et al.*, 2003; Feng *et al.*, 2004; Novotny *et al.*, 2004).

The O-PSs of *E. coli* O8 and O9a are homopolymers of mannose (Vinogradov *et al.*, 2002). O-PS synthesis takes place at the cytoplasmic face of the inner membrane on Und-P (Fig. 1.6). Synthesis is initiated by the transfer of a GlcNAc-1-P to Und-P by WecA (Rick *et al.*, 1994). In *E. coli* O8 and O9a, three mannosyltransferases (WbdC, WbdB and WbdA) are then responsible for O-PS polymerization through the addition of mannose residues to the non-reducing terminus (Kido *et al.*, 1995). Synthesis is terminated through the action of WbdD (Clarke *et al.*, 2004). In *E. coli* O8, WbdD is responsible for methylation of the nascent O-PS at the non-reducing terminus, effectively preventing further polymerization. In *E. coli* O9a, WbdD consecutively phosphorylates and methylates the O-PS. Interestingly O-PS is not synthesized in an *E. coli* O9a mutant devoid of WbdD owing to improper localization of WbdA (Clarke *et al.*, 2009). WbdD is responsible for chain-length regulation of the *E. coli* O8 and O9a polymers and addition of non-reducing terminal modifications by WbdD is also required for O-PS export across the inner membrane (Clarke *et al.*, 2004). The completed O-PS is then exported to the periplasmic face of the inner membrane through the action of an ABC transporter consisting of Wzm and Wzt (Kido *et al.*, 1995). A C-terminal extension on Wzt recognizes the terminal modifications added by WbdD, explaining why these residues are required for export (Cuthbertson *et al.*, 2005, 2007). In *K. pneumoniae* O4, O12 and *Serratia marcescens* O19, terminal Kdo residues have been identified on the O-PS (Vinogradov *et al.*, 2002, 2003) and the O-PSs of *B. parapertussis and B. bronchiseptica* terminate in a derivative of 2,3,4-triamino-2,3,4-trideoxy-L-galacturonamide (Preston *et al.*, 1999, 2006). The Wzt proteins in all of these systems contain an extended C-terminal domain analogous to the *E. coli* O8 and O9a systems indicating that a similar mechanism of coupling O-PS chain termination to export may exist in these systems (Cuthbertson *et al.*, 2010).

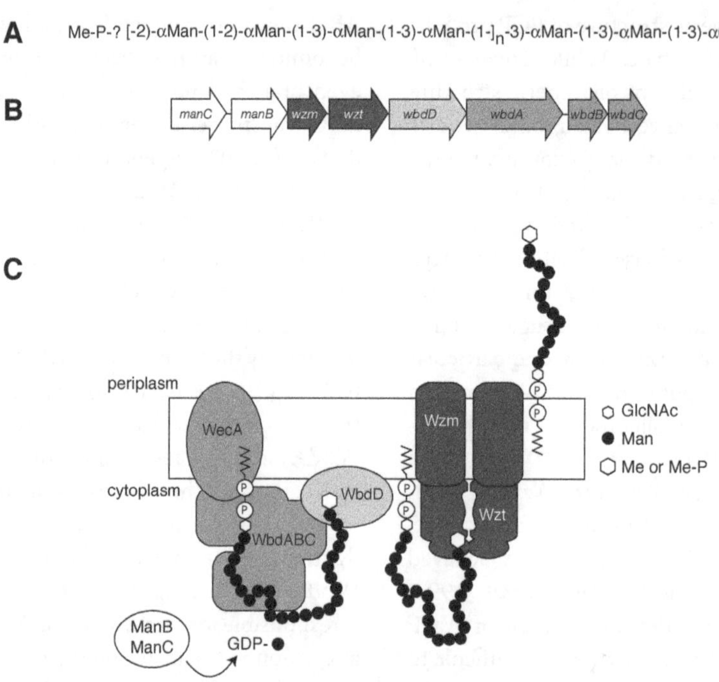

Figure 1.6 The ABC transporter-dependent pathway for O-PS biosynthesis in *E. coli* O8 and O9a. (A) Structure of the O-PS. (B) Organization of the biosynthesis locus. (C) Model for O-PS biosynthesis. GlcNAc, *N*-acetylglucosamine; Man, mannose; Me, methyl; P, phosphate.

Synthesis of D-galactan I O-PS in *K. pneumoniae* represents a second model system for O-PS synthesis and export via ABC transporters (Kos *et al.*, 2009). In this system, non-reducing terminal modifications analogous to those present on the *E. coli* O8 and O9a O-PSs are absent and O-PS chain length is determined by the stoichiometry of the biosynthesis and export apparatus. The C-terminal extension present on the *E. coli* O8 and O9a Wzt homologues is also predictably absent from the Wzt homologue of *K. pneumoniae*. The presence or absence of a C-terminal extension on Wzt serves as a diagnostic feature of O-PSs synthesized by ABC transporter-dependent pathways (Cuthbertson *et al.*, 2010). Owing to the absence of the terminal O-PS modifications and a cognate recognition domain on the transporter, the *K. pneumoniae* D-galactan I transporter is able to export structurally diverse O-PSs (Kos *et al.*, 2009) unlike the *E. coli* O8 and O9a transporters, which are specific for their cognate polysaccharide (Cuthbertson *et al.*, 2005). These differences in O-PS specificity have important implications in vaccine development.

A third type of ABC transporter involved in the export of O-PS has recently been identified in *H. pylori* (Hug *et al.*, 2010). In this transporter, the domains encoded by Wzm and Wzt are present on a single polypeptide Wzk. Wzk is more closely related to MsbA, involved in lipid A export (below) and PglK (see Chapter 4), involved in the export of an oligosaccharide destined for protein *N*-glycosylation in *C. jejuni* (Cuthbertson *et al.*, 2010; Hug and Feldman, 2010).

Synthase-dependent O-antigenic polysaccharide synthesis

There is currently only one O-PS known to be synthesized by a synthase-dependent pathway, that of *S. enterica* serovar Borreze O:54. Although not prevalent in O-PS biosynthesis, this type of pathway may be involved in the synthesis of a number of important bacterial exopolysaccharides (e.g. alginate, cellulose, poly β-D-GlcNAc) (Keiski *et al.*, 2010; Whitfield and Mainprize, 2010). The O:54 antigen is a homopolymer of *N*-acetylmannosamine (ManNAc) with alternating β1,3 and β1,4 linkages (Keenleyside *et al.*, 1994). Three plasmid-encoded genes, WbbE, WbbF and WecB, are involved in O:54 O-PS biosynthesis (Keenleyside and Whitfield, 1995). Based on the current model (Fig. 1.7), O-PS synthesis occurs on Und-P and is initiated through the transfer of a GlcNAc-1-P by WecA (Keenleyside and Whitfield, 1996). WbbE then adds the first ManNAc residue with WbbF being responsible for both polymerization and export of the nascent O-PS (Keenleyside and Whitfield, 1996; Keenleyside *et al.*, 2001; Raetz and Whitfield, 2002). WbbF shares similarity to synthases involved in the biosynthesis and export of capsular polysaccharides (see Chapter 2) and may function in a similar manner.

Ligation

Once its synthesis is complete, O-PS is ligated to a preformed molecule of lipid A core-OS at the periplasmic face of the inner membrane, through the action of the ligase enzyme WaaL (MacLachlan *et al.*, 1991; McGrath and Osborn, 1991). In some organisms, the same O-PS may be ligated to different acceptors molecules (lipid A-core or a protein acceptor) with each type of ligation requiring a different conjugating enzyme (Hug and Feldman, 2010).

WaaL is encoded in the LPS core-OS biosynthesis cluster (MacLachlan *et al.*, 1991). WaaL homologues show specificity for the core-OS (Heinrichs *et al.*, 1998a,b; Abeyrathne *et al.*, 2005; Schild *et al.*, 2005) but are able to utilize a variety of O-PSs synthesized by all three biosynthesis pathways (Kiss *et al.*, 1978; Raetz and Whitfield,

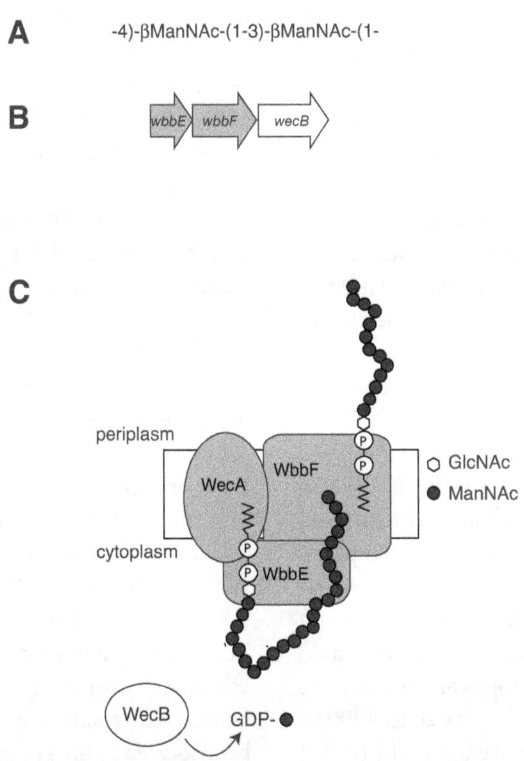

Figure 1.7 The synthase-dependent pathway for O-PS biosynthesis in *S. enterica* serovar Borreze. (A) Structure of the O-PS repeat unit. (B) Organization of the O-PS genetic locus. (C) Model for O-PS biosynthesis. GlcNAc, *N*-acetylglucosamine; ManNAc, *N*-acetylmannosamine.

2002; Meredith *et al.*, 2007). Specificity for the core-OS component may rely on factors other than WaaL (Kaniuk *et al.*, 2004).

WaaL homologues contain a predicted 12 membrane-spanning helices with a large periplasmic loop and some residues important for catalysis have been identified (Abeyrathne and Lam, 2007; Perez *et al.*, 2008; Islam *et al.*, 2010). The requirement of ATP for WaaL function is an area of debate owing to differing reports on WaaL homologues from *P. aeruginosa* and *H. pylori* (Abeyrathne and Lam, 2007; Hug *et al.*, 2010). A potential, although atypical, Walker A motif has been identified in a cytoplasmic loop of WaaL from *P. aeruginosa* while a motif corresponding to the Walker B motif has not been identified based on primary sequence (Islam *et al.*, 2010). Upon completion, the S-LPS molecule is transported to the outer face of the outer membrane via the Lpt pathway (discussed below).

O-PS modification by temperate phage

In addition to the diversity observed in O-PS structures at the level of the repeat unit, non-stoichiometric substitutions play an important role in antigenicity and diversity. Modifications include acetylation, glycosylation and an altered anomeric linkage of O-PS repeat units.

O-PS acetylation occurs in a number of bacterial species. In *S. enterica* serovar Typhimurium, O-PS acetylation by OafA (Slauch *et al.*, 1996) is thought to affect the structure and hence immunological properties of the O-PS (Slauch *et al.*, 1995). Additional phage-encoded acetyltransferases are involved in modifying O-PSs in *Shigella* and *Pseudomonas* (Verma *et al.*, 1991; Newton *et al.*, 2001). Oac from *Shigella flexneri* contains 10 transmembrane regions with residues critical for function located in both the cytoplasm and periplasm (Thanweer *et al.*, 2008). The location and mechanism of O-PS acetylation has not been determined but current models are based on those involved in peptidoglycan and alginate acetylation (see Chapter 3) (Clarke *et al.*, 2000).

In *P. aeruginosa*, three genes encoded by the D3 lysogenic phage are involved in seroconversion. Iap encodes an inhibitor of the host Wzy homologue and a second Wzy homologue specifying a different repeat-unit linkage is encoded on the phage genome. The third phage gene is involved in O-PS acetylation (Newton *et al.*, 2001).

Glycosylation of O-PS has been identified in *Salmonella* and *Shigella*. Initial studies involving the identification of biosynthetic intermediates in *Salmonella*, indicated that glycosylation occurs at the level of O-PS polymerization and was not required for repeat unit synthesis or export (Nikaido *et al.*, 1971; Nikaido and Nikaido, 1971). However, further genetic studies in *Shigella* have led to the proposed model for glycosylation that requires three phage-encoded proteins, GtrA, GtrB and GtrC (Allison and Verma, 2000). GtrB contains two transmembrane helices at its C terminus (Korres *et al.*, 2005) and is proposed to transfer glycose residues to Und-P creating a Und-P-glycose precursor different from the Und-PP-liked O-PS precursor. GtrB homologues share similarity with eukaryotic DPM1 proteins. DPM1 proteins are the catalytic subunit of the DPM synthase responsible for the synthesis and transport of dolichol-P-mannose (Beck *et al.*, 1990). ArnC, involved in L-Ara4N modification of lipid A is also related to GtrB and is responsible for the synthesis of Und-P-L-Ara4N (Breazeale *et al.*, 2005). GtrA is then thought to flip the Und-P-glycose residues across the inner membrane. GtrA proteins are predicted to contain four transmembrane segments and share structural similarity with ArnE and ArnF required for flipping Und-P-L-Ara4N residues involved in lipid A modification (Korres *et al.*, 2005; Yan *et al.*, 2007; Cuthbertson *et al.*, 2010). A third more variable protein is responsible for transferring the glycose residues to O-PS in the periplasm. A number of periplasmic glycosyltransferases have been sequenced (e.g. GtrC, GtrIc and GtrV) and while they are all predicted to contain 9–11 transmembrane helices, most do not share similarity at the amino acid level (Korres and Verma, 2004, 2006; Lehane *et al.*, 2005; Moscoso *et al.*, 2010; Ramiscal *et al.*, 2010). Topological analysis of GtrV from *S. flexneri* is consistent with nine predicted transmembrane helices and residues essential for catalysis have been located in the periplasmic loops (Korres and Verma, 2004; Ramiscal *et al.*, 2010). However, analysis of another glycosyltransferase, GtrIc, indicates that these residues may not be catalytically important for all Gtr-type transferases (Ramiscal *et al.*, 2010). In *Salmonella*, the expression of phage encoded *gtr* operons is phase variable and regulated by DNA methylation (Broadbent *et al.*, 2010). Analysis of *gtr* operons indicates that they are highly mobile, having been exchanged between phage types and are involved in O-PS synthesis/conversion in a number of *Salmonella* hosts (Villafane *et al.*, 2008). Gtr systems are not restricted to phage but can be identified in putative oligo- and polysaccharide biosynthesis loci in genome sequences (e.g. in *Xanthomonas campestris* (O-PS biosynthesis), *Legionella longbeachae, Campylobacter fetus, Streptomyces coelicolor, Aneurinibacillus thermoaerophilus* (S-layer glycosylation) (L. Cuthbertson, personal observation).

Lipopolysaccharide export

The mechanism of LPS transport from the periplasmic face of the inner membrane to its final destination in the outer membrane has received considerable attention in recent years (Ruiz *et al.*, 2006, 2009) and modifications to lipid A that occur in the periplasm and outer membrane have provided important insight into LPS export via the Lpt pathway (Ma *et al.*, 2008). Currently, seven proteins designated LptA-G are known to be involved in this process. In the current model (Fig. 1.8), the LptFGB$_2$ ABC transporter facilitates the transfer of LPS from the inner membrane to LptC (formerly YrbK), which is known to bind LPS (Sperandeo *et al.*, 2007; Ruiz *et al.*, 2008; Narita and Tokuda, 2009; Tran *et al.*, 2010). The LPS molecule is then transferred to LptA (Ma *et al.*, 2008; Suits *et al.*, 2008; Tran *et al.*, 2008), which facilitates transport across the periplasm to LptE (formerly RlpB) and LptD (formerly Imp/OstA), which form a complex in the outer membrane (Wu *et al.*, 2006). LptE has been shown to bind LPS and may serve as the outer membrane receptor (Chng *et al.*, 2010b). A number of non-*Proteobacteria* Gram-negative species produce LPS but lack the *lpt* genes for LPS transport (Sutcliffe, 2010). The mechanism of LPS transport in these bacteria is unknown.

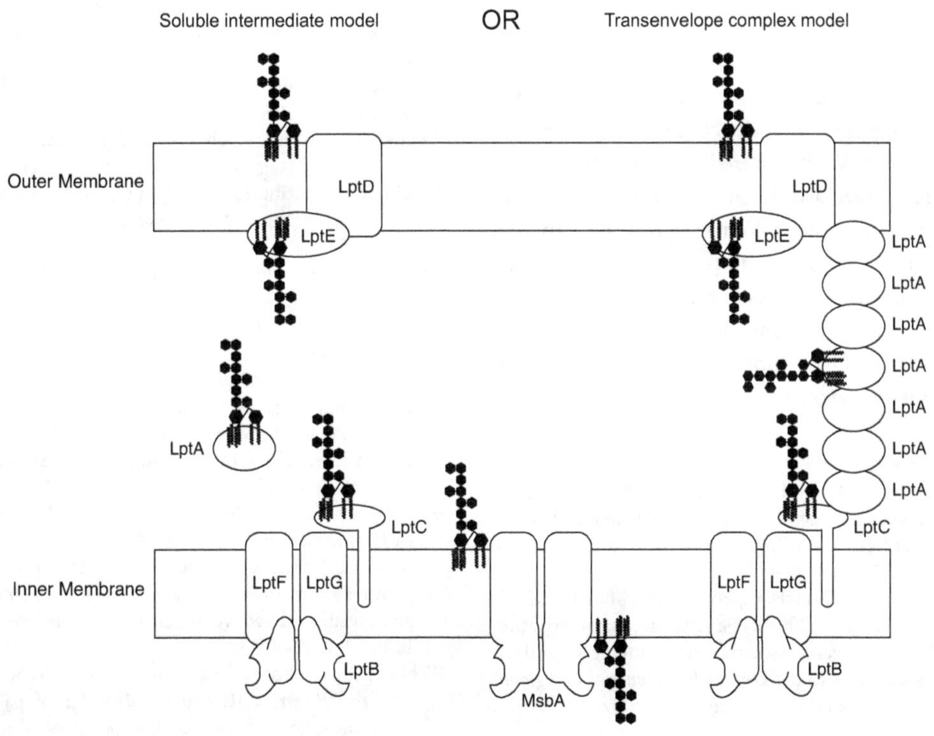

Figure 1.8 Models for the export of LPS via the Lpt pathway.

The mechanism by which LptA transfers LPS across the periplasm is not clear and there are currently two models for transport (Fig. 1.8). In one model, analogous to the transport of lipoproteins to the outer membrane, LptA serves as a chaperone for periplasmic transport similar to the role of LolA in the transport of lipoproteins (Yakushi *et al.*, 2000; Ruiz *et al.*, 2006). In the second model, LptA is thought to facilitate the formation of a transenvelope complex between the Lpt proteins present in the inner and outer membranes with LPS transport occurring through interaction sites between the inner and outer membranes known as Bayer's junctions (Mühlradt *et al.*, 1973; Leive, 1977; Bayer, 1991; Bos and Tommassen, 2004; Tefsen *et al.*, 2005b). Protein–protein interactions have been detected between LptA and LptC but not LptA and LptE, suggesting that LptD may serve as the docking site for LptA in the outer membrane (Bowyer *et al.*, 2011). Recent biochemical evidence in support of the transenvelope model has been provided (Chng *et al.*, 2010a) although the precise mechanism of transport is still not clear.

Future trends

Owing to the essential nature of LPS in most Gram-negative bacteria, it has been investigated as a potential target for antibacterial therapy. Derivatives of lipid A with altered immune stimulatory properties have also been investigated for use in vaccine development. While significant progress has been made in solving the pathways involved in LPS biosynthesis, a better understanding of the mechanisms of the conserved enzymes in lipid A, core-OS and O-PS biosynthesis and export will aid these endeavours.

References

Abeyrathne, P.D., and Lam, J.S. (2007). WaaL of *Pseudomonas aeruginosa* utilizes ATP in *in vitro* ligation of O antigen onto lipid A-core. Mol. Microbiol. *65*, 1345–1359.

Abeyrathne, P.D., Daniels, C., Poon, K.K., Matewish, M.J., and Lam, J.S. (2005). Functional characterization of WaaL, a ligase associated with linking O-antigen polysaccharide to the core of *Pseudomonas aeruginosa* lipopolysaccharide. J. Bacteriol. *187*, 3002–3012.

Albertí, S., Alvaréz, D., Merino, S., Casado, M.T., Vivanco, F., Tomás, J.M., and Benedí, V.J. (1996). Analysis of complement C3 deposition and degradation on *Klebsiella pneumoniae*. Infect. Immun. *64*, 4726–4732.

Allison, G.E., and Verma, N.K. (2000). Serotype-converting bacteriophages and O-antigen modification in *Shigella flexneri*. Trends Microbiol. *8*, 17–23.

Amor, K., Heinrichs, D.E., Frirdich, E., Ziebell, K., Johnson, R.P., and Whitfield, C. (2000). Distribution of core oligosaccharide types in lipopolysaccharides from *Escherichia coli*. Infect. Immun. *68*, 1116–1124.

Anderson, M.S., and Raetz, C.R. (1987). Biosynthesis of lipid A precursors in *Escherichia coli*. A cytoplasmic acyltransferase that converts UDP-*N*-acetylglucosamine to UDP-3-O-(R-3-hydroxymyristoyl)-*N*-acetylglucosamine. J. Biol. Chem. *262*, 5159–5169.

Anderson, M.S., Bull, H.G., Galloway, S.M., Kelly, T.M., Mohan, S., Radika, K., and Raetz, C.R. (1993). UDP-*N*-acetylglucosamine acyltransferase of *Escherichia coli*. The first step of endotoxin biosynthesis is thermodynamically unfavorable. J. Biol. Chem. *268*, 19858–19865.

Babinski, K.J., Kanjilal, S.J., and Raetz, C.R. (2002a). Accumulation of the lipid A precursor UDP-2,3-diacylglucosamine in an *Escherichia coli* mutant lacking the *lpxH* gene. J. Biol. Chem. *277*, 25947–25956.

Babinski, K.J., Ribeiro, A.A., and Raetz, C.R. (2002b). The *Escherichia coli* gene encoding the UDP-2,3-diacylglucosamine pyrophosphatase of lipid A biosynthesis. J. Biol. Chem. *277*, 25937–25946.

Bastin, D.A., Stevenson, G., Brown, P.K., Haase, A., and Reeves, P.R. (1993). Repeat unit polysaccharides of bacteria: a model for polymerization resembling that of ribosomes and fatty acid synthetase, with a novel mechanism for determining chain length. Mol. Microbiol. *7*, 725–734.

Batchelor, R.A., Haraguchi, G.E., Hull, R.A., and Hull, S.I. (1991). Regulation by a novel protein of the bimodal distribution of lipopolysaccharide in the outer membrane of *Escherichia coli*. J. Bacteriol. *173*, 5699–5704.

Batchelor, R.A., Alifano, P., Biffali, E., Hull, S.I., and Hull, R.A. (1992). Nucleotide sequences of the genes regulating O-polysaccharide chain length (*rol*) from *Escherichia coli* and *Salmonella typhimurium*: protein homology and functional complementation. J. Bacteriol. *174*, 5228–5236.

Bayer, M.E. (1991). Zones of membrane adhesion in the cryofixed envelope of *Escherichia coli*. J. Struct. Biol. *107*, 268–280.

Beck, P.J., Orlean, P., Albright, C., Robbins, P.W., Gething, M.J., and Sambrook, J.F. (1990). The *Saccharomyces cerevisiae* DPM1 gene encoding dolichol-phosphate-mannose synthase is able to complement a glycosylation-defective mammalian cell line. Mol. Cell Biol. *10*, 4612–4622.

Bishop, R.E., Gibbons, H.S., Guina, T., Trent, M.S., Miller, S.I., and Raetz, C.R. (2000). Transfer of palmitate from phospholipids to lipid A in outer membranes of gram-negative bacteria. EMBO J. *19*, 5071–5080.

Bos, M.P., and Tommassen, J. (2004). Biogenesis of the Gram-negative bacterial outer membrane. Curr. Opin. Microbiol. *7*, 610–616.

Bos, M.P., and Tommassen, J. (2005). Viability of a capsule- and lipopolysaccharide-deficient mutant of *Neisseria meningitidis*. Infect. Immun. *73*, 6194–6197.

Bowyer, A., Baardsnes, J., Ajamian, E., Zhang, L., and Cygler, M. (2011). Characterization of interactions between LPS transport proteins of the Lpt system. Biochem. Biophys. Res. Commun. *404*, 1093–1098.

Bray, D., and Robbins, P.W. (1967). The direction of chain growth in *Salmonella anatum* O-antigen biosynthesis. Biochem. Biophys. Res. Commun. *28*, 334–339.

Breazeale, S.D., Ribeiro, A.A., McClerren, A.L., Raetz, C.R. (2005). A formyltransferase required for polymyxin resistance in *Escherichia coli* and the modification of lipid A with 4-amino-4-deoxy-L-arabinose. Identification and function of UDP-4-deoxy-4-formamido-L-arabinose. J. Biol. Chem. *280*, 14154–14167.

Broadbent, S.E., Davies, M.R., and van der Woude, M.W. (2010). Phase variation controls expression of *Salmonella* lipopolysaccharide modification genes by a DNA methylation-dependent mechanism. Mol. Microbiol. *77*, 337–353.

Brozek, K.A., and Raetz, C.R. (1990). Biosynthesis of lipid A in *Escherichia coli*. Acyl carrier protein-dependent incorporation of laurate and myristate. J. Biol. Chem. *265*, 15410–15417.

Carter, J.A., Jimenez, J.C., Zaldivar, M., Alvarez, S.A., Marolda, C.L., Valvano, M.A., and Contreras, I. (2009). The cellular level of O-antigen polymerase Wzy determines chain length regulation by WzzB and WzzpHS-2 in *Shigella flexneri* 2a. Microbiology *155*, 3260–3269.

Carty, S.M., Sreekumar, K.R., and Raetz, C.R. (1999). Effect of cold shock on lipid A biosynthesis in *Escherichia coli*. Induction at 12°C of an acyltransferase specific for palmitoleoyl-acyl carrier protein. J. Biol. Chem. *274*, 9677–9685.

Chatterjee, S.N., and Chaudhuri, K. (2003). Lipopolysaccharides of *Vibrio cholerae*. I. Physical and chemical characterization. Biochim. Biophys. Acta *1639*, 65–79.

Chng, S.S., Gronenberg, L.S., and Kahne, D. (2010a). Proteins required for lipopolysaccharide assembly in *Escherichia coli* form a transenvelope complex. Biochemistry *49*, 4565–4567.

Chng, S.S., Ruiz, N., Chimalakonda, G., Silhavy, T.J., and Kahne, D. (2010b). Characterization of the two-protein complex in *Escherichia coli* responsible for lipopolysaccharide assembly at the outer membrane. Proc. Natl. Acad. Sci. U.S.A. *107*, 5363–5368.

Clarke, A.J., Strating, H., and Blackburn, N.T. (2000). Pathways for the *O*-acetylation of bacterial cell wall polysaccharides. In Glycomicrobiology, Doyle, R.J., ed. (New York, NY, Kluwer Academic/Plenum Publishers), p. 187224.

Clarke, B.R., Cuthbertson, L., and Whitfield, C. (2004). Nonreducing terminal modifications determine the chain length of polymannose O antigens of *Escherichia coli* and couple chain termination to polymer export via an ATP-binding cassette transporter. J. Biol. Chem. *279*, 35709–35718.

Clarke, B.R., Greenfield, L.K., Bouwman, C., and Whitfield, C. (2009). Coordination of polymerization, chain termination, and export in assembly of the *Escherichia coli* lipopolysaccharide O9a antigen in an ATP-binding cassette transporter-dependent pathway. J. Biol. Chem. *284*, 30662–30672.

Clementz, T., and Raetz, C.R.H. (1991). A gene coding for 3-deoxy-D-*manno*-octulosonic acid transferase in *Escherichia coli*. Identification, mapping, cloning, and sequencing. J. Biol. Chem. *266*, 9687–9696.

Coleman, J., and Raetz, C.R. (1988). First committed step of lipid A biosynthesis in *Escherichia coli*: sequence of the *lpxA* gene. J. Bacteriol. *170*, 1268–1274.

Collins, L.V., and Hackett, J. (1991). Molecular cloning, characterization, and nucleotide sequence of the *rfc* gene, which encodes an O-antigen polymerase of *Salmonella typhimurium*. J. Bacteriol. *173*, 2521–2529.

Crowell, D.N., Anderson, M.S., and Raetz, C.R. (1986). Molecular cloning of the genes for lipid A disaccharide synthase and UDP-*N*-acetylglucosamine acyltransferase in *Escherichia coli*. J. Bacteriol. *168*, 152–159.

Cunneen, M.M., and Reeves, P.R. (2008). Membrane topology of the *Salmonella enterica* serovar Typhimurium Group B O-antigen translocase Wzx. FEMS Microbiol. Lett. *287*, 76–84.

Cuthbertson, L., Kimber, M.S., and Whitfield, C. (2007). Substrate binding by a bacterial ABC transporter involved in polysaccharide export. Proc. Natl. Acad. Sci. U.S.A. *104*, 19529–19534.

Cuthbertson, L., Kos, V., and Whitfield, C. (2010). ABC transporters involved in export of cell surface glycoconjugates. Microbiol. Mol. Biol. Rev. *74*, 341–362.

Cuthbertson, L., Powers, J., and Whitfield, C. (2005). The C-terminal domain of the nucleotide-binding domain protein Wzt determines substrate specificity in the ATP-binding cassette transporter for the lipopolysaccharide O-antigens in *Escherichia coli* serotypes O8 and O9a. J. Biol. Chem. *280*, 30310–30319.

Cuthbertson, L., Mainprize, I.L., Naismith, J.H., and Whitfield, C. (2009). Pivotal roles of the outer membrane polysaccharide export and polysaccharide copolymerase protein families in export of extracellular polysaccharides in Gram-negative bacteria. Microbiol. Mol. Biol. Rev. *73*, 155–177.

Daniels, C., Vindurampulle, C., and Morona, R. (1998). Overexpression and topology of the *Shigella flexneri* O-antigen polymerase (Rfc/Wzy). Mol. Microbiol. *28*, 1211–1222.

Daniels, C., Griffiths, C., Cowles, B., and Lam, J.S. (2002). *Pseudomonas aeruginosa* O-antigen chain length is determined before ligation to lipid A core. Environ. Microbiol. *4*, 883–897.

Doerrler, W.T., Reedy, M.C., and Raetz, C.R. (2001). An *Escherichia coli* mutant defective in lipid export. J. Biol. Chem. *276*, 11461–11464.

Doerrler, W.T., Gibbons, H.S., and Raetz, C.R. (2004). MsbA-dependent translocation of lipids across the inner membrane of *Escherichia coli*. J. Biol. Chem. *279,* 45102–45109.

Eckford, P.D., and Sharom, F.J. (2010). The reconstituted *Escherichia coli* MsbA protein displays lipid flippase activity. Biochem. J. *429,* 195–203.

Fay, A., and Dworkin, J. (2009). *Bacillus subtilis* homologs of MviN (MurJ), the putative *Escherichia coli* lipid II flippase, are not essential for growth. J. Bacteriol. *191,* 6020–6028.

Feldman, M.F., Marolda, C.L., Monteiro, M.A., Perry, M.B., Parodi, A.J., and Valvano, M.A. (1999). The activity of a putative polyisoprenol-linked sugar translocase (Wzx) involved in *Escherichia coli* O antigen assembly is independent of the chemical structure of the O repeat. J. Biol. Chem. *274,* 35129–35138.

Feng, L., Senchenkova, S.N., Yang, J., Shashkov, A.S., Tao, J., Guo, H., Cheng, J., Ren, Y., Knirel, Y.A., Reeves, P.R., *et al.* (2004). Synthesis of the heteropolysaccharide O antigen of *Escherichia coli* O52 requires an ABC transporter: structural and genetic evidence. J. Bacteriol. *186,* 4510–4519.

Franco, A.V., Liu, D., and Reeves, P.R. (1998). The Wzz (Cld) protein in *Escherichia coli*: amino acid sequence variation determines O-antigen chain length specificity. J. Bacteriol. *180,* 2670–2675.

Frirdich, E., and Whitfield, C. (2005). Lipopolysaccharide inner core oligosaccharide structure and outer membrane stability in human pathogens belonging to the Enterobacteriaceae. J. Endotoxin Res. *11,* 133–144.

Frirdich, E., Lindner, B., Holst, O., and Whitfield, C. (2003). Overexpression of the *waaZ* gene leads to modification of the structure of the inner core region of *Escherichia coli* lipopolysaccharide, truncation of the outer core, and reduction of the amount of O polysaccharide on the cell surface. J. Bacteriol. *185,* 1659–1671.

Garcia Vescovi, E., Soncini, F.C., and Groisman, E.A. (1996). Mg^{2+} as an extracellular signal: environmental regulation of *Salmonella* virulence. Cell *84,* 165–174.

Garrett, T.A., Kadrmas, J.L., and Raetz, C.R. (1997). Identification of the gene encoding the *Escherichia coli* lipid A 4′-kinase. Facile phosphorylation of endotoxin analogs with recombinant LpxK. J. Biol. Chem. *272,* 21855–21864.

Gibbons, H.S., Lin, S., Cotter, R.J., and Raetz, C.R. (2000). Oxygen requirement for the biosynthesis of the S-2-hydroxymyristate moiety in *Salmonella typhimurium* lipid A. Function of LpxO, A new Fe^{2+}/alpha-ketoglutarate-dependent dioxygenase homologue. J. Biol. Chem. *275,* 32940–32949.

Gilbert, M., Karwaski, M.F., Bernatchez, S., Young, N.M., Taboada, E., Michniewicz, J., Cunningham, A.M., and Wakarchuk, W.W. (2002). The genetic bases for the variation in the lipo-oligosaccharide of the mucosal pathogen, *Campylobacter jejuni*. Biosynthesis of sialylated ganglioside mimics in the core oligosaccharide. J. Biol. Chem. *277,* 327–337.

Guerry, P., Szymanski, C.M., Prendergast, M.M., Hickey, T.E., Ewing, C.P., Pattarini, D.L., and Moran, A.P. (2002). Phase variation of *Campylobacter jejuni* 81-176 lipooligosaccharide affects ganglioside mimicry and invasiveness *in vitro*. Infect. Immun. *70,* 787–793.

Gunn, J.S. (2008). The *Salmonella* PmrAB regulon: lipopolysaccharide modifications, antimicrobial peptide resistance and more. Trends Microbiol. *16,* 284–290.

Gunn, J.S., Lim, K.B., Krueger, J., Kim, K., Guo, L., Hackett, M., and Miller, S.I. (1998). PmrA-PmrB-regulated genes necessary for 4-aminoarabinose lipid A modification and polymyxin resistance. Mol. Microbiol. *27,* 1171–1182.

Heinrichs, D.E., Monteiro, M.A., Perry, M.B., and Whitfield, C. (1998a). The assembly system for the lipopolysaccharide R2 core-type of *Escherichia coli* is a hybrid of those found in *Escherichia coli* K-12 and *Salmonella enterica*. Structure and function of the R2 WaaK and WaaL homologs. J. Biol. Chem. *273,* 8849–8859.

Heinrichs, D.E., Yethon, J.A., Amor, P.A., and Whitfield, C. (1998b). The assembly system for the outer core portion of R1 and R4-type lipopolysaccharides of *Escherichia coli*. The R1 core-specific β-glucosyltransferase provides a novel attachment site for O polysaccharides. J. Biol. Chem. *273,* 29497–29505.

Heinrichs, D.E., Yethon, J.A., and Whitfield, C. (1998c). Molecular basis for structural diversity in the core regions of the lipopolysaccharides of *Escherichia coli* and *Salmonella enterica*. Mol. Microbiol. *30,* 221–232.

Holst, O. (1999). Chemical structure of the core region of lipopolysaccharides. In Endotoxin in Health and Disease, Brade, H., Opal, S.M., Vogel, S.N., and Morrison, D.C., eds. (New York, Marcel Dekker Inc.), pp. 305–330.

Holst, O. (2002). Chemical structure of the core region of lipopolysaccharides – an update. Trends Glycosci. Glycotechnol. *14,* 87–103.

Holst, O. (2007). The structures of core regions from enterobacterial lipopolysaccharides – an update. FEMS Microbiol. Lett. *271,* 3–11.

Holst, O., and Brade, H. (1992). Chemical structure of the core region of lipopolysaccharides. In Bacterial Endotoxic Lipopolysaccharides, Morrison, D.C., and Ryan, J.L., eds. (Boca Raton, Ann Arbor, London, Tokyo, CRC Press), pp. 171–205.

Hug, I., and Feldman, M.F. (2010). Analogies and homologies in lipopolysaccharide and glycoprotein biosynthesis in bacteria. Glycobiology *21,* 138–151.

Hug, I., Couturier, M.R., Rooker, M.M., Taylor, D.E., Stein, M., and Feldman, M.F. (2010). *Helicobacter pylori* lipopolysaccharide is synthesized via a novel pathway with an evolutionary connection to protein *N*-glycosylation. PLoS Pathog. *6,* e1000819.

Hvorup, R.N., Winnen, B., Chang, A.B., Jiang, Y., Zhou, X.F., and Saier, M.H., Jr. (2003). The multidrug/oligosaccharidyl-lipid/polysaccharide (MOP) exporter superfamily. Eur. J. Biochem. *270,* 799–813.

Ikeda, M., Wachi, M., Jung, H.K., Ishino, F., and Matsuhashi, M. (1991). The *Escherichia coli mraY* gene encoding UDP-*N*-acetylmuramoyl-pentapeptide:

undecaprenyl-phosphate phospho-*N*-acetylmuramoyl-pentapeptide transferase. J. Bacteriol. *173,* 1021–1026.

Inoue, A.. Murata, Y., Takahashi, H., Tsuji, N., Fujisaki, S., and Kato, J. (2008). Involvement of an essential gene, *mviN,* in murein synthesis in *Escherichia coli.* J. Bacteriol. *190,* 7298–7301.

Islam, S.T., Taylor, V.L., Qi, M., and Lam, J.S. (2010). Membrane topology mapping of the O-antigen flippase (Wzx), polymerase (Wzy), and ligase (WaaL) from *Pseudomonas aeruginosa* PAO1 reveals novel domain architectures. MBio *1,* e00189–10.

Izquierdo, L., Merino, S., Regue, M., Rodriguez, F., and Tomas, J.M. (2003). Synthesis of a *Klebsiella pneumoniae* O-antigen heteropolysaccharide (O12) requires an ABC 2 transporter. J. Bacteriol. *185,* 1634–1641.

Jiang, X.-M., Neal, B., Santiago, F., Lee, S.J., Romana, L.K., and Reeves, P.R. (1991). Structure and sequence of the *rfb* (O antigen) gene cluster of *Salmonella* serovar Typhimurium (strain LT2). Mol. Microbiol. *5,* 695–713.

Joiner, K.A. (1988). Complement evasion by bacteria and parasites. Annu. Rev. Microbiol. *42,* 201–230.

Kabanov, D.S., and Prokhorenko, I.R. (2010). Structural analysis of lipopolysaccharides from Gram-negative bacteria. Biochemistry *75,* 383–404.

Kadam, S.K., Rehemtulla, A., and Sanderson, K.E. (1985). Cloning of *rfaG, B, I,* and *J* genes for glycosyltransferase enzymes for synthesis of the lipopolysaccharide core of *Salmonella typhimurium.* J. Bacteriol. *161,* 277–284.

Kanipes, M.I., Lin, S., Cotter, R.J., and Raetz, C.R. (2001). Ca2+-induced phosphoethanolamine transfer to the outer 3-deoxy-D-*manno*-octulosonic acid moiety of *Escherichia coli* lipopolysaccharide. A novel membrane enzyme dependent upon phosphatidylethanolamine. J. Biol. Chem. *276,* 1156–1163.

Kanipes, M.I., Holder, L.C., Corcoran, A.T., Moran, A.P., and Guerry, P. (2004). A deep-rough mutant of *Campylobacter jejuni* 81-176 is noninvasive for intestinal epithelial cells. Infect. Immun. *72,* 2452–2455.

Kanipes, M.I., Papp-Szabo, E., Guerry, P., and Monteiro, M.A. (2006). Mutation of *waaC,* encoding heptosyltransferase I in *Campylobacter jejuni 81–176,* affects the structure of both lipooligosaccharide and capsular carbohydrate. J. Bacteriol. *188,* 3273–3279.

Kanipes, M.I., Tan, X., Akelaitis, A., Li, J., Rockabrand, D., Guerry, P., and Monteiro, M.A. (2008). Genetic analysis of lipooligosaccharide core biosynthesis in *Campylobacter jejuni* 81–176. J. Bacteriol. *190,* 1568–1574.

Kaniuk, N.A., Vinogradov, E., and Whitfield, C. (2004). Investigation of the structural requirements in the lipopolysaccharide core acceptor for ligation of O antigens in the genus *Salmonella*: WaaL 'ligase' is not the sole determinant of acceptor specificity. J. Biol. Chem. *279,* 36470–36480.

Karow, M., and Georgopoulos, C. (1993). The essential *Escherichia coli msbA* gene, a multicopy suppressor of null mutations in the *htrB* gene, is related to the universally conserved family of ATP-dependent translocators. Mol. Microbiol. *7,* 69–79.

Kato, A., and Groisman, E.A. (2008). The PhoQ/PhoP regulatory network of *Salmonella enterica.* Adv. Exp. Med. Biol. *631,* 7–21.

Keenleyside, W.J., and Whitfield, C. (1995). Lateral transfer of *rfb* genes: a mobilizable ColE1-type plasmid carries the $rfb_{O:54}$ (O:54 antigen biosynthesis) gene cluster from *Salmonella enterica* serovar Borreze. J. Bacteriol. *177,* 5247–5253.

Keenleyside, W.J., and Whitfield, C. (1996). A novel pathway for O-polysaccharide biosynthesis in *Salmonella enterica* serovar Borreze. J. Biol. Chem. *271,* 28581–28592.

Keenleyside, W.J., Perry, M.B., MacLean, L.L., Poppe, C., and Whitfield, C. (1994). A plasmid-encoded $rfb_{O:54}$ gene cluster is required for biosynthesis of the O:54 antigen in *Salmonella enterica* serovar Borreze. Mol. Microbiol. *11,* 437–448.

Keenleyside, W.J., Clarke, A.J., and Whitfield, C. (2001). Identification of residues involved in catalytic activity of the inverting glycosyl transferase WbbE from *Salmonella enterica* serovar Borreze. J. Bacteriol. *183,* 77–85.

Keiski, C.L., Harwich, M., Jain, S., Neculai, A.M., Yip, P., Robinson, H., Whitney, J.C., Riley, L., Burrows, L.L., Ohman, D.E., *et al.* (2010). AlgK is a TPR-containing protein and the periplasmic component of a novel exopolysaccharide secretin. Structure *18,* 265–273.

Kelly, T.M., Stachula, S.A., Raetz, C.R., and Anderson, M.S. (1993). The *firA* gene of *Escherichia coli* encodes UDP-3-O-(R-3-hydroxymyristoyl)-glucosamine *N*-acyltransferase. The third step of endotoxin biosynthesis. J. Biol. Chem. *268,* 19866–19874.

Kido, N., Torgov, V.I., Sugiyama, T., Uchiya, K., Sugihara, H., Komatsu, T., Kato, N., and Jann, K. (1995). Expression of the O9 polysaccharide of *Escherichia coli*: sequencing of the *E. coli* O9 *rfb* gene cluster, characterization of mannosyl transferases, and evidence for an ATP-binding cassette transport system. J. Bacteriol. *177,* 2178–2187.

Kim, T.H., Sebastian, S., Pinkham, J.T., Ross, R.A., Blalock, L.T., and Kasper, D.L. (2010). Characterization of the O-antigen polymerase (Wzy) of *Francisella tularensis.* J. Biol. Chem. *285,* 27839–27849.

Kiss, P., Rinno, J., Schmidt, G., and Mayer, H. (1978). Structural studies on the immunogenic form of the enterobacterial common antigen. Eur. J. Biochem. *88,* 211–218.

Klena, J.D., Pradel, E., and Schnaitman, C.A. (1992). Comparison of lipopolysaccharide biosynthesis genes *rfaK, rfaL, rfaY,* and *rfaZ* of *Escherichia coli* K-12 and *Salmonella typhimurium.* J. Bacteriol. *174,* 4746–4752.

Klein, G., Lindner, B., Brabetz, W., Brade, H., and Raina, S. (2009). *Escherichia coli* K-12 suppressor-free mutants lacking early glycosyltransferases and late acyltransferases: minimal lipopolysaccharide structure and induction of envelope stress response. J. Biol. Chem. *284,* 15369–15389.

Korres, H., and Verma, N.K. (2004). Topological analysis of glucosyltransferase GtrV of *Shigella flexneri* by a dual reporter system and identification of a unique reentrant loop. J. Biol. Chem. *279,* 22469–22476.

Korres, H., and Verma, N.K. (2006). Identification of essential loops and residues of glucosyltransferase, V. (GtrV) of *Shigella flexneri.* Mol. Membr. Biol. *23,* 407–419.

Korres, H., Mavris, M., Morona, R., Manning, P.A., and Verma, N.K. (2005). Topological analysis of GtrA and GtrB proteins encoded by the serotype-converting cassette of *Shigella flexneri.* Biochem. Biophys. Res. Commun. *328,* 1252–1260.

Kos, V., Cuthbertson, L., and Whitfield, C. (2009). The *Klebsiella pneumoniae* O2a antigen defines a second mechanism for O antigen ATP-binding cassette transporters. J. Biol. Chem. *284,* 2947–2956.

Larue, K., Kimber, M.S., Ford, R., and Whitfield, C. (2009). Biochemical and structural analysis of bacterial O-antigen chain length regulator proteins reveals a conserved quaternary structure. J. Biol. Chem. *284,* 7395–7403.

Lee, H., Hsu, F.F., Turk, J., and Groisman, E.A. (2004). The PmrA-regulated *pmrC* gene mediates phosphoethanolamine modification of lipid A and polymyxin resistance in *Salmonella enterica.* J. Bacteriol. *186,* 4124–4133.

Lee, J.H., Lee, K.L., Yeo, W.S., Park, S.J., and Roe, J.H. (2009). SoxRS-mediated lipopolysaccharide modification enhances resistance against multiple drugs in *Escherichia coli.* J. Bacteriol. *191,* 4441–4450.

Lehane, A.M., Korres, H., and Verma, N.K. (2005). Bacteriophage-encoded glucosyltransferase GtrII of *Shigella flexneri*: membrane topology and identification of critical residues. Biochem. J. *389,* 137–143.

Lehrer, J., Vigeant, K.A., Tatar, L.D., and Valvano, M.A. (2007). Functional characterization and membrane topology of *Escherichia coli* WecA, a sugar-phosphate transferase initiating the biosynthesis of enterobacterial common antigen and O-antigen lipopolysaccharide. J. Bacteriol. *189,* 2618–2628.

Leipold, M.D., Kaniuk, N.A., and Whitfield, C. (2007a). The C-terminal Domain of the *Escherichia coli* WaaJ glycosyltransferase is important for catalytic activity and membrane association. J. Biol. Chem. *282,* 1257–1264.

Leipold, M.D., Vinogradov, E., and Whitfield, C. (2007b). Glycosyltransferases involved in biosynthesis of the outer core region of *Escherichia coli* lipopolysaccharides exhibit broader substrate specificities than is predicted from lipopolysaccharide structures. J. Biol. Chem. *282,* 26786–26792.

Leive, L. (1977). Domains involving nonrandom distribution of lipopolysaccharide in the outer membrane of *Escherichia coli.* Proc. Natl. Acad. Sci. U.S.A. *74,* 5065–5068.

Liu, D., Cole, R., and Reeves, P.R. (1996). An O-antigen processing function for Wzx (RfbX): a promising candidate for O-unit flippase. J. Bacteriol. *178,* 2102–2107.

Liu, D., and Reeves, P.R. (1994). *Escherichia coli* K12 regains its O antigen. Microbiology *140,* 49–57.

Ma, B., Reynolds, C.M., and Raetz, C.R. (2008). Periplasmic orientation of nascent lipid A in the inner membrane of an *Escherichia coli* LptA mutant. Proc. Natl. Acad. Sci. U.S.A. *105,* 13823–13828.

MacLachlan, P.R., Kadam, S.K., and Sanderson, K.E. (1991). Cloning, characterization, and DNA sequence of the *rfaLK* region for lipopolysaccharide synthesis in *Salmonella typhimurium* LT2. J. Bacteriol. *173,* 7151–7163.

Macpherson, D.F., Morona, R., Beger, D.W., Cheah, K.-C., and Manning, P.A. (1991). Genetic analysis of the *rfb* region of *Shigella flexneri* encoding the Y serotype O-antigen specificity. Mol. Microbiol. *5,* 1491–1499.

Marolda, C.L., Vicarioli, J., and Valvano, M.A. (2004). Wzx proteins involved in biosynthesis of O antigen function in association with the first sugar of the O-specific lipopolysaccharide subunit. Microbiology *150,* 4095–4105.

Marolda, C.L., Tatar, L.D., Alaimo, C., Aebi, M., and Valvano, M.A. (2006). Interplay of the Wzx translocase and the corresponding polymerase and chain length regulator proteins in the translocation and periplasmic assembly of lipopolysaccharide O antigen. J. Bacteriol. *188,* 5124–5135.

Marolda, C.L., Li, B., Lung, M., Yang, M., Hanuszkiewicz, A., Rosales, A.R., and Valvano, M.A. (2010). Membrane topology and identification of critical amino acid residues in the Wzx O-antigen translocase from *Escherichia coli* O157:H4. J. Bacteriol. *192,* 6160–6171.

Mazur, A., Krol, J.E., Marczak, M., and Skorupska, A. (2003). Membrane topology of PssT, the transmembrane protein component of the type I exopolysaccharide transport system in *Rhizobium leguminosarum* bv. trifolii strain TA1. J. Bacteriol. *185,* 2503–2511.

Mazur, A., Marczak, M., Krol, J.E., and Skorupska, A. (2005). Topological and transcriptional analysis of *pssL* gene product: a putative Wzx-like exopolysaccharide translocase in *Rhizobium leguminosarum* bv. trifolii TA1. Arch. Microbiol. *184,* 1–10.

McGrath, B.C., and Osborn, M.J. (1991). Localization of terminal steps of O-antigen synthesis in *Salmonella typhimurium.* J. Bacteriol. *173,* 649–654.

Meredith, T.C., Aggarwal, P., Mamat, U., Lindner, B., and Woodard, R.W. (2006). Redefining the requisite lipopolysaccharide structure in *Escherichia coli.* ACS Chem. Biol. *1,* 33–42.

Meredith, T.C., Mamat, U., Kaczynski, Z., Lindner, B., Holst, O., and Woodard, R.W. (2007). Modification of lipopolysaccharide with colanic acid (M-antigen) repeats in *Escherichia coli.* J. Biol. Chem. *282,* 7790–7798.

Metzger, L.E.T., and Raetz, C.R. (2010). An alternative route for UDP-diacylglucosamine hydrolysis in bacterial lipid A biosynthesis. Biochemistry *49,* 6715–6726.

Mohmammadi, T., van Dam V., Sijbrandi, R., Vernet, T., Zapun, A., Bouhss, A., Diepeveen-de Bruin M.,

Nguyen-Distèche M., de Kruijff B., and Breukink, E. (2011). Identification of FtsW as a transporter of lipid-linked cell wall precursors across the membrane. Mol. Microbiol. *47*, 1395–1406.

Moran, A.P., Rietschel, E.T., Kosunen, T.U., and Zahringer, U. (1991a). Chemical characterization of *Campylobacter jejuni* lipopolysaccharides containing *N*-acetylneuraminic acid and 2,3-diamino-2,3-dideoxy-D-glucose. J. Bacteriol. *173*, 618–626.

Moran, A.P., Zahringer, U., Seydel, U., Scholz, D., Stutz, P., and Rietschel, E.T. (1991b). Structural analysis of the lipid A component of *Campylobacter jejuni* CCUG 10936 (serotype O:2) lipopolysaccharide. Description of a lipid A containing a hybrid backbone of 2-amino-2-deoxy-D-glucose and 2,3-diamino-2,3-dideoxy-D-glucose. Eur. J. Biochem. *198*, 459–469.

Morona, R., and Van Den Bosch, L. (2003a). Lipopolysaccharide O antigen chains mask IcsA (VirG) in *Shigella flexneri*. FEMS Microbiol. Lett. *221*, 173–180.

Morona, R., and Van Den Bosch, L. (2003b). Multicopy *icsA* is able to suppress the virulence defect caused by the *wzz*(SF) mutation in *Shigella flexneri*. FEMS Microbiol. Lett. *221*, 213–219.

Morona, R., Macpherson, D.F., Van Den Bosch, L., Carlin, N.I., and Manning, P.A. (1995). Lipopolysaccharide with an altered O-antigen produced in *Escherichia coli* K-12 harbouring mutated, cloned *Shigella flexneri rfb* genes. Mol. Microbiol. *18*, 209–223.

Morona, R., Van Den Bosch, L., and Daniels, C. (2000). Evaluation of Wzz/MPA1/MPA2 proteins based on the presence of coiled- coil regions. Microbiology *146*, 1–4.

Morona, R., Daniels, C., and Van Den Bosch, L. (2003). Genetic modulation of *Shigella flexneri* 2a lipopolysaccharide O antigen modal chain length reveals that it has been optimized for virulence. Microbiology *149*, 925–939.

Moscoso, J.A., Korres, H., George, D.T., and Verma, N.K. (2010). Identification of active site residues in the *Shigella flexneri* glucosyltransferase GtrV. Mol. Membr. Biol. *27*, 104–113.

Mühlradt, P.F., Menzel, J., Golecki, J.R., and Speth, V. (1973). Outer membrane of *Salmonella*. Sites of export of newly synthsized lipopolysaccharide on the bacterial surface. Eur. J. Biochem. *35*, 471–481.

Mulford, C.A., and Osborn, M.J. (1983). An intermediate step in translocation of lipopolysaccharide to the outer membrane of *Salmonella typhimurium*. Proc. Natl. Acad. Sci. U.S.A. *80*, 1159–1163.

Muller-Loennies, S., Lindner, B., and Brade, H. (2003). Structural analysis of oligosaccharides from lipopolysaccharide (LPS) of *Escherichia coli* K12 strain W3100 reveals a link between inner and outer core LPS biosynthesis. J. Biol. Chem. *278*, 34090–34101.

Murray, G.L., Attridge, S.R., and Morona, R. (2003). Regulation of *Salmonella typhimurium* lipopolysaccharide O antigen chain length is required for virulence; identification of FepE as a second Wzz. *30*, 1425–1432.

Naito, M., Frirdich, E., Fields, J.A., Pryjma, M., Li, J., Cameron, A., Gilbert, M., Thompson, S.A., and Gaynor, E.C. (2010). Effects of sequential *Campylobacter jejuni* 81-176 lipooligosaccharide core truncations on biofilm formation, stress survival, and pathogenesis. J. Bacteriol. *192*, 2182–2192.

Narita, S., and Tokuda, H. (2009). Biochemical characterization of an ABC transporter LptBFGC complex required for the outer membrane sorting of lipopolysaccharides. FEBS Lett. *583*, 2160–2164.

Newton, G.J., Daniels, C., Burrows, L.L., Kropinski, A.M., Clarke, A.J., and Lam, J.S. (2001). Three-component-mediated serotype conversion in *Pseudomonas aeruginosa* by bacteriophage D3. Mol. Microbiol. *39*, 1237–1247.

Nikaido, H. (2003). Molecular basis of bacterial outer membrane permeability revisited. Microbiol. Mol. Biol. Rev. *67*, 593–656.

Nikaido, K., and Nikaido, H. (1971). Glucosylation of lipopolysaccharide in *Salmonella*: biosynthesis of O antigen factor 12_2. II. Structure of the lipid intermediate. J. Biol. Chem. *246*, 3912–3919.

Nikaido, H., Nikaido, K., Nakae, T., and Makela, P.H. (1971). Glucosylation of lipopolysaccharide in *Salmonella*: biosynthesis of O antigen factor 12_2. I. Over-all reaction. J. Biol. Chem. *246*, 3902–3911.

Novotny, R., Pfoestl, A., Messner, P., and Schaffer, C. (2004). Genetic organization of chromosomal S-layer glycan biosynthesis loci of *Bacillaceae*. Glycoconj. J. *20*, 435–447.

Odegaard, T.J., Kaltashov, I.A., Cotter, R.J., Steeghs, L., van der Ley, P., Khan, S., Maskell, D.J., and Raetz, C.R. (1997). Shortened hydroxyacyl chains on lipid A of *Escherichia coli* cells expressing a foreign UDP-*N*-acetylglucosamine *O*-acyltransferase. J. Biol. Chem. *272*, 19688–19696.

Olsthoorn, M.M.A., Petersen, B.O., Schlecht, S., Haverkamp, J., Bock, K., Thomas-Oates, J.E., and Holst, O. (1998). Identification of a novel core type in *Salmonella* lipopolysaccharide. Complete structural analysis of the core region of the lipopolysaccharide from *Salmonella enterica* sv. Arizonae O62. J. Biol. Chem. *273*, 3817–3829.

Papadopoulos, M., and Morona, R. (2010). Mutagenesis and chemical cross-linking suggest that Wzz dimer stability and oligomerization affect lipopolysaccharide O-antigen modal chain length control. J. Bacteriol. *192*, 3385–3393.

Parker, C.T., Gilbert, M., Yuki, N., Endtz, H.P., and Mandrell, R.E. (2008). Characterization of lipooligosaccharide-biosynthetic loci of *Campylobacter jejuni* reveals new lipooligosaccharide classes: evidence of mosaic organizations. J. Bacteriol. *190*, 5681–5689.

Patel, K.B., Furlong, S.E., and Valvano, M.A. (2010). Functional analysis of the C-terminal domain of the WbaP protein that mediates initiation of O antigen synthesis in *Salmonella enterica*. Glycobiology *20*, 1389–1401.

Paulsen, I.T., Beness, A.M., and Saier, M.H.J. (1997). Computer-based analyses of the protein constituents of transport systems catalysing export of complex carbohydrates in bacteria. Microbiology *143,* 2685–2699.

Peng, D., Hong, W., Choudhury, B.P., Carlson, R.W., and Gu, X.X. (2005). *Moraxella catarrhalis* bacterium without endotoxin, a potential vaccine candidate. Infect. Immun. *73,* 7569–7577.

Perez, J.M., McGarry, M.A., Marolda, C.L., and Valvano, M.A. (2008). Functional analysis of the large periplasmic loop of the *Escherichia coli* K-12 WaaL O-antigen ligase. Mol. Microbiol. *70,* 1424–1440.

Polissi, A., and Georgopoulos, C. (1996). Mutational analysis and properties of the *msbA* gene of *Escherichia coli,* coding for an essential ABC family transporter. Mol. Microbiol. *20,* 1221–1233.

Popoff, M.Y., and Le Minor, L. (1997). Antigenic formulas of the *Salmonella* serovars, 7th revision. World Health Organization Collaborating Centre for Reference and Research on *Salmonella* (Paris, France, Pasteur Institute).

Popoff, M.Y., Bockemuhl, J., Brenner, F.W., and Gheesling, L.L. (2001). Supplement 2000 (no. 44) to the Kauffmann-White scheme. Res. Microbiol. *152,* 907–909.

Preston A., Allen, A.G., Cadisch, J., Thomas, R., Stevens, K., Churcher, C.M., Badcock, K.L., Parkhill, J., Barrell, B., and Maskell, D.J. (1999). Genetic basis for lipopolysaccharide O-antigen biosynthesis in *Bordetellae.* Infect. Immun. *67,* 3763–3767.

Preston A., Petersen, B.O., Duus, J.Ø., Kubler-Kielb J., Ben-Menachem G., Li, J., and Vinogradov, E. (2006). Complete structures of *Bordetella bronchiseptica* and *Bordetella parapertussis* lipopolysaccharides. J. Biol. Chem. *281,* 18135–18144.

Prost, L.R., and Miller, S.I. (2008). The *Salmonellae* PhoQ sensor: mechanisms of detection of phagosome signals. Cell Microbiol. *10,* 576–582.

Raetz, C.R., and Whitfield, C. (2002). Lipopolysaccharide endotoxins. Annu. Rev. Biochem. *71,* 635–700.

Raetz, C.R., Purcell, S., Meyer, M.V., Qureshi, N., and Takayama, K. (1985). Isolation and characterization of eight lipid A precursors from a 3-deoxy-D-*manno*-octylosonic acid-deficient mutant of *Salmonella typhimurium.* J. Biol. Chem. *260,* 16080–16088.

Raetz, C.R., Reynolds, C.M., Trent, M.S., and Bishop, R.E. (2007). Lipid A modification systems in Gram-negative bacteria. Annu. Rev. Biochem. *76,* 295–329.

Raetz, C.R., Guan, Z., Ingram, B.O., Six, D.A., Song, F., Wang, X., and Zhao, J. (2009). Discovery of new biosynthetic pathways: the lipid A story. J. Lipid Res. *50 Suppl,* S103–108.

Ramiscal, R.R., Tang, S.S., Korres, H., and Verma, N.K. (2010). Structural and functional divergence of the newly identified GtrIc from its Gtr family of conserved *Shigella flexneri* serotype-converting glucosyltransferases. Mol. Membr. Biol. 27, 114–122.

Ray, B.L., and Raetz, C.R. (1987). The biosynthesis of gram-negative endotoxin. A novel kinase in *Escherichia coli* membranes that incorporates the 4′-phosphate of lipid A. J. Biol. Chem. *262,* 1122–1128.

Raymond, C.K., Sims, E.H., Kas, A., Spencer, D.H., Kutyavin, T.V., Ivey, R.G., Zhou, Y., Kaul, R., Clendenning, J.B., and Olson, M.V. (2002). Genetic variation at the O-antigen biosynthetic locus in *Pseudomonas aeruginosa.* J. Bacteriol. *184,* 3614–3622.

Reeves, P. (1995). Role of O-antigen variation in the immune response. Trends Microbiol. 3, 381–386.

Reynolds, C.M., and Raetz, C.R. (2009). Replacement of lipopolysaccharide with free lipid A molecules in *Escherichia coli* mutants lacking all core sugars. Biochemistry *48,* 9627–9640.

Reynolds, C.M., Kalb, S.R., Cotter, R.J., and Raetz, C.R. (2005). A phosphoethanolamine transferase specific for the outer 3-deoxy-D-*manno*-octulosonic acid residue of *Escherichia coli* lipopolysaccharide. Identification of the *eptB* gene and Ca2+ hypersensitivity of an *eptB* deletion mutant. J. Biol. Chem. *280,* 21202–21211.

Reynolds, C.M., Ribeiro, A.A., McGrath, S.C., Cotter, R.J., Raetz, C.R., and Trent, M.S. (2006). An outer membrane enzyme encoded by *Salmonella typhimurium lpxR* that removes the 3′-acyloxyacyl moiety of lipid A. J. Biol. Chem. *281,* 21974–21987.

Rick, P.D., Fung, L.W., Ho, C., and Osborn, M.J. (1977). Lipid A mutants of *Salmonella typhimurium.* Purification and characterization of a lipid A precursor produced by a mutant in 3-deoxy-D-*manno*-octulosonate-8-phosphate synthetase. J. Biol. Chem. *252,* 4904–4912.

Rick, P.D., Hubbard, G.L., and Barr, K. (1994). Role of the *rfe* gene in the synthesis of the O8 antigen in *Escherichia coli* K-12. J. Bacteriol. *176,* 2877–2884.

Robbins, P.W., Wright, A., and Dankert, M. (1966). Polysaccharide biosynthesis. J. Gen. Physiol. *49,* 331–346.

Robbins, P.W., Bray, D., Dankert, M., and Wright, A. (1967). Direction of chain growth in polysaccharide synthesis. Science *158,* 1536–1542.

Ruiz, N., Kahne, D., and Silhavy, T.J. (2006). Advances in understanding bacterial outer-membrane biogenesis. Nat. Rev. Microbiol. *4,* 57–66.

Ruiz, N. (2008). Bioinformatics identification of MurJ (MviN) as the peptidoglycan lipid II flippase in *Escherichia coli.* Proc. Natl. Acad. Sci. U.S.A. *105,* 15553–15557.

Ruiz, N., Gronenberg, L.S., Kahne, D., and Silhavy, T.J. (2008). Identification of two inner-membrane proteins required for the transport of lipopolysaccharide to the outer membrane of *Escherichia coli.* Proc. Natl. Acad. Sci. U.S.A. *105,* 5537–5542.

Ruiz, N., Kahne, D., and Silhavy, T.J. (2009). Transport of lipopolysaccharide across the cell envelope: the long road of discovery. Nat. Rev. Microbiol. *7,* 677–683.

Saldias, M.S., Patel, K., Marolda, C.L., Bittner, M., Contreras, I., and Valvano, M.A. (2008). Distinct functional domains of the *Salmonella enterica* WbaP transferase that is involved in the initiation reaction for synthesis of the O antigen subunit. Microbiology *154,* 440–453.

Samuel, G., and Reeves, P. (2003). Biosynthesis of O-antigens: genes and pathways involved in nucleotide sugar precursor synthesis and O-antigen assembly. Carbohydr. Res. *338*, 2503–2519.

Schild, S., Lamprecht, A.K., and Reidl, J. (2005). Molecular and functional characterization of O antigen transfer in *Vibrio cholerae*. J. Biol. Chem. *280*, 25936–25947.

Schnaitman, C.A., and Klena, J.D. (1993). Genetics of lipopolysaccharide biosynthesis in enteric bacteria. Microbiol. Rev. *57*, 655–682.

Shibayama, K., Ohsuka, S., Tanaka, T., Arakawa, Y., and Ohta, M. (1998). Conserved structural regions involved in the catalytic mechanism of *Escherichia coli* K-12 WaaO (RfaI). J. Bacteriol. *180*, 5313–5318.

Sirisena, D.M., Brozek, K.A., MacLachlan, P.R., Sanderson, K.E., and Raetz, C.R. (1992). The *rfaC* gene of *Salmonella typhimurium*. Cloning, sequencing, and enzymatic function in heptose transfer to lipopolysaccharide. J. Biol. Chem. *267*, 18874–18884.

Slauch, J.M., Mahan, M.J., Michetti, P., Neutra, M.R., and Mekalanos, J.J. (1995). Acetylation (O-factor 5) affects the structural and immunological properties of *Salmonella typhimurium* lipopolysaccharide O antigen. Infect. Immun. *63*, 437–441.

Slauch, J.M., Lee, A.A., Mahan, M.J., and Mekalanos, J.J. (1996). Molecular characterization of the *oafA* locus responsible for acetylation of *Salmonella typhimurium* O-antigen: *oafA* is a member of a family of integral membrane trans-acylases. J. Bacteriol. *178*, 5904–5909.

Smith, A.E., Kim, S.H., Liu, F., Jia, W., Vinogradov, E., Gyles, C.L., and Bishop, R.E. (2008). PagP activation in the outer membrane triggers R3 core oligosaccharide truncation in the cytoplasm of *Escherichia coli* O157:H7. J. Biol. Chem. *283*, 4332–4343.

Soldo, B., Lazarevic, V., and Karamata, D. (2002). *tagO* is involved in the synthesis of all anionic cell-wall polymers in *Bacillus subtilis* 168. Microbiology *148*, 2079–2087.

Sperandeo, P., Cescutti, R., Villa, R., Di Benedetto, C., Candia, D., Deho, G., and Polissi, A. (2007). Characterization of *lptA* and *lptB*, two essential genes implicated in lipopolysaccharide transport to the outer membrane of *Escherichia coli*. J. Bacteriol. *189*, 244–253.

Stead, C.M., Zhao, J., Raetz, C.R., and Trent, M.S. (2010). Removal of the outer Kdo from *Helicobacter pylori* lipopolysaccharide and its impact on the bacterial surface. Mol. Microbiol. *78*, 837–852.

Steeghs, L., den Hartog, R., den Boer, A., Zomer, B., Roholl, P., and van der Ley, P. (1998). Meningitis bacterium is viable without endotoxin. Nature *392*, 449–450.

Stenutz, R., Weintraub, A., and Widmalm, G. (2006). The structures of *Escherichia coli* O-polysaccharide antigens. FEMS Microbiol. Rev. *30*, 382–403.

Suits, M.D., Sperandeo, P., Deho, G., Polissi, A., and Jia, Z. (2008). Novel structure of the conserved gram-negative lipopolysaccharide transport protein A and mutagenesis analysis. J. Mol. Biol. *380*, 476–488.

Sutcliffe, I.C. (2010). A phylum level perspective on bacterial cell envelope architecture. Trends Microbiol. *18*, 464–470.

Sweet, C.R., Ribeiro, A.A., and Raetz, C.R. (2004a). Oxidation and transamination of the 3′-position of UDP-*N*-acetylglucosamine by enzymes from *Acidithiobacillus ferrooxidans*. Role in the formation of lipid a molecules with four amide-linked acyl chains. J. Biol. Chem. *279*, 25400–25410.

Sweet, C.R., Williams, A.H., Karbarz, M.J., Werts, C., Kalb, S.R., Cotter, R.J., and Raetz, C.R. (2004b). Enzymatic synthesis of lipid A molecules with four amide-linked acyl chains. LpxA acyltransferases selective for an analog of UDP-*N*-acetylglucosamine in which an amine replaces the 3′-hydroxyl group. J. Biol. Chem. *279*, 25411–25419.

Takayama, K., Qureshi, N., Mascagni, P., Nashed, M.A., Anderson, L., and Raetz, C.R. (1983). Fatty acyl derivatives of glucosamine 1-phosphate in *Escherichia coli* and their relation to lipid A. Complete structure of a diacyl GlcN-1-P found in a phosphatidylglycerol-deficient mutant. J. Biol. Chem. *258*, 7379–7385.

Tamayo, R., Choudhury, B., Septer, A., Merighi, M., Carlson, R., and Gunn, J.S. (2005). Identification of *cptA*, a PmrA-regulated locus required for phosphoethanolamine modification of the *Salmonella enterica* serovar Typhimurium lipopolysaccharide core. J. Bacteriol. *187*, 3391–3399.

Tan, L., and Darby, C. (2005). *Yersinia pestis* is viable with endotoxin composed only of lipid A. J. Bacteriol. *187*, 6599–6600.

Tefsen, B., Bos, M.P., Beckers, F., Tommassen, J., and de Cock, H. (2005a). MsbA is not required for phospholipid transport in *Neisseria meningitidis*. J. Biol. Chem. *280*, 35961–35966.

Tefsen, B., Geurtsen, J., Beckers, F., Tommassen, J., and de Cock, H. (2005b). Lipopolysaccharide transport to the bacterial outer membrane in spheroplasts. J. Biol. Chem. *280*, 4504–4509.

Thanweer, F., Tahiliani, V., Korres, H., and Verma, N.K. (2008). Topology and identification of critical residues of the *O*-acetyltransferase of serotype-converting bacteriophage, SF6, of *Shigella flexneri*. Biochem. Biophys. Res. Commun. *375*, 581–585.

Tocilj, A., Munger, C., Proteau, A., Morona, R., Purins, L., Ajamian, E., Wagner, J., Papadopoulos, M., Van Den Bosch, L., Rubinstein, J.L., *et al.* (2008). Bacterial polysaccharide co-polymerases share a common framework for control of polymer length. Nat. Struct. Mol. Biol. *15*, 130–138.

Touze, T., Tran, A.X., Hankins, J.V., Mengin-Lecreulx, D., and Trent, M.S. (2008). Periplasmic phosphorylation of lipid A is linked to the synthesis of undecaprenyl phosphate. Mol. Microbiol. *67*, 264–277.

Tran, A.X., Trent, M.S., and Whitfield, C. (2008). The LptA protein of *Escherichia coli* is a periplasmic lipid A-binding protein involved in the lipopolysaccharide export pathway. J. Biol. Chem. *283*, 20342–20349.

Tran, A.X., Dong, C., and Whitfield, C. (2010). Structure and functional analysis of LptC, a conserved membrane protein involved in the lipopolysaccharide export pathway in *Escherichia coli*. J. Biol. Chem. *285*, 33529–33539.

Trent, M.S., Pabich, W., Raetz, C.R., and Miller, S.I. (2001a). A PhoP/PhoQ-induced Lipase (PagL) that catalyses 3-*O*-deacylation of lipid A precursors in membranes of *Salmonella typhimurium*. J. Biol. Chem. *276*, 9083–9092.

Trent, M.S., Ribeiro, A.A., Lin, S., Cotter, R.J., and Raetz, C.R. (2001b). An inner membrane enzyme in *Salmonella* and *Escherichia coli* that transfers 4-amino-4-deoxy-L-arabinose to lipid A: induction on polymyxin-resistant mutants and role of a novel lipid-linked donor. J. Biol. Chem. *276*, 43122–43131.

Trent, M.S., Stead, C.M., Tran, A.X., and Hankins, J.V. (2006). Diversity of endotoxin and its impact on pathogenesis. J. Endotoxin Res. *12*, 205–223.

Valvano, M.A. (2003). Export of O-specific lipopolysaccharide. Front. Biosci. *8*, s452–471.

Verma, N.K., Brandt, J.M., Verma, D.J., and Lindberg, A.A. (1991). Molecular characterization of the *O*-acetyl transferase gene of converting bacteriophage SF6 that adds group antigen 6 to *Shigella flexneri*. Mol. Microbiol. 5, 71–75.

Villafane, R., Zayas, M., Gilcrease, E.B., Kropinski, A.M., and Casjens, S.R. (2008). Genomic analysis of bacteriophage epsilon 34 of *Salmonella enterica* serovar Anatum (15+). BMC Microbiol. *8*, 227.

Vinogradov, E., Frirdich, E., MacLean, L.L., Perry, M.B., Petersen, B.O., Duus, J.O., and Whitfield, C. (2002). Structures of lipopolysaccharides from *Klebsiella pneumoniae*. Elucidation of the structure of the linkage region between core and polysaccharide O chain and identification of the residues at the non-reducing termini of the O chains. J. Biol. Chem. *277*, 25070–25081.

Vinogradov, E., Petersen, B.O., Duus, J.O., and Radziejewska-Lebrecht, J. (2003). The structure of the polysaccharide part of the LPS from *Serratia marcescens* serotype O*19*, including linkage region to the core and the residue at the non-reducing end. Carbohydr. Res. *338*, 2757–2761.

Wang, L., and Reeves, P.R. (1994). Involvement of the galactosyl-1-phosphate transferase encoded by the *Salmonella enterica rfbP* gene in O-antigen subunit processing. J. Bacteriol. *176*, 4348–4356.

Wang, L., Liu, D., and Reeves, P.R. (1996). C-terminal half of *Salmonella enterica* WbaP (RfbP) is a galactosyl-1-phosphate transferase domain catalysing the first step of O-antigen synthesis. J. Bacteriol. *178*, 2598–2604.

Weiss, J., Hutzler, M., and Kao, L. (1986). Environmental modulation of lipopolysaccharide chain length alters the sensitivity of *Escherichia coli* to the neutrophil bactericidal/permeability-increasing protein. Infect. Immun. *51*, 594–599.

White, K.A., Kaltashov, I.A., Cotter, R.J., and Raetz, C.R. (1997). A mono-functional 3-deoxy-D-*manno*-octulosonic acid (Kdo) transferase and a Kdo kinase in extracts of *Haemophilus influenzae*. J. Biol. Chem. *272*, 16555–16563.

White, K.A., Lin, S., Cotter, R.J., and Raetz, C.R. (1999). A *Haemophilus influenzae* gene that encodes a membrane bound 3-deoxy-D-*manno*-octulosonic acid (Kdo) kinase. Possible involvement of Kdo phosphorylation in bacterial virulence. J. Biol. Chem. *274*, 31391–31400.

Whitfield, C., and Mainprize, I.L. (2010). TPR motifs: hallmarks of a new polysaccharide export scaffold. Structure *18*, 151–153.

Wilkinson, R.G., Gemski, P., and Stocker, B.A.D. (1972). Non-smooth mutants of *Salmonella typhimurium*: differentiation by phage sensitivity and genetic mapping. J. Gen. Microbiol. *70*, 527–554.

Williamson, J.M., Anderson, M.S., and Raetz, C.R. (1991). Acyl–acyl carrier protein specificity of UDP-GlcNAc acyltransferases from gram-negative bacteria: relationship to lipid A structure. J. Bacteriol. *173*, 3591–3596.

Winfield, M.D., and Groisman, E.A. (2004). Phenotypic differences between *Salmonella* and *Escherichia coli* resulting from the disparate regulation of homologous genes. Proc. Natl. Acad. Sci. U.S.A. *101*, 17162–17167.

Wösten, M.M., Kox, L.F., Chamnongpol, S., Soncini, F.C., and Groisman, E.A. (2000). A signal transduction system that responds to extracellular iron. Cell *103*, 113–125.

Wollin, R., Creeger, E.S., Rothfield, L.I., Stocker, B.A., and Lindberg, A.A. (1983). *Salmonella typhimurium* mutants defective in UDP-D-galactose:lipopolysaccharide alpha 1,6-D-galactosyltransferase. Structural, immunochemical, and enzymologic studies of *rfaB* mutants. J. Biol. Chem. *258*, 3769–3774.

Woodward, R., Yi, W., Li, L., Zhao, G., Eguchi, H., Sridhar, P.R., Guo, H., Song, J.K., Motari, E., Cai, L., *et al.* (2010). *In vitro* bacterial polysaccharide biosynthesis: defining the functions of Wzy and Wzz. Nat. Chem. Biol. *6*, 418–423.

Wu, T., McCandlish, A.C., Gronenberg, L.S., Chng, S.S., Silhavy, T.J., and Kahne, D. (2006). Identification of a protein complex that assembles lipopolysaccharide in the outer membrane of *Escherichia coli*. Proc. Natl. Acad. Sci. U.S.A. *103*, 11754–11759.

Wyckoff, T.J., Lin, S., Cotter, R.J., Dotson, G.D., and Raetz, C.R. (1998). Hydrocarbon rulers in UDP-*N*-acetylglucosamine acyltransferases. J. Biol. Chem. *273*, 32369–32372.

Yakushi, T., Masuda, K., Narita, S., Matsuyama, S., and Tokuda, H. (2000). A new ABC transporter mediating the detachment of lipid-modified proteins from membranes. Nat. Cell Biol. *2*, 212–218.

Yan, A., Guan, Z., and Raetz, C.R. (2007). An undecaprenyl phosphate-aminoarabinose flippase required for polymyxin resistance in *Escherichia coli*. J. Biol. Chem. *282*, 36077–36089.

Yethon, J.A., Gunn, J.S., Malo, D., Laroche, L., Ernst, R.K., Miller, S.I., and Whitfield, C. (2000). *Salmonella enterica* serovar Typhimurium *waaP* mutants show increased susceptibility to polymyxin and loss of virulence *in vivo*. Infect. Immun. *68*, 4485–4491.

Yoon, J.W., Minnich, S.A., Ahn, J.S., Park, Y.H., Paszczynski, A., and Hovde, C.J. (2004). Thermoregulation of the *Escherichia coli* O157:H7 pO157 *ecf* operon and lipid A myristoyl transferase activity involves intrinsically curved DNA. Mol. Microbiol. *51*, 419–435.

Young, K., Silver, L.L., Bramhill, D., Cameron, P., Eveland, S.S., Raetz, C.R., Hyland, S.A., and Anderson, M.S. (1995). The *envA* permeability/cell division gene of *Escherichia coli* encodes the second enzyme of lipid A biosynthesis. UDP-3-O-(R-3-hydroxymyristoyl)-*N*-acetylglucosamine deacetylase. J. Biol. Chem. *270*, 30384–30391.

Zhou, Z., Lin, S., Cotter, R.J., and Raetz, C.R. (1999). Lipid A modifications characteristic of *Salmonella typhimurium* are induced by NH_4VO_3 in *Escherichia coli* K12. Detection of 4-amino-4-deoxy-L-arabinose, phosphoethanolamine and palmitate. J. Biol. Chem. *274*, 18503–18514.

Biosynthesis of Capsular Polysaccharides and Exopolysaccharides

2

Anne N. Reid and Leslie Cuthbertson

Abstract

Capsular polysaccharides (CPSs) and exopolysaccharides (EPSs) enhance bacterial survival in the environment, contribute to symbiotic interactions between plants and bacteria, and mediate interactions between plant and animal pathogens and their hosts. Bacteria express a wide array of CPS and EPS structures that are assembled by one of three distinct mechanisms. The Wzy-dependent polymerization system is characterized by the synthesis of lipid-linked repeat units in the cytoplasm, and their block-wise polymerization at the periplasmic face of the inner membrane. The resulting polymer is transported across the outer membrane (in Gram-negative organisms) via a channel formed by an outer membrane polysaccharide export (OPX) protein. The ATP-binding cassette (ABC) transporter-dependent system is defined by the synthesis of full-length CPS chains in the cytoplasm, their ABC transporter-dependent export across the inner membrane, and their subsequent transport across the outer membrane, presumably via a channel formed by an OPX protein. In the synthase-dependent system, a single enzyme achieves polymer initiation, synthesis and export across the membrane. This chapter describes these modes of CPS and EPS assembly, highlighting recent findings and identifying areas where further research is warranted.

Introduction

Bacterial capsules comprise polysaccharide chains that form a coherent protective layer around some bacteria. The structures of capsular polysaccharides (CPSs) can vary widely (Fig. 2.1). More than 80 distinct capsular or K antigens have been identified in *Escherichia coli* (Orskov and Orskov, 1992), while 91 distinct CPS structures have been found in *Streptococcus pneumoniae* (Bentley *et al.*, 2006; Mavroidi *et al.*, 2007). The polymers are often acidic, can be linear or branched, composed of one or many sugars, and can be further decorated with non-carbohydrate moieties (e.g. amino acids). CPSs are distinguished from exopolysaccharides (EPSs) such as alginate and colanic acid by their degree of attachment to the bacterial surface. CPSs are firmly associated with the cell surface, while EPSs are more loosely associated and can be easily removed from the cell surface.

Biological roles and practical applications of CPSs and EPSs

Given their location on the cell surface, CPSs and EPSs are in immediate contact with the bacterium's environment. Not surprisingly, these structures play a role in the ability of a number of bacteria to survive harsh environmental conditions, to establish symbiotic interactions and in some cases to cause disease in plant or animal hosts.

The acidic nature of many CPSs and EPSs results in the formation of a hydrated layer surrounding bacterial cells, which is known to be protective against desiccation (Ophir and Gutnick, 1994). In addition, many EPSs are important components of the matrix of bacterial biofilms, and as such are important for the establishment of bacterial communities, and for the increased

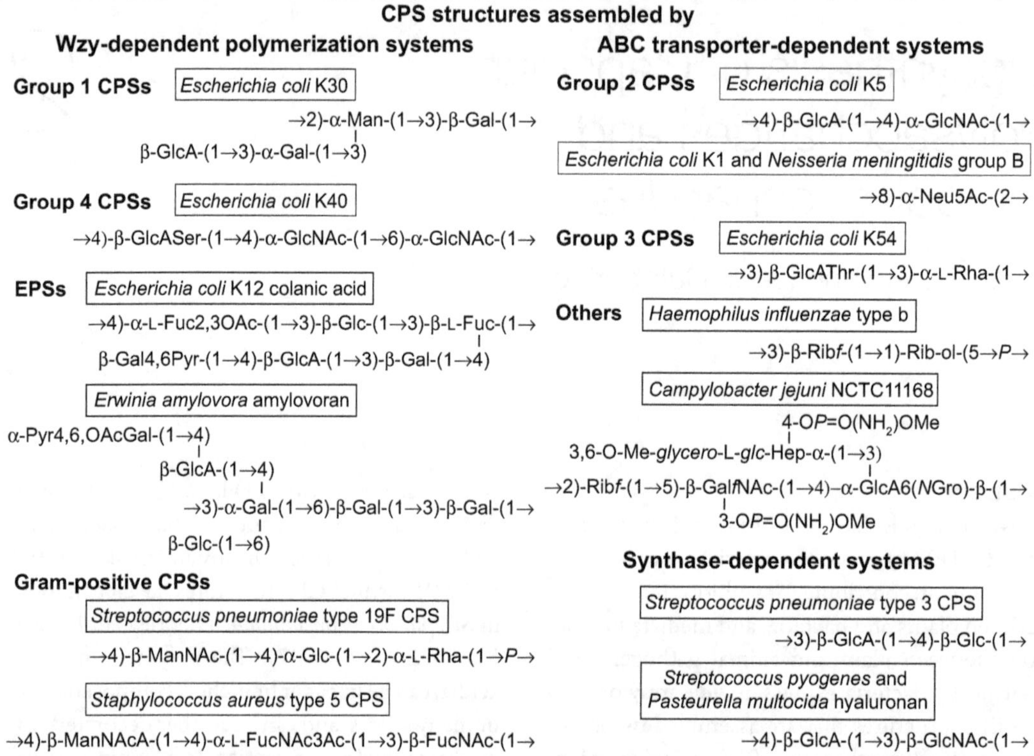

Figure 2.1 Structures of select CPSs and EPSs assembled by Wzy-, ABC transporter-, and synthase-dependent systems. References for structural data: *E. coli* K30 (Chakraborty *et al.*, 1980); *E. coli* K40 (Dengler *et al.*, 1986); colanic acid (Garegg *et al.*, 1971); amylovoran (Nimtz *et al.*, 1996); *S. pneumoniae* type 19F (Abeygunawardana *et al.*, 2000); *S. aureus* type 5 (Jones, 2005b); *E. coli* K5 (Vann *et al.*, 1981); *E. coli* K1 (McGuire and Binkley, 1964); *N. meningitidis* B (Liu *et al.*, 1971; Jennings *et al.*, 1977); *E. coli* K54 (Hofmann *et al.*, 1985); *H. influenzae* b (Jones, 2005a); *C. jejuni* NCTC11168 (St Michael *et al.*, 2002; McNally *et al.*, 2007); *S. pneumoniae* type 3 CPS (Reeves and Goebel, 1941); *S. pyogenes* (Pike, 1948; Warren and Durso, 1951; Pierce and White, 1954); *P. multocida* (Rosner *et al.*, 1992). Sugars are D-pyranosides unless otherwise indicated, and amino acids are amide-linked to the carboxyl group of glucuronic acid (GlcA). L-Fuc, L-fucose; FucNAc, *N*-acetylfucosamine; Gal, galactose; Gal*f*NAc, *N*-acetylgalactofuranose; Glc, glucose; GlcNAc, *N*-acetylglucosamine; Hep, heptose; Man, mannose; ManNAc, *N*-acetylmannosamine; ManNAcA, *N*-acetylmannuronic acid; Neu5Ac, *N*-acetylneuraminic acid; *N*Gro, aminoglycerol; OAc, *O*-acetyl group; OPO(NH_2) OMe, *O*-methyl phosphoramide; Pyr, pyruvyl group; L-Rha, L-rhamnose; Rib*f*, ribofuranose; Rib-ol, ribitol; Ser, serine; Thr, threonine. Figure modified from Reid and Szymanski (2009).

resistance of cells within these communities to a number of external stresses (Vu *et al.*, 2009; Flemming and Wingender, 2010). Some EPSs play a role in bacterial growth and motility. In the filamentous bacterium *Streptomyces coelicolor*, CslA (a cellulose synthase-like protein) is responsible for cellulose synthesis at the hyphal tip, which is important for hyphal growth and attachment (Xu *et al.*, 2008; de Jong *et al.*, 2009). *Myxococcus xanthus* synthesizes an uncharacterized slime polysaccharide that plays a role in the unique swarming motility of this organism. A pulling force is exerted at the leading pole by type IV pili, while slime secretion is believed to exert a pushing force at the trailing pole (Kaiser, 2008). A polysaccharide biosynthesis locus has been identified that is required for both social motility and fruiting body formation (Lu *et al.*, 2005).

Mutations abolishing CPS production in pathogenic bacteria can lead to impaired interactions with host tissues (adherence, colonization, and/or invasion), which ultimately leads to reduced virulence. CPSs protect bacteria from the host immune system by providing resistance to phagocytosis and in some cases, resistance to complement-mediated serum killing (Comstock and Kasper, 2006). A small number of bacteria produce CPSs whose structures mimic host molecules (e.g. the polysialic acid (PSA) capsule of *Neisseria meningitidis* and the hyaluronan capsule of *Streptococcus pyogenes*). Because the immune system does not view these molecules as foreign, the bacteria are able to evade the host immune response (Comstock and Kasper, 2006; Schneider *et al.*, 2007). In some chronic bacterial infections, the pathogens establish protective biofilms within the host. These infections are inherently difficult to treat and contain. For instance, the alginate-containing biofilm produced by *Pseudomonas aeruginosa* plays a key role in the persistence of chronic lung infections in patients suffering from cystic fibrosis (Ramsey and Wozniak, 2005; Ryder *et al.*, 2007). For further details about the role of carbohydrates in host–pathogen interactions, consult a recent review by Reid and coworkers (Reid *et al.*, 2010).

In both plant symbionts and pathogens, EPSs can mediate important bacteria–plant interactions. In *Erwinia amylovora*, a pathogen of rosaceous plant species, the EPSs amylovoran (a branched heteropolymer) and levan (a homopolymer) contribute to virulence. Wild-type strains form biofilms in xylem vessels of experimentally inoculated plants (Koczan *et al.*, 2009), that are thought to restrict water movement within the plant. Recently, amylovoran was shown to be essential for biofilm formation, and amylovoran-deficient strains were avirulent in experimental inoculations (Koczan *et al.*, 2009). Levan was not required for biofilm formation, but a levan-deficient strain showed reduced movement into plant shoots, and was not detected in xylem vessels (Koczan *et al.*, 2009). In plant symbionts such as *Rhizobium leguminosarum*, one of several rhizobial species responsible for the development of root nodules and subsequent nitrogen fixation in leguminous plants, surface polysaccharides are vital for the establishment of a successful symbiosis. In *R. leguminosarum*, EPSs appear to be involved in the development and elongation of the infection thread, and may play a role in protecting the bacterium against plant defences (Skorupska *et al.*, 2006). While the specific roles of CPSs and EPSs in many plant-associated bacteria are not well understood, it is clear that these polymers are crucial for the successful establishment of symbiosis or disease.

The surface exposure of CPSs and EPSs makes these molecules attractive targets for vaccine development (see Chapter 14). Antibodies directed against the bacterial capsule would allow opsonization and subsequent phagocytosis of the pathogen. In general, polysaccharides elicit a weak T cell-independent immune response characterized by IgM antibody production (Weintraub, 2003; Avci and Kasper, 2010). For this reason, polysaccharides are often conjugated to an immunogenic molecule, such as the carrier proteins tetanus toxoid and diphtheria toxoid. These carrier proteins help elicit a T cell-dependent immune response, which includes IgG antibody production and activation of memory cells (Weintraub, 2003; Avci and Kasper, 2010). Polysaccharide vaccines often contain many distinct CPS molecules, in an effort to provide coverage against the spectrum of CPS structures elaborated by the bacterium of interest (e.g. the polysaccharide vaccine PCV13 contains 13 antigenically distinct *S. pneumoniae* CPSs) (Weintraub, 2003; Trotter *et al.*, 2008; Duggan, 2010). Conjugate vaccines in widespread use include those against *Haemophilus influenzae* type b (Hib), *S. pneumoniae* and *N. meningitidis* (Trotter *et al.*, 2008).

Another medical application of microbial polysaccharides involves the use of microbial CPS enzymes to modify therapeutic proteins or peptides. These drugs typically suffer from short *in vivo* half-lives due to instability, proteolysis, neutralization by antibodies and/or clearance from the bloodstream. The chemical conjugation of PSA chains to clinically relevant proteins such as asparaginase (Fernandes and Gregoriadis, 1997) and insulin (Jain *et al.*, 2003) improved their circulating half-life without adversely affecting function. Enzymatic addition of PSA to therapeutic proteins has recently been achieved (Lindhout *et al.*, 2011), and remains an attractive alternative to chemical coupling.

The properties of bacterial CPSs and EPSs have been exploited by the food industry. EPSs such as xanthan (produced by *Xanthomonas campestris*) and gellan (produced by *Sphingomonas paucimobilis*) are added to foods to promote thickening, emulsification and gelling, thus improving texture and sensory qualities of foods such as dairy products, dressings and sauces (Sutherland, 2002). While these polymers are purified and subsequently added to foods, there is a push to rely on starter cultures (e.g. lactic acid bacteria) in dairy production systems to synthesize and secrete the EPSs of interest *in situ* (Welman and Maddox, 2003). In order for the latter method to be successful, a solid understanding of the factors governing EPS biosynthesis and export in these strains is crucial.

Given the importance of bacterial polysaccharides in human disease and plant symbiosis, as well as in medical and industrial applications, it is clear that a firm understanding of the processes governing the biosynthesis, export and regulation of CPS and EPS production is warranted. This information will undoubtedly lead to improved disease prevention and treatment, improved crop yields through enhanced symbiosis, and improved yields of EPSs for use in the food industry. Bacterial strains could even be genetically manipulated to produce high yields of a specific CPS or EPS structure by modifying the pool of precursors for biosynthesis, by introducing new or altering existing biosynthetic enzymes, and/or by improving transport of the resulting polymers from the cells. Complex regulatory mechanisms involved in CPS and EPS production including feedback regulation will probably complicate these endeavours, as does the sharing of precursors for CPS biosynthesis with other important molecules such as lipopolysaccharides (LPS) and peptidoglycan (PG).

Biosynthesis of CPSs and EPSs

Bacteria assemble an impressive array of CPSs and EPSs. The CPSs produced by *E. coli* have been divided into four groups based on genetic and biosynthetic features (Whitfield and Roberts, 1999; Whitfield, 2006). Despite their structural diversity, the CPSs produced by *E. coli* as well as related EPSs appear to be assembled by one of only two distinct mechanisms, described as either Wzy dependent or ATP-binding cassette (ABC) transporter dependent. Because the CPSs produced by *E. coli* have been extensively studied, they will be the primary focus of this section, though relevant examples from other bacteria will be included. The general details of these mechanisms can be found in a number of prior publications (Whitfield, 2006; Reid and Szymanski, 2009). A brief overview will be presented here, with special attention being given to recent findings that have expanded our understanding of these processes. A third biosynthetic mechanism, termed synthase dependent, has only been described in a few systems, including those for type 3 CPS assembly in *S. pneumoniae*, O antigen assembly in *Salmonella enterica* serovar Borreze O:54, and several EPSs including poly-*N*-acetylglucosamine (GlcNAc), cellulose and hyaluronan. Because of its limited description in CPS and EPS systems, this mechanism will be briefly presented here.

Wzy-dependent polymerization mechanism

A Wzy-dependent polymerization mechanism is used to assemble *E. coli* CPSs belonging to groups 1 and 4, EPSs produced by a number of other bacteria, as well as some *E. coli* LPS O antigens (Fig. 2.2). In fact, many of the early steps of Wzy-dependent CPS and EPS assembly are inferred from studies of O-antigen assembly (see Chapter 1).

Assembly of lipid-linked repeat units

Assembly of Wzy-dependent CPSs and EPSs begins in the cytoplasm, where pools of nucleotide-activated sugars are used to build a single repeat unit of the polymer. The first reaction is catalysed by an initiating glycosyltransferase, which transfers a sugar-1-phosphate to undecaprenol-phosphate, the carrier lipid upon which repeat units are built. The initiating glycosyltransferase enzyme can be either a polyisoprenyl-P hexose-1-P transferase (e.g. WbaP), which transfers a hexose-1-P to the carrier lipid, or a polyisoprenyl-P *N*-acetyl-hexosamine- (HexNAc-) 1-P transferase (e.g. WecA), which transfers a HexNAc-1-P to the lipid acceptor. The initiating transferases are large integral membrane proteins with multiple domains. The role(s) of some of these domains

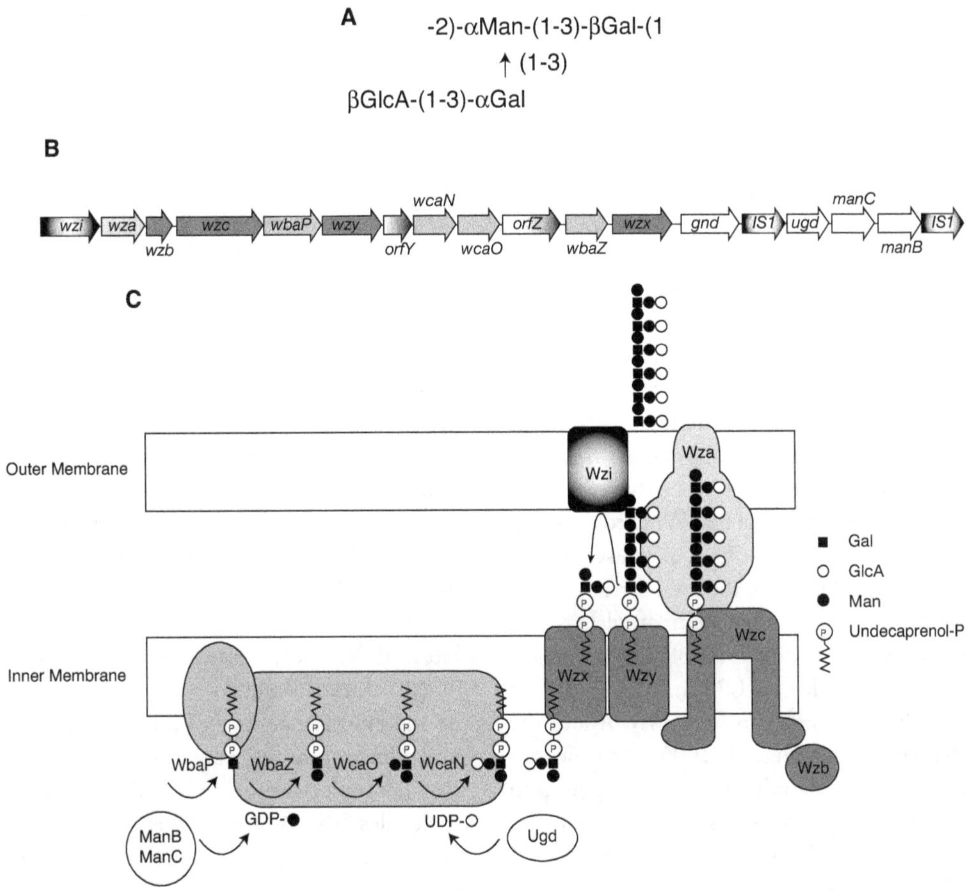

Figure 2.2 The Wzy-dependent mechanism for CPS biosynthesis in *E. coli* K30. (A) Structure of the CPS repeat unit. (B) Organization of the CPS biosynthetic locus. (C) Model for CPS biosynthesis. See text for details. Man, mannose; Gal, galactose; GDP, guanosine diphosphate; GlcA, glucuronic acid; UDP, uridine diphosphate.

have begun to be elucidated, and include regions responsible for substrate binding, transferase activity, membrane localization of the enzyme, and release of the reaction product (Wang *et al.*, 1996; Amer and Valvano, 2000, 2001; Anderson *et al.*, 2000; Pelosi *et al.*, 2005; Lehrer *et al.*, 2007; Saldias *et al.*, 2008; Patel *et al.*, 2010). The periplasmic domain of a WbaP-type initiating transferase was shown to influence *S. enterica* O-antigen chain length (Saldias *et al.*, 2008), highlighting the fact that these enzymes may harbour multiple activities. As discussed below, there is mounting evidence to suggest that initiating glycosyltransferases may serve as a central point for regulation of CPS biosynthesis (Whitfield, 2006; Xayarath and Yother, 2007).

After the initiating reaction has occurred, soluble or membrane-bound glycosyltransferases act sequentially to complete the repeat unit structure.

Transport of repeat units across the inner membrane

Completed repeat units are transferred from the cytoplasmic to the periplasmic face of the inner membrane, where they are added to the growing polymer chain. Chromosomal mutations implicate the Wzx protein in this transbilayer movement (Liu *et al.*, 1996; Rick *et al.*, 2003). Wzx proteins (often called flippases) are large inner membrane proteins with 12 transmembrane domains (TMDs) (Cunneen and Reeves, 2008; Islam *et al.*, 2010; Marolda *et al.*, 2010). Though

Wzx proteins share limited sequence similarity (largely confined to Pfam domain PF01943) (Marolda *et al.*, 2006; Cunneen and Reeves, 2008), they appear to have relaxed specificity, as Wzx proteins can complement a *wzx* defect in a heterologous system. However, this cross-complementation is only successful if both the native and heterologous systems employ the same class (WecA or WbaP) of initiating transferase (Feldman *et al.*, 1999; Marolda *et al.*, 2004, 2006), and if the heterologous Wzx protein's native Wzy and Wzz proteins are absent (Marolda *et al.*, 2006). To date, these cross-complementation experiments have only been carried out in systems where the repeat unit is initiated by a WecA-type enzyme. These observations suggest that Wzx proteins may recognize either the initiating transferase enzyme or the first sugar bound to the lipid carrier. This is consistent with the finding that incomplete repeat units (GlcNAc-PP-undecaprenol) can be incorporated into *E. coli* O16 LPS, in a process that requires the initiating transferase WecA and the Wzx flippase (Feldman *et al.*, 1999). Transport of soluble GlcNAc-PP-undecaprenol analogues into everted membrane vesicles also revealed that Wzx from the enterobacterial common antigen (ECA) assembly system exhibits specificity towards pyrophosphoryl-linked GlcNAc and an unsaturated alpha-isoprene unit (Rick *et al.*, 2003).

While these studies clearly implicate Wzx in moving repeat units across the inner membrane and suggest means used to impart substrate specificity, they do not shed light on the mechanism used to transfer polar (and in some cases charged) substrates across a hydrophobic membrane. Data from studies using soluble GlcNAc-PP-undecaprenol analogues and everted membrane vesicles suggested that Wzx allows facilitated diffusion across the membrane (Rick *et al.*, 2003). Similar findings were recently reported for the FtsW-dependent transport of the PG precursor lipid II in membrane vesicles and in a model membrane system (Mohammadi *et al.*, 2011). Nuclear magnetic resonance studies revealed that interaction of specific peptides (called polyisoprenol recognition sequence peptides) with polyisoprenoid lipids such as undecaprenol-P can alter membrane structure (Zhou and Troy, 2003, 2005). The interaction of multiple peptides with a single polyisoprenoid lipid led to the formation of a membrane channel (Zhou and Troy, 2003, 2005). While there is no report of such peptide sequences in Wzx, such membrane alterations may represent a means by which Wzx proteins could facilitate repeat unit translocation across the membrane. Topology mapping studies of Wzx have noted the presence of polar residues within certain membrane-spanning domains (Cunneen and Reeves, 2008; Islam *et al.*, 2010; Marolda *et al.*, 2010). In one of these studies, site-specific replacement of polar residues within the TMDs and on either face of the membrane revealed that the charge at these positions was critical for protein function, leading the authors to propose that charged residues at these positions could allow passage of lipid-linked repeat units across the membrane (Marolda *et al.*, 2010).

Repeat unit polymerization and chain length determination

At the periplasmic face of the inner membrane, individual repeat units are polymerized to generate high molecular mass (HMM) polysaccharide molecules. The Wzy protein is implicated in this process, catalysing growth at the reducing end of the polymer by transferring the polysaccharide chain onto the incoming repeat unit. Deletion of *wzy* in O-antigen systems leads to the generation of LPS molecules containing a single O-antigen repeat unit (Daniels *et al.*, 1998; Bengoechea *et al.*, 2002; Tao *et al.*, 2004; Cheng *et al.*, 2007; Nakhamchik *et al.*, 2007). Similarly, in the *E. coli* K30 CPS system, mutation of *wzy* leads to the expression of single K repeat units on lipid A-core (termed K_{LPS}) (Drummelsmith and Whitfield, 1999). This ligation of K antigens to lipid A-core is a phenomenon unique to *E. coli*, and its biological significance, if any, is not known. In contrast, *wzx* and *wzy* deletion mutants in *S. pneumoniae* were only obtained in the presence of suppressor mutations that abrogate repeat unit biosynthesis (Xayarath and Yother, 2007).

Wzy proteins are large integral inner membrane proteins that share limited sequence homology. Reporter fusion studies of Wzy proteins from O-antigen systems have yielded topological maps for Wzy proteins that indicate either 12 (*Shigella flexneri*; Daniels et al., 1998) or 14 (*P. aeruginosa*;

Islam *et al.*, 2010) TMDs, and two large periplasmic loops (Daniels *et al.*, 1998; Islam *et al.*, 2010). It has been suggested that the observed differences between Wzy proteins may reflect inherent differences in the substrate specificities of these enzymes (Islam *et al.*, 2010). The mechanism used by Wzy to polymerize repeat units is not currently known, but may involve the $HX_{10}G$ 'polymerase motif' (or the modified $RX_{10}G$ motif) that is present in many proteins known to interact with an undecaprenol-PP-linked polysaccharide (e.g. Wzy, Wzx, WaaL (O-antigen ligase), Wzz) (Schild *et al.*, 2005; Abeyrathne and Lam, 2007; Islam *et al.*, 2010). Wzy proteins must minimally interact with both the growing polysaccharide chain and an incoming lipid-linked repeat unit, and catalyse a transfer reaction in the periplasm. Wzy proteins share a conserved periplasmic domain (Wzy_C, pfam04921) with O-antigen ligases (WaaL) and PilO/TfpO/PglL homologues that transfer glycan to pilin subunits indicating that these proteins and their mechanisms of action are evolutionarily conserved (Hug and Feldman, 2011). While limited sequence similarities among Wzx and Wzy proteins has frustrated attempts to pinpoint catalytic residues and functional domains, this variability has been exploited for the development of PCR-based detection methods for pathogens in food, environmental and biological samples (Wang *et al.*, 2010; Fratamico *et al.*, 2011; Lorusso *et al.*, 2011).

In O-antigen polysaccharide biosynthesis, chain length is determined by one or several Wzz protein(s) (see Chapter 1). In systems lacking a Wzz protein, chain length is uncontrolled, and short chains have a higher frequency of occurrence than long chains. Wzz proteins do not influence chain length in CPS and EPS systems (Dodgson *et al.*, 1996; Amor and Whitfield, 1997; Drummelsmith and Whitfield, 1999), but the synthesis of these polymers shows a clear preference for HMM material. It has been proposed that chain length in these systems is determined in part by the reversible phosphorylation of Wzc, a protein tyrosine kinase whose precise role in CPS and EPS biosynthesis has proven difficult to elucidate. Both Wzz and Wzc proteins are members of the polysaccharide co-polymerase (PCP) family, where Wzz proteins are designated PCP-1 and Wzc proteins are designated PCP-2 (Morona *et al.*, 2000b). Both Wzz and Wzc share a similar topology that includes two transmembrane domains and a large periplasmic loop (Morona *et al.*, 1995, 2000b; Wugeditsch *et al.*, 2001; Doublet *et al.*, 2002). Wzc proteins have tyrosine autokinase activity, harbouring Walker A and B ATP-binding motifs and a cytoplasmic C-terminal tyrosine-rich region that is subject to phosphorylation (Morona *et al.*, 2000a, 2003; Niemeyer and Becker, 2001; Wugeditsch *et al.*, 2001; Doublet *et al.*, 2002; Grangeasse *et al.*, 2002; Bender *et al.*, 2003; Nakar and Gutnick, 2003). In Gram-positive organisms (and a few Gram-negative organisms), Wzc homologues comprise an integral membrane protein and a separate soluble cytosolic protein (Guidolin *et al.*, 1994; Sau *et al.*, 1997).

Deletion of *wzc* or impairment of autokinase activity by mutation of catalytic residues or deletion of the C-terminal tyrosine residues causes loss or reduction of HMM CPS and EPS (Becker *et al.*, 1995; Drummelsmith and Whitfield, 1999; Morona *et al.*, 2000a, 2003; Vincent *et al.*, 2000; Cieslewicz *et al.*, 2001; Niemeyer and Becker, 2001; Wugeditsch *et al.*, 2001; Paiment *et al.*, 2002; Bender *et al.*, 2003; Nakar and Gutnick, 2003; Minic *et al.*, 2007; Obadia *et al.*, 2007). The precise number of tyrosine residues present in the C-terminal domain of Wzc proteins varies between homologues, but information to date suggests Wzc proteins require a minimum level of phosphorylation for proper function, rather than phosphorylation of specific residues (Morona *et al.*, 2000a, 2003; Paiment *et al.*, 2002).

As evidenced by *in vivo* and *in vitro* studies, Wzc proteins are capable of transphosphorylation reactions (meaning one Wzc protein can phosphorylate the C-terminal tyrosine residues on another Wzc protein) (Wugeditsch *et al.*, 2001; Grangeasse *et al.*, 2002). Phosphorylation of Wzc from the *E. coli* colanic acid system (Wzc_{CA}) was suggested to occur in a two-step process that involved autophosphorylation of Wzc at a tyrosine residue upstream of the C-terminal tyrosine-rich domain (Y569) followed by transphosphorylation of the C-terminal tyrosine residues of an adjacent Wzc protein (Grangeasse *et al.*, 2002). Structural analysis as well as molecular dynamics simulation coupled with biochemical analysis of the Wzc homologue, Etk, suggested that intramolecular phosphorylation of the upstream tyrosine residue causes a conformational change that frees up the active site for phosphorylation of the C-terminal tyrosine residues (Lee *et al.*, 2008; Lu *et al.*, 2009). However, structural analysis of the soluble C-terminal domain of Wzc_{CA} did not support intramolecular phosphorylation of this internal tyrosine residue, and rather suggested that Y569 plays an important role in nucleotide binding (Bechet *et al.*, 2010). This may help to explain why mutation of the corresponding upstream tyrosine residue in Wzc from *E. coli* K30 did not affect CPS production (Paiment *et al.*, 2002) and why not all Wzc homologues harbour an upstream tyrosine residue corresponding to Y569 (Grangeasse *et al.*, 2002; Cuthbertson *et al.*, 2009).

The three-dimensional structure of full-length Wzc has been reported, and revealed a tetramer resembling a molar tooth, with individual cytoplasmic root domains (containing the tyrosine-rich domain and the kinase motifs) and a large periplasmic crown region that appears to include sites for protein interaction (Collins *et al.*, 2006). While the subunits were heterogeneously phosphorylated in this structure, phosphorylation was not required for multimer formation (Collins *et al.*, 2006). These structural data confirmed and extended previous genetic and biochemical studies that suggested Wzc multimerization (Wugeditsch *et al.*, 2001; Doublet *et al.*, 2002; Grangeasse *et al.*, 2002; Paiment *et al.*, 2002). High resolution crystal structures have also been obtained of the soluble cytoplasmic domains of several Wzc homologues (Lee *et al.*, 2008; Olivares-Illana *et al.*, 2008; Bechet *et al.*, 2010). The structures of non-phosphorylated forms of both *Staphylococcus aureus* CapB (fused to a portion of the CapA transmembrane protein) and the cytoplasmic domain of *E. coli* Wzc_{CA} revealed an octameric oligomerization state, and identified an EX_2RX_2R motif important for interaction between subunits (Olivares-Illana *et al.*, 2008; Bechet *et al.*, 2010) as well as for kinase activity and polymer export (Bechet *et al.*, 2010). Mutagenesis studies also highlighted important roles for residues within the RK cluster of Wzc_{CA} in kinase activity, nucleotide binding and colanic acid expression (Bechet *et al.*, 2010). The octameric oligomerization state was lost when phosphorylation-competent forms of these proteins were analysed (Olivares-Illana *et al.*, 2008; Bechet *et al.*, 2010), which is consistent with the monomeric state observed in the structure of phosphorylated Etk (Lee *et al.*, 2008). These structures also show a C-terminal tyrosine residue from each monomer bound in the active site of the adjacent monomer, lending support to the notion of intermolecular phosphorylation between Wzc proteins (Bechet *et al.*, 2010). Discrepancies between the oligomerization state observed in the lower resolution structure of full-length Wzc (Collins *et al.*, 2006) versus the high resolution structures of the cytoplasmic domains of Wzc (Olivares-Illana *et al.*, 2008; Bechet *et al.*, 2010) will probably be resolved once a high-resolution structure is obtained for the full-length form of the protein.

Most Wzy-dependent CPS and EPS systems also include a member of the low molecular weight phosphotyrosine protein phosphatase (LMW PTP) family (named Wzb) that is capable of removing phosphate residues from Wzc (Grangeasse *et al.*, 1998; Vincent *et al.*, 1999, 2000; Wugeditsch *et al.*, 2001; Preneta *et al.*, 2002; Klein *et al.*, 2003; Nakar and Gutnick, 2003). While phosphorylation of Wzc is essential for CPS and EPS expression, an active Wzb phosphatase is also required (Bugert and Geider, 1995; Vincent *et al.*, 2000; Wugeditsch *et al.*, 2001; Nakar and Gutnick, 2003), which led to the proposal that CPS and EPS biosynthesis and/or export requires Wzc cycling between phosphorylated and dephosphorylated forms (Wugeditsch *et al.*, 2001; Paiment *et al.*, 2002). LMW PTP proteins are widely distributed in Nature, and employ a catalytic mechanism that relies on nucleophilic attack by a conserved cysteine residue (Zhang *et al.*, 1994). The three-dimensional crystal structure of Wzb from *E. coli* K30 has been solved, and confirms the use of a catalytic mechanism similar to that of other LMW PTP enzymes (Hagelueken *et al.*, 2009). While the Wzb phosphatase is absent from Gram-positive organisms that assemble Wzy-dependent CPSs, these systems harbour a member of the polymerase and histidinol phosphatase (PHP) family (e.g. CpsB in *Streptococcus* spp.) (Bender and Yother, 2001; Morona *et al.*, 2002). The crystal structure of CpsB from *S. pneumoniae* is consistent with these phosphatases employing a mechanism that is dependent on metal ions and arginine residues in the active site, and nucleophilic attack by a metal-bound water molecule (Hagelueken *et al.*, 2009). It has been suggested that PHP phosphatases may modulate the kinase activity of the Wzc homologues in addition to catalysing dephosphorylation of the C-terminal tyrosine residues (Bender and Yother, 2001; Morona *et al.*, 2002). Similar to the phenotype of a *wzb* deletion mutant, deletion of *cpsB* leads to a reduction in CPS production (Morona *et al.*, 2000a, 2006; Cieslewicz *et al.*, 2001; Bender *et al.*, 2003). Purified CpsB was able to dephosphorylate Wzc from *E. coli* K30, revealing that both classes of phosphatases are able to act on a shared substrate (Hagelueken *et al.*, 2009).

While it is clear that phosphorylation of Wzc is important for CPS and EPS expression, the outcome of this reversible phosphorylation is still not known. Asides from the sequence and topological similarities between Wzz and Wzc proteins, there is little direct evidence to support a role for Wzc in CPS and EPS chain length determination. This is due in part to the fact that mutation of *wzc* often yields an acapsular phenotype, caused perhaps by feedback mechanisms designed to prevent polymer synthesis when the final stages of assembly and/or transport are compromised. Mutation of *wzc* in *Streptococcus agalactiae* led to the expression of CPS molecules that were approximately half the typical molecular mass (Cieslewicz *et al.*, 2001), while expression of a phosphorylation-null Wzc mutant narrowed the size distribution of colanic acid (Obadia *et al.*, 2007). Both Wzz and Wzc proteins harbour a Pro-Gly rich domain just upstream of and within the second transmembrane domain. Mutations in this domain in Wzz (Daniels and Morona, 1999) and in several Wzc homologues involved in EPS production (Becker and Puhler, 1998; Dams-Kozlowska and Kaplan, 2007; Obadia *et al.*, 2007) led to the expression of polymer with altered chain length and/or molecular mass. The use of the *E. coli* colanic acid *wza*, *wzb* and *wzc* genes to complement mutations in *Acinetobacter venetianus* resulted in the expression of an EPS of increased molecular mass (Dams-Kozlowska *et al.*, 2008). In several systems where deletion of *wzc* causes loss of the predominant HMM CPS or EPS, an increase in the production of shorter polymers is noted (Drummelsmith and Whitfield, 1999; Niemeyer and Becker, 2001; Wugeditsch *et al.*, 2001; Bender *et al.*, 2003). Whether this reflects a defect in chain-length regulation or is the by-product of an inefficient polymerization and/or export machinery is not clear. While Wzc may be involved in setting polymer chain length in a manner analogous to Wzz in O-antigen systems (perhaps in conjunction with the initiating transferase, as discussed above), the involvement of reversible phosphorylation and the varied phenotypes obtained from mutations of *wzc* suggest this protein may serve a multifaceted role in CPS and EPS production. In fact, several lines of evidence suggest a link between Wzc and initiating transferase activity. Mutation of *wzc* (*epsC* or *epsD*) in *Streptococcus thermophilus* led to decreased initiating transferase (EpsE) activity (Minic *et al.*, 2007). A similar decrease in galactosyltransferase (WbaP) activity was observed in isolated membranes of *E. coli* K30 *wza*, *wzc* and *wzy* mutants (Reid, 2005). To further highlight the complexity of Wzc activity, *in vitro* experiments have shown that Wzc can phosphorylate uridine diphosphate- (UDP-) sugar dehydrogenases in *E. coli* K-12 (Grangeasse *et al.*, 2003; Lacour *et al.*, 2008), *Bacillus subtilis* (Mijakovic *et al.*, 2003) and *S. aureus* (Soulat *et al.*, 2006). In addition, *E. coli* K-12 harbours a Wzc/Wzb pair (referred to as Etk/Etp, or Wzc_{22min}/Wzb_{22min}) that influences the organism's heat shock response by modulating the phosphorylation level of the RpoH sigma factor and the RseA anti-sigma factor (Klein *et al.*, 2003). Finally, it has recently become clear that Wzc interacts with the outer membrane channel Wza in Gram-negative organisms, and that these interactions are crucial for polymer export (discussed below).

ABC transporter-dependent mechanism

The ABC transporter-dependent mechanism is used to assemble *E. coli* group 2 and 3 CPSs, including the PSA capsule of *E. coli* K1, the *N*-acetylheparosan capsule of *E. coli* K5, as well as CPSs from a number of other organisms (e.g. *N. meningitidis*, *Campylobacter jejuni* and *H. influenzae*) (Fig. 2.3). Unlike the block-wise synthesis

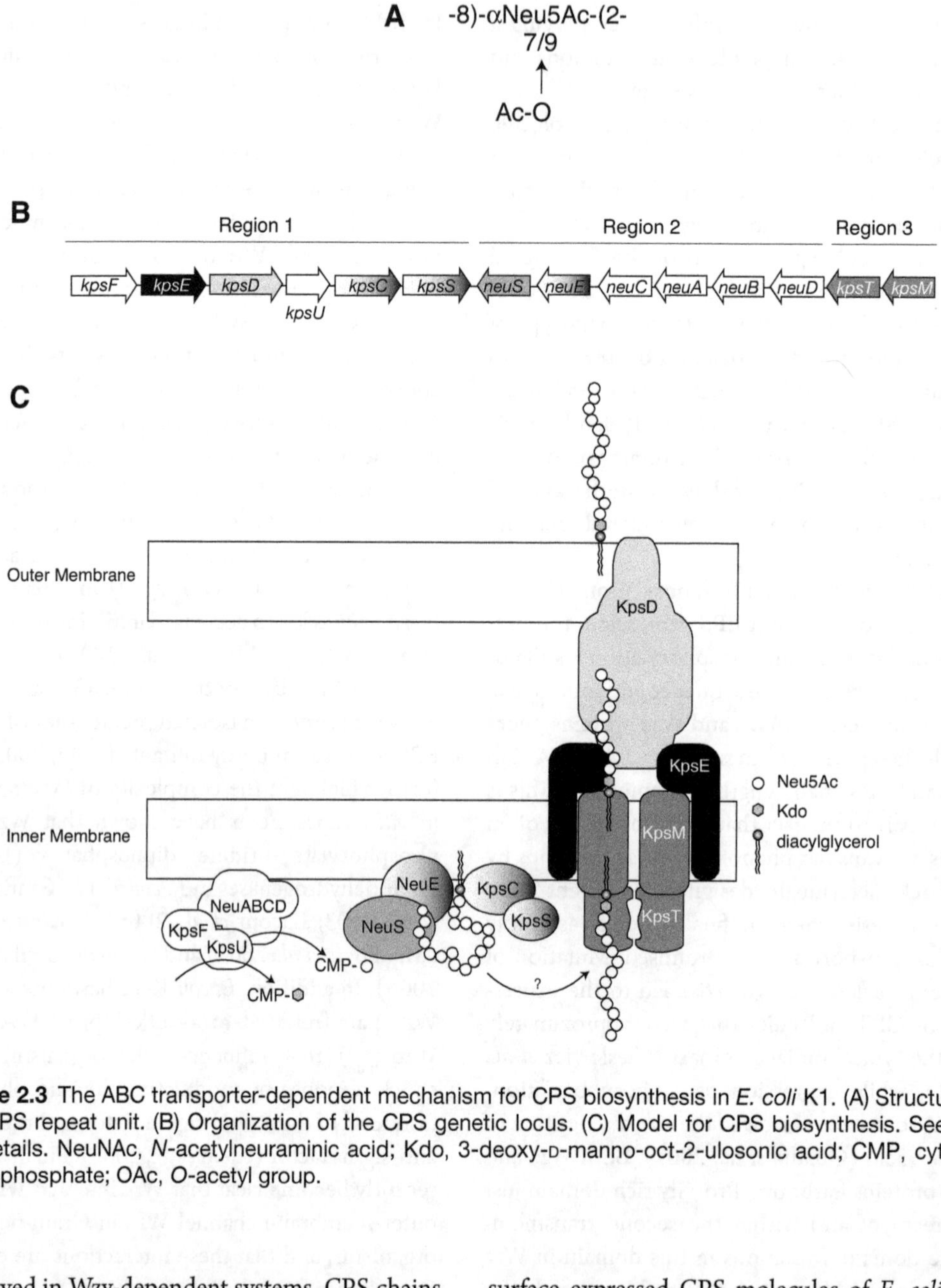

Figure 2.3 The ABC transporter-dependent mechanism for CPS biosynthesis in *E. coli* K1. (A) Structure of the CPS repeat unit. (B) Organization of the CPS genetic locus. (C) Model for CPS biosynthesis. See text for details. NeuNAc, *N*-acetylneuraminic acid; Kdo, 3-deoxy-D-manno-oct-2-ulosonic acid; CMP, cytidine monophosphate; OAc, *O*-acetyl group.

observed in Wzy-dependent systems, CPS chains assembled by ABC transporter-dependent systems are fully synthesized in the cytoplasm prior to transport across the cell envelope.

Cytoplasmic polymer assembly

The early steps of ABC transporter-dependent polymer synthesis are poorly understood. Both the details of the initiation reaction(s) and the identity of the lipid carrier remain unclear. A 1,2-diacylglycerol moiety has been identified on surface expressed CPS molecules of *E. coli* and *N. meningitidis* (Gotschlich *et al.*, 1981; Schmidt and Jann, 1982; Tzeng *et al.*, 2005), as well as on polymer accumulating in the cytoplasm as a result of a *lipA* and/or *lipB* mutation in *N. meningitidis* NMB (Tzeng *et al.*, 2005). However, these studies do not preclude the involvement of a second lipid carrier during early steps of biosynthesis. Early genetic studies implicated LipA and LipB (homologues of KpsC and KpsS in *E. coli*) in lipid-modification of the PSA CPS expressed by

N. meningitidis (Bronner *et al.*, 1993; Frosch and Muller, 1993; Roberts, 1996; Rigg *et al.*, 1998). However, comparable studies in *E. coli* K1 and *N. meningitidis* B revealed the accumulation of lipid-modified CPS in *lipA/kpsC* and *lipB/kpsS* mutants (Cieslewicz and Vimr, 1996; Tzeng *et al.*, 2005).

It has been suggested that a 3-deoxy-D-manno-oct-2-ulosonic acid (Kdo) residue may link the lipid to the polymer in at least some ABC transporter-dependent systems. Group 2 CPS loci in *E. coli* encode enzymes involved in cytidine monophosphate (CMP)-Kdo biosynthesis (KpsF and KpsU) (Finke *et al.*, 1989; Pazzani *et al.*, 1993; Rosenow *et al.*, 1995), and mutation of these genes reduces the levels of CPS produced. However, these genes are not essential for group 2 CPS biosynthesis (Cieslewicz and Vimr, 1996, 1997), and equivalent genes are absent from the loci of some ABC transporter-dependent CPS systems (e.g. group 3 CPSs of *E. coli* and the CPS of *Mannheimia haemolytica* A1). It is not clear whether homologous genes in LPS biosynthesis loci could compensate for the absence of these genes in some systems (Whitfield, 2006).

The minimum requirements for *de novo* synthesis of *E. coli* K1 CPS include CMP-*N*-acetylneuraminic acid (Neu5Ac), a polysialyltransferase (NeuS), an accessory membrane protein (NeuE) and KpsC, while addition of KpsS to this system increased polymer yields (Andreishcheva and Vann, 2006). The requirement of NeuE for *de novo* CPS biosynthesis raises the possibility that NeuE may somehow initiate polymer synthesis. However, this is inconsistent with the finding that *neuE* is not essential for polymer synthesis, as a mutant in *E. coli* K1 accumulated intracellular polymer (Vimr and Steenbergen, 1993; Vimr *et al.*, 2004).

While the process of polymer initiation remains difficult to unravel, the membrane-associated glycosyltransferases responsible for PSA CPS elongation have been extensively studied. These enzymes catalyse the sequential addition of sugars to the non-reducing end of the polysaccharide chain (Bliss and Silver, 1996). Studies using purified glycosyltransferases from *E. coli* K1 and *N. meningitidis* PSA CPS systems have confirmed transferase activity on exogenous acceptors (Steenbergen and Vimr, 1990; Steenbergen *et al.*, 1992; Cho and Troy, 1994; Freiberger *et al.*, 2007; Willis *et al.*, 2008), and revealed that efficient polymer elongation requires a primer of sialic acid on the acceptor (Steenbergen and Vimr, 1990; Ferrero *et al.*, 1991; Chao *et al.*, 1999; Freiberger *et al.*, 2007; Willis *et al.*, 2008). Polysialyltransferase reaction intermediates were not detected *in vivo* (Steenbergen and Vimr, 2003), suggesting these enzymes employ a processive mechanism. Such intermediates were however apparent from *in vitro* experiments (Freiberger *et al.*, 2007; Vionnet and Vann, 2007), but these data do not preclude processivity in the enzyme's native environment. The glycosyltransferases involved in the biosynthesis of the *N*-acetylheparosan capsule of *E. coli* K5 have also been studied in purified form. The biosynthesis of K5 CPS is achieved by the alternating activities of KfiC, a glucuronic acid (GlcA) transferase, and KfiA, a GlcNAc transferase (Sugiura *et al.*, 2010). The uridine diphosphate-glucose (UDP-Glc) dehydrogenase enzyme KfiD catalyses the formation of UDP-GlcA from UDP-Glc. The last biosynthetic protein encoded by the K5 CPS locus is KfiB. This protein is proposed to be an accessory protein (Sugiura *et al.*, 2010), but its exact role in CPS biosynthesis has not been determined.

Polymer translocation across the inner membrane

As is the case with Wzy-dependent CPS molecules, some CPSs assembled by ABC transporter-dependent systems exhibit a fairly defined size range (e.g. chains of PSA in *E. coli* K1 range in length from 160–230 Neu5Ac residues) (Pelkonen *et al.*, 1988). The mechanism(s) governing chain length in this system is not known. In several O-antigen assembly systems, details concerning the termination of chain elongation and the initiation of polymer export by the ABC transporter have come to light (see Chapter 1). Transfer of a phosphate and/or methyl group to the non-reducing end of the O8 and O9a polysaccharides terminates chain elongation (Clarke *et al.*, 2004). These modifications are in turn recognized by an extended C-terminal domain of the nucleotide binding domain (NBD) component of the ABC transporter (Cuthbertson *et al.*, 2007). However, equivalent terminal modifications have not been detected on CPS molecules, and KpsT (NBD domain of the ABC transporter) from *E. coli* K1 does not have the extended domain involved in recognition of terminal modifications (Cuthbertson *et al.*, 2005, 2007). The D-galactan I system in *Klebsiella pneumoniae* represents a second model system for O-antigen chain length regulation in an ABC transporter system. Unlike the O8 and O9a systems of *E. coli*, the Wzt homologue in *K. pneumoniae* lacks an extended C-terminal domain and the O antigen lacks an analogous terminal modification (Vinogradov *et al.*, 2002; Kos *et al.*, 2009). In this system, O-antigen chain length is determined by the stoichiometry of the biosynthesis and export apparatus. Similar to ABC transporters in CPS systems, the *K. pneumoniae* D-galactan I transporter is also able to export structurally unrelated O antigens (Kos *et al.*, 2009). These similarities raise the possibility that chain length may be regulated in a similar manner in *K. pneumonaie* D-galactan I and ABC transporter-dependent CPS systems (Cuthbertson *et al.*, 2010).

Lipid modification of the CPS was once believed to initiate export, but the intracellular accumulation of lipid-modified polymer in some mutants (e.g. *kpsC* or *kpsS*) argues against this. Given the likelihood that proteins involved in CPS biosynthesis and export form a multienzyme complex (discussed below), it remains possible that such mutations disrupt or alter complex formation, and that lipid-modification may be sufficient to initiate export in the native system. The *E. coli* K1 biosynthesis locus encodes an *O*-acetyltransferase (NeuD) that is proposed to acetylate sialic acid (Neu5Ac) prior to its nucleotide-activation and subsequent incorporation into CPS chains (Steenbergen *et al.*, 2006). In Group B *Streptococcus*, under certain conditions, the *O*-acetyl group added by NeuD can spontaneously migrate from Neu5Ac position 7 to positions 8 or 9 (Lewis *et al.*, 2004, 2006). Incorporation of an 8-*O*-acetyl-Neu5Ac residue into the growing K1 CPS chain would block further chain elongation (as the subunits are α-2,8-linked and terminate PSA biosynthesis (Vimr and Steenbergen, 2006). However, terminal Neu5Ac residues acetylated at the 8 position have not been reported to date in a PSA system. Sialic acid-specific *O*-acetyltransferases have also been described in *N. meningitidis* C, which produces an α-2,9-linked PSA (Bergfeld *et al.*, 2009). However, the α-2,8-linked PSA of group B meningococci is not acetylated (Jennings *et al.*, 1977; Orskov *et al.*, 1979), arguing against *O*-acetylation as a conserved means of chain termination in PSA CPS systems.

Following chain elongation, polysaccharide export across the inner membrane is achieved by a transporter belonging to the ABC-2 family. The TMD of the transporter comprises a homodimer of KpsM, while a homodimer of KpsT makes up the NBD of the transporter (Pigeon and Silver, 1994; Bliss and Silver, 1996). Deletion of either of the genes encoding these proteins leads to intracellular polymer accumulation (Kroncke *et al.*, 1990; Pavelka *et al.*, 1994; Pigeon and Silver, 1994). Complementation studies have shown that KpsM, KpsT, and KpsMT are functionally interchangeable between systems that assemble polymers with different structures (Roberts *et al.*, 1986, 1988; Pearce and Roberts, 1995; Ward and Inzana, 1997; Bacon *et al.*, 2001; Lo *et al.*, 2001; Cuthbertson *et al.*, 2005), suggesting the transporter does not recognize its native polysaccharide. This is in contrast to the LPS O-antigen export apparatus (comprising Wzm and Wzt), where Wzt (NBD) proteins are not functionally interchangeable (Cuthbertson *et al.*, 2005). The actual means by which the ABC transporter for CPS expression accommodates both a large hydrophilic polysaccharide and a hydrophobic lipid anchor are not known. It is conceivable that the KpsMT transporter may function in a manner analogous to the MsbA ABC transporter for lipid A-core export, which appears to move the hydrophilic portion of the substrate through the lumen of the transporter, while dragging the hydrophobic anchor through the lipid bilayer (Reyes and Chang, 2005; Whitfield, 2006).

Synthase-dependent mechanism

Synthases are processive glycosyltransferases that are responsible for the formation of some biologically important CPSs and EPSs including cellulose, hyaluronan, alginate and poly-GlcNAc (Fig. 2.4). Conserved domain searches show the presence

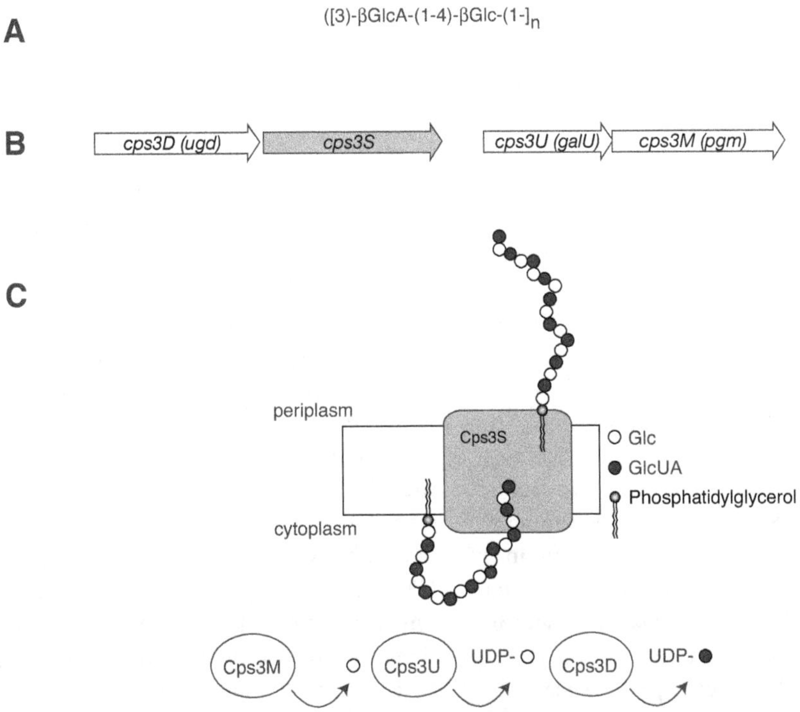

Figure 2.4 The synthase-dependent pathway for CPS biosynthesis in *S. pneumoniae* type 3. (A) Structure of the CPS repeat unit. (B) Organization of the CPS genetic locus. (C) Model for CPS biosynthesis. See text for details. Glc, glucose; GlcA, glucuronic acid; UDP, uridine diphosphate.

of one of two related conserved domains, the CESA_like domain and the GT2_HAS domain, in characterized synthase enzymes. A search for these domains in sequenced bacterial genomes reveals that these protein domains are widespread in Gram-positive and Gram-negative bacteria, indicating that while the synthase-dependent pathway of CPS biosynthesis is perhaps less well characterized than the Wzy- and ABC transporter-dependent mechanisms, it is no less important to bacterial physiology.

In *S. pneumoniae,* more than 90 different CPS serotypes have been identified. Although most of these follow a Wzy-dependent mechanism of biosynthesis, the types 3 and 37 CPSs follow a synthase-dependent mechanism of synthesis and export (Bentley *et al.,* 2006). The *S. pneumoniae* type 3 CPS contains a disaccharide repeat unit of GlcA and Glc (Reeves and Goebel, 1941). Despite the disaccharide repeating unit, the *S. pneumoniae* type 3 synthase (Cps3S) contains a single glycosyltransferase domain and is thought to be responsible for polymer initiation, elongation and export (Dillard *et al.,* 1995). While details of the initiation process are lacking, synthesis is initiated on a phosphatidylglycerol carrier lipid (Cartee *et al.,* 2005a) with glucose as the predominant initiating residue (Forsee *et al.,* 2006, 2009a). CPS polymerization occurs at the non-reducing terminus (Cartee *et al.,* 2000; Forsee *et al.,* 2009a) and there are thought to be two distinct kinetic phases involved (Cartee *et al.,* 2005b; Forsee *et al.,* 2006). In the first phase, chain initiation occurs and the lipid-linked oligosaccharide is loosely associated with the enzyme. In the second phase, the substrate is tightly bound and the enzyme is in a fully processive state. CPS chain length is controlled by the concentrations of the UDP–sugar substrates (Ventura *et al.,* 2006; Forsee *et al.,* 2009b). The mechanism of CPS export by the Cps3S synthase is unknown but Cps3S contains four predicted transmembrane domains, one at the N terminus and three at the C terminus, which may play a role (Dillard *et al.,* 1995).

While the hyaluronan CPSs of *S. pyogenes* and *Pasteurella multocida* are both polymerized by a synthase-type glycosyltransferase, the enzymes are very different (Weigel and DeAngelis, 2007). Details on the mechanism of hyaluronan biosynthesis in both species are lacking but available data indicate that this probably occurs by very different mechanisms. HasA of *S. pyogenes* contains one glycosyltransferase domain with two transmembrane segments on either side (Heldermon *et al.*, 2001). HasA of *P. multocida* contains two glycosyltransferase domains and lacks transmembrane domains (Jing and DeAngelis, 2000). While the details of initiation are unknown, hyaluronan polymerization is thought to occur at the reducing end in *S. pyogenes* (Tlapak-Simmons *et al.*, 2005) but at the non-reducing terminus in *P. multocida* (DeAngelis, 1999). Hyaluronan chain length is controlled by UDP–sugar substrates in *Streptococcus zooepidemicus* (Chen *et al.*, 2009) similar to the type 3 CPS of *S. pneumoniae*, indicating a possible common mechanism of CPS chain-length regulation in the synthase-dependent pathway. While HasA of *S. pyogenes* is thought to mediate polymer export, the HasA protein of *P. multocida* is located within a CPS cluster closely resembling ABC transporter-dependent clusters like that of *E. coli* K1 and as such, export of hyaluronan in *P. multocida* is probably mediated by the ABC transporter via a mechanism similar to other ABC transporter-dependent CPSs (Chung *et al.*, 1998). In one instance, an ABC transporter has been implicated in the export of hyaluronan in *S. pyogenes* (Ouskova *et al.*, 2004). However, this transporter is most closely related to ABC transporters involved in cobalt transport and is conserved in other species of *Streptococcus* lacking the hyaluronan biosynthesis genes (L. Cuthbertson, personal observation).

Surface expression of CPSs and EPSs

Transport across the outer membrane

In Gram-negative organisms, the polymerized CPS or EPS molecules must cross the outer membrane before reaching the cell surface. Most CPS or EPS systems include one of two distinct outer membrane proteins believed to serve this function. These are exemplified by the *E. coli* Wza protein involved in the Wzy-dependent polymerization of K30 CPS, and the *E. coli* KpsD proteins involved in the ABC transporter-dependent assembly of K1 and K5 CPS. Wza and KpsD belong to the outer membrane auxiliary (OMA) protein family (Paulsen *et al.*, 1997), which has recently been renamed the outer membrane polysaccharide export (OPX) family to reflect the role of these proteins in polymer extrusion (Cuthbertson *et al.*, 2009). Primary sequence similarity between Wza and KpsD proteins is poor and largely limited to the N-terminal region (McNulty *et al.*, 2006), where both proteins share a polysaccharide biosynthesis/export (PES) motif (Pfam02563) (Dong *et al.*, 2006). Despite this, secondary structure analysis revealed that OPX proteins share a significant degree of structural similarity (Cuthbertson *et al.*, 2009).

Wza is an acylated outer membrane lipoprotein (Paulsen *et al.*, 1997; Drummelsmith and Whitfield, 1999). Early studies revealed that Wza forms SDS-stable aggregates that resemble a ring with a central pore (Drummelsmith and Whitfield, 2000; Nesper *et al.*, 2003). Single particle analyses later revealed that Wza forms an octamer, whose structure is organized as a tetramer of dimers (Beis *et al.*, 2004). The structure of the Wza octamer was further refined by crystallographic studies that revealed a narrow α-helical barrel in the outer membrane and a series of larger stacked rings in the periplasm (Dong *et al.*, 2006). The conserved PES domain is located within the innermost ring of this complex. The structure is sealed in the periplasmic end, suggesting the complex can exist in both open and closed conformations. A channel lined with hydrophilic amino acids runs along the interior of the structure, and is believed to form a water-filled channel that allows passage of the CPS or EPS chains. Such a channel would preclude polymer binding by Wza, and is consistent with observations that heterologous Wza proteins can function in cross-complementation studies in systems that export drastically different polymers (Reid and Whitfield, 2005).

Deletion of *wza* from *E. coli* K30 causes loss of HMM CPS and the production of shorter K_{LPS}

molecules (Drummelsmith and Whitfield, 2000), as seen in both *wzc* and *wzb* deletion mutants (Drummelsmith and Whitfield, 1999; Wugeditsch *et al.*, 2001; Nesper *et al.*, 2003). The lack of periplasmic accumulation of HMM CPS in these strains is consistent with the presence of a feedback mechanism that would prevent extensive repeat unit biosynthesis and polymerization when export of the CPS or EPS is not possible. Accumulation of HMM CPS does occur in a strain that expresses a derivative of Wza that can not be acylated (Nesper *et al.*, 2003), suggesting the expression of this non-functional Wza protein is sufficient to circumvent the feedback mechanism, even though its improper localization prevents polymer export.

Early electron microscopy studies revealed that sites of nascent CPS biosynthesis coincided with zones of membrane adhesion (termed Bayer's junctions), areas where the inner and outer membrane were in close apposition (Bayer and Thurow, 1977; Kroncke *et al.*, 1990). Interactions between Wzc at the inner membrane and Wza at the outer membrane were first suggested by *in vivo* cross-linking studies (Nesper *et al.*, 2003) and by genetic studies that showed that Wzc from the colanic acid system required its cognate Wza protein to complement a *wzc* defect in *E. coli* K30 (Reid and Whitfield, 2005). The structure of the Wza–Wzc complex was later obtained by single particle analysis, and revealed a molecular scaffold spanning the periplasm, bridging the inner and outer membranes (Collins *et al.*, 2007). The co-crystal revealed a Wza octamer whose innermost ring was opened by a large conformational change in the PES domain. The broadening observed in this co-crystal structure relative to the shape of the individual Wza and Wzc structures suggests that periplasmic polymer is able to enter the Wza–Wzc complex via openings in the side of the structure. While it is clear that Wzc interactions with Wza are crucial for polymer export, the requirement for Wzc homologues in Gram-positive CPS systems where export across an outer membrane is not required indicates this is not the only function of Wzc. Of note, Gram-positive Wzc homologues differ from their Gram-negative counterparts by having shorter periplasmic domains and fewer terminal tyrosine residues.

The outer membrane protein KpsD has to date only been found in CPS systems that employ ABC transporter-dependent biosynthesis. Unlike Wza, KpsD is not acylated (Pazzani *et al.*, 1993; Wunder *et al.*, 1994). Proper localization of KpsD to the outer membrane depends on the presence of other members of the CPS biosynthesis and export apparatus (Arrecubieta *et al.*, 2001), which may explain early results that indicated KpsD was located in the periplasm (Silver *et al.*, 1987; Wunder *et al.*, 1994). As with Wza, heat-labile and SDS-stable aggregates of KpsD have been observed (McNulty *et al.*, 2006). Structural information for the KpsD multimer is not yet available, but will be required to elucidate the mechanism of polymer export. While ABC transporter-dependent systems for CPS assembly do not have a Wzc homologue, they do have a dedicated member of the PCP family (PCP-3) (Morona *et al.*, 2000b, 2009), also called membrane fusion proteins (MFP) or membrane periplasmic auxiliary proteins (MPA-2) (Paulsen *et al.*, 1997). The MFP components of bacterial efflux pumps are proposed to span the periplasm, linking activities at the inner membrane to those at the outer membrane (Misra and Bavro, 2009). The PCP-3 protein KpsE is a periplasmic protein anchored in the inner membrane by a single TMD and an amphipathic α-helix (Arrecubieta *et al.*, 2001; Phoenix *et al.*, 2001). KpsE may exist as a dimer (Arrecubieta *et al.*, 2001), and genetic studies have provided indirect evidence to suggest KpsE interacts with the outer membrane protein KpsD (Arrecubieta *et al.*, 2001), thus bridging the inner and outer membranes. Until structural data are available to support the existence of this interaction, it remains to be seen whether KpsE–KpsD interactions form a scaffold similar to that formed by Wzc-Wza.

While most Wzy-dependent and ABC transporter-dependent systems are associated with a Wza-type and KpsD-type protein, respectively, exceptions have been noted. For instance, *H. influenzae* (Kroll *et al.*, 1990) and *N. meningitidis* (Frosch *et al.*, 1992) express CPSs assembled by ABC transporter-dependent systems, but contain an OPX protein that most closely resembles Wza (Cuthbertson *et al.*, 2009). Phylogenetic analysis based on the sequence of the PES domain showed that these Wza-like proteins cluster together (Cuthbertson *et al.*, 2009). Unlike Wza from *E. coli* K30, these OPX proteins function with an ABC transporter and a PCP-3 protein to mediate polymer export, and can complement a KpsD mutation when expressed with their cognate ABC transporter and PCP protein (Silver *et al.*, 2001).

Details surrounding export of synthase-dependent polymers such as those involved in the biosynthesis of cellulose, alginate and poly-GlcNAc are lacking. However, recent structural studies on AlgK, a protein involved in alginate production, indicate that a conserved protein apparatus is involved in export of synthase-dependent polymers across the outer membrane in Gram-negative bacteria. AlgK is an outer membrane lipoprotein containing multiple tetratricopeptide-like repeats (Keiski *et al.*, 2010). It is required for proper localization of the outer membrane protein AlgE (Rehm *et al.*, 1994), similar to the requirement of KpsE for proper localization of KpsD in *E. coli* group 2 CPS biosynthesis (below). In the current model for alginate biosynthesis, AlgK and AlgE are thought to fulfil the roles of the OPX and PCP proteins from other polysaccharide biosynthesis systems (discussed below). In other synthase systems, AlgK and AlgE proteins are fused into one polypeptide (e.g. PgaA, BcsC and PelB).

It has become clear that at least some of the proteins involved in CPS and EPS biosynthesis and export interact, and that these interactions are important for polymer expression. This chapter has already described interactions between Wzc–Wza and KpsE–KpsD, as well as the possibility of interactions between Wzx, Wzy and WbaP/WecA. Interactions between biosynthetic enzymes in *E. coli* K5 have been suggested by subcellular localization studies and studies of *in vitro* enzyme activity. KfiA required KfiC for membrane association, while KfiC required both KfiA and KfiB, and KfiB required both KfiA and KfiC for membrane localization (Hodson *et al.*, 2000). The presence of KfiA was required for KfiC activity *in vitro*, though KfiA functions in the absence of KfiC (Sugiura *et al.*, 2010). Using a LexA-based system for detecting protein interactions, homodimerization of several biosynthetic enzymes from *E. coli* K1 (NeuD, NeuA, NeuC and NeuB) was detected, as well as heterodimerization between NeuD and NeuB (Daines and Silver, 2000). Several lines of evidence support the existence of coordinated polymer biosynthesis and export. In *E. coli* K5, subcellular localization studies suggested interactions between proteins involved in polymer biosynthesis (KfiA, KfiC and KfiD) and those involved in export (KpsM, KpsT) (Rigg *et al.*, 1998). Those same biosynthetic enzymes were also found associated with KpsS and KpsC (Rigg *et al.*, 1998), proteins whose role in CPS expression remains unclear. Proper localization of KpsD was impaired in *kfiC* and *kpsE* mutants (Arrecubieta *et al.*, 2001), suggesting either direct or indirect interactions between these proteins. Membrane preparations from *kpsT*-deficient strains were not able to support *in vitro* addition of sialic acid onto endogenous acceptors, suggesting that transport defects impact polymer biosynthesis (Vimr *et al.*, 1989; Bronner *et al.*, 1993). Chemical cross-linking studies in *E. coli* K5 also revealed the existence of protein complexes that included KpsS, KpsD, KpsE, KpsM, KpsT and KfiA (McNulty *et al.*, 2006). Protein interactions in *E. coli* K1 have been probed using a bacterial two-hybrid system. In this system, a strong interaction was detected between KpsC and the polysialyltransferase NeuS, between KpsC and KpsE, as well as between KpsC proteins themselves, suggesting that KpsC could couple polymer synthesis and export (Steenbergen and Vimr, 2008). Finally, K1 CPS accumulating in the cytoplasm of a *kpsC* mutant is sensitive to digestion by a plasmid-encoded endosialidase, while nascent CPS in the wild-type strain is protected (Steenbergen and Vimr, 2008). This is consistent with concerted biosynthesis and export in the wild-type K1 system. Similar studies in *E. coli* K5 revealed that the polymer in this strain is sensitive to degradation by a plasmid-encoded lyase (Hudson *et al.*, 2009). The authors suggest, among other explanations, that this discrepancy may be due in part to differences in the biosynthesis of these polymers, as K1 is a homopolymer elongated by a single glycosyltransferase and K5 is a heteropolymer whose synthesis requires the alternating activities of two glycosyltransferases (Hudson *et al.*, 2009).

The question of how protein complexes for CPS and EPS biosynthesis and export would traverse the stress-bearing PG layer has not been addressed to date. Multienzyme complexes such as those described above may require the creation of transient holes in the PG layer. While some envelope-spanning structures such as protein secretion systems are associated with a lytic transglycosylase believed to create such openings in the PG layer (Scheurwater and Burrows, 2011), CPS and EPS loci described to date do not encode such an enzyme. It remains possible that this function is carried out by lytic transglycosylases encoded elsewhere on the chromosome. In *E. coli* K5, components of the CPS biosynthesis and export apparatus, as well as nascent polysaccharide, appear to localize to the cell poles (McNulty *et al.*, 2006). It is conceivable that the unique properties of the cell poles (e.g. metabolically inert PG) (Burman *et al.*, 1983; de Pedro *et al.*, 1997) may be important for the insertion of such envelope-spanning structures.

Surface-association of CPS and EPS molecules

While both CPSs and EPSs are assembled by common mechanisms, we distinguish these polymers by their degree of attachment to the cell surface. EPSs are loosely associated with the surface, and can easily be disrupted by routine laboratory manipulations such as centrifugation and resuspension of cell pellets. CPSs are more firmly anchored, and resist such manipulations. At least some CPSs are covalently linked to the outer membrane via attachment to a lipid anchor (e.g. 1,2-diacylglycerol phosphate anchor on PSA CPS of *E. coli* and *N. meningitidis*) (Gotschlich *et al.*, 1981; Schmidt and Jann, 1982; Tzeng *et al.*, 2005). However, it is not clear if all ABC transporter-dependent CPSs are anchored in this manner. The terminal molecule on group 1 CPSs has yet to be conclusively identified, though genetic studies suggested that the Wzi protein might play a role in surface attachment of *E. coli* K30 CPS (Rahn *et al.*, 2003). The *wzi* gene has been found in all group 1 CPS loci of *E. coli* and *K. pneumoniae* examined to date (Rahn *et al.*, 1999), and homologues of *wzi* have also been found in *Acinetobacter*, *Providencia*, *Serratia* and *Shewanella* (Rahn *et al.*, 2003; Bushell *et al.*, 2010). It remains difficult to identify the terminal moiety on long, hydrophilic polysaccharide chains, as isolation procedures must be gentle enough to preserve the linkage, and analytical methods sensitive enough to characterize a small fraction of a very large molecule. Advances in lipid isolation and characterization methodologies will hopefully allow these questions to be resolved in the near future.

In Gram-positive organisms, polymerized CPS molecules at the outer face of the inner membrane can be covalently linked to the PG layer (De Cueninck *et al.*, 1982; Yeung and Mattingly, 1983a,b; Sorensen *et al.*, 1990; Deng *et al.*, 2000), or alternatively anchored to the membrane via a phosphatidylglycerol moiety (Bender *et al.*, 2003; Cartee *et al.*, 2005a).

Origins and impact of CPS variability

It has long been appreciated that CPS expression must be modulated to ensure a balance between the protective function of the capsule and exposure of outer membrane structures such as adhesins required for substrate binding and/or colonization. Temperature provides transcriptional regulation for some CPSs and EPSs, and may serve to ensure polymer production when suitable environmental conditions are present (e.g. 37°C for human pathogens). The expression of group 2 CPS molecules in *E. coli* and many other organisms is thermoregulated, and optimal CPS expression occurs at or near 37°C, while CPS expression is negligible at growth temperatures below 20°C (Cieslewicz and Vimr, 1996; Simpson *et al.*, 1996; Rowe *et al.*, 2000). Expression of the *kps* gene cluster is driven by two temperature-regulated promoters (Prf1 and Prf3), whose activities are influenced by a number of accessory proteins including SlyA, H-NS, BipA and IHF (Rowe *et al.*, 2000; Corbett *et al.*, 2007). The interplay of these factors and the conditions governing their expression and activities are beyond the scope of this chapter, and have recently been reviewed elsewhere (Ferrero and Aparicio, 2010). Temperature also influences expression of EPSs such as colanic acid (Yeh and Chen, 2004; Whitfield, 2006; Navasa *et al.*, 2009) and amylovoran (Smirnova *et al.*, 2001), as these polysaccharides are only produced at environmentally relevant temperatures (below 30°C). In contrast, group 1 CPSs (e.g. *E. coli* K30) appear to be constitutively expressed (Rahn and Whitfield, 2003). In the latter system, the Rcs (regulator of capsule synthesis) two-component regulatory system affects the availability of sugar–nucleotide precursors, and as such indirectly affects the amount of CPS produced (Rahn and Whitfield, 2003). The Rcs system is now known to affect a wide array of cellular processes, the details of which can be found in a recent review (Clarke, 2010).

In addition to turning CPS expression on and off, many bacteria harbour the means to modify CPS chains. This can have the effect of preventing recognition by host immune molecules, as well as altering binding to and/or invasion of host tissues. Properties of PSA capsules such as degree of *O*-acetylation have been shown to vary, and affect virulence properties such as cell binding, invasion and resistance to the host immune response (Bergfeld *et al.*, 2007; King *et al.*, 2007). In *E. coli* K1, acetylation of Neu5Ac residues within a PSA chain is carried out by NeuO (Steenbergen *et al.*, 2006), whose expression is phase-variable (Bergfeld *et al.*, 2007). This leads to the production of CPS chains with *O*-acetyl groups at different positions and frequencies along the polysaccharide chain (Bergfeld *et al.*, 2007; King *et al.*, 2007). In *C. jejuni*, genes containing hypervariable regions are clustered within CPS, lipo-oligosaccharide and flagellar glycosylation loci (Parkhill *et al.*, 2000; Gundogdu *et al.*, 2007). Nuclear magnetic resonance spectroscopy analysis of *C. jejuni* CPSs confirmed the phase-variability of several CPS modifications (e.g. phosphoramidate) (Szymanski *et al.*, 2003). Phase-variable CPS modifications may account for reports of antigenic variability of *Campylobacter* CPSs (Mills *et al.*, 1991; Aspinall *et al.*, 1992).

In *Bacteroides fragilis* the expression of eight CPS loci (PSA-PSH) is regulated through DNA inversions and transcriptional antitermination, and contributes to the antigenic diversity of this bacterium (Krinos *et al.*, 2001; Chatzidaki-Livanis *et al.*, 2009, 2010). In addition, *B. fragilis* also expresses a phase-variable extracellular polysaccharide (Chatzidaki-Livanis *et al.*, 2008) and encodes for the synthesis of an additional two CPSs not yet observed (Patrick *et al.*, 2010). Among three sequenced strains of *B. fragilis* examined, 28 different CPS loci were found (Patrick *et al.*, 2010), representing an extraordinary degree of diversity and highlighting the importance of antigenic diversity to the physiology of this organism.

Given that CPS molecules form the basis of serological typing schemes for the identification of many bacteria, variability in the expression (on/off), amount and properties (length, modifications) of the polymer can lead to erroneous results. For instance, the expression of phase-variable modifications in *C. jejuni* NCTC 11168 altered the strain's antibody reactivity (Szymanski *et al.*, 2003). In order to circumvent these issues, efforts are under way to develop typing schemes that rely on differences in gene sequences rather than antibody reactivity.

Future directions

Despite the decades of research that have gone into studying CPS and EPS biosynthesis and surface expression, many questions remain unanswered to this day. In Wzy-dependent polysaccharide assembly systems, the mechanisms used to 'flip' repeat units across the inner membrane and to effect their polymerization in the periplasm remain unknown. Improved methods for the overexpression and purification of large integral membrane proteins will be required to obtain these proteins in their active forms in quantities sufficient for biochemical and/or crystallographic studies. Unanswered questions in ABC transporter-dependent CPS systems include the mechanism of polymer initiation, the nature of the lipid anchor(s) and the means of chain termination/initiation of polymer export. The development of suitable methods for the analysis of lipid-linked intermediates (see Chapter 9) may help to address whether a single lipid anchor is used, and whether this lipid is 1,2-diacylglycerol phosphate for all CPSs assembled by this mechanism. Such methods could also solve the question of how group 1 CPSs are retained on the cell surface. Crystallographic studies of Wza and the Wza–Wzc pair have greatly improved our understanding of CPS transport across the Gram-negative outer membrane. Comparable structural data will be required to elucidate the nature of the KpsD channel, and to determine whether the KpsD–KpsE pair form a transenvelope scaffold similar to Wza–Wzc. It is conceivable that structural data may someday be able to define the players and the nature of the protein interactions in the multienzyme complexes for CPS and EPS expression.

References

Abeygunawardana, C., Williams, T.C., Sumner, J.S., and Hennessey, J.P., Jr. (2000). Development and validation of an NMR-based identity assay for bacterial polysaccharides. Anal. Biochem. *279*, 226–240.

Abeyrathne, P.D., and Lam, J.S. (2007). WaaL of *Pseudomonas aeruginosa* utilizes ATP in *in vitro* ligation of O antigen onto lipid A-core. Mol. Microbiol. *65*, 1345–1359.

Amer, A.O., and Valvano, M.A. (2000). The N-terminal region of the *Escherichia coli* WecA (Rfe) protein, containing three predicted transmembrane helices, is required for function but not for membrane insertion. J. Bacteriol. *182*, 498–503.

Amer, A.O., and Valvano, M.A. (2001). Conserved amino acid residues found in a predicted cytosolic domain of the lipopolysaccharide biosynthetic protein WecA are implicated in the recognition of UDP-N-acetylglucosamine. Microbiology *147*, 3015–3025.

Amor, P.A., and Whitfield, C. (1997). Molecular and functional analysis of genes required for expression of group IB K antigens in *Escherichia coli*: characterization of the *his*-region containing gene clusters for multiple cell-surface polysaccharides. Mol. Microbiol. *26*, 145–161.

Anderson, M.S., Eveland, S.S., and Price, N.P. (2000). Conserved cytoplasmic motifs that distinguish sub-groups of the polyprenol phosphate:N-acetylhexosamine-1-phosphate transferase family. FEMS Microbiol. Lett. *191*, 169–175.

Andreishcheva, E.N., and Vann, W.F. (2006). Gene products required for *de novo* synthesis of polysialic acid in *Escherichia coli* K1. J. Bacteriol. *188*, 1786–1797.

Arrecubieta, C., Hammarton, T.C., Barrett, B., Chareonsudjai, S., Hodson, N., Rainey, D., and Roberts, I.S. (2001). The transport of group 2 capsular polysaccharides across the periplasmic space in *Escherichia coli*. Roles for the KpsE and KpsD proteins. J. Biol. Chem. *276*, 4245–4250.

Aspinall, G.O., McDonald, A.G., Raju, T.S., Pang, H., Mills, S.D., Kurjanczyk, L.A., and Penner, J.L. (1992). Serological diversity and chemical structures of *Campylobacter jejuni* low-molecular-weight lipopolysaccharides. J. Bacteriol. *174*, 1324–1332.

Avci, F.Y., and Kasper, D.L. (2010). How bacterial carbohydrates influence the adaptive immune system. Ann. Rev. Immunol. *28*, 107–130.

Bacon, D.J., Szymanski, C.M., Burr, D.H., Silver, R.P., Alm, R.A., and Guerry, P. (2001). A phase-variable capsule is involved in virulence of *Campylobacter jejuni* 81–176. Mol. Microbiol. *40*, 769–777.

Bayer, M.E., and Thurow, H. (1977). Polysaccharide capsule of *Escherichia coli*: microscope study of its size, structure, and sites of synthesis. J. Bacteriol. *130*, 911–936.

Bechet, E., Gruszczyk, J., Terreux, R., Gueguen-Chaignon, V., Vigouroux, A., Obadia, B., Cozzone, A.J., Nessler, S., and Grangeasse, C. (2010). Identification of structural and molecular determinants of the tyrosine-kinase Wzc and implications in capsular polysaccharide export. Mol. Microbiol. *77*, 1315–1325.

Becker, A., and Puhler, A. (1998). Specific amino acid substitutions in the proline-rich motif of the *Rhizobium meliloti* ExoP protein result in enhanced production of low-molecular-weight succinoglycan at the expense of high-molecular-weight succinoglycan. J. Bacteriol. *180*, 395–399.

Becker, A., Niehaus, K., and Puhler, A. (1995). Low-molecular-weight succinoglycan is predominantly produced by *Rhizobium meliloti* strains carrying a mutated ExoP protein characterized by a periplasmic N-terminal domain and a missing C-terminal domain. Mol. Microbiol. *16*, 191–203.

Beis, K., Collins, R.F., Ford, R.C., Kamis, A.B., Whitfield, C., and Naismith, J.H. (2004). Three-dimensional structure of Wza, the protein required for translocation of group 1 capsular polysaccharide across the outer membrane of *Escherichia coli*. J. Biol. Chem. *279*, 28227–28232.

Bender, M.H., and Yother, J. (2001). CpsB is a modulator of capsule-associated tyrosine kinase activity in *Streptococcus pneumoniae*. J. Biol. Chem. *276*, 47966–47974.

Bender, M.H., Cartee, R.T., and Yother, J. (2003). Positive correlation between tyrosine phosphorylation of CpsD and capsular polysaccharide production in *Streptococcus pneumoniae*. J. Bacteriol. *185*, 6057–6066.

Bengoechea, J.A., Pinta, E., Salminen, T., Oertelt, C., Holst, O., Radziejewska-Lebrecht, J., Piotrowska-Seget, Z., Venho, R., and Skurnik, M. (2002). Functional characterization of Gne (UDP-N-acetylglucosamine-4-epimerase), Wzz (chain length determinant), and Wzy (O-antigen polymerase) of *Yersinia enterocolitica* serotype O:8. J. Bacteriol. *184*, 4277–4287.

Bentley, S.D., Aanensen, D.M., Mavroidi, A., Saunders, D., Rabbinowitsch, E., Collins, M., Donohoe, K., Harris, D., Murphy, L., Quail, M.A., *et al.* (2006). Genetic analysis of the capsular biosynthetic locus from all 90 pneumococcal serotypes. PLoS Genet. *2*, e31.

Bergfeld, A.K., Claus, H., Vogel, U., and Muhlenhoff, M. (2007). Biochemical characterization of the polysialic acid-specific O-acetyltransferase NeuO of *Escherichia coli* K1. J. Biol. Chem. *282*, 22217–22227.

Bergfeld, A.K., Claus, H., Lorenzen, N.K., Spielmann, F., Vogel, U., and Muhlenhoff, M. (2009). The polysialic acid-specific O-acetyltransferase OatC from *Neisseria meningitidis* serogroup C evolved apart from other bacterial sialate O-acetyltransferases. J. Biol. Chem. *284*, 6–16.

Bliss, J.M., and Silver, R.P. (1996). Coating the surface: a model for expression of capsular polysialic acid in *Escherichia coli* K1. Mol. Microbiol. *21*, 221–231.

Bronner, D., Sieberth, V., Pazzani, C., Smith, A., Boulnois, G., Roberts, I., Jann, B., and Jann, K. (1993). Synthesis of the K5 (group II) capsular polysaccharide in transport-deficient recombinant *Escherichia coli.* FEMS Microbiol. Lett. *113,* 279–284.

Bugert, P., and Geider, K. (1995). Molecular analysis of the *ams* operon required for exopolysaccharide synthesis of *Erwinia amylovora.* Mol. Microbiol. *15,* 917–933.

Burman, L.G., Raichler, J., and Park, J.T. (1983). Evidence for diffuse growth of the cylindrical portion of the *Escherichia coli* murein sacculus. J. Bacteriol. *155,* 983–988.

Bushell, S.R., Lou, H., Wallat, G.D., Beis, K., Whitfield, C., and Naismith, J.H. (2010). Crystallization and preliminary diffraction analysis of Wzi, a member of the capsule export and assembly pathway in *Escherichia coli.* Acta Crystallogr. Sect. F Struct. Biol. Cryst. Commun. *66,* 1621–1625.

Cartee, R.T., Forsee, W.T., Schutzbach, J.S., and Yother, J. (2000). Mechanism of type 3 capsular polysaccharide synthesis in *Streptococcus pneumoniae.* J. Biol. Chem. *275,* 3907–3914.

Cartee, R.T., Forsee, W.T., Bender, M.H., Ambrose, K.D., and Yother, J. (2005a). CpsE from type 2 *Streptococcus pneumoniae* catalyses the reversible addition of glucose-1-phosphate to a polyprenyl phosphate acceptor, initiating type 2 capsule repeat unit formation. J. Bacteriol. *187,* 7425–7433.

Cartee, R.T., Forsee, W.T., and Yother, J. (2005b). Initiation and synthesis of the *Streptococcus pneumoniae* type 3 capsule on a phosphatidylglycerol membrane anchor. J. Bacteriol. *187,* 4470–4479.

Chakraborty, A.K., Friebolin, H., and Stirm, S. (1980). Primary structure of the *Escherichia coli* serotype K30 capsular polysaccharide. J. Bacteriol. *141,* 971–972.

Chao, C.F., Chuang, H.C., Chiou, S.T., and Liu, T.Y. (1999). On the biosynthesis of alternating alpha-2,9/alpha-2,8 heteropolymer of sialic acid catalysed by the sialyltransferase of *Escherichia coli* Bos-12. J. Biol. Chem. *274,* 18206–18212.

Chatzidaki-Livanis, M., Coyne, M.J., Roche-Hakansson, H., and Comstock, L.E. (2008). Expression of a uniquely regulated extracellular polysaccharide confers a large-capsule phenotype to *Bacteroides fragilis.* J. Bacteriol. *190,* 1020–1026.

Chatzidaki-Livanis, M., Coyne, M.J., and Comstock, L.E. (2009). A family of transcriptional antitermination factors necessary for synthesis of the capsular polysaccharides of *Bacteroides fragilis.* J. Bacteriol. *191,* 7288–7295.

Chatzidaki-Livanis, M., Weinacht, K.G., and Comstock, L.E. (2010). Trans locus inhibitors limit concomitant polysaccharide synthesis in the human gut symbiont *Bacteroides fragilis.* Proc. Natl. Acad. Sci. U.S.A. *107,* 11976–11980.

Chen, W.Y., Marcellin, E., Hung, J., and Nielsen, L.K. (2009). Hyaluronan molecular weight is controlled by UDP-N-acetylglucosamine concentration in *Streptococcus zooepidemicus.* J. Biol. Chem. *284,* 18007–18014.

Cheng, J., Liu, B., Bastin, D.A., Han, W., Wang, L., and Feng, L. (2007). Genetic characterization of the *Escherichia coli* O66 antigen and functional identification of its *wzy* gene. J. Microbiol. *45,* 69–74.

Cho, J.W., and Troy, F.A., 2nd (1994). Polysialic acid engineering: synthesis of polysialylated neoglycosphingolipids by using the polysialyltransferase from neuroinvasive *Escherichia coli* K1. Proc. Natl. Acad. Sci. U.S.A. *91,* 11427–11431.

Chung, J.Y., Zhang, Y., and Adler, B. (1998). The capsule biosynthetic locus of *Pasteurella multocida* A:1. FEMS Microbiol. Lett. *166,* 289–296.

Cieslewicz, M., and Vimr, E. (1996). Thermoregulation of *kpsF,* the first region 1 gene in the *kps* locus for polysialic acid biosynthesis in *Escherichia coli* K1. J. Bacteriol. *178,* 3212–3220.

Cieslewicz, M., and Vimr, E. (1997). Reduced polysialic acid capsule expression in *Escherichia coli* K1 mutants with chromosomal defects in *kpsF.* Mol. Microbiol. *26,* 237–249.

Cieslewicz, M.J., Kasper, D.L., Wang, Y., and Wessels, M.R. (2001). Functional analysis in type Ia group B *Streptococcus* of a cluster of genes involved in extracellular polysaccharide production by diverse species of streptococci. J. Biol. Chem. *276,* 139–146.

Clarke, B.R., Cuthbertson, L., and Whitfield, C. (2004). Nonreducing terminal modifications determine the chain length of polymannose O antigens of *Escherichia coli* and couple chain termination to polymer export via an ATP-binding cassette transporter. J. Biol. Chem. *279,* 35709–35718.

Clarke, D.J. (2010). The Rcs phosphorelay: more than just a two-component pathway. Future Microbiol. *5,* 1173–1184.

Collins, R.F., Beis, K., Clarke, B.R., Ford, R.C., Hulley, M., Naismith, J.H., and Whitfield, C. (2006). Periplasmic protein–protein contacts in the inner membrane protein Wzc form a tetrameric complex required for the assembly of *Escherichia coli* group 1 capsules. J. Biol. Chem. *281,* 2144–2150.

Collins, R.F., Beis, K., Dong, C., Botting, C.H., McDonnell, C., Ford, R.C., Clarke, B.R., Whitfield, C., and Naismith, J.H. (2007). The 3D structure of a periplasm-spanning platform required for assembly of group 1 capsular polysaccharides in *Escherichia coli.* Proc. Natl. Acad. Sci. U.S.A. *104,* 2390–2395.

Comstock, L.E., and Kasper, D.L. (2006). Bacterial glycans: key mediators of diverse host immune responses. Cell *126,* 847–850.

Corbett, D., Bennett, H.J., Askar, H., Green, J., and Roberts, I.S. (2007). SlyA and H-NS regulate transcription of the *Escherichia coli* K5 capsule gene cluster, and expression of *slyA* in *Escherichia coli* is temperature-dependent, positively autoregulated, and independent of H-NS. J. Biol. Chem. *282,* 33326–33335.

Cunneen, M.M., and Reeves, P.R. (2008). Membrane topology of the *Salmonella enterica* serovar Typhimurium Group B O-antigen translocase Wzx. FEMS Microbiol. Lett. *287,* 76–84.

Cuthbertson, L., Powers, J., and Whitfield, C. (2005). The C-terminal domain of the nucleotide-binding domain protein Wzt determines substrate specificity in the ATP-binding cassette transporter for the lipopolysaccharide O-antigens in *Escherichia coli* serotypes O8 and O9a. J. Biol. Chem. *280,* 30310–30319.

Cuthbertson, L., Kimber, M.S., and Whitfield, C. (2007). Substrate binding by a bacterial ABC transporter involved in polysaccharide export. Proc. Natl. Acad. Sci. U.S.A. *104,* 19529–19534.

Cuthbertson, L., Mainprize, I.L., Naismith, J.H., and Whitfield, C. (2009). Pivotal roles of the outer membrane polysaccharide export and polysaccharide copolymerase protein families in export of extracellular polysaccharides in Gram-negative bacteria. Microbiol. Mol. Biol. Rev. *73,* 155–177.

Cuthbertson, L., Kos, V., and Whitfield, C. (2010). ABC transporters involved in export of cell surface glycoconjugates. Microbiol. Mol. Biol. Rev. *74,* 341–362.

Daines, D.A., and Silver, R.P. (2000). Evidence for multimerization of Neu proteins involved in polysialic acid synthesis in *Escherichia coli* K1 using improved LexA-based vectors. J. Bacteriol. *182,* 5267–5270.

Dams-Kozlowska, H., and Kaplan, D.L. (2007). Protein engineering of Wzc to generate new emulsan analogs. Appl. Environ. Microbiol. *73,* 4020–4028.

Dams-Kozlowska, H., Sainath, N., and Kaplan, D.L. (2008). Construction of a chimeric gene cluster for the biosynthesis of apoemulsan with altered molecular weight. Appl. Microbiol. Biotechnol. *78,* 677–683.

Daniels, C., and Morona, R. (1999). Analysis of *Shigella flexneri wzz* (Rol) function by mutagenesis and cross-linking: Wzz is able to oligomerize. Mol. Microbiol. *34,* 181–194.

Daniels, C., Vindurampulle, C., and Morona, R. (1998). Overexpression and topology of the *Shigella flexneri* O-antigen polymerase (Rfc/Wzy). Mol. Microbiol. *28,* 1211–1222.

De Cueninck, B.J., Shockman, G.D., and Swenson, R.M. (1982). Group B, type III streptococcal cell wall: composition and structural aspects revealed through endo-N-acetylmuramidase-catalysed hydrolysis. Infect. Immun. 35, 572–581.

DeAngelis, P.L. (1999). Molecular directionality of polysaccharide polymerization by the *Pasteurella multocida* hyaluronan synthase. J. Biol. Chem. *274,* 26557–26562.

Deng, L., Kasper, D.L., Krick, T.P., and Wessels, M.R. (2000). Characterization of the linkage between the type III capsular polysaccharide and the bacterial cell wall of group B *Streptococcus.* J. Biol. Chem. *275,* 7497–7504.

Dengler, T., Jann, B., and Jann, K. (1986). Structure of the serine-containing capsular polysaccharide K40 antigen from *Escherichia coli* O8:K40:H9. Carbohydr. Res. *150,* 233–240.

Dillard, J.P., Vandersea, M.W., and Yother, J. (1995). Characterization of the cassette containing genes for type 3 capsular polysaccharide biosynthesis in *Streptococcus pneumoniae.* J. Exp. Med. *181,* 973–983.

Dodgson, C., Amor, P., and Whitfield, C. (1996). Distribution of the *rol* gene encoding the regulator of lipopolysaccharide O-chain length in *Escherichia coli* and its influence on the expression of group I capsular K antigens. J. Bacteriol. *178,* 1895–1902.

Dong, C., Beis, K., Nesper, J., Brunkan-Lamontagne, A.L., Clarke, B.R., Whitfield, C., and Naismith, J.H. (2006). Wza the translocon for *E. coli* capsular polysaccharides defines a new class of membrane protein. Nature *444,* 226–229.

Doublet, P., Grangeasse, C., Obadia, B., Vaganay, E., and Cozzone, A.J. (2002). Structural organization of the protein-tyrosine autokinase Wzc within *Escherichia coli* cells. J. Biol. Chem. *277,* 37339–37348.

Drummelsmith, J., and Whitfield, C. (1999). Gene products required for surface expression of the capsular form of the group 1 K antigen in *Escherichia coli* (O9a:K30). Mol. Microbiol. *31,* 1321–1332.

Drummelsmith, J., and Whitfield, C. (2000). Translocation of group 1 capsular polysaccharide to the surface of *Escherichia coli* requires a multimeric complex in the outer membrane. EMBO J. *19,* 57–66.

Duggan, S.T. (2010). Pneumococcal polysaccharide conjugate vaccine (13-valent, adsorbed) [prevenar 13(R)]. Drugs *70,* 1973–1986.

Feldman, M.F., Marolda, C.L., Monteiro, M.A., Perry, M.B., Parodi, A.J., and Valvano, M.A. (1999). The activity of a putative polyisoprenol-linked sugar translocase (Wzx) involved in *Escherichia coli* O antigen assembly is independent of the chemical structure of the O repeat. J. Biol. Chem. *274,* 35129–35138.

Fernandes, A.I., and Gregoriadis, G. (1997). Polysialylated asparaginase: preparation, activity and pharmacokinetics. Biochim. Biophys. Acta *1341,* 26–34.

Ferrero, M.A., and Aparicio, L.R. (2010). Biosynthesis and production of polysialic acids in bacteria. Appl. Microbiol. Biotechnol. *86,* 1621–1635.

Ferrero, M.A., Luengo, J.M., and Reglero, A. (1991). H.p.l.c. of oligo(sialic acids). Application to the determination of the minimal chain length serving as exogenous acceptor in the enzymic synthesis of colominic acid. Biochem. J. *280,* 575–579.

Finke, A., Roberts, I., Boulnois, G., Pzzani, C., and Jann, K. (1989). Activity of CMP-2-keto-3-deoxyoctulosonic acid synthetase in *Escherichia coli* strains expressing the capsular K5 polysaccharide implication for K5 polysaccharide biosynthesis. J. Bacteriol. *171,* 3074–3079.

Flemming, H.C., and Wingender, J. (2010). The biofilm matrix. Nat. Rev. Microbiol. *8,* 623–633.

Forsee, W.T., Cartee, R.T., and Yother, J. (2006). Role of the carbohydrate binding site of the *Streptococcus pneumoniae* capsular polysaccharide type 3 synthase in the transition from oligosaccharide to polysaccharide synthesis. J. Biol. Chem. *281,* 6283–6289.

Forsee, W.T., Cartee, R.T., and Yother, J. (2009a). Characterization of the lipid linkage region and chain length of the cellubiuronic acid capsule of *Streptococcus pneumoniae.* J. Biol. Chem. *284,* 11826–11835.

Forsee, W.T., Cartee, R.T., and Yother, J. (2009b). A kinetic model for chain length modulation of *Streptococcus pneumoniae* cellubiuronan capsular polysaccharide by nucleotide sugar donor concentrations. J. Biol. Chem. *284,* 11836–11844.

Fratamico, P.M., Bagi, L.K., Cray, W.C., Narang, N., Yan, X., Medina, M., and Liu, Y. (2011). Detection by multiplex real-time polymerase chain reaction assays and isolation of Shiga toxin-producing *Escherichia coli* serogroups O*26,* O*45,* O*103,* O*111,* O*121,* and O145 in ground beef. Foodborne Pathog. Dis. *8,* 601–607.

Freiberger, F., Claus, H., Gunzel, A., Oltmann-Norden, I., Vionnet, J., Muhlenhoff, M., Vogel, U., Vann, W.F., Gerardy-Schahn, R., and Stummeyer, K. (2007). Biochemical characterization of a *Neisseria meningitidis* polysialyltransferase reveals novel functional motifs in bacterial sialyltransferases. Mol. Microbiol. *65,* 1258–1275.

Frosch, M., and Muller, A. (1993). Phospholipid substitution of capsular polysaccharides and mechanisms of capsule formation in *Neisseria meningitidis.* Mol. Microbiol. *8,* 483–493.

Frosch, M., Muller, D., Bousset, K., and Muller, A. (1992). Conserved outer membrane protein of *Neisseria meningitidis* involved in capsule expression. Infect. Immun. *60,* 798–803.

Garegg, P.J., Lindberg, B., Onn, T., and Sutherland, I.W. (1971). Comparative structural studies on the M-antigen from *Salmonella typhimurium, Escherichia coli* and *Aerobacter cloacae.* Acta Chem. Scand. *25,* 2103–2108.

Gotschlich, E.C., Fraser, B.A., Nishimura, O., Robbins, J.B., and Liu, T.Y. (1981). Lipid on capsular polysaccharides of gram-negative bacteria. J. Biol. Chem. *256,* 8915–8921.

Grangeasse, C., Doublet, P., Vincent, C., Vaganay, E., Riberty, M., Duclos, B., and Cozzone, A.J. (1998). Functional characterization of the low-molecular-mass phosphotyrosine-protein phosphatase of *Acinetobacter johnsonii.* J. Mol. Biol. *278,* 339–347.

Grangeasse, C., Doublet, P., and Cozzone, A.J. (2002). Tyrosine phosphorylation of protein kinase Wzc from *Escherichia coli* K12 occurs through a two-step process. J. Biol. Chem. *277,* 7127–7135.

Grangeasse, C., Obadia, B., Mijakovic, I., Deutscher, J., Cozzone, A.J., and Doublet, P. (2003). Autophosphorylation of the *Escherichia coli* protein kinase Wzc regulates tyrosine phosphorylation of Ugd, a UDP-glucose dehydrogenase. J. Biol. Chem. *278,* 39323–39329.

Guidolin, A., Morona, J.K., Morona, R., Hansman, D., and Paton, J.C. (1994). Nucleotide sequence analysis of genes essential for capsular polysaccharide biosynthesis in *Streptococcus pneumoniae* type 19F. Infect. Immun. *62,* 5384–5396.

Gundogdu, O., Bentley, S.D., Holden, M.T., Parkhill, J., Dorrell, N., and Wren, B.W. (2007). Re-annotation and re-analysis of the *Campylobacter jejuni* NCTC11168 genome sequence. BMC Genomics *8,* 162.

Hagelueken, G., Huang, H., Mainprize, I.L., Whitfield, C., and Naismith, J.H. (2009). Crystal structures of Wzb of *Escherichia coli* and CpsB of *Streptococcus pneumoniae,* representatives of two families of tyrosine phosphatases that regulate capsule assembly. J. Mol. Biol. *392,* 678–688.

Heldermon, C., DeAngelis, P.L., and Weigel, P.H. (2001). Topological organization of the hyaluronan synthase from *Streptococcus pyogenes.* J. Biol. Chem. *276,* 2037–2046.

Hodson, N., Griffiths, G., Cook, N., Pourhossein, M., Gottfridson, E., Lind, T., Lidholt, K., and Roberts, I.S. (2000). Identification that KfiA, a protein essential for the biosynthesis of the *Escherichia coli* K5 capsular polysaccharide, is an alpha -UDP-GlcNAc glycosyltransferase. The formation of a membrane-associated K5 biosynthetic complex requires KfiA, KfiB, and KfiC. J. Biol. Chem. *275,* 27311–27315.

Hofmann, P., Jann, B., and Jann, K. (1985). Structure of the amino acid-containing capsular polysaccharide (K54 antigen) from *Escherichia coli* O6:K54:H10. Carbohydr. Res. *139,* 261–271.

Hudson, T., Goldrick, M., and Roberts, I.S. (2009). The *Escherichia coli* K5 capsule is not synthesized in a protected compartment within the cytoplasm. J. Bacteriol. *191,* 1716–1718.

Hug, I., and Feldman, M.F. (2011). Analogies and homologies in lipopolysaccharide and glycoprotein biosynthesis in bacteria. Glycobiol. *21,* 138–151.

Islam, S.T., Taylor, V.L., Qi, M., and Lam, J.S. (2010). Membrane topology mapping of the O-antigen flippase (Wzx), polymerase (Wzy), and ligase (WaaL) from *Pseudomonas aeruginosa* PAO1 reveals novel domain architectures. MBio *1,* e00189–10.

Jain, S., Hreczuk-Hirst, D.H., McCormack, B., Mital, M., Epenetos, A., Laing, P., and Gregoriadis, G. (2003). Polysialylated insulin: synthesis, characterization and biological activity *in vivo.* Biochim. Biophys. Acta *1622,* 42–49.

Jennings, H.J., Bhattacharjee, A.K., Bundle, D.R., Kenny, C.P., Martin, A., and Smith, I.C. (1977). Structures of the capsular polysaccharides of *Neisseria meningitidis* as determined by 13C-nuclear magnetic resonance spectroscopy. J. Infect. Dis. *136* (Suppl.), S78–S83.

Jing, W., and DeAngelis, P.L. (2000). Dissection of the two transferase activities of the *Pasteurella multocida* hyaluronan synthase: two active sites exist in one polypeptide. Glycobiol. *10,* 883–889.

Jones, C. (2005a). NMR assays for carbohydrate-based vaccines. J. Pharm. Biomed. Anal. *38,* 840–850.

Jones, C. (2005b). Revised structures for the capsular polysaccharides from *Staphylococcus aureus* Types 5 and 8, components of novel glycoconjugate vaccines. Carbohydr. Res. *340,* 1097–1106.

de Jong, W., Wosten, H.A., Dijkhuizen, L., and Claessen, D. (2009). Attachment of *Streptomyces coelicolor* is mediated by amyloidal fimbriae that are anchored to the cell surface via cellulose. Mol. Microbiol. *73,* 1128–1140.

Kaiser, D. (2008). *Myxococcus*-from single-cell polarity to complex multicellular patterns. Annu. Rev. Genet. *42,* 109–130.

Keiski, C.L., Harwich, M., Jain, S., Neculai, A.M., Yip, P., Robinson, H., Whitney, J.C., Riley, L., Burrows, L.L., Ohman, D.E., *et al.* (2010). AlgK is a TPR-containing protein and the periplasmic component of a novel exopolysaccharide secretin. Structure *18,* 265–273.

King, M.R., Steenbergen, S.M., and Vimr, E.R. (2007). Going for baroque at the *Escherichia coli* K1 cell surface. Trends Microbiol. *15,* 196–202.

Klein, G., Dartigalongue, C., and Raina, S. (2003). Phosphorylation-mediated regulation of heat shock response in *Escherichia coli.* Mol. Microbiol. *48,* 269–285.

Koczan, J.M., McGrath, M.J., Zhao, Y., and Sundin, G.W. (2009). Contribution of *Erwinia amylovora* exopolysaccharides amylovoran and levan to biofilm formation: implications in pathogenicity. Phytopathology *99,* 1237–1244.

Kos, V., Cuthbertson, L., and Whitfield, C. (2009). The *Klebsiella pneumoniae* O2a antigen defines a second mechanism for O antigen ATP-binding cassette transporters. J. Biol. Chem. *284,* 2947–2956.

Krinos, C.M., Coyne, M.J., Weinacht, K.G., Tzianabos, A.O., Kasper, D.L., and Comstock, L.E. (2001). Extensive surface diversity of a commensal microorganism by multiple DNA inversions. Nature *414,* 555–558.

Kroll, J.S., Loynds, B., Brophy, L.N., and Moxon, E.R. (1990). The *bex* locus in encapsulated *Haemophilus influenzae*: a chromosomal region involved in capsule polysaccharide export. Mol. Microbiol. *4,* 1853–1862.

Kroncke, K.D., Boulnois, G., Roberts, I., Bitter-Suermann, D., Golecki, J.R., Jann, B., and Jann, K. (1990). Expression of the *Escherichia coli* K5 capsular antigen: immunoelectron microscopic and biochemical studies with recombinant *E. coli.* J. Bacteriol. *172,* 1085–1091.

Lacour, S., Bechet, E., Cozzone, A.J., Mijakovic, I., and Grangeasse, C. (2008). Tyrosine phosphorylation of the UDP-glucose dehydrogenase of *Escherichia coli* is at the crossroads of colanic acid synthesis and polymyxin resistance. PLoS One 3, e3053.

Lee, D.C., Zheng, J., She, Y.M., and Jia, Z. (2008). Structure of *Escherichia coli* tyrosine kinase Etk reveals a novel activation mechanism. EMBO J. *27,* 1758–1766.

Lehrer, J., Vigeant, K.A., Tatar, L.D., and Valvano, M.A. (2007). Functional characterization and membrane topology of *Escherichia coli* WecA, a sugar-phosphate transferase initiating the biosynthesis of enterobacterial common antigen and O-antigen lipopolysaccharide. J. Bacteriol. *189,* 2618–2628.

Lewis, A.L., Nizet, V., and Varki, A. (2004). Discovery and characterization of sialic acid O-acetylation in group B *Streptococcus.* Proc. Natl. Acad. Sci. U.S.A. *101,* 11123–11128.

Lewis, A.L., Hensler, M.E., Varki, A., and Nizet, V. (2006). The group B streptococcal sialic acid O-acetyltransferase is encoded by *neuD,* a conserved component of bacterial sialic acid biosynthetic gene clusters. J. Biol. Chem. *281,* 11186–11192.

Lindhout, T., Iqbal, U., Willis, L., Reid, A., Li, J., Liu, X., Moreno, M., and Wakarchuk, W. (2011). Site-specific enzymatic polysialylation of therapeutic proteins using bacterial enzymes. Proc. Natl. Acad. Sci. USA. *108,* 7397–7402.

Liu, D., Cole, R.A., and Reeves, P.R. (1996). An O-antigen processing function for Wzx (RfbX): a promising candidate for O-unit flippase. J. Bacteriol. *178,* 2102–2107.

Liu, T.Y., Gotschlich, E.C., Dunne, F.T., and Jonssen, E.K. (1971). Studies on the meningococcal polysaccharides. II. Composition and chemical properties of the group B and group C polysaccharide. J. Biol. Chem. *246,* 4703–4712.

Lo, R.Y., McKerral, L.J., Hills, T.L., and Kostrzynska, M. (2001). Analysis of the capsule biosynthetic locus of *Mannheimia (Pasteurella) haemolytica* A1 and proposal of a nomenclature system. Infect. Immun. *69,* 4458–4464.

Lorusso, V., Dambrosio, A., Quaglia, N.C., Parisi, A., Lasalandra, G., Mula, G., Virgilio, S., Lucifora, G., Dario, M., and Normanno, G. (2011). Development of a multiplex PCR for rapid detection of verocytotoxin-producing *Escherichia coli* O26 in raw milk and ground beef. J. Food Protect. *74,* 13–17.

Lu, A., Cho, K., Black, W.P., Duan, X.Y., Lux, R., Yang, Z., Kaplan, H.B., Zusman, D.R., and Shi, W. (2005). Exopolysaccharide biosynthesis genes required for social motility in *Myxococcus xanthus.* Mol. Microbiol. *55,* 206–220.

Lu, T., Tan, H., Lee, D., Chen, G., and Jia, Z. (2009). New insights into the activation of *Escherichia coli* tyrosine kinase revealed by molecular dynamics simulation and biochemical analysis. Biochemistry *48,* 7986–7995.

Mavroidi, A., Aanensen, D.M., Godoy., Skovsted, I.C., Kaltoft, M.S., Reeves, P.R., Bentley, S.D., Spratt, B.G. (2007). Genetic relatedness of the *Streptococcus pneumoniae* capsular biosynthetic loci. J. Bacteriol. *189,* 7841–7855.

McGuire, E.J., and Binkley, S.B. (1964). The structure and chemistry of colominic acid. Biochemistry 3, 247–251.

McNally, D.J., Lamoureux, M.P., Karlyshev, A.V., Fiori, L.M., Li, J., Thacker, G., Coleman, R.A., Khieu, N.H., Wren, B.W., Brisson, J.R., *et al.* (2007). Commonality and biosynthesis of the O-methyl phosphoramidate capsule modification in *Campylobacter jejuni.* J. Biol. Chem. *282,* 28566–28576.

McNulty, C., Thompson, J., Barrett, B., Lord, L., Andersen, C., and Roberts, I.S. (2006). The cell surface expression of group 2 capsular polysaccharides in *Escherichia coli*: the role of KpsD, RhsA and a multi-protein complex at the pole of the cell. Mol. Microbiol. *59,* 907–922.

Marolda, C.L., Vicarioli, J., and Valvano, M.A. (2004). Wzx proteins involved in biosynthesis of O antigen function in association with the first sugar of the O-specific lipopolysaccharide subunit. Microbiology *150,* 4095–4105.

Marolda, C.L., Tatar, L.D., Alaimo, C., Aebi, M., and Valvano, M.A. (2006). Interplay of the Wzx translocase and the corresponding polymerase and chain length regulator proteins in the translocation and periplasmic assembly of lipopolysaccharide O antigen. J. Bacteriol. *188,* 5124–5135.

Marolda, C.L., Li, B., Lung, M., Yang, M., Hanuszkiewicz, A., Rosales, A.R., and Valvano, M.A. (2010). Membrane topology and identification of critical amino acid residues in the Wzx O-antigen translocase from *Escherichia coli* O157:H4. J. Bacteriol. *192,* 6160–6171.

Mijakovic, I., Poncet, S., Boel, G., Maze, A., Gillet, S., Jamet, E., Decottignies, P., Grangeasse, C., Doublet, P., Le Marechal, P., *et al.* (2003). Transmembrane modulator-dependent bacterial tyrosine kinase activates UDP-glucose dehydrogenases. EMBO J. *22,* 4709–4718.

Mills, S.D., Kurjanczyk, L.A., Shames, B., Hennessy, J.N., and Penner, J.L. (1991). Antigenic shifts in serotype determinants of *Campylobacter coli* are accompanied by changes in the chromosomal DNA restriction endonuclease digestion pattern. J. Med. Microbiol. *35,* 168–173.

Minic, Z., Marie, C., Delorme, C., Faurie, J.M., Mercier, G., Ehrlich, D., and Renault, P. (2007). Control of EpsE, the phosphoglycosyltransferase initiating exopolysaccharide synthesis in *Streptococcus thermophilus,* by EpsD tyrosine kinase. J. Bacteriol. *189,* 1351–1357.

Misra, R., and Bavro, V.N. (2009). Assembly and transport mechanism of tripartite drug efflux systems. Biochim. Biophys. Acta *1794,* 817–825.

Mohammadi, T., van Dam, V., Sijbrandi, R., Vernet, T., Zapun, A., Bouhss, A., Diepeveen-de Bruin, M., Nguyen-Disteche, M., de Kruijff, B., and Breukink, E. (2011). Identification of FtsW as a transporter of lipid-linked cell wall precursors across the membrane. EMBO J. *30,* 1425–1432.

Morona, R., van den Bosch, L., and Manning, P.A. (1995). Molecular, genetic, and topological characterization of O-antigen chain length regulation in *Shigella flexneri*. J. Bacteriol. *177,* 1059–1068.

Morona, J.K., Paton, J.C., Miller, D.C., and Morona, R. (2000a). Tyrosine phosphorylation of CpsD negatively regulates capsular polysaccharide biosynthesis in *Streptococcus pneumoniae.* Mol. Microbiol. *35,* 1431–1442.

Morona, J.K., Morona, R., Miller, D.C., and Paton, J.C. (2002). *Streptococcus pneumoniae* capsule biosynthesis protein CpsB is a novel manganese-dependent phosphotyrosine-protein phosphatase. J. Bacteriol. *184,* 577–583.

Morona, J.K., Morona, R., Miller, D.C., and Paton, J.C. (2003). Mutational analysis of the carboxy-terminal (YGX)4 repeat domain of CpsD, an autophosphorylating tyrosine kinase required for capsule biosynthesis in *Streptococcus pneumoniae.* J. Bacteriol. *185,* 3009–3019.

Morona, J.K., Morona, R., and Paton, J.C. (2006). Attachment of capsular polysaccharide to the cell wall of *Streptococcus pneumoniae* type 2 is required for invasive disease. Proc. Natl. Acad. Sci. U.S.A. *103,* 8505–8510.

Morona, R., Van Den Bosch, L., and Daniels, C. (2000b). Evaluation of Wzz/MPA1/MPA2 proteins based on the presence of coiled-coil regions. Microbiology *146 (Pt 1),* 1–4.

Morona, R., Purins, L., Tocilj, A., Matte, A., and Cygler, M. (2009). Sequence–structure relationships in polysaccharide co-polymerase (PCP) proteins. Trends Biochem. Sci. *34,* 78–84.

Nakar, D., and Gutnick, D.L. (2003). Involvement of a protein tyrosine kinase in production of the polymeric bioemulsifier emulsan from the oil-degrading strain *Acinetobacter lwoffii* RAG-1. J. Bacteriol. *185,* 1001–1009.

Nakhamchik, A., Wilde, C., and Rowe-Magnus, D.A. (2007). Identification of a Wzy polymerase required for group IV capsular polysaccharide and lipopolysaccharide biosynthesis in *Vibrio vulnificus.* Infect. Immun. *75,* 5550–5558.

Navasa, N., Rodriguez-Aparicio, L., Martinez-Blanco, H., Arcos, M., and Ferrero, M.A. (2009). Temperature has reciprocal effects on colanic acid and polysialic acid biosynthesis in *E. coli* K92. Appl. Microbiol. Biotechnol. *82,* 721–729.

Nesper, J., Hill, C.M., Paiment, A., Harauz, G., Beis, K., Naismith, J.H., and Whitfield, C. (2003). Translocation of group 1 capsular polysaccharide in *Escherichia coli* serotype K30. Structural and functional analysis of the outer membrane lipoprotein Wza. J. Biol. Chem. *278,* 49763–49772.

Niemeyer, D., and Becker, A. (2001). The molecular weight distribution of succinoglycan produced by *Sinorhizobium meliloti* is influenced by specific tyrosine phosphorylation and ATPase activity of the cytoplasmic domain of the ExoP protein. J. Bacteriol. *183,* 5163–5170.

Nimtz, M., Mort, A., Domke, T., Wray, V., Zhang, Y., Qiu, F., Coplin, D., and Geider, K. (1996). Structure of amylovoran, the capsular exopolysaccharide from the fire blight pathogen *Erwinia amylovora.* Carbohydr. Res. *287,* 59–76.

Obadia, B., Lacour, S., Doublet, P., Baubichon-Cortay, H., Cozzone, A.J., and Grangeasse, C. (2007). Influence of tyrosine-kinase Wzc activity on colanic acid production in *Escherichia coli* K12 cells. J. Mol. Biol. *367,* 42–53.

Olivares-Illana, V., Meyer, P., Bechet, E., Gueguen-Chaignon, V., Soulat, D., Lazereg-Riquier, S., Mijakovic, I., Deutscher, J., Cozzone, A.J., Laprevote, O., *et al.* (2008). Structural basis for the regulation mechanism of the tyrosine kinase CapB from *Staphylococcus aureus.* PLoS Biol. *6,* e143.

Ophir, T., and Gutnick, D.L. (1994). A role for exopolysaccharides in the protection of microorganisms from desiccation. Appl. Environ. Microbiol. *60,* 740–745.

Orskov, F., and Orskov, I. (1992). *Escherichia coli* serotyping and disease in man and animals. Can. J. Microbiol. *38,* 699–704.

Orskov, F., Orskov, I., Sutton, A., Schneerson, R., Lin, W., Egan, W., Hoff, G.E., and Robbins, J.B. (1979). Form variation in *Escherichia coli* K1: determined by O-acetylation of the capsular polysaccharide. J. Exp. Med. *149,* 669–685.

Ouskova, G., Spellerberg, B., and Prehm, P. (2004). Hyaluronan release from *Streptococcus pyogenes*: export by an ABC transporter. Glycobiology *14,* 931–938.

Paiment, A., Hocking, J., and Whitfield, C. (2002). Impact of phosphorylation of specific residues in the tyrosine autokinase, Wzc, on its activity in assembly of group 1 capsules in *Escherichia coli*. J. Bacteriol. *184*, 6437–6447.

Parkhill, J., Wren, B.W., Mungall, K., Ketley, J.M., Churcher, C., Basham, D., Chillingworth, T., Davies, R.M., Feltwell, T., Holroyd, S., *et al.* (2000). The genome sequence of the food-borne pathogen *Campylobacter jejuni* reveals hypervariable sequences. Nature *403*, 665–668.

Patel, K.B., Furlong, S.E., and Valvano, M.A. (2010). Functional analysis of the C-terminal domain of the WbaP protein that mediates initiation of O antigen synthesis in *Salmonella enterica*. Glycobiol. *20*, 1389–1401.

Patrick, S., Blakely, G.W., Houston, S., Moore, J., Abratt, V.R., Bertalan, M., Cerdeno-Tarraga, A.M., Quail, M.A., Corton, N., Corton, C., *et al.* (2010). Twenty-eight divergent polysaccharide loci specifying within- and amongst-strain capsule diversity in three strains of *Bacteroides fragilis*. Microbiology *156*, 3255–3269.

Paulsen, I.T., Beness, A.M., and Saier, M.H., Jr. (1997). Computer-based analyses of the protein constituents of transport systems catalysing export of complex carbohydrates in bacteria. Microbiology *143 (Pt 8)*, 2685–2699.

Pavelka, M.S., Jr., Hayes, S.F., and Silver, R.P. (1994). Characterization of KpsT, the ATP-binding component of the ABC transporter involved with the export of capsular polysialic acid in *Escherichia coli* K1. J. Biol. Chem. *269*, 20149–20158.

Pazzani, C., Rosenow, C., Boulnois, G.J., Bronner, D., Jann, K., and Roberts, I.S. (1993). Molecular analysis of region 1 of the *Escherichia coli* K5 antigen gene cluster: a region encoding proteins involved in cell surface expression of capsular polysaccharide. J. Bacteriol. *175*, 5978–5983.

Pearce, R., and Roberts, I.S. (1995). Cloning and analysis of gene clusters for production of the *Escherichia coli* K10 and K54 antigens: identification of a new group of *serA*-linked capsule gene clusters. J. Bacteriol. *177*, 3992–3997.

de Pedro, M.A., Quintela, J.C., Holtje, J.V., and Schwarz, H. (1997). Murein segregation in *Escherichia coli*. J. Bacteriol. *179*, 2823–2834.

Pelkonen, S., Hayrinen, J., and Finne, J. (1988). Polyacrylamide gel electrophoresis of the capsular polysaccharides of *Escherichia coli* K1 and other bacteria. J. Bacteriol. *170*, 2646–2653.

Pelosi, L., Boumedienne, M., Saksouk, N., Geiselmann, J., and Geremia, R.A. (2005). The glucosyl-1-phosphate transferase WchA (Cap8E) primes the capsular polysaccharide repeat unit biosynthesis of *Streptococcus pneumoniae* serotype 8. Biochem. Biophys. Res. Commun. *327*, 857–865.

Phoenix, D.A., Brandenburg, K., Harris, F., Seydel, U., Hammerton, T., and Roberts, I.S. (2001). An investigation into the membrane-interactive potential of the *Escherichia coli* KpsE C-terminus. Biochem. Biophys. Res. Comm. *285*, 976–980.

Pierce, W.A., Jr., and White, A.G. (1954). Hyaluronic acid formation by *Streptococcus pyogenes*. Proc. Soc. Exp. Biol. Med. *87*, 50–54.

Pigeon, R.P., and Silver, R.P. (1994). Topological and mutational analysis of KpsM, the hydrophobic component of the ABC transporter involved in the export of polysialic acid in *Escherichia coli* K1. Mol. Microbiol. *14*, 871–881.

Pike, R.M. (1948). Hyaluronidase and hyaluronic acid of group A streptococci. Am. J. Med. *4*, 468.

Preneta, R., Jarraud, S., Vincent, C., Doublet, P., Duclos, B., Etienne, J., and Cozzone, A.J. (2002). Isolation and characterization of a protein-tyrosine kinase and a phosphotyrosine-protein phosphatase from *Klebsiella pneumoniae*. Comp. Biochem. Physiol. B, Biochem. Mol. Biol. *131*, 103–112.

Rahn, A., and Whitfield, C. (2003). Transcriptional organization and regulation of the *Escherichia coli* K30 group 1 capsule biosynthesis (*cps*) gene cluster. Mol. Microbiol. *47*, 1045–1060.

Rahn, A., Drummelsmith, J., and Whitfield, C. (1999). Conserved organization in the *cps* gene clusters for expression of *Escherichia coli* group 1 K antigens: relationship to the colanic acid biosynthesis locus and the *cps* genes from *Klebsiella pneumoniae*. J. Bacteriol. *181*, 2307–2313.

Rahn, A., Beis, K., Naismith, J.H., and Whitfield, C. (2003). A novel outer membrane protein, Wzi, is involved in surface assembly of the *Escherichia coli* K30 group 1 capsule. J. Bacteriol. *185*, 5882–5890.

Ramsey, D.M., and Wozniak, D.J. (2005). Understanding the control of *Pseudomonas aeruginosa* alginate synthesis and the prospects for management of chronic infections in cystic fibrosis. Mol. Microbiol. *56*, 309–322.

Reeves, R.E., and Goebel, W.F. (1941). Chemoimmunological studies on the soluble specific substance of pneumococcus. J. Biol. Chem. *139*, 511–519.

Rehm, B.H., Boheim, G., Tommassen, J., and Winkler, U.K. (1994). Overexpression of AlgE in *Escherichia coli*: subcellular localization, purification, and ion channel properties. J. Bacteriol. *176*, 5639–5647.

Reid, A.N. (2005). The role of the tyrosine autokinase, Wzc, in Group 1 capsule assembly in *Escherichia coli* O9a:K30. Department of Molecular and Cellular Biology (Guelph, University of Guelph), pp. 290.

Reid, A.N., and Whitfield, C. (2005). Functional analysis of conserved gene products involved in assembly of *Escherichia coli* capsules and exopolysaccharides: evidence for molecular recognition between Wza and Wzc for colanic acid biosynthesis. J. Bacteriol. *187*, 5470–5481.

Reid, A.N., and Szymanski, C.M. (2009). Biosynthesis and assembly of capsular polysaccharides. In Microbial Glycobiology. Structures, Relevance and Applications, Moran, A.P., Holst, O., Brennan, P.J., and von Itzstein, M., eds. (London, UK, Academic Press), pp. 351–373.

Reid, C.W., Fulton, K.M., and Twine, S.M. (2010). Never take candy from a stranger: the role of the bacterial glycome in host–pathogen interactions. Future Microbiol. *5*, 267–288.

Reyes, C.L., and Chang, G. (2005). Structure of the ABC transporter MsbA in complex with ADP.vanadate and lipopolysaccharide. Science *308*, 1028–1031.

Rick, P.D., Barr, K., Sankaran, K., Kajimura, J., Rush, J.S., and Waechter, C.J. (2003). Evidence that the *wzxE* gene of *Escherichia coli* K-12 encodes a protein involved in the transbilayer movement of a trisaccharide-lipid intermediate in the assembly of enterobacterial common antigen. J. Biol. Chem. *278*, 16534–16542.

Rigg, G.P., Barrett, B., and Roberts, I.S. (1998). The localization of KpsC, S and T, and KfiA, C and D proteins involved in the biosynthesis of the *Escherichia coli* K5 capsular polysaccharide: evidence for a membrane-bound complex. Microbiology *144*, 2905–2914.

Roberts, I., Mountford, R., High, N., Bitter-Suermann, D., Jann, K., Timmis, K., and Boulnois, G. (1986). Molecular cloning and analysis of genes for production of K5, K7, K*12*, and K92 capsular polysaccharides in *Escherichia coli*. J. Bacteriol. *168*, 1228–1233.

Roberts, I.S. (1996). The biochemistry and genetics of capsular polysaccharide production in bacteria. Ann. Rev. Microbiol. *50*, 285–315.

Roberts, I.S., Mountford, R., Hodge, R., Jann, K.B., and Boulnois, G.J. (1988). Common organization of gene clusters for production of different capsular polysaccharides (K antigens) in *Escherichia coli*. J. Bacteriol. *170*, 1305–1310.

Rosenow, C., Roberts, I.S., and Jann, K. (1995). Isolation from recombinant *Escherichia coli* and characterization of CMP-Kdo synthetase, involved in the expression of the capsular K5 polysaccharide (K-CKS). FEMS Microbiol. Lett. *125*, 159–164.

Rosner, H., Grimmecke, H.D., Knirel, Y.A., and Shashkov, A.S. (1992). Hyaluronic acid and a (1–4)-beta-D-xylan, extracellular polysaccharides of *Pasteurella multocida* (Carter type A) strain 880. Carbohydr. Res. *223*, 329–333.

Rowe, S., Hodson, N., Griffiths, G., and Roberts, I.S. (2000). Regulation of the *Escherichia coli* K5 capsule gene cluster: evidence for the roles of H-NS, BipA, and integration host factor in regulation of group 2 capsule gene clusters in pathogenic *E. coli*. J. Bacteriol. *182*, 2741–2745.

Ryder, C., Byrd, M., and Wozniak, D.J. (2007). Role of polysaccharides in *Pseudomonas aeruginosa* biofilm development. Curr. Opin. Microbiol. *10*, 644–648.

Saldias, M.S., Patel, K., Marolda, C.L., Bittner, M., Contreras, I., and Valvano, M.A. (2008). Distinct functional domains of the *Salmonella enterica* WbaP transferase that is involved in the initiation reaction for synthesis of the O antigen subunit. Microbiology *154*, 440–453.

Sau, S., Bhasin, N., Wann, E.R., Lee, J.C., Foster, T.J., and Lee, C.Y. (1997). The *Staphylococcus aureus* allelic genetic loci for serotype 5 and 8 capsule expression contain the type-specific genes flanked by common genes. Microbiology *143 (Pt 7)*, 2395–2405.

Scheurwater, E.M., and Burrows, L.L. (2011). Maintaining network security: how macromolecular structures cross the peptidoglycan layer. FEMS Microbiol. Lett. *318*, 1–9.

Schild, S., Lamprecht, A.K., and Reidl, J. (2005). Molecular and functional characterization of O antigen transfer in *Vibrio cholerae*. J. Biol. Chem. *280*, 25936–25947.

Schmidt, M.A., and Jann, K. (1982). Phospholipid substitution of capsular (K) polysaccharide antigens from *Escherichia coli* causing extraintestinal infections. FEMS Microbiol. Lett. *14*, 69–74.

Schneider, M.C., Exley, R.M., Ram, S., Sim, R.B., and Tang, C.M. (2007). Interactions between *Neisseria meningitidis* and the complement system. Trends Microbiol. *15*, 233–240.

Silver, R.P., Aaronson, W., and Vann, W.F. (1987). Translocation of capsular polysaccharides in pathogenic strains of *Escherichia coli* requires a 60-kilodalton periplasmic protein. J. Bacteriol. *169*, 5489–5495.

Silver, R.P., Prior, K., Nsahlai, C., and Wright, L.F. (2001). ABC transporters and the export of capsular polysaccharides from gram-negative bacteria. Res. Microbiol. *152*, 357–364.

Simpson, D.A., Hammarton, T.C., and Roberts, I.S. (1996). Transcriptional organization and regulation of expression of region 1 of the *Escherichia coli* K5 capsule gene cluster. J. Bacteriol. *178*, 6466–6474.

Skorupska, A., Janczarek, M., Marczak, M., Mazur, A., and Krol, J. (2006). Rhizobial exopolysaccharides: genetic control and symbiotic functions. Microb. Cell Fact. *5*, 7.

Smirnova, A., Li, H., Weingart, H., Aufhammer, S., Burse, A., Finis, K., Schenk, A., and Ullrich, M.S. (2001). Thermoregulated expression of virulence factors in plant-associated bacteria. Arch. Microbiol. *176*, 393–399.

Sorensen, U.B., Henrichsen, J., Chen, H.C., and Szu, S.C. (1990). Covalent linkage between the capsular polysaccharide and the cell wall peptidoglycan of *Streptococcus pneumoniae* revealed by immunochemical methods. Microb. Pathog. *8*, 325–334.

Soulat, D., Jault, J.M., Duclos, B., Geourjon, C., Cozzone, A.J., and Grangeasse, C. (2006). *Staphylococcus aureus* operates protein-tyrosine phosphorylation through a specific mechanism. J. Biol. Chem. *281*, 14048–14056.

St Michael, F., Szymanski, C.M., Li, J., Chan, K.H., Khieu, N.H., Larocque, S., Wakarchuk, W.W., Brisson, J.R., and Monteiro, M.A. (2002). The structures of the lipooligosaccharide and capsule polysaccharide of *Campylobacter jejuni* genome sequenced strain NCTC 11168. Eur. J. Biochem. *269*, 5119–5136.

Steenbergen, S.M., and Vimr, E.R. (1990). Mechanism of polysialic acid chain elongation in *Escherichia coli* K1. Mol. Microbiol. *4*, 603–611.

Steenbergen, S.M., and Vimr, E.R. (2003). Functional relationships of the sialyltransferases involved in expression of the polysialic acid capsules of *Escherichia coli* K1 and K92 and *Neisseria meningitidis* groups B or C. J. Biol. Chem. *278,* 15349–15359.

Steenbergen, S.M., and Vimr, E.R. (2008). Biosynthesis of the *Escherichia coli* K1 group 2 polysialic acid capsule occurs within a protected cytoplasmic compartment. Mol. Microbiol. *68,* 1252–1267.

Steenbergen, S.M., Wrona, T.J., and Vimr, E.R. (1992). Functional analysis of the sialyltransferase complexes in *Escherichia coli* K1 and K92. J. Bacteriol. *174,* 1099–1108.

Steenbergen, S.M., Lee, Y.C., Vann, W.F., Vionnet, J., Wright, L.F., and Vimr, E.R. (2006). Separate pathways for O acetylation of polymeric and monomeric sialic acids and identification of sialyl O-acetyl esterase in *Escherichia coli* K1. J. Bacteriol. *188,* 6195–6206.

Sugiura, N., Baba, Y., Kawaguchi, Y., Iwatani, T., Suzuki, K., Kusakabe, T., Yamagishi, K., Kimata, K., Kakuta, Y., and Watanabe, H. (2010). Glucuronyltransferase activity of KfiC from *Escherichia coli* strain K5 requires association of KfiA: KfiC and KfiA are essential enzymes for production of K5 polysaccharide, N-acetylheparosan. J. Biol. Chem. *285,* 1597–1606.

Sutherland, I. (2002). A sticky business. Microbial polysaccharides: current products and future trends. Microbiol. Today *29,* 70–71.

Szymanski, C.M., Michael, F.S., Jarrell, H.C., Li, J., Gilbert, M., Larocque, S., Vinogradov, E., and Brisson, J.R. (2003). Detection of conserved N-linked glycans and phase-variable lipooligosaccharides and capsules from *Campylobacter* cells by mass spectrometry and high resolution magic angle spinning NMR spectroscopy. J. Biol. Chem. *278,* 24509–24520.

Tao, J., Feng, L., Guo, H., Li, Y., and Wang, L. (2004). The O-antigen gene cluster of *Shigella boydii* O11 and functional identification of its *wzy* gene. FEMS Microbiol. Lett. *234,* 125–132.

Tlapak-Simmons, V.L., Baron, C.A., Gotschall, R., Haque, D., Canfield, W.M., and Weigel, P.H. (2005). Hyaluronan biosynthesis by class I streptococcal hyaluronan synthases occurs at the reducing end. J. Biol. Chem. *280,* 13012–13018.

Trotter, C.L., McVernon, J., Ramsay, M.E., Whitney, C.G., Mulholland, E.K., Goldblatt, D., Hombach, J., and Kieny, M.P. (2008). Optimising the use of conjugate vaccines to prevent disease caused by *Haemophilus influenzae* type b, *Neisseria meningitidis* and *Streptococcus pneumoniae.* Vaccine *26,* 4434–4445.

Tzeng, Y.L., Datta, A.K., Strole, C.A., Lobritz, M.A., Carlson, R.W., and Stephens, D.S. (2005). Translocation and surface expression of lipidated serogroup B capsular polysaccharide in *Neisseria meningitidis.* Infect. Immun. *73,* 1491–1505.

Vann, W.F., Schmidt, M.A., Jann, B., and Jann, K. (1981). The structure of the capsular polysaccharide (K5 antigen) of urinary-tract-infective *Escherichia coli* 010:K5:H4. A polymer similar to desulfo-heparin. Eur. J. Biochem. *116,* 359–364.

Ventura, C.L., Cartee, R.T., Forsee, W.T., and Yother, J. (2006). Control of capsular polysaccharide chain length by UDP–sugar substrate concentrations in *Streptococcus pneumoniae.* Mol. Microbiol. *61,* 723–733.

Vimr, E., and Steenbergen, S.M. (1993). Mechanisms of polysialic acid assembly in *Escherichia coli* K1: a paradigm for microbes and mammals. In Polysialic acid: from microbes to men, Roth, J., Rutishauser, U., and Troy, F.A., eds (Basel, Switzerland, Birkhauser Verlag), pp. 73–91.

Vimr, E.R., and Steenbergen, S.M. (2006). Mobile contingency locus controlling *Escherichia coli* K1 polysialic acid capsule acetylation. Mol. Microbiol. *60,* 828–837.

Vimr, E.R., Aaronson, W., and Silver, R.P. (1989). Genetic analysis of chromosomal mutations in the polysialic acid gene cluster of *Escherichia coli* K1. J. Bacteriol. *171,* 1106–1117.

Vimr, E.R., Kalivoda, K.A., Deszo, E.L., and Steenbergen, S.M. (2004). Diversity of microbial sialic acid metabolism. Microbiol. Mol. Biol. Rev. *68,* 132–153.

Vincent, C., Doublet, P., Grangeasse, C., Vaganay, E., Cozzone, A.J., and Duclos, B. (1999). Cells of *Escherichia coli* contain a protein-tyrosine kinase, Wzc, and a phosphotyrosine-protein phosphatase, Wzb. J. Bacteriol. *181,* 3472–3477.

Vincent, C., Duclos, B., Grangeasse, C., Vaganay, E., Riberty, M., Cozzone, A.J., and Doublet, P. (2000). Relationship between exopolysaccharide production and protein-tyrosine phosphorylation in gram-negative bacteria. J. Mol. Biol. *304,* 311–321.

Vinogradov, E., Frirdich, E., MacLean, L.L., Perry, M.B., Petersen, B.O., Duus, J.O., and Whitfield, C. (2002). Structures of lipopolysaccharides from *Klebsiella pneumoniae.* Elucidation of the structure of the linkage region between core and polysaccharide O chain and identification of the residues at the non-reducing termini of the O chains. J. Biol. Chem. *277,* 25070–25081.

Vionnet, J., and Vann, W.F. (2007). Successive glycosyltransfer of sialic acid by *Escherichia coli* K92 polysialyltransferase in elongation of oligosialic acceptors. Glycobiol. *17,* 735–743.

Vu, B., Chen, M., Crawford, R.J., and Ivanova, E.P. (2009). Bacterial extracellular polysaccharides involved in biofilm formation. Molecules *14,* 2535–2554.

Wang, L., Liu, D., and Reeves, P.R. (1996). C-terminal half of *Salmonella enterica* WbaP (RfbP) is the galactosyl-1-phosphate transferase domain catalysing the first step of O-antigen synthesis. J. Bacteriol. *178,* 2598–2604.

Wang, Q., Ruan, X., Wei, D., Hu, Z., Wu, L., Yu, T., Feng, L., and Wang, L. (2010). Development of a serogroup-specific multiplex PCR assay to detect a set of *Escherichia coli* serogroups based on the identification of their O-antigen gene clusters. Mol. Cell. Probes *24,* 286–290.

Ward, C.K., and Inzana, T.J. (1997). Identification and characterization of a DNA region involved in the export of capsular polysaccharide by *Actinobacillus pleuropneumoniae* serotype 5a. Infect. Immun. *65,* 2491–2496.

Warren, G.H., and Durso, J.G. (1951). The existence of hyaluronidase substrate factors in microbiological culture media. Science *113*, 359.

Weigel, P.H., and DeAngelis, P.L. (2007). Hyaluronan synthases: a decade-plus of novel glycosyltransferases. J. Biol. Chem. *282*, 36777–36781.

Weintraub, A. (2003). Immunology of bacterial polysaccharide antigens. Carbohydr. Res. *338*, 2539–2547.

Welman, A.D., and Maddox, I.S. (2003). Exopolysaccharides from lactic acid bacteria: perspectives and challenges. Trends Biotechnol. *21*, 269–274.

Whitfield, C. (2006). Biosynthesis and assembly of capsular polysaccharides in *Escherichia coli*. Ann. Rev. Biochem. *75*, 39–68.

Whitfield, C., and Roberts, I.S. (1999). Structure, assembly and regulation of expression of capsules in *Escherichia coli*. Mol. Microbiol. *31*, 1307–1319.

Willis, L.M., Gilbert, M., Karwaski, M.F., Blanchard, M.C., and Wakarchuk, W.W. (2008). Characterization of the alpha-2,8-polysialyltransferase from *Neisseria meningitidis* with synthetic acceptors, and the development of a self-priming polysialyltransferase fusion enzyme. Glycobiology *18*, 177–186.

Wugeditsch, T., Paiment, A., Hocking, J., Drummelsmith, J., Forrester, C., and Whitfield, C. (2001). Phosphorylation of Wzc, a tyrosine autokinase, is essential for assembly of group 1 capsular polysaccharides in *Escherichia coli*. J. Biol. Chem. *276*, 2361–2371.

Wunder, D.E., Aaronson, W., Hayes, S.F., Bliss, J.M., and Silver, R.P. (1994). Nucleotide sequence and mutational analysis of the gene encoding KpsD, a periplasmic protein involved in transport of polysialic acid in *Escherichia coli* K1. J. Bacteriol. *176*, 4025–4033.

Xayarath, B., and Yother, J. (2007). Mutations blocking side-chain assembly, polymerization, or transport of a Wzy-dependent *Streptococcus pneumoniae* capsule are lethal in the absence of suppressor mutations and can affect polymer transfer to the cell wall. J. Bacteriol. *189*, 3369–3381.

Xu, H., Chater, K.F., Deng, Z., and Tao, M. (2008). A cellulose synthase-like protein involved in hyphal tip growth and morphological differentiation in *Streptomyces*. J. Bacteriol. *190*, 4971–4978.

Yeh, J.Y., and Chen, J. (2004). Production of slime polysaccharide by EHEC and STEC as well as the influence of culture conditions on slime production in *Escherichia coli* O157:H7. Lett. Appl. Microbiol. *38*, 488–492.

Yeung, M.K., and Mattingly, S.J. (1983a). Biosynthesis of cell wall peptidoglycan and polysaccharide antigens by protoplasts of type III group B *Streptococcus*. J. Bacteriol. *154*, 211–220.

Yeung, M.K., and Mattingly, S.J. (1983b). Isolation and characterization of type III group B streptococcal mutants defective in biosynthesis of the type-specific antigen. Infect. Immun. *42*, 141–151.

Zhang, Z.Y., Wang, Y., Wu, L., Fauman, E.B., Stuckey, J.A., Schubert, H.L., Saper, M.A., and Dixon, J.E. (1994). The Cys(X)$_5$Arg catalytic motif in phosphoester hydrolysis. Biochemistry *33*, 15266–15270.

Zhou, G.P., and Troy, F.A., 2nd (2003). Characterization by NMR and molecular modeling of the binding of polyisoprenols and polyisoprenyl recognition sequence peptides: 3D structure of the complexes reveals sites of specific interactions. Glycobiology *13*, 51–71.

Zhou, G.P., and Troy, F.A., 2nd (2005). NMR studies on how the binding complex of polyisoprenol recognition sequence peptides and polyisoprenols can modulate membrane structure. Curr. Protein Pept. Sci. *6*, 399–411.

Control of Lytic Transglycosylase Activity Within Bacterial Cell Walls

3

John M. Pfeffer, Patrick J. Moynihan, Chelsea A. Clarke, Chris Vandenende and Anthony J. Clarke

Abstract

Lytic transglycosylases are an important class of bacterial enzymes that act on peptidoglycan with the same substrate specificity as lysozyme. Unlike the latter enzymes however, the lytic transglycosylases are not hydrolases, but instead cleave the glycosidic linkage between *N*-acetylmuramyl and *N*-acetylglucosaminyl residues with the concomitant formation of a 1,6-anydromuramyl product. They are ubiquitous in bacteria which produce a complement of different forms that are responsible for creating space within the peptidoglycan sacculus for its biosynthesis and recycling, cell division, and the insertion of cell envelope-spanning structures, such as flagella and secretion systems. Given their catastrophic autolytic potential, the activity of lytic transglycosylases must be tightly controlled within the producing cells. Three modes of control at the enzymatic level have been identified: the modification of substrate, membrane association and complex formation, and the production of proteinaceous inhibitors. These modes of control and their potential as new targets for antibacterials are discussed.

Introduction

The rigid bacterial cell wall component peptidoglycan (PG; murein) is a macromolecular heteropolymer composed of glycan chains of alternating *N*-acetylglucosamine (GlcNAc) and *N*-acetylmuramic acid (MurNAc). These glycan chains are cross-linked by peptides associated with the lactyl moiety of MurNAc (Fig. 3.1A). The heteropolymer forms a mesh-like sacculus that surrounds the bacterial cell to confer shape and, more importantly, support and strength to resist the internal turgor pressure imparted by the cytoplasm (reviewed in Vollmer *et al.*, 2008). Maintaining the integrity of the PG sacculus is thus essential to cell viability and its importance is reflected by the number of important therapeutic antibiotics that target PG biosynthesis which include the glycopeptide vancomycin and the β-lactams.

The PG sacculus is not a static structure, but rather it is continually expanded and turned over as bacteria grow and divide (reviewed in Höltje, 1998; Handelsman, 2006). The metabolism of PG also involves the creation of sites for the insertion of appendages, such as pili and flagella, and the creation of pores for membrane-associated secretory and uptake systems. A key class of enzymes responsible for cleaving PG to accommodate these requirements are the lytic transglycosylases (LTs). The LTs lyse PG with the same substrate specificity as the muramidases (EC 3.2.1.17), that is the β-1,4-glycosidic bond between MurNAc and GlcNAc. However, unlike the muramidases, the LTs are not hydrolases, but instead utilize the C-6 hydroxyl group of the substrate MurNAc residue to cleave PG with the concomitant formation of an intramolecular 1,6-anhydromuramyl reaction product (Höltje *et al.*, 1975) (Fig. 3.1).

LTs appear to be ubiquitous amongst the eubacteria, except for the wall-less organisms, such as the mycoplasmas. Based on a proteomic analysis, a variety of LTs have been recognized and a classification scheme has been developed that organizes these hypothetical proteins into

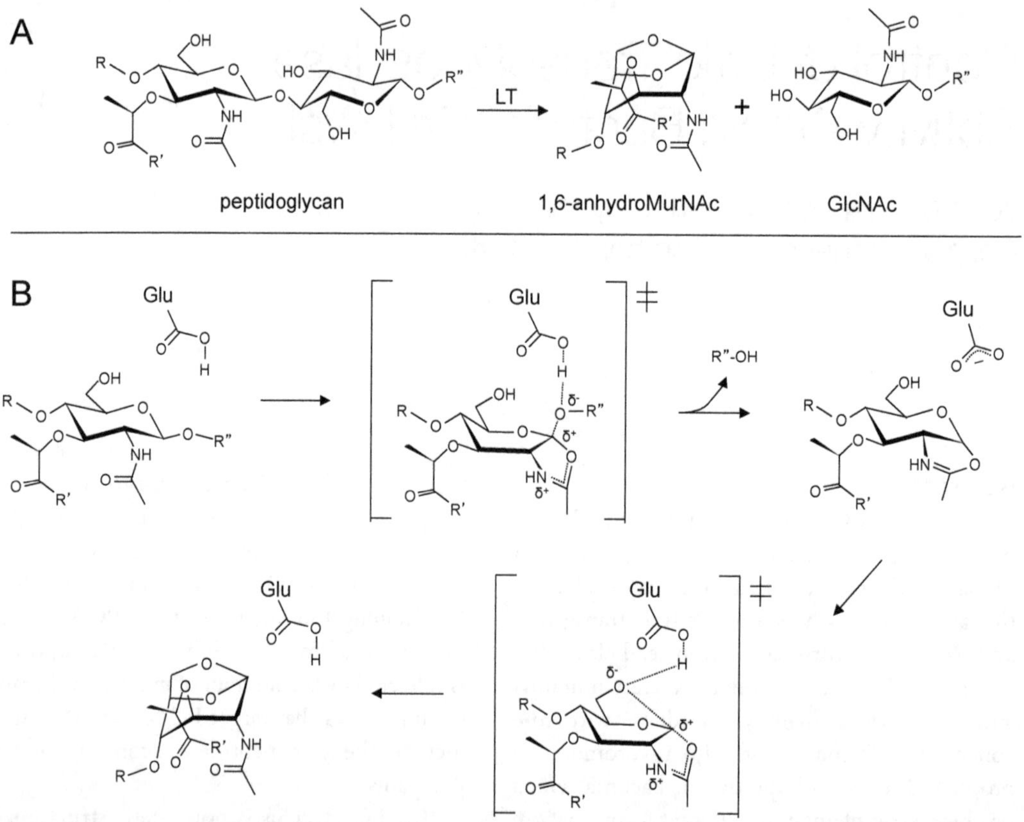

Figure 3.1 Reaction and proposed mechanism catalysed by LTs. (A) The LTs catalyse the cleavage of peptidoglycan between MurNAc and GlcNAc residues with the formation of 1,6-anhydromuramyl residues. (B) Proposed reaction mechanism of LTs. The catalytic Glu serves initially as an acid to protonate the glycosidic linkage to be cleaved leading to the formation of a muramyl oxazolinium-ion intermediate, and then as a base to abstract the C-6 hydroxyl proton of the oxazolinium ion promoting its collapse and concomitant formation of the 1,6-anhydromuramyl product. R and R′, GlcNAc; R′, stem peptide.

four distinct families, each with unique consensus motifs (Blackburn and Clarke, 2001). Family 1 is a superfamily comprising five subfamilies, each sharing some limited sequence similarity to the goose-type lysozymes. Both classes of these PG active enzymes are members of Family GH23 of the CAZy classification system for glycoside hydrolases (www.CAZy.org) (Henrissat and Davies, 1997). Families 2 (GH102) and 3 (GH103) comprise enzymes with sequence similarity to membrane-bound lytic transglycosylases A (MltA) and B (MltB) from *Escherichia coli*, respectively. A more recent analysis of the database suggests that, like family 1, family 3 also could be further divided into subfamilies (Reid *et al.*, 2006). Family 4 enzymes are primarily of bacteriophage origin and they belong to CAZy family GH104.

In addition to being ubiquitous amongst bacteria, a given species appears to produce a complement of LTs from the different families. For example, *E. coli* produces seven unique LTs, one from each of families 1A (Slt70), 1B (MltF), 1C (MltC), 1D (MltD), 1E (EmtA), 2 (MltA) and 3 (MltB). The genome of the opportunistic pathogen *Pseudomonas aeruginosa* appears to encode a variety of LTs, but it lacks genes for homologues of family 1C and 1E enzymes, and instead produces four MltB isozymes in addition to the other LTs produced by *E. coli* (Blackburn and Clarke, 2001). The apparent redundancy in the production of LTs is similar to the situation with the production of the penicillin-binding proteins (PBPs), enzymes responsible for the final stages of PG biosynthesis. As with the PBPs, 'functional equivalency' amongst the LTs appears to permit the inactivation of one or more without causing a lethal effect, at least *in vitro* (Blackburn and Clarke, 2002; Heidrich *et al.*, 2002).

Structure and function relationship of LTs

The three-dimensional structures of five LTs, representing families 1A (*E. coli* Slt70 (Thunnissen *et al.*, 1994a; 1995), 2 (*E. coli* MltA (van Straaten *et al.*, 2005), *Neisseria gonorrhoeae* MltA (Powell *et al.*, 2006)), 3 [*E. coli* Slt35, a naturally occurring product of MltB (van Asselt *et al.*, 1999)], and 4 [bacteriophage lambda LT (Leung *et al.*, 2001)] have been solved by X-ray crystallography. The family 1A, 3 and 4 LTs are mostly α-helical and, despite lacking similarity in both primary and overall tertiary structure, their catalytic domains are formed of a 'lysozyme' fold. As a beta-barrel, MltA (family 2) adopts a unique structure that is more similar to an endoglucanase V fold. Nonetheless, the active site clefts of Slt70, MltA and MltB possess four substrate binding subsites (labelled −2, −1, +1, and +2) that accommodate the residues of a tetrasaccharide stretch of a PG strand (Thunnissen *et al.*, 1995b; van Asselt *et al.*, 1999, 2000; van Straaten *et al.*, 2005; Powell *et al.*, 2006). This subsite architecture is thus consistent with the exo-activity of these LTs, as they catalyse the release of disaccharide anhydromuropeptides from the ends of PG strands. As endo-acting enzymes that cleave glycosidic linkages within PG strands, the family 4 LTs, accommodate hexasaccharide substrates within their active site clefts (Leung *et al.*, 2001).

The MurNAc and GlcNAc residues of a PG metabolite are seen to bind the −1 and +1 subsites, respectively, accounting for the reaction specificity of LT function. Protein engineering and kinetic studies with family 3 LTs from *P. aeruginosa* have led to the identification of important structural requirements of both binding ligands and amino acid residues on the enzymes for substrate binding. In particular, extensive interactions between amino acid residues comprising subsite −1 and both the lactyl moiety of bound MurNAc and its associated peptide are critical for the proper orientation of substrate for lytic activity (Reid *et al.*, 2004, 2006).

Irrespective of its length, the active site of an LT involves a single catalytic acidic residue positioned between subsites +1 and −1. Based on inhibition studies (Reid *et al.*, 2004) and theoretical considerations (Thunnissen *et al.*, 1994; van Asselt *et al.*, 1999), this residue is proposed to function as an acid/base catalyst to affect glycosidic bond cleavage involving anchimeric assistance (Fig. 3.1B). Initially acting as a general acid, the catalytic residue donates its proton to the glycosidic oxygen of the scissile bond to promote bond cleavage. The resulting oxocarbenium ion transition state is stabilized by the formation of a relatively stable oxazolinium intermediate involving the *N*-acetyl group of the muramyl residue at subsite 1. The deprotonated catalytic residue then acts as a general base to abstract a proton from the C-6 OH of MurNAc at subsite +1 thereby promoting intramolecular nucleophilic attack at C-1 to collapse the oxazoline intermediate with the concomitant formation of the 1,6-anhydro reaction product.

Biological functions

As endogenous PG-cleaving enzymes, LTs are viewed as space-making enzymes for bacteria to accommodate a variety of physiological functions. They are essential for the expansion of the PG sacculus to permit cell growth by creating sites for the insertion of new PG precursors (reviewed in Höltje, 1998). To ensure efficiency, this process of cleavage and re-synthesis involves the turnover and recycling of PG metabolites, where released 1,6-anhydromuropeptides are transported back into the cytoplasm for reuse. LTs also contribute to the formation of septa and the separation of dividing cells (Heidrich *et al.*, 2002) and they participate in endospore germination. Finally, given that the primary function of PG is to maintain the integrity of the cytoplasmic membrane, the space between neighbouring PG strands is necessarily short. Hence, LTs are required to create sites or pores for the insertion of protein complexes that extend through the PG sacculus, such as secretion systems, flagella, and pili (reviewed in Koraimann, 2003).

In addition to physiological functions, LTs have been suggested to contribute to pathogenesis (reviewed in Cloud-Hansen *et al.*, 2006). The production of LTs from different animal and plant pathogens, including *Haemophilus influenzae* (Shen *et al.*, 2005), *Neisseria meningitidis* (Sun *et al.*, 2000), *Shigella flexneri* (Bartoleschi *et al.*, 2002), *Erwinia amylovora* (Zhao *et al.*, 2005), and *Pseudomonas syringae* (Boch *et al.*, 2002) have been shown to be up-regulated during host infection. As well, reaction products of LTs have been demonstrated to elicit pathobiological effects. PG fragments in general elicit many of the general symptoms of bacterial infections, such as pyrogenicity, somnogenicity, and loss of appetite (reviewed in Clarke *et al.*, 2000), as well as the stimulation of the NOD1 and NOD2 intracellular receptors (reviewed in Inohara *et al.*, 2005). In addition, the LT product GlcNAc-anhMurNAc-tetrapeptide has been identified as the cytotoxin responsible for the pathology of both *Bordetella pertussis* (Goldman *et al.*, 1982; Cookson *et al.*, 1989) and *N. gonorrhoeae* (Melly *et al.*, 1984) infections. As many bacteria release this PG metabolite, it is probable that other pathobiological roles will eventually be ascribed to it.

Control of LT activity

LTs represent a class of autolysins in view of their ability to cause cellular lysis if their activity is allowed to proceed unchecked. Hence, it is essential that a bacterium has mechanisms in place to control its endogenous autolysins to remain viable. Until recently, little was known about the control of LTs, and indeed all autolysins in general. However, an understanding is emerging that now recognizes three different levels of control that operate in different cells. These involve the chemical modification of the substrate, the 'compartmentalization' and physical association of the enzymes, and the production of specific inhibitors.

Chemical modification of PG

A relatively common form of chemically modifying the PG sacculus to control LT activity is through O-acetylation. The O-acetylation of PG occurs at the C-6 hydroxyl group of MurNAc (Fig. 3.2). This modification precludes the activity of LTs because they require a free C-6 hydroxyl for the formation of the 1,6-anhydromuramyl reaction product (Fig. 3.1) (Blackburn and Clarke, 2002). The O-acetylation of PG has been found in many bacterial species, including a large number of known human pathogens (Table 3.1). The extent of modification varies with species and strain and typically ranges between 15% and 70%. The age of a bacterial culture also appears to influence PG O-acetylation. For example, increases in O-acetylation of 10–40% were observed with cultures of *Enterococcus faecalis* entering stationary phase and a further 10–16% when cells become viable but non-culturable (VBNC) (Pfeffer *et al.*, 2006). Similar observations have been made with *Bacillus anthracis* where the levels of PG O-acetylation of these cells growing in batch culture increases with time (Laaberki *et al. 2011*). However, this level drops in spores, which may reflect the production of high levels of new PG with the formation of spore cortex.

Early biochemical evidence indicated that the O-acetylation of PG is a maturation event, occurring after the assembly of PG precursors into the existing sacculus (reviewed in Clarke and Dupont, 1992; Clarke *et al.*, 2000). Thus, the source of acetate for this modification must be transported from the cytoplasm to the PG sacculus through the cytoplasmic membrane. At least two distinct families of membrane bound *O*-acetyltransferases have been identified for peptidoglycan O-acetylation. Mutation of the *oatA* gene in *S. aureus*, encoding a putative *O*-acetyltransferase, significantly decreases *O*-acetyl PG (Bera *et al.*, 2005). Based on genomic analyses, the existence of OatA for PG O-acetylation appears to be confined to Gram-positive bacteria, and the activity of homologues of *S. aureus* OatA have been found in *Streptococcus pneumoniae*, *E. faecalis* (Hébert *et al.*, 2007), and *Lactococcus lactis* (Veiga *et al.*, 2007), in addition to other species of *Staphylococcus* (Bera *et al.*, 2006). Although OatA has not been characterized biochemically characterized, it is classified as a member of the membrane-bound *O*-acetyltransferase (MBOAT) family of acyltransferases and thus as such it is predicted to be

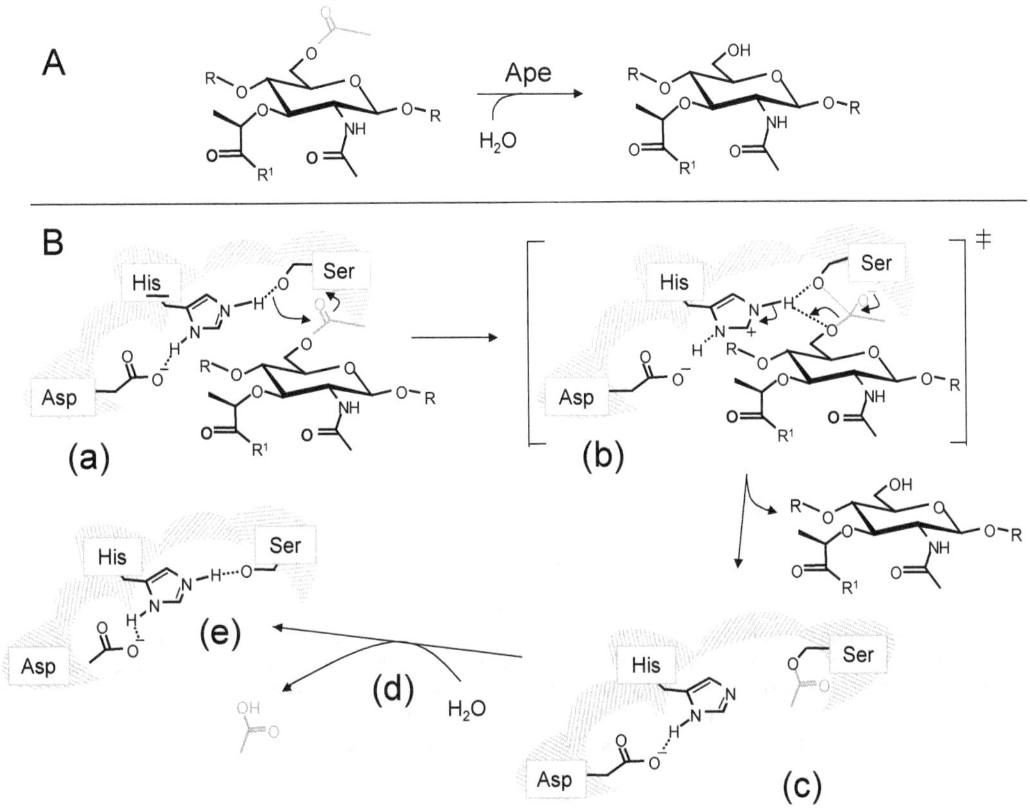

Figure 3.2 Reaction and proposed mechanism catalysed by *O*-acetylPG esterase (Ape). (A) Ape catalyses the specific hydrolysis of acetate from the C-6 position of *O*-acetylmuramyl residues in PG. (B) The proposed reaction mechanism of Ape involving a catalytic triad of Asp, His and Ser residues in which the nucleophilic Ser attacks the ester linkage (A) leading to the formation of a tetrahedral transition state (B) which collapses with the release of PG and the generation of an acylated enzyme intermediate (C). The reaction pathway is then reversed using water to release the bound acetate (D) and return of the free enzyme (E). R, GlcNAc; R′, stem peptide.

an integral membrane protein. Its mode of action is proposed to involve the transfer of acetate from cytoplasmic pools of acetyl-CoA to acceptor sites on PG following its translocation across the cytoplasmic membrane (Fig. 3.3) (Clarke *et al.*, 2000). Recent work with *B. anthracis* suggests that paralogues of OatA exist which, while functioning to O-acetylate PG, are distantly related based on sequence alignments (Laaberki *et al.*, 2011). Consequently, these paralogues are named OatB.

Gram-negative bacteria do not appear to encode homologues of OatA or OatB. Rather, they produce a second family of enzymes, called Pat, for the O-acetylation of their PG, which are encoded in OAP (O-acetylation of peptidoglycan) gene clusters. OAP clusters have been identified in a number of species, including the human pathogens *N. gonorrhoeae, N. meningitidis, Helicobacter pylori*, and species of *Proteus* (Weadge *et al.*, 2005). These proteins are homologues of Alg proteins involved in the O-acetylation of alginate, an exopolysaccharide secreted by *P. aeruginosa* (Franklin and Ohman, 1996). Phenotypic studies with *N. gonorrhoeae* demonstrated that loss of its PatA, a hypothetical integral membrane protein, results in a decrease in O-acetylation (Dillard and Hackett, 2005). More recently, the second gene product of the OAP cluster, PatB, was shown to function as an *O*-acetyltransferase and its activity was dependent upon functional PatA (Moynihan and Clarke, 2010). Thus, PatA and PatB are postulated to function as a two-component system

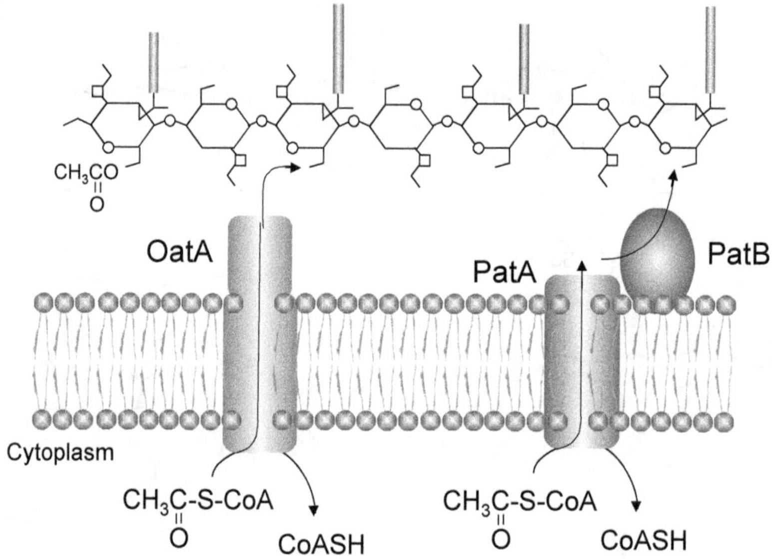

Figure 3.3 Proposed reaction pathways for the O-acetylation of PG. The integral membrane protein OatA of Gram-positive bacteria translocates acetate from cytoplasmic pools of acetyl-CoA through the cytoplasmic membrane to the outside of the cell where it is transferred directly to muramyl residues in PG. In Gram-negative bacteria and a limited number of Gram-positives, PatA functions like OatA to translocate acetate across the cytoplasmic membrane to the periplasm where it is accepted by PatB for its transfer to PG.

for PG O-acetylation (Fig. 3.3). As a putative integral membrane protein, PatA would function like OatA or OatB to translocate acetate across the cytoplasmic membrane and then PatB, which is localized to the periplasm (Moynihan and Clarke, 2010), would receive the acetate and transfer it to PG. Interestingly, in addition to being present in Gram-negative bacteria that O-acetylate PG, OAP clusters are present in some Gram-positive bacteria, notably species of *Bacillus* (Laaberki *et al.*, 2011). Thus, with *B. anthracis,* both Oat and Pat systems operate to O-acetylate its PG.

In addition to the acetyltransferase system, OAP clusters also encode *O*-acetyl-PG esterases (Ape). These enzymes hydrolyse the controlling *O*-acetyl group from muramyl residues of PG (Weadge and Clarke, 2006) (Fig. 3.2A). They belong to the GDSL superfamily of serine esterases which use a catalytic triad of Ser, His and Asp residues to catalyse the release of acetate from PG (Fig. 3.2B) (Weadge and Clarke, 2007). Thus, the Gram-negative bacteria balance the expression and activities of their Pat and Ape enzymes to provide sites for cleavage by the LTs as required. How the activity of the Ape esterases is localized to permit site-specific cleavage of PG is currently unknown, but it is possible that they comprise the 'murosome' complexes of enzymes responsible for localized degradation and synthesis (discussed further below).

Localization and association of LTs

Physical association and separation presents a second level of control of LT activity. Many of the LTs are membrane-bound lipoproteins, possessing a lipidation sequence at their N-terminus. Studies with both *E. coli and P. aeruginosa* indicate that these LTs are localized to the inner leaflet of the outer-membrane of these Gram-negative bacteria (Höltje, 1998; Blackburn and Clarke, 2002) thereby physically confining the enzymes away from their PG substrate and precluding their indiscriminate lytic activity. To act on their substrate, these LTs would have to be drawn physically into close proximity of the PG strands that need to be cleaved and this is thought to be accomplished through the formation of complexes with the PBPs. These PG biosynthetic enzymes are peripheral membrane proteins associated with the outer (periplasmic) leaflet of the cytoplasmic membrane. Experimental evidence provided by affinity chromatography (Romeis and Höltje, 1994; von Rechenberg *et al.*, 1996; Vollmer *et al.*, 1999; Jennings *et al.*, 2002; Figge *et al.*, 2004; Divakaruni *et al.*, 2005; Höppner *et al.*, 2005; Bertsche *et al.*, 2006; Legaree and Clarke, 2008), cross-linking (Satta *et al.*, 1994; Bhardwaj and Day, 1997; Alaedini and Day, 1999), and surface-plasmon resonance (Legaree and Clarke, 2008) studies have been obtained to support the existence of such associations and the partnerships of specific LTs with specific PBPs. Hence, these 'murosome' complexes would serve to fuse the cytoplasmic and outer membranes together and sandwich the PG sacculus between them. Murosomes would thus facilitate the appropriate juxtaposition of both the cleaving and biosynthetic activities and the physical constraints imposed would thus guarantee that LT-catalysed lysis is coupled with the biosynthetic transglycosylase activity of the PBPs to either extend the cell length or form division sites.

LTs have also been found to be associated with protein complexes that constitute secretory systems (Höppner *et al.*, 2005). In some cases, the LT appears to be fused to other membrane proteins to direct their specific function. For example, *E. coli* MltF is bimodal and possesses a C-terminal LT module and an N-terminal module with strong amino acid sequence identity (Scheurwater and Clarke, 2008) and three-dimensional structural similarity (Madoori and Thunnissen, 2010) to amino acid-binding proteins associated with ABC transport systems (Fig. 3.4). The recombinant protein fragments representing the respective modules have been overproduced both together and individually (Scheurwater and Clarke, 2008). Only the C-terminal LT module was found to bind PG and catalyse PG lysis. For the localized insertion of these secretory and transport systems, only limited lytic activity would be required to permit the extrusion of the complete secretion complex of proteins through the PG layer. The physical association or fusion of the LT responsible for creating the pore with the secretion apparatus would thus permit the appropriate coordination and localization of this required lytic activity while precluding catastrophic autolysis.

Inhibitors of LTs

The proteinaceous inhibitor of vertebrate lysozymes (Ivy) is produced by a collection of Gram-negative bacteria as a stress response to PG damage (Deckers *et al.*, 2004). Recently, we demonstrated that the production Ivy, and its paralogue Ivy2 produced by species of Pseudomonads, correlates exclusively with Gram-negative bacteria that do not O-acetylate their PG (Clarke *et al.*, 2010). Moreover, both proteins were shown experimentally to be potent inhibitors of MltB from *P. aeruginosa*. A 3-fold molar excess of either Ivy is sufficient to inhibit MltB by 50%, and complete inhibition is achieved with a 10-fold molar excess of the proteins (Clarke *et al.*, 2010).

Ivy was first identified as the product of the *ykfE* gene in *E. coli* (Deckers *et al.*, 2004), and homologues of Ivyc (Ivy from *E. coli*) have been identified on the chromosome of 27 other Gram-negative bacteria, each possessing the CKPHDC consensus sequence (Abergel *et al.*, 2007; Clarke *et al.*, 2010). Structural studies with both Ivyc and the homologous Ivyp1 from *P. aeruginosa* revealed that the consensus sequence forms a rigid loop which is locked by both a disulfide bridge between the two Cys residues and a salt bridge between the Lys and Asp residues, and that this rigid loop protrudes from each of the inhibitors to occlude lysozyme's active site (Abergel *et al.*, 2007; Yum *et al.*, 2009). Paralogues of Ivyc and IvyP1 were identified in species of *Pseudomonas*, including *P. aeruginosa*, but these proteins lack the consensus loop sequence of Ivyc and consequently they do not inhibit lysozyme (Abergel *et al.*, 2007; Clarke *et al.*, 2010).

Despite being identified originally as inhibitors of lysozyme, the Ivy proteins (both Ivy1 and Ivy2) are not exported out of the producing bacteria to the external milieu where the lysozyme of host innate immune systems would be. Instead, they are localized to the periplasm (Deckers *et al.*, 2004; Abergel *et al.*, 2007; Clarke *et al.*, 2010). This strongly suggests that their physiological function is not to inhibit the exogenous lysozyme, but rather to inhibit any LTs that have dissociated

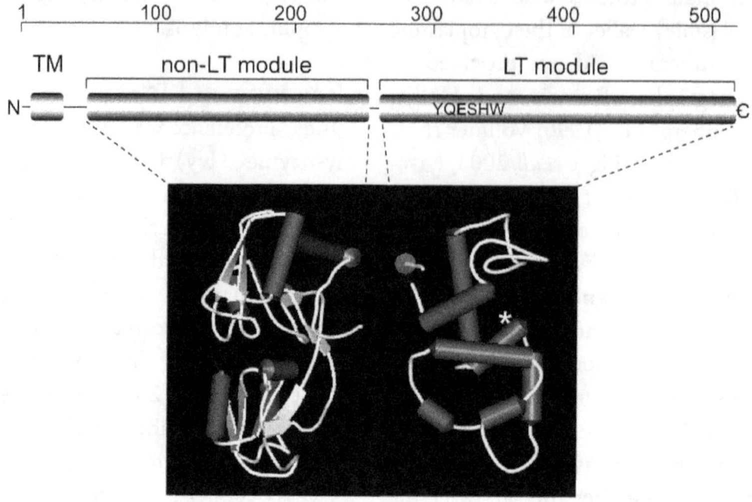

Figure 3.4 Structure of *E. coli* MltF. MltF is bimodal and possibly a bifunctional enzyme comprising an N-terminal transmembrane domain (TM) followed by a module that is similar to periplasmic proteins associated with ABC transporters and then a C-terminal LT module. The consensus sequence of MltF homologues involving the catalytic Glu residue is presented. The three-dimensional structures predicted by PHYRE are of residues 1–270 threaded onto the periplasmic LAO-binding protein from *S. enterica* (PDB 1LST) and residues 283–518 onto *E. coli* Slt70 (PDB 1QSA). The asterisk denotes the putative single catalytic Glu314 residue which is positioned appropriately at the bottom of a groove that runs through the middle of the module.

from either the murosomes or secretion/transport complexes. The production of Ivy would thus compensate for the lack of PG O-acetylation by directly inhibiting LT activity.

The discovery of Ivy provided the impetus to search for other 'lysozyme inhibitors' produced by bacteria. This has led to the identification of homologues and paralogues of small proteins produced by Gram-negative bacteria that are distantly related to Ivy but have apparent inhibitory activity against the different classes of lysozymes (Fig. 3.5). Thus, periplasmic lysozyme inhibitor C (PliC) from *Salmonella enterica* and its homologues (named membrane-bound lysozyme inhibitor, MliC) from both *E. coli* and *P. aeruginosa* were demonstrated experimentally to inhibit the class C hen egg-white lysozyme (Callewaert *et al.*, 2008). More recent studies led to the identification of inhibitors with apparent specificity for G-type (Vanderkelen *et al.*, 2011) and I-type (Van Herreweghe *et al.*, 2010) lysozymes. Interestingly, with homologues of the G-type inhibitors, the proteins from both *E. coli* and *S. enterica* were found to inhibit the goose-type lysozymes, but the protein isolated from *Bordetella avium* was devoid of this activity despite its high sequence similarity (Vanderkelen *et al.*, 2011). This situation is thus analogous to that with the Ivy proteins produced by the Pseudomonads, which likewise have strong sequence similarity to other Ivy proteins but lack the capacity to inhibit lysozyme. Also like Ivy, (i) each of these other lysozyme inhibitors are localized specifically to the periplasm by their producing cells, either as soluble proteins or associated with the inner leaflet of the outer membrane (Callewaert *et al.*, 2008), and (ii) with the exception of *Proteus penneri* and *Providencia rettgeri*, the production of these inhibitors would appear to be confined to Gram-negative bacteria that do not O-acetylate PG. Given these observations, we predict that, like Ivy, the true physiological function of each of these other periplasmic proteins is to inhibit and thereby control the LTs.

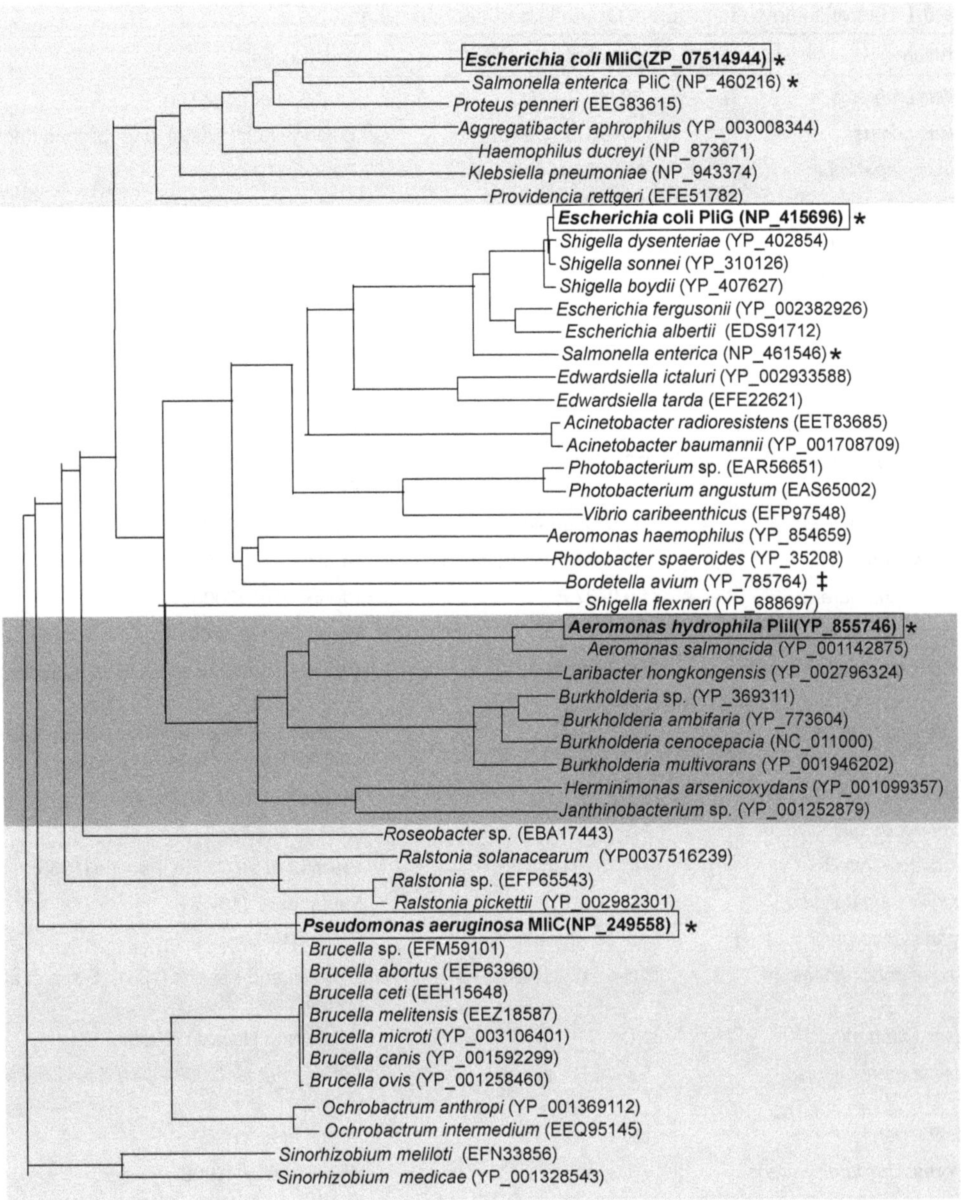

Figure 3.5 Phylogenic analysis of proteinaceous inhibitors of lysozymes distantly related to Ivy. The sequences of hypothetical and known homologs of *E. coli* MliC, *P. aeruginosa* MliC, *E. coli* PliG, and *Aeromonas hydrophila* PliI were aligned by CLUSTALW and presented as the phylogram. The asterisks denote those proteins that have been demonstrated experimentally to inhibit specific lysozymes (C, chicken, G, goose, I, invertebrate); (‡) denotes the protein devoid of inhibitory activity.

Concluding remarks and future directions

The LTs represent an important class of enzymes essential for the viability of Eubacteria that produce PG (Scheurwater and Clarke, 2008). With the potential to cause autolysis if the activity of endogenous LTs is not controlled, it would appear that bacteria have developed redundant systems to direct appropriate lytic activity at the enzymatic level. Although limited experimental evidence

Table 3.1 Bacteria known to produce O-acetylated peptidoglycan

Bacterium	Concentration (%)[1]	References
Bacillus anthracis	33–58	Laaberki *et al.* (2011)
Bacillus cereus	18–65 (3 strains)	Weadge *et al.* (2005) (unpublished data)
Bacillus megaterium	ND	Perkins (1965)
Bacillus subtilis	22 – 37 (4 strains)	Laaberki *et al.* (2011) (unpublished data)
Bacillus thuringiensis	ND	Unpublished data
Bacteroides fragilis	69.4	Weadge *et al.* (2005)
Bacteroides thetaiotamicron	27.0	Weadge *et al.* (2005)
Bradyrhizobium japonicum	18.5	Weadge *et al.* (2005)
Campylobacter jejuni	56–63 (2 strains)	Weadge *et al.* (2005)
Chromobacterium violaceum	14.7	Weadge *et al.* (2005)
Enterococcus feacalis	19–72 (5 strains)	Pfeffer *et al.* (2006), Hebert *et al.* (2007), Emirian *et al.* (2009)
Enterococcus faecium	5–57 (2 strains)	Pfeffer *et al.* (2006), Patti *et al.* (2008)
Enterococcus hirae	26.2–68.3 (4 strains)	Abrams (1958), Pfeffer *et al.* (2006)
Enterococcus durans	25.2–37.8	Pfeffer *et al.* (2006)
Helicobacter pylori	45.6	Weadge *et al.* (2005)
Lactobacillus acidophilus	60–70 (2 strains)	Coyette and Ghuysen (1970), Kleerebezem *et al.* (2010)
Lactobacillus fermentum	ND	Logardt and Neujahr (1975), Kleerebezem *et al.* (2010)
Lactobacillus plantarum	ND	Kleerebezem *et al.* (2010)
Macrococcus caseolyticus	ND	Bera *et al.* (2006)
Micrococcus luteus	ND (two strains)	Brumfitt *et al.* (1958), Perkins (1965)
Moraxella glucidolytica	ND	Martin *et al.* (1973)
Morganella morganii	43–50 (three strains)	Clarke (1993)
Neisseria gonorrhoeae	15–52 (20 strains)	Blundell and Perkins (1981), Swim *et al.* (1983)
Neisseria lactamica	ND	Dillard and Hackett (2005)
Neisseria meningitidis	32–40 (11 strains)	Antignac *et al.* (2003), Dillard and Hackett (2005)
Neisseria perflava	ND	Martin *et al.* (1973)
Photorhabdus luminescens	65.6	Weadge *et al.* (2005)
Proteus mirabilis	20–71 (18 strains)	Martin and Gmeiner (1979), Blundell and Perkins (1981), Dupont and Clarke (1991), Clarke (1993)
Proteus myxofaciens	52.8	Clarke (1993)
Proteus penneri	36.0	Clarke (1993)
Proteus vulgaris	29.4–50 (five strains)	Fleck *et al.* (1971), Clarke (1993)
Providencia alcalifaciens	42	Clarke (1993)
Providencia heinbachae	33.6	Clarke (1993)
Providencia rettgeri	36.6–42.1 (two strains)	Clarke (1993)
Providencia rustigianii	40.9	Clarke (1993)
Providencia stuartii	39.4–68 (nine strains)	Clarke (1993), Payie *et al.* (1995)
Pseudomonas alcaligenes	ND	Martin *et al.* (1973)
Ruminococcus flavefaciens	33.4	Weadge *et al.* (2005)

Table 3.1 continued

Bacterium	Concentration (%)[1]	References
Staphylococcus aureus	35–90 (8 strains)	Ghuysen and Strominger (1963), Burghaus *et al.* (1983), Snowden *et al.* (1989), MacDonald and Beveridge (2002), Bera *et al.* (2005)
Staphylococcus epidermidis	ND	Bera *et al.* (2006)
Staphylococcus haemolyticus	ND	Bera *et al.* (2006)
Staphylococcus hyicus	ND	Bera *et al.* (2006)
Staphylococcus lugdunensis	ND	Bera *et al.* (2006)
Staphylococcus saccharolyticus	ND	Bera *et al.* (2006)
Staphylococcus saprophyticus	ND	Bera *et al.* (2006)
Streptococcus pneumoniae	ND (2 strains)	Crisostomo *et al.* (2006)

[1]% of acetylation relative to MurNAc content in isolated PG.

ND, not determined.

has been obtained with only a few bacterial species, it is assumed that all bacteria organize their LTs within murosome complexes involving the PBPs. To further protect the cells from random lysis resulting from unassociated LTs, the cells appear to block the enzymes from acting by either modifying the substrate through O-acetylation or through the production of periplasmic inhibitors. With *P. rettgeri* and *P. penneri*, it seems that all three forms of control are operative further exemplifying the requirement to control LT activity. It follows then that perturbation of any of these systems may provide an opportunity to control the undesired growth of bacterial pathogens. For example, compounds developed to inhibit PG O-acetylation or prevent the association of the periplasmic inhibitors may provide new leads to the development of novel classes of antibiotics. However, more detailed investigations on the respective systems will be required to validate these potential opportunities.

References

Abergel, C., Monchois, V., Byrne, D., Chenivesse, S., Lembo, F., Lazzaroni, J.-C., and Claverie, J.-M. (2007). Structure and evolution of the Ivy protein family, unexpected lysozyme inhibitors in Gram-negative bacteria. Proc. Natl. Acad. Sci. U.S.A. *104,* 6394–6399.

Abrams, A. (1958). *O*-Acetyl groups in the cell wall of *Streptococcus faecalis*. J. Biol. Chem. *230,* 949–959.

Alaedini, A., and Day, R.A. (1999). Identification of two penicillin-binding multienzyme complexes in *Haemophilus influenzae*. Biochem. Biophys. Res. Commun. *264,* 191–195.

Antignac, A., Rousselle, J.C., Namane, A., Labigne, A., Taha, M.K., and Boneca, I.G. (2003). Detailed structural analysis of the peptidoglycan of the human pathogen *Neisseria meningitidis*. J. Biol. Chem. *278,* 31521–31528.

van Asselt, E.J., Dijkstra, A.J., Kalk, K.H., Takacs, B., Keck, W., and Dijkstra, B.W. (1999a). Crystal structure of *Escherichia coli* lytic transglycosylase Slt35 reveals a lysozyme-like catalytic domain with an EF-hand. Structure 7, 1167–1180.

van Asselt, E.J., Thunnissen, A.-M.W.H., and Dijkstra, B.W. (1999b). High resolution crystal structures of the *Escherichia coli* lytic transglycosylase Slt70 and its complex with a peptidoglycan fragment. J. Mol. Biol. *291,* 877–898.

van Asselt, E.J., Kalk, K.H., and Dijkstra, B.W. (2000). Crystallographic studies of the interactions of *Escherichia coli* lytic transglycosylase Slt35 with peptidoglycan. Biochemistry *39,* 1924–1934.

Bartoleschi, C., Pardini, M.C., Scaringi, C., Martino, M.C., Pazzani, C., and Bernardini, M.L. (2002). Selection of *Shigella flexneri* candidate virulence genes specifically induced in bacteria resident in host cell cytoplasm. Cell. Microbiol. *4,* 613–626.

Bera, A., Herbert, S., Jakob, A., Vollmer, W., and Götz, F. (2005). Why are pathogenic staphylococci so lysozyme resistant? The peptidoglycan *O*-acetyltransferase OatA is the major determinant for lysozyme resistance of *Staphylococcus aureus*. Mol. Microbiol. *55,* 778–787.

Bera. A., Biswas, R., Herbert, S., and Götz, F. (2006). The presence of peptidoglycan *O*-acetyltransferase in various staphylococcal species correlates with lysozyme resistance and pathogenicity. Infect Immun. *74,* 4598–4604.

Bertsche, U., Kast,T., Wolf, B., Fraipont, C., Aarsman, M.E.G. Kannenberg, K., von Reschenberg, M., Nguyen-Distèche, M., den Blaauwen, T., Höltje, J.-V., and Vollmer. W. (2006). Interaction between two murein (peptidoglycan) synthases, PBP 3 and PBP 1B, in *Escherichia coli*. Mol. Microbiol. *61*, 675–690.

Bhardwaj, S., and Day, R.A. (1997). Detection of intracellular protein–protein interactions: penicillin interactive proteins and morphogene proteins. Tech. Protein Chem. *VIII*, 469–480.

Blackburn, N.T., and Clarke, A.J. (2001). Identification of four families of peptidoglycan lytic transglycosylases. J. Mol. Evol. *52*, 78–84.

Blackburn, N.T., and Clarke, A.J. (2002). Characterization of soluble and membrane-bound family 3 lytic transglycosylases from *Pseudomonas aeruginosa*. Biochemistry *41*, 1001–1013.

Blundell, J.K., and Perkins, H.R. (1981). Effects of β-lactam antibiotics on peptidoglycan synthesis in growing *Neisseria gonorrhoeae*, including changes in the degree of *O*-acetylation. J. Bacteriol. *147*, 633–641.

Boch, J., Joardar, V., Gao, L., Robertson, T.L., Lim, M., and Kunkel, B.N. (2002). Identification of *Pseudomonas syringae* pv. tomato genes induced during infection of *Arabidopsis thaliana*. Mol. Microbiol. *44*, 73–88.

Brumfitt, W., Wardlaw, A.C., and Park, J.T. (1958). Development of lysozyme resistance in *Micrococcus lysodeikticus* and its association with an increased *O*-acetyl content in the cell wall. Nature *181*, 1783–1784.

Burghaus, P., Johannsen, L., Naumann, D., Labischinski, H., Bradaczek, H., and Giesbrecht, P. (1983). The influence of different antibiotics on the degree of O-acetylation of staphylococcal cell walls. In The Target of Penicillin. The Murein Sacculus of Bacterial Cell Walls Architecture and Growth, Hakenbeck, R., Höltje, J.-V., and Labischinski, H., eds. (New York, Walter de Gruyter and Co.) pp. 317–322.

Callewaert, L., Aertsen, A., Deckers, D., Vanoirbeek, K.G., Vanderkelen, L., Van Herreweghe, J.M., Masschalck, B., Nakimbugwe, D., Robben, J., and Michiels, C.W. (2008). A new family of lysozyme inhibitors contributing to lysozyme tolerance in gram-negative bacteria. PLoS Pathog. *4*, e1000019.

Clarke, A.J. (1993). Extent of peptidoglycan *O*-acetylation in the tribe *Proteeae*. J. Bacteriol. *175*, 4550–4553.

Clarke, A.J., and Dupont, C. (1992). O-Acetylated peptidoglycan: its occurrence, pathobiological significance, and biosynthesis. Can. J. Microbiol. *38*, 85–91.

Clarke, A.J., Strating, H., and Blackburn, N.T. (2000). Pathways for the O-acetylation of bacterial cell wall polysaccharides. In Glycomicrobiology, Doyle, R.J., ed. (New York, Plenum Publishing Co. Ltd.), pp. 187–223.

Clarke, C.A., Scheurwater, E.M., and Clarke, A.J. (2010). The vertebrate lysozyme inhibitor Ivy functions to inhibit the activity of lytic transglycosylases. J. Biol. Chem. *285*, 14843–14847.

Cloud-Hansen, K.A., Peterson, S.B., Stabb, E.V., Goldman, W.E., McFall-Ngai, M.J., and Handelsman, J. (2006). Breaching the great wall: peptidoglycan and microbial interactions. Nat. Rev. Microbiol. *4*, 710–716.

Cookson, B.T., Cho, H.L., Herwaldt, L.A., and Goldman, W.E. (1989). Biological activities and chemical composition of purified tracheal cytotoxin of *Bordetella pertussis*. Infect. Immun. *57*, 2223–2229.

Coyette, J., and Ghuysen, J.M. (1970). Structure of the walls of *Lactobacillus acidophilus* strain 63 AM Gasser. Biochemistry *9*, 2935–2943.

Crisóstomo, M.I., Vollmer, W., Kharat, A.S., Inhülsen, S., Gehre, F., Buckenmaier, S., and Tomasz, A. (2006). Attenuation of penicillin resistance in a peptidoglycan *O*-acetyl transferase mutant of *Streptococcus pneumoniae*. Mol. Microbiol. *61*, 1497–1509.

Deckers, D., Masschalck, B., Aertsen, A., Callewaert, L., Van Tiggelen, C.G., Atanassova, M., and Michiels, C.W. (2004). Periplasmic lysozyme inhibitor contributes to lysozyme resistance in *Escherichia coli*. Cell Mol. Life Sci. *61*, 1229–1237.

Dillard, J.P., and Hackett, K.T. (2005). Mutations affecting peptidoglycan acetylation in *Neisseria gonorrhoeae* and *Neisseria meningitidis*. Infect. Immun. *73*, 5697–5705.

Divakaruni, A.V., Loo, R.R., Xie, Y., Loo, J.A., and Gober, J.W. (2005). The cell-shape protein MreC interacts with extracytoplasmic proteins including cell wall assembly complexes in *Caulobacter crescentus*. Proc. Natl. Acad. Sci. U.S.A. *102*, 18602–18607.

Emirian, A., Fromentin, S., Eckert, C., Chau, F., Dubost, L., Delepierre, M., Gutmann, L., Arthur, M., and Mesnage, S. (2009). Impact of peptidoglycan O-acetylation on autolytic activities of the *Enterococcus faecalis* *N*-acetylglucosaminidase AtlA and *N*-acetylmuramidase AtlB. FEBS Lett. *583*, 3033–3038.

Figge, R.M., Divakaruni, A.V., and Gober, J.W. (2004). MreB, the cell shape-determining bacterial actin homologue, co-ordinates cell wall morphogenesis in *Caulobacter crescentus*. Mol. Microbiol. *51*, 1321–1332.

Fleck, J., Mock, M., Minck, R., and Ghuysen, J.M. (1971). The cell envelope in *Proteus vulgaris* P 18. Isolation and characterization of the peptidoglycan component. Biochim. Biophys. Acta *233*, 489–503.

Franklin, M.J., and Ohman, D.E. (1996). Identification of *algI* and *algJ* in the *Pseudomonas aeruginosa* alginate biosynthetic gene cluster which are required for alginate O acetylation. J. Bacteriol. *178*, 2186–2195.

Ghuysen, J.M., and Strominger, J.L. 1963. Structure of the cell wall of *S. aureus* strain Copenhagen. II. Separation and structure of disaccharides. Biochemistry 2, 1119–1125.

Goldman, W.E., Klapper, D.G., and Baseman, J.B. (1982). Detection, isolation, and analysis of a released *Bordetella pertussis* product toxic to cultured tracheal cells. Infect. Immun. *36*, 782–794.

Hébert. L., Courtin, P., Torelli, R., Sanguinetti, M., Chapot-Chartier, M.P., Auffray, Y., and Benachour, A. (2007). *Enterococcus faecalis* constitutes an unusual bacterial model in lysozyme resistance. Infect Immun. *75*, 5390–5398.

Heidrich, C., Ursinus, A., Berger, J., Schwarz, H., and Höltje, J.-V. (2002). Effects of multiple deletions of murein hydrolases on viability, septum cleavage, and sensitivity to large toxic molecules in *Escherichia coli*. J. Bacteriol. *184*, 6093–6099.

Henrissat, B., and Davies, G.J. (1997). Structural and sequence-based classification of glycoside hydrolases. Curr. Op. Struct. Biol. *7*, 637–644.

Höltje, J.-V. (1998). Growth of the stress-bearing and shape-maintaining murein sacculus of *Escherichia coli*. Microbiol. Mol. Biol. Rev. *62*, 181–203.

Höltje, J.-V., Mirelman, D., Sharon, N., and Schwarz, U. (1975). Novel type of murein transglycosylase in *Escherichia coli*. J. Bacteriol. *124*, 1067–1076.

Höppner, C., Carle, A., Sivanesan, D., Hoeppner, S.,and Baron, C. (2005). The putative lytic transglycosylase VirB1 from *Brucella suis* interacts with the type IV secretion system core components VirB8, VirB9 and VirB11. Microbiology *151*, 3469–3482.

Inohara, N., Chamaillard, M., McDonald, C., and Nunez, G. NOD-LRR proteins: role in host–microbial interactions and inflammatory disease. Annu. Rev. Biochem. *74*, 355–383.

Jennings, G.T., Savino, S., Marchetti, E., Arico, B., Kast, T., Baldi, L., Ursinus, A., Höltje, J.-V., Nicholas, R.A., Rappuoli, R., and Grandi, G. (2002). GNA33 from *Neisseria meningitidis* serogroup B encodes a membrane-bound lytic transglycosylase (MltA). Eur. J. Biochem. *269*, 3722–3731.

Kleerebezem, M., Hols, P., Bernard, E., Rolain, T., Zhou, M., Siezen, R.J., and Bron, P.A. (2010). The extracellular biology of the lactobacilli. FEMS Microbiol. Rev. *34*, 199–230.

Koraimann, G. (2003). Lytic transglycosylases in macromolecular transport systems of Gram-negative bacteria. Cell. Mol. Life Sci. *60*, 2371–2388.

Laaberki, M.H., Pfeffer, J.M., Clarke, A.J., and Dworkin, J. (2011). O-Acetylation of peptidoglycan is required for proper cell separation and S-layer anchoring in *Bacillus anthracis*. J. Biol. Chem. *286*, 5278–5288.

Legaree, B.A., and A.J. Clarke (2008). Interaction of penicillin-binding protein 2 with soluble lytic transglycosylase B1 in *Pseudomonas aeruginosa*. J. Bacteriol. *190*, 6922–6926.

Leung, A.K.-W., Duewel, H.S., Honek, J.F., and Berghuis, A.M. (2001). Crystal structure of the lytic transglycosylase from bacteriophage lambda in complex with hexa-*N*-acetylchitohexaose, Biochemistry *40*, 5665–5673.

Logardt, I.M., and Neujahr, H.Y. (1975). Lysis of modified walls from *Lactobacillus fermentum*. J. Bacteriol. *124*, 73–77.

Madoori, P.K., and Thunnissen, A.M. (2010). Purification, crystallization and preliminary X-ray diffraction analysis of the lytic transglycosylase MltF from *Escherichia coli*. Acta Crystallogr. Sect. F Struct. Biol. Cryst. Commun. *66*, 534–538.

Martin, H.H., and Gmeiner, J. (1979). Modification of peptidoglycan structure by penicillin-action in cell walls of *Proteus mirabilis*. Eur. J. Biochem. *95*, 487–495.

Martin, J.-P., Fleck, J., Mock, M., and Ghuysen, J.-M. (1973). The wall peptidoglycans of *Neisseria perflava, Moraxella glucidolytica, Pseudomonas alcaligenes* and *Proteus vulgaris* strain P18. Eur. J. Biochem. *38*, 301–306.

Melly, M.A., McGee, Z.A., and Rosenthal, R.S. (1984). Ability of monomeric peptidoglycan fragments from *Neisseria gonorrhoeae* to damage human fallopian tube mucosa. J. Infect. Dis. *149*, 378–386.

Moynihan, P.J., and Clarke, A.J. (2010). O-Acetylation of peptidoglycan in Gram-negative bacteria: Identification and characterization of peptidoglycan *O*-acetyltransferase in *Neisseria gonorrhoeae*. J. Biol. Chem. *285*, 13264–13273.

Patti, G.J., Chen, J., Schaefer, J., and Gross, M.L. (2008). Characterization of structural variations in the peptidoglycan of vancomycin-susceptible *Enterococcus faecium*: understanding glycopeptide-antibiotic binding sites using mass spectrometry. J. Am. Soc. Mass Spectrom. *19*, 1467–1475.

Payie, K.G., Rather, P.N., and Clarke, A.J. (1995). Contribution of gentamicin 2′-*N*-acetyltransferase to the O-acetylation of peptidoglycan in *Providencia stuartii*. J. Bacteriol. *177*, 4303–4310.

Perkins, H.R. (1965). The action of hot formamide on bacterial cell walls. Biochem. J. *95*, 876–882.

Pfeffer, J.M., Strating, H., Weadge, J.T., and Clarke, A.J. (2006). Peptidoglycan O-acetylation and autolysin profile of *Enterococcus faecalis* in the viable but nonculturable state. J. Bacteriol. *188*, 902–908.

Powell, A.J., Liu, Z.J., Nicholas, R.A., and Davies, C. (2006). Crystal structures of the lytic transglycosylase MltA from *N. gonorrhoeae* and *E. coli*: insights into interdomain movements and substrate binding. J. Mol. Biol. *359*, 122–136.

von Rechenberg, M., Ursinus, A., and Höltje, J.-V. (1996). Affinity chromatography as a means to study multienzyme complexes involved in murein synthesis. Microb. Drug Resist. *2*, 155–157.

Reid, C.W., Blackburn, N.T., Legaree, B.A., Auzanneau, F.I., and Clarke, A.J. (2004). Inhibition of membrane-bound lytic transglycosylase B by NAG-thiazoline. FEBS Lett. *574*, 73–79.

Reid, C.W., Blackburn, N.T., and Clarke, A.J. (2006). Role of arginine residues in the active site of the membrane-bound lytic transglycosylase B from *Pseudomonas aeruginosa*. Biochemistry *45*, 2129–2138.

Romeis, T., and Höltje, J.-V. (1994). Specific interaction of penicillin-binding proteins 3 and 7/8 with soluble lytic transglycosylase in *Escherichia coli*. J. Biol. Chem. *269*, 21603–21607.

Satta, G., Fontana, R., and Canepari, P. (1994). The two-competing site (TCS) model for cell shape regulation in bacteria: the envelope as an integration point for the regulatory circuits of essential physiological events. Adv. Microb. Physiol. *36*, 281–245.

Shen, K., Antalis, P., Gladitz, J., Sayeed, S., Ahmed, A., Yu, S., Hayes, J., Johnson, S., Dice, B., Dopico, R., Keefe, R., Janto, B., Chong, W., Goodwin, J., Wadowsky, R.M., Erdos, G., Post, J.C., Ehrlich, G.D., and Hu, F.Z. (2005). Identification, distribution and expression of novel genes in 10 clinical isolates of nontypeable *Haemophilus influenzae*. Infect. Immun. *73*, 3479–3491.

Scheurwater, E.M., and Clarke, A.J. (2008). The C-terminal domain of *Escherichia coli* YfhD functions as a lytic transglycosylase. J. Biol. Chem. *283*, 8363–8373.

Snowden, M.A., Perkins, H.R., Wyke, A.W., Hayes, M.V., and Ward, J.B. (1989). Cross-linking and *O*-acetylation of newly synthesized peptidoglycan in *Staphylococcus aureus* H. J. Gen. Micro. *135*, 3015–3022.

van Straaten, K.E., Dijkstra, B.W., Vollmer, W., and Thunnissen, A.-M.W.H. (2005). Crystal structure of MltA from *Escherichia coli* reveals a unique lytic transglycosylase fold. J. Mol. Biol. *352*, 1068–1082.

Sun, Y.H., Bakshi, S., Chalmers, R., and Tang, C.M. (2000). Functional genomics of *Neisseria meningitidis* pathogenesis. Nature Med. *6*, 1269–1273.

Swim, S.C., Gfell, M.A., Wilde, C.E. 3rd, and Rosenthal, R.S. (1983). Strain distribution in extents of lysozyme resistance and O-acetylation of gonococcal peptidoglycan determined by high-performance liquid chromatography. Infect Immun. *42*, 446–452.

Thunnissen, A.-M.W.H., Dijkstra, A.J., Kalk, K.H., Rozeboom, H.J., Engel, H., Keck, W., and Dijkstra, B.W. (1994). Doughnut-shaped structure of a bacterial muramidase revealed by X-ray crystallography. Nature *367*, 750–753.

Thunnissen, A.-M.W.H., Isaacs, N., and Dijkstra, B.W. (1995a). The catalytic domain of a bacterial lytic transglycosylase defines a novel class of lysozymes. Proteins *22*, 245–258.

Thunnissen, A.-M.W.H., Rozeboom, H.J., Kalk, K.H., and Dijkstra, B.W. (1995b). Structure of the 70-kDa soluble lytic transglycosylase complexed with bulgecin A. Implications for the enzymatic mechanism. Biochemistry *34*, 12729–12737.

Vanderkelen, L., Van Herreweghe, J.M., Vanoirbeek, K.G., Baggerman, G., Myrnes, B., Declerck, P.J., Nilsen, I.W., Michiels, C.W., and Callewaert, L. (2011). Identification of a bacterial inhibitor against g-type lysozyme. Cell. Mol. Life Sci. *68*, 1053–1064.

Van Herreweghe, J.M., Vanderkelen, L., Callewaert, L., Aertsen, A., Compernolle, G., Declerck, P.J., and Michiels, C.W. (2010). Lysozyme inhibitor conferring bacterial tolerance to invertebrate type lysozyme Cell. Mol. Life Sci. *67*, 1177–1188.

Veiga. P., Bulbarela-Sampieri, C., Furlan, S., Maisons, A., Chapot-Chartier, M.P., Erkelenz, M., Mervelet, P., Noirot, P., Frees, D., Kuipers, O.P., Kok, J., Gruss, A., Buist, G., and Kulakauskas, S. (2007). SpxB regulates O-acetylation-dependent resistance of *Lactococcus lactis* peptidoglycan to hydrolysis. J. Biol. Chem. *282*, 19342–19354.

Vollmer, W., von Rechenberg, M., and Höltje, J.-V. (1999). Demonstration of molecular interactions between the murein polymerase PBP1B, the lytic transglycosylase MltA, and the scaffolding protein MipA of *Escherichia coli*. J. Biol. Chem. *274*, 6726–6734.

Vollmer, W., Blanot, D., and de Pedro, M.A. (2008). Peptidoglycan structure and architecture. FEMS Microbiol Rev. *32*, 149–167.

Weadge, J.T., and Clarke, A.J. (2006). Identification and characterization of O-acetylpeptidoglycan esterase: a novel enzyme discovered in *Neisseria gonorrhoeae*. Biochemistry *45*, 839–851.

Weadge, J.T., and Clarke, A.J. (2007). *Neisseria gonorrheae O*-acetylpeptidoglycan esterase, a serine esterase with a Ser-His-Asp catalytic triad. Biochemistry *46*, 4932–4941.

Weadge, J.T., Pfeffer, J.M., and Clarke, A.J. (2005). Identification of a new family of enzymes with potential *O*-acetylpeptidoglycan esterase activity in both Gram-positive and Gram-negative bacteria. BMC Microbiol. *5*, 49.

Yum, S., Kim, M.J., Xu, Y., Jin, X.L., Yoo, H.Y., Park, J.W., Gong, J.H., Choe, K.M., Lee, B.L., and Ha, N.C. (2009). Structural basis for the recognition of lysozyme by MliC, a periplasmic lysozyme inhibitor in Gram-negative bacteria. Biochem. Biophys. Res. Commun. *378*, 244–248.

Zhao, Y., Blumer, S.E., and Sundin, G.W. (2005). Identification of *Erwinia amylovora* genes induced during infection of immature pear tissue. J. Bacteriol. *187*, 8088–8103.

Bacterial Flagellar Glycosylation Systems – Opportunities and Applications

4

Susan M. Twine and Susan M. Logan

Abstract

The biosynthesis and assembly of the flagellar apparatus has been the subject of extensive studies over many decades. More recently, glycosylation of major structural protein, the flagellin, has been shown to be an important component of numerous flagellar systems in both Archaea and Bacteria; playing either an integral role in assembly and for a number of bacterial pathogens a role in virulence. Increasingly, it is apparent that bacteria elaborate a structurally diverse array of flagellin-modifying glycans. This chapter focuses firstly upon reviewing recent research on the structural diversity in Gram-positive and Gram-negative flagellar glycosylation systems. In the second part, the ways in which flagellin glycosylation and associated biosynthetic pathways can be exploited are discussed.

Introduction

In recent years, the abundance of examples of the presence of prokaryotic glycoproteins has clearly established that this form of post-translational modification is in fact an integrated part of prokaryotic biology. While extensive structural characterization of many of these glycoproteins has been completed, our understanding of the biological significance and utility of these novel glycoconjugates remains in its infancy. However, some of the most extensively studied glycoproteins are the bacterial flagellins. These proteins form the flagellar filament and impart motility on cells via the sophisticated flagellar rotary motor (Macnab, 2003). It has been demonstrated in recent years for a number of polarly flagellated bacteria, that the major flagellin structural proteins bear O-linked glycans. Of significance, is the fact that glycosylated flagellins appear to be commonly produced by bacterial pathogens, yet the glycan structures produced by distinct species are unique. The body of work describing flagellar glycosylation has been reviewed elsewhere (Logan, 2006) and so the focus of this chapter will be on more recent advances, as well as potential applications and opportunities, which are emerging from our understanding of these unique flagellar glycosylation systems.

Tools and techniques

Mass spectrometry (MS) has proved to be an essential tool in the elucidation of glycan structure. 'Bottom-up' analysis involves proteolytic cleavage of flagellin proteins and analysis of glycopeptides by collision-assisted dissociation tandem MS (CAD–MS/MS) or electron transfer dissociation tandem MS (ETD–MS/MS). These analyses allow the mapping of sites of glycosylation, determination of glycan linkage as well as a more detailed characterization of glycan structure. In addition to the more traditional 'bottom-up' analysis of a glycoprotein, recent development of a 'top-down' strategy incorporating electrospray mass spectrometry (ES–MS) for the analysis of intact bacterial glycoproteins has greatly facilitated flagellin structural elucidation. Top-down analysis has been used extensively for flagellar structural analysis (Schirm *et al.*, 2005; Twine *et al.*, 2008, 2009). The process allows determination of the glycoprotein mass while the formation of characteristic glycan oxonium ions in the low mass range also provides information for preliminary structural assignment using accurate mass and fragmentation patterns. While these MS techniques are critical for structural assignment of flagellar glycans, they are usually not sufficient. The sugars found on bacterial flagellins are often novel in structure with unique stereochemistry, and so NMR is required for structural determination of each of the glycan components. The precise structural assignment of numerous novel flagellar glycans has been completed following analysis of either purified glycopeptide (Thibault *et al.*, 2001; Twine *et al.*, 2008), purified glycan (Voisin *et al.*, 2005) or nucleotide-activated sugar purified from the metabolome of the bacterial cell (Soo *et al.*, 2004; McNally *et al.*, 2006, 2007).

The development of focused metabolomics techniques in recent years has greatly facilitated the structural elucidation of novel glycan moieties and corresponding biosynthetic pathways (Soo *et al.*, 2004). The application of novel capillary electrophoresis mass spectrometry (CE-ESMS) and precursor ion scanning methods in metabolomics studies of *Helicobacter pylori* and *Campylobacter jejuni* has provided insight into the role of genes from the flagellin glycosylation loci of these two bacteria. In addition, the process has been successfully employed to purify biosynthetic intermediates for precise structural analysis (McNally *et al.*, 2006, 2007; Logan *et al.*, 2009).

A number of chemical approaches have recently been used to identify bacterial glycoproteins and these studies have been extensively reviewed by Dube *et al.* (2011). Metabolic oligosaccharide engineering (MOE) is one such approach which has been utilized recently for glycoprotein discovery (Koenigs *et al.*, 2009). In this process slightly modified monosaccharide residues containing a chemical handle (e.g. ketone, azide, thiol or alkene functional groups) are incorporated into cellular glycans following uptake into the bacterial cell and utilization by glycan biosynthetic pathways. Once incorporated into cellular glycans the chemical handle can be tagged for visualization and enrichment. Metabolic incorporation of an azide sugar derivative into a flagellar glycan has actually been demonstrated (Liu *et al.*, 2009). Chemienzymatic synthesis was used to generate 6-deoxyAltNac4Naz, an azido containing derivative of a pseudaminic acid (5,7-diacetamido-3,5,7,9-tetradeoxy-L-glycero-α-L-manno-nonulosonic acid, Pse), Pse5Ac7Ac metabolic precursor. *C. jejuni* cells fed this precursor were then shown to replace Pse5Ac7Ac with Pse5Ac7NAz on the flagellin protein. Moreover, these azide-containing glycoproteins which were assembled into a flagellar filament on the cell surface, could be tagged via Staudinger ligation with biotin.

Flagellar glycan structures – an update

Gram-negative bacteria

Campylobacter spp.

An extensive body of work has been completed at both the structural and genetic level describing the process of glycosylation in this species, and this has been previously reviewed (Logan, 2006). *C. jejuni* flagellins have been shown to be glycosylated with derivatives of the sialic acid-like sugar pseudaminic acid (Thibault *et al.*, 2001; Logan *et al.*, 2002; McNally *et al.*, 2006). More recently, reports have demonstrated that in addition to these novel Pse monosaccharides, some strains of *C. jejuni* and *Campylobacter coli* can also glycosylate their flagellins with derivatives of a second sialic acid-like monosaccharide, legionaminic acid (5,7-diacetamido-3,5,7,9-tetradeoxy-D-*glycero*-β-D-*galacto*-nonulosonic acid, Leg) (Logan *et al.*, 2009). The diverse Pse and Leg derivatives which have now been identified on *Campylobacter* flagellin are listed in Table 4.1 and include replacement of an acetamido with an acetamidino group (Pse5Ac7Am, Leg5Am7Ac) as well as variations in N-acyl groups (e.g. 2,3-di-*O*-methylglyceric acid).

Aeromonas spp.

In a similar fashion to *Campylobacter* spp., the flagellins of *Aeromonas* spp. are also glycosylated with nonulosonate sugars (Schirm *et al.*, 2005). A novel nonulosonate sugar derivative of mass 373 Da on *Aeromonas caviae* UU51 flagellin was described using 'top-down' analysis and the

Table 4.1 Flagellar glycan structures

Organism	Type of flagella	No. of glycosylation sites	Glycan structure	Glycan component mass (Da) (monosaccharide masses)
Gram-negative				
Aeromonas caviae Sch3	Polar, complex	6–8 sites	Pse5Ac7Ac	316
	Laf, complex	ND	ND	
Aeromonas caviae UU 51	Polar, complex	FlaA 6 sites	Pse5Ac7Ac8Ac	373
	Lateral, complex	FlaB 7 sites	ND	
		ND		
Campylobacter jejuni, *Campylobacter coli*	Bipolar, complex	16–19 sites	Pse5Ac7Am	315
			Pse5Ac7Ac	316
			Pse8OAc	358
			Pse (2,3-di-O-methylglyceric acid derivatives)	390/391
			Pse5Pr7Pr	408
			Pse5Ac7AmGlnAc	485
			Leg5Am7Ac	315
			Leg5AmNMe7Ac	329
Helicobacter pylori 1061	Polar, complex, sheathed	FlaA 7sites, FlaB 10 sites	Pse5Ac7Ac	316
Helicobacter pylori NCTC11637	Polar, complex, sheathed	FlaA 7 sites, FlaB 10 sites	Pse5Ac7Ac	316
Pseudomonas aeruginosa PAK (a-type)	Polar, simple	2 sites	Heterogeneous glycan with rhamnose linking monosaccharide and novel trisaccharide capping	Oligosaccharide masses range from 729 to 1597 Da (146–variable oligosaccharide-176–146–174)
Pseudomonas aeruginosa JJ692 (a-type)	Polar, simple	2 sites	Rhamnose	146
Pseudomonas aeruginosa PAO1 (b-type)	Polar, simple	2 sites	Unassigned	355 (146–80–129)
Pseudomonas aeruginosa 1244 (b-type)	Polar, simple	2 sites	Unassigned	355 (146–80–129)
Pseudomonas syringea	Polar, simple	6 sites	Di- and tri-saccharide rhamnosyl residues with capping 4,6-dideoxy-4(3-hydroxybutanamido)2-O-methylglucose	538 (146–146–246)

Table 4.1 continued

Organism	Type of flagella	No. of glycosylation sites	Glycan structure	Glycan component mass (Da) (monosaccharide masses)
Gram-positive				
Clostridium botulinum[1]	–	–	Leg5GluNMe7Ac	417
			Di-*N*-acetylHexA	258
			Tri-*N*-acetylHexA	300
			Putative Leg5Ac7Ac	316
Clostridium difficile 630	–	–	HexNac, PO4, Asp	398 (203–98–115)
C. difficile (recent clinical isolates)[2]	–	–	Heterogeneous glycan with linking HexNAc	203
				146
				192
				160
Listeria monocytogenes	Peritrichous, simple	6 sites	β-GlcNAc	203

[1]See Table 4.2 for detailed descriptions of flagellin glycan diversity between strains.

[2]See Fig. 4.1 for structural details.

ND, not determined.

fragmentation pattern indicated that this sugar was a Pse derivative. More recently a 'bottom up' analysis of flagellar glycopeptides identified the glycan on *A. caviae* Sch3 polar flagella as Pse5Ac7Ac based on MS fragmentation patterns. In addition, complementation experiments with *C. jejuni pseI* and *H. pylori pseB* biosynthetic genes showed that these genes were able complement their *A. caviae* homologues providing further evidence that *A. caviae* was able to biochemically synthesize Pse5Ac7Ac (Tabei *et al.*, 2009). In contrast to *C. jejuni* and *Aeromonas hydrophila* where a dedicated genetic island was shown to be responsible for polar flagellar glycan production (Gryllos *et al.*, 2001; Szymanski *et al.*, 2003), the genes involved in synthesis of the flagellar glycan in *A. caviae* are also involved in LPS biosynthesis, and Pse5Ac7Ac is a component of both surface glycoconjugate structures (Tabei *et al.*, 2009). A report describing glycosylation of both polar and lateral flagella from *A. hydrophila* AH-3 was recently published although it was not determined if unique glycans were found on the respective flagellin proteins (Canals *et al.*, 2007).

Pseudomonas spp.

The opportunistic pathogen *Pseudomonas aeruginosa* possesses a polar flagellum which has been shown to be glycosylated. The flagellin protein is modified with one of two glycans described as a-type or b-type (Brimer and Montie, 1998; Schirm *et al.*, 2004b; Verma *et al.*, 2006). Both a- and b-type flagellin proteins have been characterized and while both have distinct glycan structures each glycan has been shown to contain an O-linked deoxyhexose residue as the linking sugar (Verma *et al.*, 2006). In *P. aeruginosa* PAK a-type flagellin the unique glycan was shown to be linked to the protein backbone via a rhamnose (Rha) moiety (Schirm *et al.*, 2004a). Recent work by Lindhout *et al.* (2009) has demonstrated in *P. aeruginosa* PAO1 (b-type) that only L-Rha and not D-Rha is incorporated into the flagellar filament glycan and the nucleotide-activated sugar precursor TDP-L-Rha is shared between LPS biosynthesis and flagellar glycan biosynthesis. MS analysis of flagellin from a second b-type strain of *P. aeruginosa* (strain 1244) revealed an identical b-type glycan to that found on PAO1 flagellin. The mass of the glycan is 355 Da and is linked to the flagellin protein via a Rha moiety (S. Twine and S. Logan, unpublished observation). In contrast, the pilin protein from this strain has been shown to be glycosylated with an LPS O-antigen repeat unit which is quite distinct in structure (Castric *et al.*, 2001).

In the plant pathogen *Pseudomonas syringae*, Rha-linked glycans have also been identified at six serine residues of the flagellin protein (Taguchi *et al.*, 2006). Glycan structures from two pathovars have recently been determined and shown to be composed of di-,tri- or tetrasaccharides of Rha residues capped with 4,6-dideoxy-4-(3-hydroxybutanamido)-2-*O*-methylglucose (Takeuchi *et al.*, 2007; Konishi *et al.*, 2009). Analysis of the flagellin protein of *P. syringae* pv tabaci 6605 revealed that the Rha content of the glycan was solely L-Rha, while *P. syringae* pv glycinea race 4 flagellin contained both L-Rha and D-Rha at a ratio of 4:1. The role of these glycans in the biology of these important plant pathogens is discussed below.

Gram-positive organisms

In contrast to the significant number of Gram-negative species which have been shown to produce glycosylated flagellins, a very limited number of Gram-positive species were described in a review (Logan, 2006). Of these, only the glycan from *Listeria monocytogenes* flagellin had been characterized and shown to be composed of *N*-acetylglucosamine in O-linkage at six sites on each flagellin monomer (Schirm *et al.*, 2004b). Recently, however novel flagellar glycan structures have been identified and characterized from two significant bacteria – *Clostridium botulinum* and *Clostridium difficile*.

C. botulinum is an anaerobic, spore-forming rod that produces a potent neurotoxin, botulinum neurotoxin (BoNT) that is the causative agent of botulism. The extreme toxicity and potential bioterrorism threat have directed much of the *C. botulinum* research focus towards BoNT. There were, until recently, no published reports describing *C. botulinum* flagellin. To date, several studies have examined *C. botulinum* flagellins at the DNA, protein sequence and post-translational level (Paul *et al.*, 2007b; Twine *et al.*, 2008). Aberrant migration of purified flagellin monomers on SDS-PAGE indicated that flagellin proteins from strains with identical flagellin gene sequences had markedly different molecular masses from the predicted mass. This not only suggested post-translational modification, but some diversity in flagellin post-translational modification between different *C. botulinum* strains (Paul *et al.*, 2007b). A top-down MS characterization of flagellins from group I and II strains confirmed that *C. botulinum* flagellins were modified with a diverse array of glycan moieties, with observed glycan masses ranging from 258 Da to 417 Da (Twine *et al.*, 2008). Tandem mass spectrometry showed the glycan fragment ion patterns to be characteristic of nonulosonic acids or *N*-acetyl hexuronic acids. A glycopeptide harbouring a glycan of 417 Da from a Group I strain (strain Alberta) was isolated in sufficient quantity for NMR structural studies and the glycan from this strain was shown to be a novel legionaminic acid derivative, 7-acetamido-5-(*N*-methyl-glutam-4-yl)-3,5,7,9-tetradeoxy-D-*glycero*-α-D-*galacto*-legionaminic acid (αLeg5GluNMe7Ac) (Twine *et al.*, 2008). This was the first report of a nonulosonic acid type sugar modifying a protein from a Gram-positive bacterium. Comparative genomic indexing of *C. botulinum* strains by DNA microarray identified the flagellar glycosylation island locus as a highly variable region of the *C. botulinum* genome that may have evolved independently from much of the genetic complement. This work also linked differences in the glycosylation profiles of flagellins from various strains to the gene content of the flagellar glycosylation island, dividing the screened strains into six groups, based on the gene content of their respective flagellar glycosylation island (Carter *et al.*, 2009).

C. difficile is an emerging opportunistic pathogen and the leading cause of antibiotic-associated diarrhoea and pseudomembranous colitis in humans. The recent emergence of the hypervirulent NAP1/027 strain in hospitals globally has resulted in increased severity of infection and mortality rates. While studies have shown that two toxins, TcdA and TcdB, are responsible for severe tissue damage and consequent manifestation of disease, little is known about other putative virulence factors, such a flagellas. It has been reported that purified *C. difficile* flagellin monomers from various strains migrate at a molecular weight greater than that predicted from the translated DNA sequence, but flagellin monomers showed no reactivity with glycan staining kits (Tasteyre *et al.*, 2000). Mass spectrometry studies of purified flagellin from strain CD630 showed the protein to be modified in O-linkage at up to seven sites with a glycan of 398 Da. Accurate mass measurement of the glycan, combined with multistage mass spectrometry (also known as MS^n) fragmentation experiments, provided evidence that the modification comprised a *N*-acetylhexosamine (HexNAc) linked to a methylated aspartic acid via a phosphate bond. In contrast, the flagellins isolated from hypervirulent strains, including strains isolated from recent *C. difficile* outbreaks, were shown to be modified with a heterogeneous glycan containing up to four different monosaccharide residues with masses of 204 Da (putative HexNAc), 146 Da (putative deoxyhexose), 160 Da (putative methylated deoxyhexose), and 192 Da (putative heptose) which formed glycan chains of various length, as detailed in Table 4.1 and Fig. 4.1 (Twine *et al.*, 2009; Hitchen *et al.*, 2010). Although variation in glycan composition was observed amongst strains, all of the *C. difficile* flagellins so far examined were modified with a glycan attached in O-linkage via a HexNAc moiety (Twine *et al.*, 2009). Examination of the flagellar glycosylation loci from multiple *C. difficile* genome sequences revealed a conserved glycosyltransferase gene lying proximal to the flagellin structural gene. Disruption of this gene in strain CD630 resulted in complete loss of flagellin glycosylation, in addition to subsequent inability to assemble the unglycosylated flagellin into flagellar filaments. Correspondingly, bacterial cells were no longer motile. It remains to be determined if the extended flagellar glycan chains present on

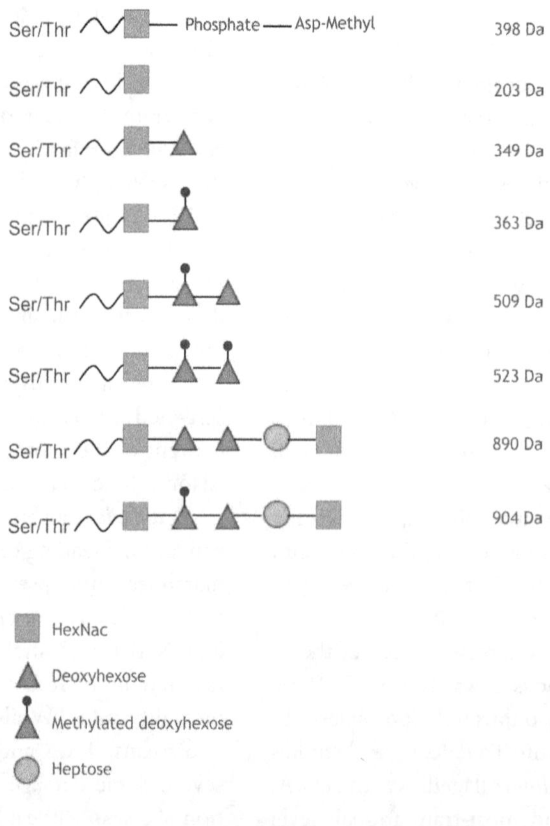

Figure 4.1 Schematic depiction of the diversity of glycan moieties (and corresponding masses) identified to date modifying flagellins of *C. difficile* (from Twine *et al.*, 2009; Hitchen *et al.*, 2010).

hypervirulent, clinical isolates play an additional role in host pathogen interactions of these *C. difficile* strains.

Biological role(s) of flagellar glycans

While the list of bacterial species producing glycosylated flagellins continues to grow (Hayakawa *et al.*, 2009) our understanding of the role of glycosylation in biological systems remains limited. Of significance however, is the identification of flagellar glycans on many bacterial pathogens. This has stimulated a considerable body of work to explore the link between glycan presence and the ability to cause disease. A number of examples of biological roles for flagellar glycans have been provided to date.

Probably the most extensively studied group of organisms with respect to a biological role of the flagellar glycan are members of the epsilon division of Proteobacteria. The flagellin structural proteins (FlaA, FlaB) of *C. jejuni* and *H. pylori* cannot assemble into filaments unless the protein monomers are glycosylated (Thibault *et al.*, 2001; Schirm *et al.*, 2003).

Autoagglutination (Agg) of *C. jejuni* cells has been shown to be associated with flagellin (Misawa and Blaser, 2000) and Guerry and coworkers demonstrated that when the flagellar glycan composition of *C. jejuni* 81-176 was altered (loss of PseAm or PseAc) cells failed to autoagglutinate although motility was unaffected (Guerry *et al.*, 2006). Flagellar glycans appear to mediate Agg and microcolony formation by interactions with other flagellar glycans on adjacent bacteria. Studies have recently demonstrated that 5 of the 19 sites of glycosylation on *Campylobacter* flagellin are critical for this Agg phenotype, while three other sites of glycosylation are critical for structural subunit–subunit interactions within the filament (Ewing *et al.*, 2009). The flagellar glycans on *C. jejuni* have also been shown to contribute to virulence. The loss of Pse5Am7Ac from flagellin resulted in reduced adherence and invasion of intestinal epithelial cells as well as attenuation in a ferret diarrhoeal disease model (Guerry *et al.*, 2006). A comparative phylogenomics study of *C. jejuni* strains demonstrated that isolates fell into two distinct clades; livestock-associated and non-livestock-associated. A group of five genes from the flagellar glycosylation locus were identified as being associated with the livestock clade (Champion *et al.*, 2005). Recent studies characterizing these genes have revealed that they are responsible for the biosynthesis of unique Leg derivatives and the presence of these novel sugars in livestock isolates provides an explanation for why some strains of *C. jejuni* are more probable than others to colonize chickens (Howard *et al.*, 2009).

H. pylori flagella are covered with a membranous sheath and so the role of flagellar glycans in host–pathogen interactions is less clear. However, recent work demonstrated that flagellar hyperglycosylation promoted enhanced interaction between bacteria and a human gastric epithelial cells (AGS) cells thereby accelerating host cellular responses. CagA phosphorylation and NF-κB activation occurred more efficiently following infection of AGS cells with mutant strains which were producing elevated levels of O-linked glycans on flagellin proteins. This increase in amount of glycosylation was shown to result in bacterial cells which were hypermotile, which presumably improved the colonization ability of *H. pylori* (Asakura *et al.*, 2010).

In contrast to these bacterial species, glycosylation of flagellin from *Pseudomonas* spp. is not a prerequisite for assembly into functional flagellar filaments. In the human pathogen *P. aeruginosa*, a mutant strain of PAO1 (b-type) which was unable to glycosylate its flagellin, exhibited a reduction in virulence in the murine model of burn wound infection although the organism retained full motility (Arora *et al.*, 2005). The molecular basis of this attenuation in virulence is currently not defined.

Flagellins from the plant pathogenic bacteria *P. syringae* pv tabaci and pv glycinea have been reported to induce hypersensitive (HS) cell death in non-host plants. As the flagellin amino acid sequences in the two pathovars studied to date are identical it appears that the glycans on the respective flagellins play an important part in stimulating plant HS responses to each pathogenic strain (Taguchi *et al.*, 2003a,b; Takeuchi *et al.*, 2003). In addition it was shown that the glycan structures of *P. syringae* pv tabaci flagellin play a role in flagellar filament suprastructure and glycosylation defective mutants were shown to be defective in swimming ability in a highly viscous medium. Flagellar filaments appeared to aggregate into bundles at the cell surface when glycosylation was abolished which was in distinct contrast to cells where flagellins were glycosylated and filaments remained as distinct entities (Taguchi *et al.*, 2008).

Recent genetic analysis of *C. difficile* (Gram-positive, phylum Firmicutes) has demonstrated that, in a similar fashion to *C. jejuni* and *H. pylori*, glycosylation is required for flagellar assembly in this group of organisms (Twine *et al.*, 2008). It remains to be established if flagellin glycosylation is required for flagellar assembly and subsequent motility in *C. botulinum.*

Applications and opportunities

The utility of flagellar glycans as biomarkers for strain detection/disease diagnosis as well as the potential of the glycosylation biosynthetic pathways as targets for therapeutic intervention are beginning to be explored.

Flagellar glycosylation pathways as therapeutic targets

It is becoming increasingly clear that the ability of pathogenic bacteria to synthesize glycoproteins is often directly linked to their ability to cause disease and as described above there is a growing body of evidence that glycosylation of flagellin is critical for virulence of a number of significant pathogens. Bacterial glycans including flagellar glycans have unique structures that distinguish them from eukaryotes and as such make attractive targets for therapeutic intervention. To validate the flagellar glycosylation systems as targets, inactivation of flagellar glycosylation genes led to pathogens which had lost their ability to adhere to and invade host cells, initiate infection and colonize the host (Goon *et al.*, 2003; Schirm *et al.*, 2003; Arora *et al.*, 2005). As a consequence of these studies, the enzymes of the glycan pathways can now be considered as attractive targets for therapeutic intervention.

For *H. pylori* where the flagellin protein is glycosylated with the novel sialic acid-like nonulosonate sugar Pse5Ac7Ac, recent studies have focused on defining the biosynthetic pathway for this unique prokaryotic sugar. An elegant study by Schoenhofen and coworkers (Schoenhofen *et al.*, 2006b) characterized six *H. pylori* biosynthetic enzymes previously identified by genetic studies and reconstituted the complete biosynthesis of Pse5Ac7Ac from UDP-GlcNAc in a single reaction combining the six enzymes. This *in vitro* enzymatic synthesis of Pse5Ac7Ac set the groundwork for small molecule inhibitor screening and design of inhibitors of the Pse pathway enzymes. Indeed, McNally *et al* (McNally *et al.*, 2008) established that cytidine monophosphate CMP-Pse5Ac7Ac is a potent inhibitor of PseB, the first enzyme in the pathway, where it binds to the substrate binding site and acts as a competitive inhibitor. STD-NMR, CORCEMA calculations and molecular modelling revealed that the nucleotide group of CMP-Pse is the key binding epitope. Such information can be exploited in future rational design of a cell permeable molecule inhibitor which targets PseB. The biosynthetic pathway leading to Leg5Ac7Ac, the related nine carbon sugar found on flagellins of *C. jejuni* has recently been biochemically characterized and found to proceed via unique guanosine diphosphate (GDP)-linked intermediates (Schoenhofen *et al.*, 2009).

Structural studies of a number of Pse pathway enzymes (PseC, PseB, PseG) has also been completed (Ishiyama *et al.*, 2006; Schoenhofen *et al.*, 2006a; Rangarajan *et al.*, 2009) in addition to detailed characterization of enzyme reaction mechanisms (Chou *et al.*, 2005; Liu and Tanner, 2006; Morrison *et al.*, 2008).

With the characterization of these flagellin glycoprotein biosynthesis pathways complete, strategies to identify small molecule lead compounds which inhibit the respective biosynthetic enzymes can now be explored. Such strategies would include library screening, structure-based design, transition state analogue design, and fragment-based screening.

Flagellar glycans as biomarkers/typing tools

Bacterial surface structures have been exploited in numerous bacterial typing schemes, with two of the most extensive being the O (lipopolysaccharide) and H (flagellar) typing schemes of *Escherichia coli*. Surface-exposed or shed components may also have unique features that can be exploited in bacterial detection or forensic strain differentiation. As a result of the terrorist attacks in the United States in 2001, there has been a renewed interest in detection of potential bioterrorism agents, such as *C. botulinum* and the associated BoNT. In particular, techniques capable of detecting *C. botulinum* or BoNT, discriminating BoNT type or subtype as well as forensic differentiation of individual strains are much sought after. Detection of BoNT is challenging, owing to the low levels that are typically required to cause illness and currently the mouse bioassay is the most widely accepted method for confirming BoNT presence. This bioassay is time consuming, requires specialized personnel and animal handling facilities, and is usually limited to reference laboratories. If routine screening of potential sources of contamination in the food chain, or rapid response to intentional contamination is a key priority, alternative detection or differentiation methods will be required.

Recent studies identified *C. botulinum* flagellin as a novel biomarker that, in contrast to BoNT, is produced by cells in relatively large amounts and contains both sequence and post-translational diversity. Furthermore, flagellin proteins have been identified as components of BoNT preparations (for example culture supernatants) and it has been suggested that *C. botulinum* flagellin could be exploited as a surrogate for detection of BoNT. The flagellin structural gene *flaA* was identified as a target for rapid PCR-based identification of *C. botulinum* strains (Paul *et al.*, 2007a,b). Gene sequencing of *flaA* in multiple *C. botulinum* strains resulted in division into 13 flagellar types and correlated with strain epidemiology. Comparative genomic indexing of *C. botulinum* strains by DNA microarray identified the flagellar glycosylation locus as a highly variable region of the *C. botulinum* genome which contained gene targets suitable for use in the proposed multilocus sequence typing protocol (Carter *et al.*, 2009).

At the protein level, top-down MS studies of purified *C. botulinum* flagellins showed that two marker ions were readily detectable for glycosylated flagellin. One marker ion (doubly charged m/z 512) was detected in every *C. botulinum* strain screened and corresponded to the C-terminal flagellin peptide. This peptide, arising from non-specific collision-induced cleavage at a C-terminal proline, is specific for *C. botulinum*. Flagellin sequences of other Gram-positive organisms were examined with respect to their C-terminal peptides, and were shown to have a proline residue close to the C-terminus of the protein sequence. However, the C-terminal peptide sequences differed sufficiently between species such that the C-terminal flagellin peptide could be distinguished by MS and used to identify Gram-positive bacterial species (Table 4.3). The second marker ion from *C. botulinum* flagellin was specific for the modifying glycan species, and varied between strains (Tables 4.1 and 4.2). Thus, from rapid, top-down MS of purified flagellins, two marker ions identify a protein as belonging to *C. botulinum* and in addition aid in strain differentiation. Using top-down liquid chromatography–mass spectrometry (LC–MS), *C. botulinum*-specific flagellin marker ions were readily detectable in toxin positive culture supernatant, illustrated in Fig. 4.2 (Twine *et al.*, 2008). Marker ions characteristic of *C. botulinum* flagellin were observed at m/z 418.2, and 512.3 and MS/MS fragmentation patterns of these ions confirmed them to be *C. botulinum* flagellin marker ions (Fig. 4.2). The utility of this discovery as a surrogate marker for detection of BoNT or for strain differentiation has yet to be exploited. It is possible that monoclonal antibodies could be developed to specifically recognize unique carbohydrate epitopes, and these could be used in a high-throughput ELISA-type screening assay. Detection and differentiation at the protein level has the potential to offer complementary information from that offered by more traditional nucleic acid-based approaches which are limited to primary sequence information and do not incorporate the additional discriminatory potential offered by post-translational modifications.

Table 4.2 Diversity of *Clostridium botulinum* flagellin proteins determined by mass spectrometry

Strain	FlaVR	Region	Source	MW[1]	Exact protein mass[2]	Marker ions[3]	FGI[4]
Group 1							
FE9909ACS	3	AB, Canada	Faeces	29.4	32.30, 31.89, 31.40	512.3^{2+}, 418.1^{+}	3
FE303A1YO	3	ON, Canada	Faeces	29.4	32.30, 31.88, 31.46	512.3^{2+}, 418.1^{+}	3
MUL0109ASA	1	Gulf of Kuwait, Kuwait	Mullet fish	29.80	32.63, 32.60, 32.27, 32.24, 31.88	512.3^{2+}, 418.1^{+}	1
FE0205A1AK	1	AB, Canada	Faeces	29.59	32.32, 32.35, 32.28	512.3^{2+}, 301.1^{+}, 259.1^{+}	2
Type B							
PA9508B	2	QC, Canada	Paté	29.6	32.92, 32.88, 32.49	512.3^{2+}, 259.1^{+}	1
FE9904BMT	4	ON, Canada	Faeces	29.4	32.96, 32.80, 32.47	512.3^{2+}, 299.1^{+}, (317.1^{+})	ND
Type AB							
FE9504ACG	1	QC, Canada	Faeces	29.7	32.36	512.3^{2+}, 259.1^{+}	1
Type F							
Langeland	3	Denmark	Liver paste	29.4	32.32, 31.91, 31.49	512.3^{2+}, 418.1^{+}	3
H461297F	1	WI, USA	Honey	29.8	32.27, 32.31, 32.35	512.3^{2+}, 301.1^{+}, 259.1^{+}	2
Group II							
Type B							
17B	9	Pacific, USA	Sediment	29.1	32.18	512.3^{2+}, 697.2^{+}	ND
Kap-B3		CA, USA	Kapchunka	29.1	32.89	512.3^{2+}	ND
Type E							
Bennett	8	NL, Canada	Gastric fluid	52.1	52.15	512.3^{2+}	ND
Gordon	10	QC, Canada	Clinical	nd	52.15	512.3^{2+}	ND
Type F							
610F	9	OR, USA	Salmon	29.3	32.18, 32.20, 32.22	512.3^{2+}, 697.2^{+}	–

Flagellin from a diverse array of *C. botulinum* strains was purified from cultured cells by mechanical shearing (Paul *et al.*, 2007) and accurate mass of the protein determined by infusion into QTOF2 mass spectrometer (Waters). Marker ions denote labile protein ions observed during tandem mass spectrometry experiments using intact protein. 512.3^{2+} ion was observed as a labile marker ion in all *C. botulinum* flagellin surveyed.

[1]Predicted molecular mass from DNA sequence (Paul *et al.*, 2007).

[2]Protein molecular mass of intact clostridial flagellins, determined by mass spectrometry. Intact proteins were infused into the QTOF2 mass spectrometer and the ion profile recorded. Exact protein mass was determined using maximum entropy algorithm.

[3]Ions observed during tandem mass spectrometry of multiple charged intact protein ions. Data originally published in Twine *et al.* (2008).

[4]FGI refers to the flagellar glycosylation island division, based upon microarray comparisons of the strains (Carter *et al.*, 2009).

Table 4.3 Predicted flagellin C-terminal peptide backbone ion phase cleavages

Strain	Accession number[1]	Predicted non-specific C-terminal peptide cleavage[2]	Theoretical monoisotopic mass[3]
C. botulinum strain Alberta[4]		PQGVLQLLR	1022.6237
C. botulinum strain Bennet		PQGVLQLLR	1022.6237
C. difficile strain BI-1		PQGVLQLLG	923.5440
C. difficile strain 630		PQGVLQLLG	923.5440
Clostridium tetani E88	NP_782313	pqgilqllr	1036.6393
Clostridium novyi NT Flagellin_C	YP_877971	PQDALRVLEGIK	1337.7667
Clostridium thermocellum ATCC 27405	ABN53439	pqgvlqllg	923.5440
Bacillus subtilis subsp. *subtilis* str. 168	NP_391416	pqnvlqllr	1079.6451
Campylobacter jejuni subsp. *jejuni* NCTC 11168	NP_282485	No proline at C-terminus of protein	–

[1]Accession number according to the NCBI.

[2]C-terminal peptide fragmentation predicted non-specific cleavage site N-terminal to proline.

[3]Theoretical peptide monoisotopic mass, calculated using BioLynx (Waters).

[4]All group I and II strains have identical C-terminal cleavage peptides.

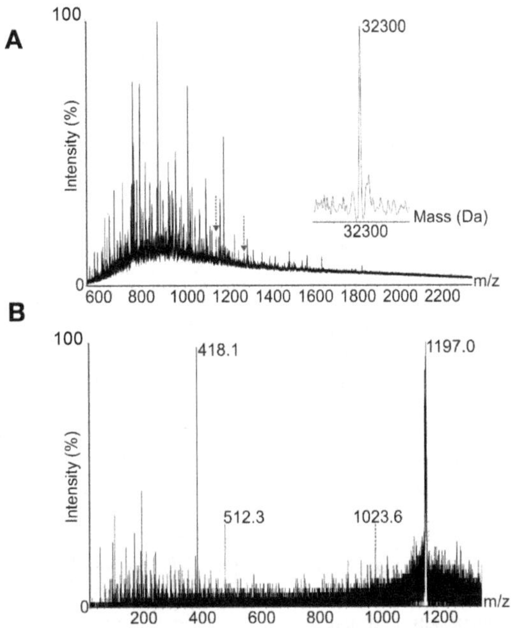

Figure 4.2 Mass spectrometry detection of flagellin-associated marker ions in crude preparations of *Clostridium botulinum* strain Alberta BoNT. (A) Electrospray mass spectrum of crude BoNT preparation. Inset shows the deconvoluted mass profile, indicating an intact protein mass of 32300 Da. The arrows indicate protein ions that were selected for tandem mass spectrometry. (B) Tandem mass spectrometry analyses of multiply charged protein ion at *m/z* 1197.0. Similar results were obtained when analysis was carried out on the multiple charged protein ion at *m/z* 1346.7. Ions at *m/z* 418.2, 512.3 and 1023.6 were observed in the tandem mass spectrum, characteristic of flagellin glycan and peptide markers from *C. botulinum* strain Alberta flagellin. Specifically, ions at *m/z* 512.3 and 1023.6 corresponded to the double and single charged forms of the conserved C-terminal flagellin peptide (PQGVLQLLR), whilst *m/z* 418.2 corresponded to a labile glycan modification of flagellin.

Conclusion

It has been clearly demonstrated in recent years that a significant number of bacteria, many of which are pathogens, produce flagellins which are glycosylated in O-linkage with novel glycan moieties. The diversity in glycan structure which has been revealed through comprehensive MS and NMR analyses probably reflects distinct biological roles for these unique flagellar glycans and offers potential for development of diagnostic and typing biomarkers. The future holds significant promise in uncovering novel and exciting biological roles for these prominent cell surface prokaryotic glycans as well as exploring the utility of the glycosylation systems in novel therapeutic strategies against a number of important bacterial pathogens.

Acknowledgements

The authors would like to thank all members of their laboratories (past and present) for their contributions to flagellar glycosylation research.

References

Arora, S.K., Neely, A.N., Blair, B., Lory, S., and Ramphal, R. (2005). Role of motility and flagellin glycosylation in the pathogenesis of *Pseudomonas aeruginosa* burn wound infections. Infect. Immun. *73,* 4395–4398.

Asakura, H., Churin, Y., Bauer, B., Boettcher, J.P., Bartfeld, S., Hashii, N., Kawasaki, N., Mollenkopf, H.J., Jungblut, P.R., Brinkmann, V., *et al.* (2010). *Helicobacter pylori* HP0518 affects flagellin glycosylation to alter bacterial motility. Mol. Microbiol. *78,* 1130–1144.

Brimer, C.D., and Montie, T.C. (1998). Cloning and comparison of *fliC* genes and identification of glycosylation in the flagellin of *Pseudomonas aeruginosa* a-type strains. J. Bacteriol. *180,* 3209–3217.

Canals, R., Vilches, S., Wilhelms, M., Shaw, J.G., Merino, S., and Tomas, J.M. (2007). Non-structural flagella genes affecting both polar and lateral flagella-mediated motility in *Aeromonas hydrophila.* Microbiology *153,* 1165–1175.

Carter, A.T., Paul, C.J., Mason, D.R., Twine, S.M., Alston, M.J., Logan, S.M., Austin, J.W., and Peck, M.W. (2009). Independent evolution of neurotoxin and flagellar genetic loci in proteolytic *Clostridium botulinum.* BMC Genomics *10,* 115.

Castric, P., Cassels, F.J., and Carlson, R.W. (2001). Structural characterization of the *Pseudomonas aeruginosa* 1244 pilin glycan. J. Biol. Chem. *276,* 26479–26485.

Champion, O.L., Gaunt, M.W., Gundogdu, O., Elmi, A., Witney, A.A., Hinds, J., Dorrell, N., and Wren, B.W. (2005). Comparative phylogenomics of the food-borne pathogen *Campylobacter jejuni* reveals genetic markers predictive of infection source. Proc. Natl. Acad. Sci. U.S.A. *102,* 16043–16048.

Chou, W.K., Dick, S., Wakarchuk, W.W., and Tanner, M.E. (2005). Identification and characterization of NeuB3 from *Campylobacter jejuni* as a pseudaminic acid synthase. J. Biol. Chem. *43,* 35922–35928.

Dube, D.H., Champasa, K., and Wang, B. (2011). Chemical tools to discover and target bacterial glycoproteins. Chem. Commun. (Camb.) *47,* 87–101.

Ewing, C.P., Andreishcheva, E., and Guerry, P. (2009). Functional characterization of flagellin glycosylation in *Campylobacter jejuni* 81-176. J. Bacteriol. *191,* 7086–7093.

Goon, S., Kelly, J.F., Logan, S.M., Ewing, C.P., and Guerry, P. (2003). Pseudaminic acid, the major modification on *Campylobacter* flagellin, is synthesized via the Cj1293 gene. Mol. Microbiol. *50,* 659–671.

Gryllos, I., Shaw, J.G., Gavin, R., Merino, S., and Tomas, J.M. (2001). Role of flm locus in mesophilic Aeromonas species adherence. Infect. Immun. *69,* 65–74.

Guerry, P., Ewing, C.P., Schirm, M., Lorenzo, M., Kelly, J., Pattarini, D., Majam, G., Thibault, P., and Logan, S.M. (2006). Changes in flagellin glycosylation affect Campylobacter autoagglutination and virulence. Mol. Microb. *60,* 299–311.

Hayakawa, J., Kondoh, Y., and Ishizuka, M. (2009). Cloning and characterization of flagellin genes and identification of flagellin glycosylation from thermophilic Bacillus species. Biosci. Biotechnol. Biochem. *73,* 1450–1452.

Hitchen, P.G., Twigger, K., Valiente, E., Langdon, R.H., Wren, B.W., and Dell, A. (2010). Glycoproteomics: a powerful tool for characterizing the diverse glycoforms of bacterial pilins and flagellins. Biochem. Soc. Trans. *38,* 1307–1313.

Howard, S.L., Jagannathan, A., Soo, E.C., Hui, J.P., Aubry, A.J., Ahmed, I., Karlyshev, A., Kelly, J.F., Jones, M.A., Stevens, M.P., *et al.* (2009). *Campylobacter jejuni* glycosylation island important in cell charge, legionaminic acid biosynthesis, and colonization of chickens. Infect. Immun. *77,* 2544–2556.

Ishiyama, N., Creuzenet, C., Miller, W.L., Demendi, M., Anderson, E.M., Harauz, G., Lam, J.S., and Berghuis, A.M. (2006). Structural studies of FlaA1 from *Helicobacter pylori* reveal the mechanism for inverting 4,6-dehydratase activity. J. Biol. Chem. *281,* 24489–24495.

Koenigs, M.B., Richardson, E.A., and Dube, D.H. (2009). Metabolic profiling of *Helicobacter pylori* glycosylation. Mol. Biosyst. *5,* 909–912.

Konishi, T., Taguchi, F., Iwaki, M., Ohnishi-Kameyama, M., Yamamoto, M., Maeda, I., Nishida, Y., Ichinose, Y., Yoshida, M., and Ishii, T. (2009). Structural characterization of an O-linked tetrasaccharide from *Pseudomonas syringae* pv. tabaci flagellin. Carbohydr. Res. *344,* 2250–2254.

Lindhout, T., Lau, P.C., Brewer, D., and Lam, J.S. (2009). Truncation in the core oligosaccharide of lipopolysaccharide affects flagella-mediated motility in

Pseudomonas aeruginosa PAO1 via modulation of cell surface attachment. Microbiology *155*, 3449–3460.

Liu, F., and Tanner, M.E. (2006) PseG of pseudaminic acid biosynthesis: a UDP–sugar hydrolase as a masked glycosyltransferase. J. Biol. Chem. *281*, 20902–20909.

Liu, F., Aubry, A.J., Schoenhofen, I.C., Logan, S.M., and Tanner, M.E. (2009). The engineering of bacteria bearing azido-pseudaminic acid-modified flagella. Chembiochemistry *10*, 1317–1320.

Logan, S.M. (2006). Flagellar glycosylation – a new component of the motility repertoire? Microbiology *152*, 1249–1262.

Logan, S.M., Kelly, J.F., Thibault, P., Ewing, C.P., and Guerry, P. (2002). Structural heterogeneity of carbohydrate modifications affects serospecificity of Campylobacter flagellins. Mol. Microbiol. *46*, 587–597.

Logan, S.M., Hui, J.P., Vinogradov, E., Aubry, A.J., Melanson, J.E., Kelly, J.F., Nothaft, H., and Soo, E.C. (2009). Identification of novel carbohydrate modifications on *Campylobacter jejuni* 11168 flagellin using metabolomics-based approaches. FEBS J. *276*, 1014–1023.

Macnab, R.M. (2003). How bacteria assemble flagella. Annu. Rev. Microbiol. *57*, 77–100.

McNally, D.J., Hui, J.P., Aubry, A.J., Mui, K.K., Guerry, P., Brisson, J.R., Logan, S.M., and Soo, E.C. (2006). Functional characterization of the flagellar glycosylation locus in *Campylobacter jejuni* 81-176 using a focused metabolomics approach. J. Biol. Chem. *281*, 18489–18498.

McNally, D.J., Aubry, A.J., Hui, J.P., Khieu, N.H., Whitfield, D., Ewing, C.P., Guerry, P., Brisson, J.R., Logan, S.M., and Soo, E.C. (2007). Targeted metabolomics analysis of *Campylobacter coli* VC167 reveals legionaminic acid derivatives as novel flagellar glycans. J. Biol. Chem. *282*, 14463–14475.

McNally, D.J., Schoenhofen, I.C., Houliston, R.S., Khieu, N.H., Whitfield, D.M., Logan, S.M., Jarrell, H.C., and Brisson, J.R. (2008). CMP-pseudaminic acid is a natural potent inhibitor of PseB, the first enzyme of the pseudaminic acid pathway in *Campylobacter jejuni* and *Helicobacter pylori*. ChemMedChem. *3*, 55–59.

Misawa, N., and Blaser, M.J. (2000). Detection and characterization of autoagglutination activity by *Campylobacter jejuni*. Infect. Immun. *68*, 6168–6175.

Morrison, J.P., Schoenhofen, I.C., and Tanner, M.E. (2008). Mechanistic studies on PseB of pseudaminic acid biosynthesis: a UDP-N-acetylglucosamine 5-inverting 4,6-dehydratase. Bioorg. Chem. *36*, 312–320.

Paul, C.J., Tran, S., Tam, K.J., and Austin, J.W. (2007a) A unique restriction site in the flaA gene allows rapid differentiation of group I and group II *Clostridium botulinum* strains by PCR-restriction fragment length polymorphism analysis. J. Food Prot. *70*, 2133–2139.

Paul, C.J., Twine, S.M., Tam, K.J., Mullen, J.A., Kelly, J.F., Austin, J.W., and Logan, S.M. (2007b) Flagellin diversity in *Clostridium botulinum* groups I and II: a new strategy for strain identification. Appl. Environ. Microbiol. *73*, 2963–2975.

Rangarajan, E.S., Proteau, A., Cui, Q., Logan, S.M., Potetinova, Z., Whitfield, D., Purisima, E.O., Cygler, M., Matte, A., Sulea, T., *et al.* (2009). Structural and functional analysis of *Campylobacter jejuni* PseG: a UDP–sugar hydrolase from the pseudaminic acid biosynthetic pathway. J. Biol. Chem. *284*, 20989–21000.

Schirm, M., Soo, E.C., Aubry, A.J., Austin, J., Thibault, P., and Logan, S.M. (2003). Structural, genetic and functional characterization of the flagellin glycosylation process in *Helicobacter pylori*. Mol. Microbiol. *48*, 1579–1592.

Schirm, M., Arora, S.K., Verma, A., Vinogradov, E., Thibault, P., Ramphal, R., and Logan, S.M. (2004a) Structural and genetic characterization of glycosylation of type a flagellin in *Pseudomonas aeruginosa*. J. Bacteriol. *186*, 2523–2531.

Schirm, M., Kalmokoff, M., Aubry, A., Thibault, P., Sandoz, M., and Logan, S.M. (2004b) Flagellin from *Listeria monocytogenes* is glycosylated with beta-O-linked N-acetylglucosamine. J. Bacteriol. *186*, 6721–6727.

Schirm, M., Schoenhofen, I.C., Logan, S.M., Waldron, K. C., and Thibault, P. (2005). Identification of unusual bacterial glycosylation by tandem mass spectrometry analyses of intact proteins. Anal. Chem. *77*, 7774–7782.

Schoenhofen, I.C., Lunin, V.V., Julien, J.P., Li, Y., Ajamian, E., Matte, A., Cygler, M., Brisson, J.R., Aubry, A., Logan, S.M., *et al.* (2006a) Structural and functional characterization of PseC, an aminotransferase involved in the biosynthesis of pseudaminic acid, an essential flagellar modification in *Helicobacter pylori*. J. Biol. Chem. *281*, 8907–8916.

Schoenhofen, I.C., McNally, D.J., Brisson, J.R., and Logan, S.M. (2006b) Elucidation of the CMP-pseudaminic acid pathway in *Helicobacter pylori*: synthesis from UDP-N-acetylglucosamine by a single enzymatic reaction. Glycobiology *16*, 8C–14C.

Schoenhofen, I.C., Vinogradov, E., Whitfield, D.M., Brisson, J.R., and Logan, S.M. (2009). The CMP-legionaminic acid pathway in Campylobacter: biosynthesis involving novel GDP-linked precursors. Glycobiology *19*, 715–725.

Soo, E.C., Aubry, A.J., Logan, S.M., Guerry, P., Kelly, J.F., Young, N.M., and Thibault, P. (2004). Selective detection and identification of sugar nucleotides by CE-electrospray–MS and its application to bacterial metabolomics. Anal. Chem. *76*, 619–626.

Szymanski, C.M., Logan, S.M., Linton, D., and Wren, B.W. (2003). Campylobacter – a tale of two protein glycosylation systems. Trends Microbiol. *11*, 233–238.

Tabei, S.M., Hitchen, P.G., Day-Williams, M.J., Merino, S., Vart, R., Pang, P.C., Horsburgh, G.J., Viches, S., Wilhelms, M., Tomas, J.M., *et al.* (2009). An *Aeromonas caviae* genomic island is required for both O-antigen lipopolysaccharide biosynthesis and flagellin glycosylation. J. Bacteriol. *191*, 2851–2863.

Taguchi, F., Shimizu, R., Ikeda, Y., Inagaki, Y., Toyoda, K., Shiraishi, T., and Ichinose, Y. (2003a) Post-translational modifcation of flagellin determines the specific induction of HR. Plant Physiol. Biochem. *44*, 342–349.

Taguchi, F., Shimizu, R., Nakajima, R., Toyoda, K., Shiraishi, T., and Ichinose, Y. (2003b) Differential effects of flagellins from *Pseudomonas syringea* pvs. tabaci, tomato and glycinea on plant defence response. Plant Physiol. Biochem. *41,* 165–174.

Taguchi, F., Takeuchi, K., Katoh, E., Murata, K., Suzuki, T., Marutani, M., Kawasaki, T., Eguchi, M., Katoh, S., Kaku, H., *et al.* (2006). Identification of glycosylation genes and glycosylated amino acids of flagellin in *Pseudomonas syringae* pv. tabaci. Cell Microbiol. *8,* 923–938.

Taguchi, F., Shibata, S., Suzuki, T., Ogawa, Y., Aizawa, S., Takeuchi, K., and Ichinose, Y. (2008). Effects of glycosylation on swimming ability and flagellar polymorphic transformation in *Pseudomonas syringae* pv. tabaci 6605. J. Bacteriol. *190,* 764–768.

Takeuchi, K., Taguchi, F., Inagaki, Y., Toyoda, K., Shiraishi, T., and Ichinose, Y. (2003). Flagellin glycosylation island in *Pseudomonas syringea* pv glycinea and its role in host specificity. J. Bacteriol. *185,* 6658–6665.

Takeuchi, K., Ono, H., Yoshida, M., Ishii, T., Katoh, E., Taguchi, F., Miki, R., Murata, K., Kaku, H., and Ichinose, Y. (2007). Flagellin glycans from two pathovars of *Pseudomonas syringae* contain rhamnose in D and L configurations in different ratios and modified 4-amino-4,6-dideoxyglucose. J. Bacteriol. *189,* 6945–6956.

Tasteyre, A., Barc, M.C., Karjalainen, T., Dodson, P., Hyde, S., Bourlioux, P., and Borriello, P. (2000). A *Clostridium difficile* gene encoding flagellin. Microbiology *146,* 957–966.

Thibault, P., Logan, S.M., Kelly, J.F., Brisson, J.R., Ewing, C.P., Trust, T.J., and Guerry, P. (2001). Identification of the carbohydrate moieties and glycosylation motifs in *Campylobacter jejuni* flagellin. J. Biol. Chem. *276,* 34862–34870.

Twine, S.M., Paul, C.J., Vinogradov, E., McNally, D.J., Brisson, J.R., Mullen, J.A., McMullin, D.R., Jarrell, H.C., Austin, J.W., Kelly, J.F., *et al.* (2008). Flagellar glycosylation in *Clostridium botulinum.* FEBS J. *275,* 4428–4444.

Twine, S.M., Reid, C.W., Aubry, A., McMullin, D.R., Fulton, K.M., Austin, J., and Logan, S.M. (2009). Motility and flagellar glycosylation in *Clostridium difficile.* J. Bacteriol. *191,* 7050–7062.

Verma, A., Schirm, M., Arora, S.K., Thibault, P., Logan, S.M., and Ramphal, R. (2006). Glycosylation of b-Type flagellin of *Pseudomonas aeruginosa*: structural and genetic basis. J. Bacteriol. *188,* 4395–4403.

Voisin, S., Houliston, R.S., Kelly, J., Brisson, J.R., Watson, D., Bardy, S.L., Jarrell, K.F., and Logan, S.M. (2005). Identification and characterization of the unique N-linked glycan common to the flagellins and S-layer glycoprotein of *Methanococcus voltae.* J. Biol. Chem. *280,* 16586–16593.

Bacterial N-linked Glycosylation Systems: Current Knowledge and Future Perspectives

5

Jon Cuccui, Rebecca H. Langdon, Madeleine G. Moule and Brendan W. Wren

Abstract

Once thought to be restricted to eukaryotes and archaea, N-linked glycosylation has now been discovered in prokaryotes. Over the past decade, our understanding of bacterial N-linked glycosylations systems and their abundance has been expanding. This type of protein modification was first demonstrated in *Campylobacter jejuni*, a human gut pathogen, and we now know that N-linked glycosylation also exists in other ε-proteobacteria ranging from the deep-sea vent *Nitratiruptor* spp. and *Sulfurovum* spp. to sulfate-reducing δ-proteobacteria. A greater understanding of these systems is necessary in order to comprehend the evolutionary reasons for their development and maintenance. In addition, this knowledge may also be exploited for glycoengineering purposes to produce cheaper subunit vaccines as well as humanized proteins.

Introduction

N-linked glycosylation is the most frequent post-translational modification of secretory proteins in eukaryotes and has been studied for over 60 years. Deficiencies in this glycosylation system have been linked to a variety of human disorders including neurological, liver and intestinal dysfunctions (Aebi and Hennet, 2001). Until recently it was thought that this system was unique to eukaryotic cells, and that bacteria were capable of O-linked glycosylation only of cell surface structures such as pili and flagella. The discovery of an N-linked glycosylation system in the bacterium *Campylobacter jejuni*, a common cause of acute gastroenteritis in humans, was a major finding that challenged this concept. Consequently, several additional bacterial glycosylation systems have been discovered, indicating that this post-translational modification is abundant in all domains of life. As our understanding of bacterial *pgl* systems has developed, the importance and potential applications of this novel process has become apparent.

Discovery and characterization of the first bacterial N-linked glycosylation system in *C. jejuni*

In 1999, Szymanski *et al.* described the cloning of a 16-kb gene cluster adjacent to the lipopolysaccharide (LPS) biosynthetic locus of *C. jejuni*. Surprisingly, this locus did not play a role in LPS core or O-antigen assembly and mutagenesis of the locus resulted in the disappearance of several immunodominant proteins in Western blots, suggesting the presence of a general glycosylation system (Szymanski *et al.*, 1999). Genes within the locus were subsequently named *pgl* for *p*rotein *gl*ycosylation. In 2000, Parkhill *et al.* published the full genome sequence of *C. jejuni* strain NCTC11168 and confirmed that the *pgl* locus was independent of the LPS and O-linked flagellin glycosylation loci, further suggesting the presence of a new general glycosylation system (Parkhill *et al.*, 2000). The glycan modification synthesized by the *pgl* locus was structurally determined by mass spectrometry and nuclear magnetic resonance (NMR) spectroscopy as a heptasaccharide – GalNAc$_5$GlcBac, where Bac at the reducing end is bacillosamine (2,4-diacetamido-2,4,6-trideoxyglucose) (Wacker *et al.*, 2002; Young *et al.*, 2002). These studies helped to confirm the *C. jejuni* Pgl locus as the first N-linked general glycosylation system to be found in a bacterial species.

The key enzyme in the newly discovered pathway that initially suggested the potential presence of an N-linked general protein glycosylation system in *C. jejuni* was a single oligosaccharyltransferase termed PglB that has significant homology to Stt3p. This has previously been found exclusively in eukaryotes and in archaea and genetic and biochemical studies in yeast have demonstrated that Stt3p is an essential component of the oligosaccharyltransferase complex, an integral part of the machinery of macromolecular synthesis and is crucial for the proper functioning of all eukaryotic cells (Zufferey *et al.*, 1995; Silberstein and Gilmore, 1996; Knauer and Lehle, 1999). In eukaryotes, Stt3p has been shown to play a central role in the eight-member oligosaccharyltransferase complex and is evolutionarily conserved (Yan and Lennarz, 2002). Stt3p homologues, including PglB, contain a conserved C-terminal catalytic domain (W**WD**YG) where the second and third amino acids are essential and the third residue is variable (Yan and Lennarz, 2002).

Analysis of the 16-kb *C. jejuni pgl* locus revealed sequence similarity to glycosyltransferases and enzymes required for sugar modification, which allowed an initial model for the pathway to be developed. Further analysis involving mutagenesis of each of the genes in the 16-kb locus, in combination with structural analyses, confirmed the role of individual genes in the locus (Linton *et al.*, 2005). Some components of the *C. jejuni* Pgl pathway have similarity to the O-linked glycosylation pathway of pilin in *Neisseria meningitidis* and to other systems that involve the synthesis of bacillosamine derivatives (Power *et al.*, 2000). As in *N. meningitides,* bacillosamine (Bac) is the initiating sugar in the synthesis of the *C. jejuni* glycan and is converted from UDP-HexNAc by the sequential modification of PglF (dehydratase), PglE (aminotransferase) and PglD (acetyltransferase). PglC then attaches the Bac residue to the lipid carrier undecaprenolpyrophosphate and PglA adds an α-1,3-linked GalNAc to Bac (Virji, 1999; Glover *et al.*, 2005). In *C. jejuni,* PglH and PglJ are both involved in transferring the four additional α-1,4-linked GalNAc residues. *In vitro* assays have demonstrated that the first α 1,4-GalNAc is transferred by PglJ whilst the remaining three are transferred by PglH (Glover *et al.*, 2005). The final glycosyltransferase, PglI, adds the branched glucose in *C. jejuni.* Once assembled, the entire heptasaccharide is flipped across the inner membrane into the periplasm by the ABC transporter, PglK (Kelly *et al.*, 2006) (Fig. 5.1). This assembly process is reminiscent of other bacterial glycoconjugate biosynthetic pathways such as those described for LPS and capsule polysaccharides and is also analogous to the eukaryotic N-glycosylation pathway where the assembled sugars are flipped from the cytoplasm into the lumen of the endoplasmic reticulum.

Characterization of the first glycoproteins isolated from *C. jejuni*, PEB3 and AcrA, confirmed that the *pgl* locus encoded an N-linked general protein glycosylation system where the glycan was attached to an asparagine residue within the consensus sequence N-X-S/T in which X could be any amino acid except for proline (Wacker *et al.*, 2002). Further studies revealed that a negatively charged side group was necessary at the Asn –2 position, a requirement that is not seen in eukaryotic systems, thus extending the acceptor sequon to D/E-X-N-X-S/T, where X could be any amino acid except for proline (Kowarik *et al.*, 2006). The number of proteins carrying the glycan modification is increasing as *C. jejuni* proteome analysis continues. There are at least 38 glycoproteins in *C. jejuni* NCTC11168 and more than 50 glycoproteins have been identified from *C. jejuni* strain HB93–13 (Young *et al.*, 2002; Scott *et al.*, 2010). Based on the abundance of potential acceptor sites, it is predicted that this number could be as high as 150 (Nothaft, 2008).

Potential role(s) of N-linked glycosylation in *C. jejuni*?

Ablation of PglB function by mutagenesis was found to have multiple effects on the bacterium (summarized in Table 5.1), which is not surprising given that over 50 proteins are modified by this system. One of the first hypotheses offered for the role of N-linked glycosylation in *C. jejuni* was

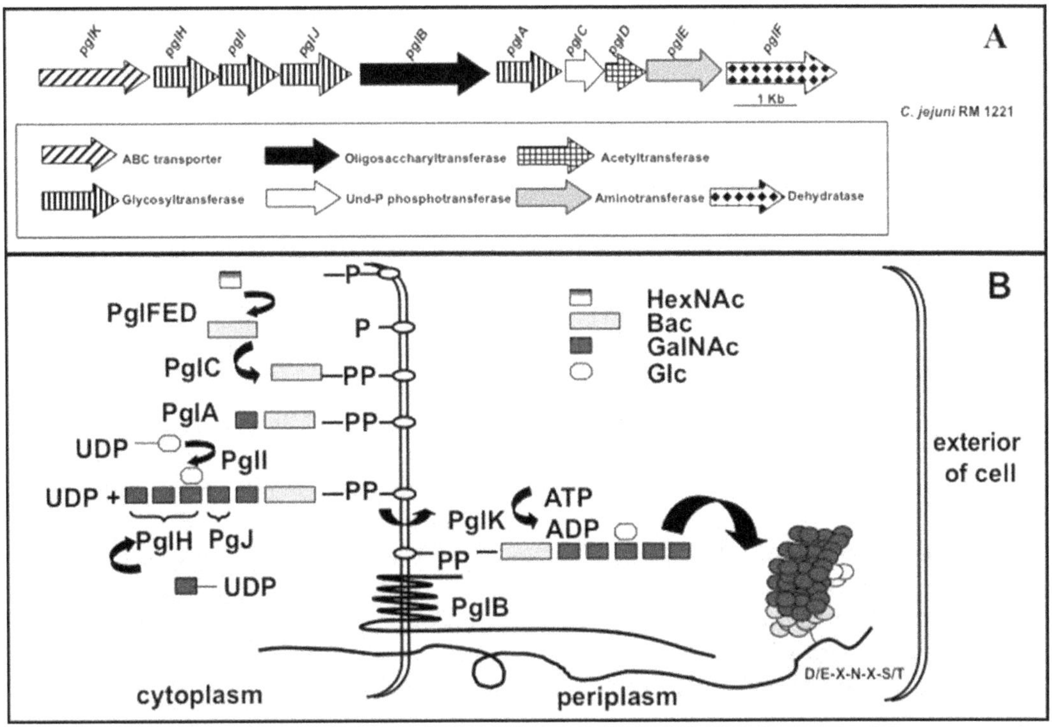

Figure 5.1 Schematic showing *pgl* locus in *C. jejuni* strain R1221 (A) and *N*-glycan biosynthesis pathway in *C. jejuni* (B). Glycan is built on an undecaprenyl pyrophosphate anchor then flipped across the inner membrane by PglK before being transferred to proteins that carry the D/E-X-N-S/T sequon by the OSTase PglB.

Table 5.1 Summary of the pleiotropic effects of *pglB* mutants

Phenotype observed in *C. jejuni pglB* deficient mutants	Reference
Reduced cell invasion	Szymanski *et al.* (2002)
Reduced mouse colonization	Szymanski *et al.* (2002)
Reduced chicken colonization	Karlyshev *et al.* (2004)
Altered host cell cytokine production	van Sorge *et al.* (2009)
Loss of natural competence	Larsen *et al.* (2004)
Reduced salt stress tolerance	Nothaft *et al.* (2009)

that the sugar structure shielded proteins from the host immune system (Szymanski *et al.*, 1999). In strain 81-176 the *pglB* mutant showed a reduced ability to invade INT407 cells and to colonize mice (Szymanski *et al.*, 2002). In addition, a protein designated VirB10 (known to be involved in DNA uptake) was dysfunctional and resulted in a loss of natural competence in this strain (Larsen *et al.*, 2004). In strain *C. jejuni* NCTC11168 a PglB deficient mutant was found to have reduced ability to colonize chickens (Karlyshev *et al.*, 2004). Also, the interaction with macrophage galactose type lectins on immature human dendritic cells is affected in a *pglB* mutant resulting in altered cytokine production (van Sorge *et al.*, 2009). More recently, it has been proposed that *C. jejuni* NCTC11168 generates free oligosaccharides via the N-linked pathway and that these are produced to maintain osmotic balance. Generation of free oligosaccharide was found to be reduced in the presence of high salt and sucrose concentrations (Nothaft *et al.*, 2009).

Due to the pleiotropic nature of the effects associated with the abolition of this pathway, the exact role of the N-linked glycosylation machinery and why *C. jejuni* has specifically retained this locus is uncertain.

However, in direct contrast to the O-linked flagella modification, the capsular polysaccharide and the LOS biosynthetic pathways the Pgl locus is absolutely conserved among *C. jejuni* strains. This evolutionary selective pressure suggests that it plays a central, but not vital, role for the organism.

Subsequent identification of PglB orthologues in delta proteobacteria and other epsilon proteobacteria

Although the *pgl* system identified in *C. jejuni* was initially thought to be unique, it is now evident that *pgl* orthologues exist in an increasing number of epsilon- and delta-proteobacteria. The central enzyme PglB has been the focus of searches to identify orthologous protein glycosylation systems in other bacteria. A combination of approaches including degenerate PCR, genome sequencing, and BLAST searches of available databases have led to the identification of over 40 bacterial *pglB* orthologues (Nothaft and Szymanski, 2010). These are all found in either epsilon- or delta-proteobacteria and cover genera isolated from diverse environments. Unsurprisingly, many additional *Campylobacter* species have *pglB* orthologues that are within loci presumed to encode protein glycosylation systems. These generally constitute orthologues of the *C. jejuni pgl* locus, however there is some variation in gene order and gene composition (Nothaft and Szymanski, 2010). The *pgl* system in *C. lari* was recently the second *Campylobacter* N-linked glycosylation system to be characterized (Schwarz *et al.*, 2011). The *C. lari pgl* locus is similar to that of *C. jejuni*, with the omission of a *pglI* orthologue. This is reflected in the structure of the glycan generated by *C. lari*, which is a linear hexasaccharide resembling the *C. jejuni* glycan minus a branching glucose. Although *C. lari pglB* is more closely related to *C. jejuni pglB* than many of the other bacterial *pglB* orthologues, the *C. lari* PglB enzyme has a more relaxed specificity in terms of the protein acceptor sequon, and was found to transfer glycan to sequons lacking a negatively charged amino acid at the –2 position (Schwarz *et al.*, 2011).

Of additional interest is the identification of *pgl* loci in *C. concisus* and *C. curvus* which contain two divergent *pglB* orthologues. The *pgl* loci identified in the genomes from these organisms consist of genes predicted to encode proteins involved in protein glycosylation, including a *pglB* orthologue (designated *pglB1*). There is also a second putative *pglB* orthologues, *pglB2*, immediately upstream of the putative *pgl* region (Fig. 5.2). The significance of this is currently under investigation. A third *Campylobacter* species, *C. gracilis*, also has two *pglB* orthologues, but in this case the first *pglB* orthologue is found within a putative *pgl* locus, whereas the second *pglB* is located at a distinct location on the chromosome and does not appear to be part of a glycosylation locus.

In addition to the *pglB* orthologues in *Campylobater* species, *pglB*s have also been identified in some species from the closely related *Helicobacter* genus, from *Wolinella succinogenes* and from deep-sea vent proteobacteria. The genome sequence *of W. succinogenes* strain DMZ1740 provided the first non-*Campylobacter pgl* system to be identified (Baar *et al.*, 2003) and is phylogenetically located between the *Campylobacter* and *Helicobacter* genera. The *W. succinogenes pgl* locus is larger than the corresponding locus in *C. jejuni*. In addition to orthologues of *pglDEF*, *pglKHIJ* and *pglBAC*, the *W. succinogenes* locus has five additional open reading frames that are predicted to encode proteins with putative roles in polysaccharide synthesis. The *W. succinogenes pgl* locus can be functionally expressed in *E. coli* together with a suitable *C. jejuni* acceptor protein to produce recombinant glycoprotein (see subheading '*C. jejuni* PglB demonstrates relaxed sugar substrate specificity' for details of this system). Mass spectrometry analysis of the recombinant glycoprotein revealed that the glycan consisted of an Asn-linked Bacillosamine residue linked to a chain of three unknown residues of 216 Da, a branching glucose and a terminal HexNAc (Hitchen *et al.*, 2010).

The *pgl* systems identified from *Helicobacter*

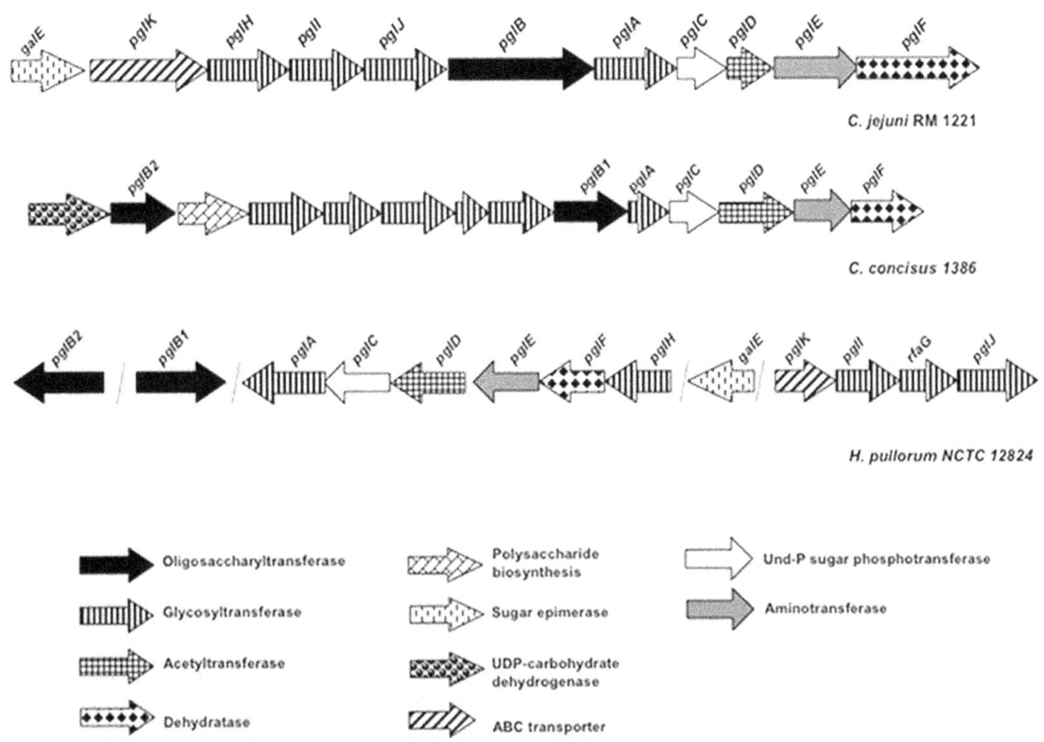

Figure 5.2 Genetic organization of *pglB* and associated genes clusters *Campylobacter* and *Helicobacter* species. The arrangement seen in *C. jejuni* RM 1221 is representative of *C. coli*, *C. lari* and *C. upsaliensis*; the arrangement in *C. concisus* 1386 is representative of *C. curvus* and the arrangement of *H. pullorum* is similar to *H. canadensis* and *H. winghamensis*.

spp. invariably include two *pglB* orthologues, although the arrangement of the *Helicobacter* genomes is such that the *pgl* orthologues are not situated in a distinct locus, but are scattered throughout the chromosome (Fig. 5.2). Investigation of protein glycosylation in *H. pullorum* has shown that one of the PglB orthologues from this organism, PglB1, has a similar function to *C. jejuni* PglB (Jervis *et al.*, 2010). By using the recombinant *C. jejuni pgl* system in *E. coli* it was shown that a mutation in *C. jejuni pglB* was partially restored by complementation with *H. pullorum pglB1*. though the specificity of these enzymes differs in the recognition of the glycosylation site, as *C. jejuni* PglB will glycosylate all four acceptor sites of the *C. jejuni* glycoprotein Cj0114, whereas only two of the sites serve as acceptors for *H. pullorum* PglB1. No transfer was detected when complementation was repeated with *H. pullorum* PglB2. Further *in vitro* analysis of the *H. pullorum pgl* system using a short peptide acceptor revealed that this organism synthesizes an unusual linear pentasaccharide composed of an Asn-linked HexNAc linked to a residue of 216 Da, two residues of 217 Da and a second, terminal HexNAc (Jervis *et al.*, 2010). Although this is a very different structure to the glycan in *C. jejuni* (Young *et al.*, 2002), *pglA, pglJ, pglH* and *pglDEF* orthologues in *H. pullorum* were all shown to be involved in synthesis of the glycan. A role for PglB2 has not yet been identified.

One of the most interesting recent discoveries is the presence of *pglB* orthologues in the genomes of bacteria found in deep-sea vents. There has been an increased interest in microorganisms in sea environments, and combined with high throughput genome sequencing, this has led to an upsurge in the amount of available sequence data from these organism. The first examples of *pglB* orthologues from deep-sea vent bacteria were from *Nitratiruptor tergacus* SB155–2 and *Sulfurovum lithotrophicum* NBC37–1 (Nakagawa *et al.*, 2007). The *pglB*s in these two genomes are found in loci that putatively encode proteins involved in polysaccharide biosynthesis, although they are quite divergent from the *Campylobacter* and *Helicobacter* counterparts (Langdon *et al.*, 2009). This suggests that the glycans synthesized by these bacteria are also different. Recent findings suggest that PglB from *N. tergacus* can transfer glycan with a reducing end Gal residue, which is not transferred by *C. jejuni* PglB (Langdon *et al.*, in preparation).

A putative N-linked glycosylation locus has also been discovered in the sulphate reducing delta-proteobacterium *Desulfovibrio desulfuricans* strain G20. A clue to the presence of N-linked glycosylation in this class of bacteria was first found in 2005 when analysis of the draught genome of *D. desulfuricans* (Joint Genome Institute, CA, USA) revealed a putative oligosaccharyltransferase (OSTase) (Szymanski and Wren, 2005). The *D. desulfuricans* G20 PglB orthologue (termed PglBDd), more closely related to its eukaryotic and archael counterparts, was determined to have a slightly different catalytic domain of WWDWG, rather than the more commonly occurring WWDYG. PglBDd was recently cloned and functionally transferred into *E. coli* where it was shown to be capable of glycosylating the *C. jejuni* acceptor protein AcrA (Ielmini and Feldman, 2010). Subtle differences were noted between *C. jejuni* PglB (PglBCj) and PglBDd. AcrA is di-glycosylated by the *C. jejuni* heptasaccharide when the entire *pgl* locus is expressed in *E. coli*. This is because despite the fact that there are five N-X-S/T motifs within the protein, only two carry the necessary negatively charged residue at the –2 position. It was found that when PglBDd was expressed with AcrA, the sequon corresponding to N_{274}ST was occupied. Therefore PglBDd lacks a requirement for a negatively charged residue at the –2 position of the glycosylation sequon N-X-S/T, instead the shorter recognition sequon preferred by eukaryotic (OSTases) is used. There is further work to be done to functionally characterize this novel OSTase in terms of sugar substrate specificity, because it was also found that PglBDd did not transfer *Pseudomonas aeruginosa* O11 and *E. coli* O16 antigens, both of which are substrates for PglBCj (Ielmini and Feldman, 2010). However, the finding that OSTases also exist in the delta proteobacteria and that they probably have altered sequon recognition sequences as well as sugar specificities is intriguing as they may expand the utility of Pgls in glycoengineering (see below). In addition, PglB orthologues have been identified from several more deep-sea bacteria and this most probably represents the tip of the iceberg. What was first thought to be a unique characteristic of a human pathogen is turning out to be a much more abundant modification present in a large range of bacteria that occupy diverse niches.

Other bacterial N-linked glycosylation systems

The gamma-proteobacterium and human respiratory pathogen *Haemophilus influenzae* has recently been reported to have an N-linked glycosylation system unlike any other previously described in bacteria. Grass *et al.* described the glycosylation of a high molecular weight adhesin known as HMW1 that is related to proteins that mediate attachment to human epithelial cells (Grass *et al.* 2003). This protein was found to form a complex with HMW1C in the cytoplasm, which was necessary for glycosylation. This post translational modification is necessary for tethering the HMW1 to the bacterial surface and prevents premature protein degradation during secretion (Grass *et al.*, 2003). The glycosylation sites on HMW1 were identified by examining tryptic digests of the protein. Subsequent MS/MS analysis found a total of 31 asparagine-modified sites, of which 30 were consistent with the N-X-S/T consensus sequon, where X could be any amino acid except for proline. An additional asparagine at position 5 within the DTTFNVER peptide was also found to be glycosylated. A total of 47 hexose units made up of hexose or dihexose sugars were found to be attached to these 31 sequons. This is the first occasion in either eukaryotes or prokaryotes whereby the N-linked carbohydrate modification did not contain N-acetylated sugars (Gross *et al.*, 2008). The glycosyltransferase responsible for transferring glucose and galactose to the asparagines sites on HMW1 adhesin was recently found to be HMW1C. The *H. influenzae* system is much simpler than the previously described bacterial N-linked glycosylation systems and has two major differences. Firstly, the modification occurs in the cytoplasm rather than the periplasm (Grass *et al.*, 2010). Secondly, an oligosaccharyl transferase such as PglB is not required and instead UDP-Glc or UDP-Gal are directly linked to HMW1 by the HMW1C glycosyltransferase, without the requirement of a lipid-linked intermediate (Grass *et al.*, 2010).

C. jejuni PglB demonstrates relaxed sugar substrate specificity

Using the recombinant *E. coli* system, Linton *et al.* carried out mutagenesis of each of the cloned genes within the *pgl* locus of *C. jejuni* to investigate their roles. An observation from this study was that glycans of different lengths could be transferred to the *C. jejuni* reporter protein PEB3. In addition, mutation of *pgl* genes involved in the synthesis of bacillosamine resulted in the synthesis and transfer of a glycan with a reducing end GalNAc. These results indicated that PglB had alternative sugar substrate specificities as well as the native *C. jejuni* N-linked heptasaccharide structure $GalNAc_5GlcBac$ (Linton *et al.*, 2005). At the same time, Feldman *et al.* (2005) demonstrated that expression of *C. jejuni* PglB (CjPglB) alone facilitated the transfer of alternative glycans, including *E. coli* O7 and O16 and *Pseudomonas aeruginosa* O11, to the *C. jejuni* reporter glycoprotein AcrA. Although the native substrate for CjPglB is one unit composed of seven sugar residues, the enzyme was capable of transferring O-antigen chains of increasing length, resulting in the observation of a protein-linked ladder. Another striking finding was that the linkages between the reducing end sugars could vary and yet still remain a substrate for *C. jejuni* PglB. In fact, the reducing end sugar could be branched. *E. coli* O7 has GlcNAc at the reducing end which is β1,3 linked to a galactose residue. *E. coli* O16 also has a GlcNAc at the reducing end with the addition of a branching α-1,6-linked glucose. In the polysaccharide the GlcNAc residue is attached via an α1,3 link to rhamnose. Finally, *P. aeruginosa* O11 has a reducing end FucNAc, which is β1,3 linked to another FucNAc (Feldman *et al.*, 2005). The successful transfer of all of these polysaccharides demonstrated the versatility of CjPglB. However, structural limitations for CjPglB have been observed. Polysaccharides with β1,4 linkages between the first two sugars are not accepted as substrates (Chen *et al.*, 2007) and further studies showed that only sugars with a C-2 position acetamido group modification at the reducing end could be accepted by CjPglB. For example, the *Shigella dysenteriae* O1 antigen (reducing end sugar GlcNAc) was a suitable substrate for CjPglB, whereas the *Salmonella* Typhimurium O4 antigen (reducing end sugar Gal) was not (Wacker *et al.*, 2006). We have recently shown that QuiNAc is also a substrate for the CjPglB (unpublished data). A summary of reducing end sugar substrates tested is shown in Table 5.2.

Table 5.2 *C. jejuni* PglB substrate specificity tested

Reducing end sugar	*C. jejuni* PglB substrate?
GalNAc	Yes
FucNAc	Yes
GlcNAc	Yes
Bacillosamine	Yes
QuiNAc	Yes
Gal	No
Glc	No

GalNAc, *N*-acetylgalactosamine, FucNAc, *N*-acetylfucosamine, GlcNAc, *N*-acetylglucosamine, Bac, bacillosamine (2,4-diacetamido-2,4,6-trideoxyglucose); QuiNAc 2-acetamido-2,6-dideoxy-D-glucose; Gal, galactose; Glc, glucose.

Bacterial N-linked glycosylation – an unexpected application

During the initial characterization of the N-linked glycan from *C. jejuni* it was difficult to purify the native glycoprotein in sufficient yield and purity to allow structural characterization by mass spectroscopy. Therefore, an attempt was made to individually clone the glycosylation system and an acceptor glycoprotein into *E. coli.* Surprisingly, there was sufficient expression to allow the acceptor glycoprotein (PEB3) to be purified for full structural analysis (Wacker *et al.*, 2002). This marked the first instance in which a recombinant glycoprotein had been expressed in *E. coli* and established a potential new era in glycoengineering.

Expansion of proteins that can be modified by the *C. jejuni* PglB systems

An important question still remained in terms of the exploitation of the CjPglB. Could any protein be modified to carry the cloned glycan? CjPglB can transfer oligosaccharide from a lipid-pyrophosphate donor to asparagine (N) residues within a conserved consensus sequence on target glycoproteins. By grafting the D/E-Y-N-X-S/T acceptor sequons onto proteins such as the cholera toxin-binding domain and the green fluorescent protein, it was found that these non-campylobacter proteins could be *N*-glycosylated (Kowarik *et al.*, 2006). It appears that an important requirement for PglB is that for successful transfer the D/E-Y-N-X-S/T sequon must be located in a surface-exposed, flexible region of the folded protein (Kowarik *et al.*, 2006). These studies demonstrate the possibility of using a *pglB* expressing *E. coli* strain to convert any given protein to a glycoprotein by directing its secretion to the periplasm, where the glycosylation reaction occurs, and grafting in an accessible D/E-Y-N-X-S/T sequon.

The technological advances of grafting glycosylation sequons to target proteins coupled with the relaxed specificity paves the way towards coupling almost any protein with a suitable glycan with significant implications for glycoengineering and glycoconjugate vaccine production (Fig. 5.3).

Application of N-linked glycosylation in glycoconjugate vaccine production

As our understanding of the mechanism(s) behind bacterial protein glycosylation has increased, so has the potential to exploit this process to produce engineered glycoconjugate proteins. This process has been termed Protein Glycan Coupling Technology (PGCT). One potential use of PGCT is to optimize the production of glycoconjugate vaccines. It has been established for decades that vaccines consisting of carbohydrates conjugated to a protein carrier are more protective and longer lasting than either component alone, and thus this type of vaccine candidate is highly desirable (Avery and Goebel, 1931; Svenson *et al.*, 1979; Astronomo and Burton, 2010). The increased protection imparted by glycoconjugate vaccines is attributed to the ability of these vaccines to induce a T-cell-dependent immune response, in contrast to free polysaccharides which are thought to induce a transient T-cell independent response characterized primarily by IgM production (Snapper and Mond, 1996; Weintraub, 2003). However, production of these conjugate vaccines is severely limited by the technical difficulties of chemical conjugation, which is a multi-step expensive process prone to variability in yields owing to protein loss at each stage of the conjugation process (Frasch, 2009). Adapting the *C. jejuni pglB* glycosylation system to attach specific immunogenic bacterial sugars onto protein carriers provides a potential solution to the problem by producing glycoconjugates *in vivo* in *E. coli,* thus avoiding the difficulties of chemical conjugation altogether.

The bacterial N-linked glycosylation system can probably be widely applied, opening up the possibility of a range of new potential vaccines. However, there are some limitations to PGCT that must be taken into consideration when choosing a suitable candidate. First and foremost, as described above PglB has thus far only been demonstrated to be capable of transferring polysaccharides with a reducing end sugar that contains an acetamido group at the C-2 position. This means that polysaccharides from bacteria such as *E. coli, C. jejuni, Pseudomanas aeruginosa, Shigella dysenteriae,* and *Francisella tularensis* have reducing end sugars that are suitable substrates

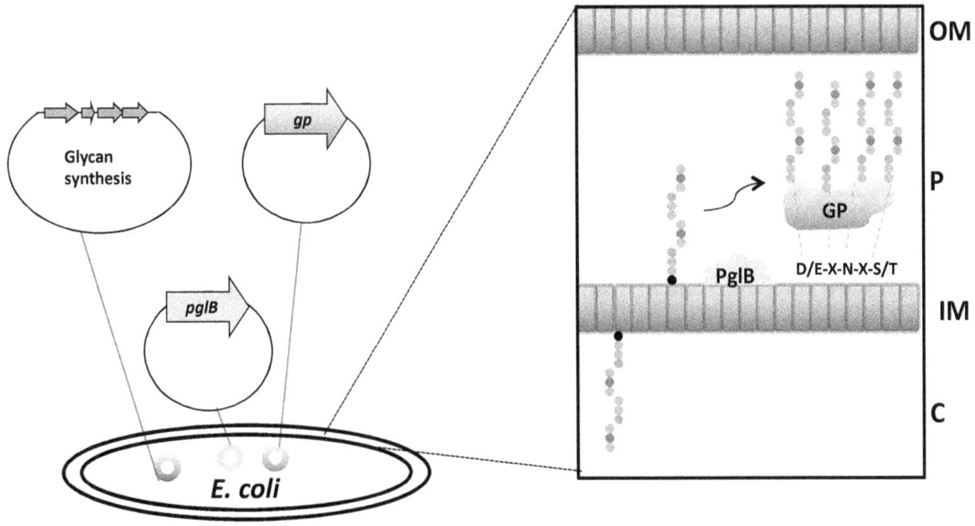

Figure 5.3 Gycoconjugate generation model using *E. coli* as the host organism. Three plasmids coding for a polysaccharide of choice, PglB and an acceptor protein (GP) are transformed into *E. coli* and co expressed prior to purification based on tag sequence carried on GP. C, cytoplasm; IM, inner membrane; P, periplasm; OM, outer membrane.

(Wacker *et al.*, 2006). However, other bacterial species will require further advancements in this technology to modify substrate specificity in order to allow conjugation to occur *in vivo*. Potential solutions to this problem include the identification of PglB proteins from other bacteria with different substrate specificity, molecular modification of the polysaccharides to make them suitable for bacterial conjugation, or even exploration of the O-linked bacterial glycosylation system as an alternative to N-linked glycosylation (Faridmoayer *et al.*, 2007).

In addition, in order to produce a candidate vaccine, a suitable carrier protein to which the bacterial sugars are attached must be carefully chosen. In the chemically conjugated vaccines currently available for *H. influenzae, Streptococcus pneumonia,* and bacterial meningitis, the carrier proteins used are bacterial toxins from tetanus and diphtheria that were chosen for their ability to produce an enhanced immune response (Dagan *et al.*, 2010). However, these proteins are currently not substrates for the *C. jejuni* PglB protein, and are thus unsuitable for bacterial conjugation. Initial work with the CjPglB system used the acceptor protein AcrA, a periplasmic protein thought to be part of an efflux pump system. However this protein is not an ideal vaccine carrier candidate because it would not be predicted to produce and optimal T-cell-dependent immune response (Feldman *et al.*, 2005). More recently, it has been suggested that modifying immunogenic carrier proteins such as bacterial toxins to serve as PglB acceptors would produce more effective vaccines. Bioconjugation of bacterial sugars has been successfully accomplished using exotoxin A from *Pseudomonas aeruginosa* modified to contain two N-glycosylation sites (Ihssen *et al.*, 2010). Exotoxin A has previously been demonstrated to be an effective carrier protein for vaccines produces against *Shigella* species, and thus the modified form would be an ideal candidate for a bioconjugate vaccine (Taylor *et al.*, 1993).

Recently, the first glycoconjugate vaccine produced in *E. coli* has successfully completed a phase I clinical trial in human volunteers by GlycoVaxyn AG, a Swiss biotechnology company that exploits PGCT (press release, 8 October 2010). This provides the first evidence that biologically conjugated glycoprotein vaccines may become commercially available in the foreseeable future, and provides motivation to continue to research in this area. The GlycoVaxyn vaccine is directed against *Shigella dysenteria O1,* and uses the exotoxin

A carrier protein (Ihssen *et al.*, 2010). Future challenges include demonstration of the ability of this vaccine to protect against subsequent infection, and developing additional vaccines for bacterial pathogens that are adaptable for PGCT. Future work may include adapting PGCT to include a wider range of pathogens, GlycoVaxyn have already developed a *Staphylococcus aureus* vaccine based on a surface polysaccharide demonstrating that the system can be extended to Gram-positive pathogens as well (press release, 5 May 2010). In addition, it has recently been demonstrated that the bacterial conjugation system can be modified to accept eukaryotic *N*-glycoproteins (Schwarz *et al.*, 2010). This further expands the potential applications of this technology beyond vaccine production against bacterial targets. In the future, PGCT could be used to produce large quantities of human glycoproteins for both laboratory and biomedical purposes.

Concluding remarks

Despite only discovered a decade ago great strides have been made in functionally characterizing bacterial N-linked glycosylation. Unsurprisingly, the *C. jejuni* system is best understood, but the precise role of N-linked glycosylation in this and other bacteria remains unclear. Of particular interest is the origin and evolution of PglBs and the associated N-linked glycosylation system. It is evident that they are almost exclusively found among epsilon- and delta- proteobacteria. However, although it is found in a broad spectrum of this subdivision of bacteria, it is not found in all species as most *Helicobacter* species lack *pglB* orthologues. In addition, some species have two *pglB*s, either adjacent or scattered on the chromosome. Interestingly, the glycosylation system appears common in bacteria from deep-sea vents, where the majority of bacteria are from the epsilon-proteobacteria subdivision (Nakagawa *et al.*, 2007). This points to an ancient origin of these systems, and given that archaea are also often found in extreme environments such as deep-sea vents, it is tempting to speculate a distant gene transfer event of this system from archaea to the epsilon- or possibly delta-proteobacteria.

Beyond the diverse PglBs, the adjacent glycan biosynthetic genes enables the functional characterization of genetic loci relating to an overall glycan structure to be determined by modern spectroscopy analyses. The ability to determine the functions of cassettes of genes relating to a given glycan structure has been demonstrated for *C. jejuni* (Fig. 5.1) and could have applications in the field of synthetic biology where 'plug and play' genetic systems are the mainstay for this burgeoning research area.

Further cheap and rapid methods to generate glycoconjugate vaccines where potentially any glycan could be coupled to any protein will herald a new era in vaccinology. Other PGCT applications include the generation of glycoproteins to treat human with deficiencies in N-linked glycoprotein generation. The field of glycoengineering is on a voyage of discovery with many more bacterial N-linked glycosylation systems to be characterized and exploited. Perhaps the true legacy for *C. jejuni* may not be as an important gastrointestinal pathogen, but as the trailblazer of a rapidly expanding basic and applied research discipline.

References

Aebi, M., and Hennet, T. (2001). Congenital disorders of glycosylation: genetic model systems lead the way. Trends Cell. Biol. *11*, 136–141.

Astronomo, R.D., and Burton, D.R. (2010). Carbohydrate vaccines: developing sweet solutions to sticky situations? Nat. Rev. Drug Discov. 9, 308–324.

Avery, O.T., and Goebel, W.F. (1931). Chemo-immunological studies on conjugated carbohydrate-proteins: V. the immunological specificity of an antigen prepared by combining the capsular polysaccharide of Type III Pneumococcus with foreign protein. J. Exp. Med. *54*, 437–447.

Baar, C., Eppinger, M., Raddatz, G., Simon, J., Lanz, C., Klimmek, O., Nandakumar, R., Gross, R., Rosinus, A., Keller, H., *et al.* (2003). Complete genome sequence and analysis of *Wolinella succinogenes*. Proc. Natl. Acad. Sci. U.S.A. *100*, 11690–11695.

Chen, M.M., Glover, K.J., and Imperiali, B. (2007). From peptide to protein: comparative analysis of the substrate specificity of N-linked glycosylation in *C. jejuni*. Biochemistry *46*, 5579–5585.

Dagan, R., Poolman, J., and Siegrist, C.A. (2010). Glycoconjugate vaccines and immune interference: A review. Vaccine *28*, 5513–5523.

Faridmoayer, A., Fentabil, M.A., Mills, D.C., Klassen, J.S., and Feldman, M.F. (2007). Functional characterization of bacterial oligosaccharyltransferases involved in O-linked protein glycosylation. J. Bacteriol. *189,* 8088–8098.

Feldman, M.F., Wacker, M., Hernandez, M., Hitchen, P.G., Marolda, C.L., Kowarik, M., Morris, H.R., Dell, A., Valvano, M.A., and Aebi, M. (2005). Engineering N-linked protein glycosylation with diverse O antigen lipopolysaccharide structures in *Escherichia coli.* Proc. Natl. Acad. Sci. U.S.A. *102,* 3016–3021.

Frasch, C.E. (2009). Preparation of bacterial polysaccharide–protein conjugates: analytical and manufacturing challenges. Vaccine *27,* 6468–6470.

Glover, K.J., Weerapana, E., and Imperiali, B. (2005). *In vitro* assembly of the undecaprenylpyrophosphate-linked heptasaccharide for prokaryotic N-linked glycosylation. Proc. Natl. Acad. Sci. U.S.A. *102,* 14255–14259.

Grass, S., Buscher, A.Z., Swords, W.E., Apicella, M.A., Barenkamp, S.J., Ozchlewski, N., and St Geme, J.W., 3rd (2003). The *Haemophilus influenzae* HMW1 adhesin is glycosylated in a process that requires HMW1C and phosphoglucomutase, an enzyme involved in lipooligosaccharide biosynthesis. Mol. Microbiol. *48,* 737–751.

Grass, S., Lichti, C.F., Townsend, R.R., Gross, J., and St Geme, J.W., 3rd. (2010). The *Haemophilus influenzae* HMW1C protein is a glycosyltransferase that transfers hexose residues to asparagine sites in the HMW1 adhesin. PLoS Pathog. *6,* e1000919.

Gross, J., Grass, S., Davis, A.E., Gilmore-Erdmann, P., Townsend, R.R., and St Geme, J.W., 3rd. (2008). The *Haemophilus influenzae* HMW1 adhesin is a glycoprotein with an unusual N-linked carbohydrate modification. J. Biol. Chem. *283,* 26010–26015.

Hitchen, P.G., Twigger, K., Valiente, E., Langdon, R.H., Wren, B.W., and Dell, A. (2010). Glycoproteomics: a powerful tool for characterizing the diverse glycoforms of bacterial pilins and flagellins. Biochem. Soc. Trans. *38,* 1307–1313.

Ielmini, M.V., and Feldman, M.F. (2010). *Desulfovibrio desulfuricans* PglB homolog possesses oligosaccharyltransferase activity with relaxed glycan specificity and distinct protein acceptor sequence requirements. Glycobiology *6,* 734–742.

Ihssen, J., Kowarik, M., Dilettoso, S., Tanner, C., Wacker, M., and Thony-Meyer, L. (2010). Production of glycoprotein vaccines in *Escherichia coli.* Microb. Cell Fact *9,* 61.

Jervis, A.J., Langdon, R., Hitchen, P., Lawson, A.J., Wood, A., Fothergill, J.L., Morris, H.R., Dell, A., Wren, B., and Linton, D. (2010). Characterization of N-linked protein glycosylation in *Helicobacter pullorum.* J. Bacteriol. *192,* 5228–5236.

Karlyshev, A.V., Everest, P., Linton, D., Cawthraw, S., Newell, D.G., and Wren, B.W. (2004). The *Campylobacter jejuni* general glycosylation system is important for attachment to human epithelial cells and in the colonization of chicks. Microbiology *150,* 1957–1964.

Kelly, J., Jarrell, H., Millar, L., Tessier, L., Fiori, L.M., Lau, P.C., Allan, B., and Szymanski, C.M. (2006). Biosynthesis of the N-linked glycan in *Campylobacter jejuni* and addition onto protein through block transfer. J. Bacteriol. *188,* 2427–2434.

Knauer, R., and Lehle, L. (1999). The oligosaccharyltransferase complex from *Saccharomyces cerevisiae.* Isolation of the OST6 gene, its synthetic interaction with OST3, and analysis of the native complex. J. Biol. Chem. *274,* 17249–17256.

Kowarik, M., Numao, S., Feldman, M.F., Schulz, B.L., Callewaert, N., Kiermaier, E., Catrein, I., and Aebi, M. (2006). N-linked glycosylation of folded proteins by the bacterial oligosaccharyltransferase. Science *314,* 1148–1150.

Langdon, R.H., Cuccui, J., and Wren, B.W. (2009). N-linked glycosylation in bacteria: an unexpected application. Future Microbiol. *4,* 401–412.

Larsen, J.C., Szymanski, C., and Guerry, P. (2004). N-linked protein glycosylation is required for full competence in *Campylobacter jejuni* 81–176. J. Bacteriol. *186,* 6508–6514.

Linton, D., Dorrell, N., Hitchen, P.G., Amber, S., Karlyshev, A.V., Morris, H.R., Dell, A., Valvano, M.A., Aebi, M., and Wren, B.W. (2005). Functional analysis of the *Campylobacter jejuni* N-linked protein glycosylation pathway. Mol. Microbiol. *55,* 1695–1703.

Nakagawa, S., Takaki, Y., Shimamura, S., Reysenbach, A.L., Takai, K., and Horikoshi, K. (2007). Deep-sea vent epsilon-proteobacterial genomes provide insights into emergence of pathogens. Proc. Natl. Acad. Sci. U.S.A. *104,* 12146–12150.

Nothaft, H., Amber, S., Aebi, M., and Szymanski, C.M. (2008). *Campylobacter,* 3rd edition. (Washington, D.C., ASM), pp. 447–469

Nothaft, H., Liu, X., McNally, D.J., Li, J., and Szymanski, C.M. (2009). Study of free oligosaccharides derived from the bacterial N-glycosylation pathway. Proc. Natl. Acad. Sci. U.S.A. *106,* 15019–15024.

Nothaft, H., and Szymanski, C.M. (2010). Protein glycosylation in bacteria: sweeter than ever. Nat. Rev. Microbiol. *8,* 765–778.

Parkhill, J., Wren, B.W., Mungall, K., Ketley, J.M., Churcher, C., Basham, D., Chillingworth, T., Davies, R.M., Feltwell, T., Holroyd, S., *et al.* (2000). The genome sequence of the food-borne pathogen *Campylobacter jejuni* reveals hypervariable sequences. Nature *403,* 665–668.

Power, P.M., Roddam, L.F., Dieckelmann, M., Srikhanta, Y.N., Tan, Y.C., Berrington, A.W., and Jennings, M.P. (2000). Genetic characterization of pilin glycosylation in *Neisseria meningitidis.* Microbiology *146 (Pt 4),* 967–979.

Schwarz, F., Huang, W., Li, C., Schulz, B.L., Lizak, C., Palumbo, A., Numao, S., Neri, D., Aebi, M., and Wang, L.X. (2010). A combined method for producing homogeneous glycoproteins with eukaryotic N-glycosylation. Nat Chem Biol *6,* 264–266.

Schwarz, F., Lizak, C., Fan, Y.Y., Fleurkens, S., Kowarik, M., and Aebi, M. (2011). Relaxed acceptor site specificity of bacterial oligosaccharyltransferase *in vivo.* Glycobiology *21,* 45–54.

Scott, N.E., Parker, B.L., Connolly, A.M., Paulech, J., Edwards, A.V., Crossett, B., Falconer, L., Kolarich, D.,

Djordjevic, S.P., Hojrup, P., *et al.* (2010). Simultaneous glycan-peptide characterization using hydrophilic interaction chromatography and parallel fragmentation by CID, HCD and ETD–MS applied to the N-linked glycoproteome of *Campylobacter jejuni*. Mol Cell Proteomics.

Silberstein, S., and Gilmore, R. (1996). Biochemistry, molecular biology, and genetics of the oligosaccharyltransferase. Faseb J *10*, 849–858.

Snapper, C.M., and Mond, J.J. (1996). A model for induction of T cell-independent humoral immunity in response to polysaccharide antigens. J. Immunol. *157*, 2229–2233.

van Sorge, N.M., Bleumink, N.M., van Vliet, S.J., Saeland, E., van der Pol, W.L., van Kooyk, Y., and van Putten, J.P. (2009). N-glycosylated proteins and distinct lipooligosaccharide glycoforms of *Campylobacter jejuni* target the human C-type lectin receptor MGL. Cell Microbiol. *11*, 1768–1781.

Svenson, S.B., Nurminen, M., and Lindberg, A.A. (1979). Artificial Salmonella vaccines: O-antigenic oligosaccharide–protein conjugates induce protection against infection with *Salmonella typhimurium*. Infect. Immun. *25*, 863–872.

Szymanski, C.M., and Wren, B.W. (2005). Protein glycosylation in bacterial mucosal pathogens. Nat. Rev. Microbiol. *3*, 225–237.

Szymanski, C.M., Yao, R., Ewing, C.P., Trust, T.J., and Guerry, P. (1999). Evidence for a system of general protein glycosylation in *Campylobacter jejuni*. Mol. Microbiol. *32*, 1022–1030.

Szymanski, C.M., Burr, D.H., and Guerry, P. (2002). Campylobacter protein glycosylation affects host cell interactions. Infect. Immun. *70*, 2242–2244.

Taylor, D.N., Trofa, A.C., Sadoff, J., Chu, C., Bryla, D., Shiloach, J., Cohen, D., Ashkenazi, S., Lerman, Y., Egan, W., *et al.* (1993). Synthesis, characterization, and clinical evaluation of conjugate vaccines composed of the O-specific polysaccharides of *Shigella dysenteriae* type 1, *Shigella flexneri* type 2a, and *Shigella sonnei* (*Plesiomonas shigelloides*) bound to bacterial toxoids. Infect. Immun. *61*, 3678–3687.

Virji, M. (1999). Glycans in meningococcal pathogenesis. Biochem. Soc. Trans. *27*, 498–507.

Wacker, M., Feldman, M.F., Callewaert, N., Kowarik, M., Clarke, B.R., Pohl, N.L., Hernandez, M., Vines, E.D., Valvano, M.A., Whitfield, C., *et al.* (2006). Substrate specificity of bacterial oligosaccharyltransferase suggests a common transfer mechanism for the bacterial and eukaryotic systems. Proc. Natl. Acad. Sci. U.S.A. *103*, 7088–7093.

Wacker, M., Linton, D., Hitchen, P.G., Nita-Lazar, M., Haslam, S.M., North, S.J., Panico, M., Morris, H.R., Dell, A., Wren, B.W., *et al.* (2002). N-linked glycosylation in *Campylobacter jejuni* and its functional transfer into *E. coli*. Science *298*, 1790–1793.

Weintraub, A. (2003). Immunology of bacterial polysaccharide antigens. Carbohydr. Res. *338*, 2539–2547.

Yan, Q., and Lennarz, W.J. (2002). Studies on the function of oligosaccharyl transferase subunits. Stt3p is directly involved in the glycosylation process. J. Biol. Chem. *277*, 47692–47700.

Young, N.M., Brisson, J.R., Kelly, J., Watson, D.C., Tessier, L., Lanthier, P.H., Jarrell, H.C., Cadotte, N., St Michael, F., Aberg, E., *et al.* (2002). Structure of the N-linked glycan present on multiple glycoproteins in the Gram-negative bacterium, *Campylobacter jejuni*. J. Biol. Chem. *277*, 42530–42539.

Zufferey, R., Knauer, R., Burda, P., Stagljar, I., te Heesen, S., Lehle, L., and Aebi, M. (1995). STT3, a highly conserved protein required for yeast oligosaccharyl transferase activity *in vivo*. Embo J. *14*, 4949–4960.

Bacterial Lipoarabinomannan: Structure to Biogenesis

6

Arun K. Mishra, Sarah M. Batt, Luke J. Alderwick, Klaus Futterer and Gurdyal Singh Besra

Abstract

Lipoarabinomannan is an amphipathic lipoglycan found in the cell wall of most Actinomycetes. The majority of bacteria from the suborder *Corynebacterineae,* including *Mycobacterium tuberculosis, Mycobacterium smegmatis* and *Corynebacterium glutamicum*, and from genus *Rhodococcus, Gordonia* and *Amycolatopsis*; all possess lipoarabinomannan and related glycoconjugates, such as lipomannan and phosphatidyl-*myo*-inositol mannosides. In addition to their physiological function in these microorganisms, these glycoconjugates play a key immunomodulatory role for pathogenic bacteria during infection. Herein, we report the work from this laboratory and several others, which has led to the biochemical characterization of key enzymes involved in the biogenesis of these complex glycoconjugates.

Introduction

There are a number of aerobic actinomycetes found in soil, freshwater and marine life, of which a few are pathogenic to humans. These aerobic actinomycetes are subdivided into the 'nocardioform actinomycetes' which include mycobacteria, nocardia, rhodococcus, and corynebacteria. All members of the nocardioform actinomycetes possess an unusual lipid rich cell wall which is unique to the order *Actinomycetes* (Brennan and Nikaido, 1995). The majority of our current knowledge about actinomycetes cell wall architecture comes from pioneering studies on mycobacterial strains. The cell wall of these actinomycetes is basically composed of an arabinogalactan–peptidoglycan complex (Daffe *et al.*, 1990; McNeil *et al.*, 1990, 1991; Besra *et al.*, 1995; Brennan, 2003; Dover *et al.*, 2004), connected to which are the unusually long chain mycolic acids (Hoffmann *et al.*, 2008). In addition, distinct glycolipids, phospho-*myo*-inositolmannosides (PIMs), lipomannan (LM) and lipoarabinomannan (LAM) are also found in the cell wall of mycobacteria and related species (Hill and Ballou, 1966; Hunter *et al.*, 1986; Besra *et al.*, 1997; Briken *et al.*, 2004; Morita *et al.*, 2004). PIM comprises a membrane-anchored phosphatidyl-*myo*-inositiol (PI), which has been decorated in a precise order with mannose sugars on the inositol group. Further mannosylation yields LM and addition of arabinose affords LAM (Chatterjee *et al.*, 1991; Besra *et al.*, 1997). In addition to their physiological function, these complex glycoconjugates play a key role in modulating the host response during infection (Schlesinger *et al.*, 1994; Chatterjee and Khoo, 1998; Nigou *et al.*, 2002; Maeda *et al.*, 2003). In this chapter, we focus on recent advances in the structure and biogenesis of LAM and other glycoconjugates from mycobacteria, and also revisit the available literature on other actinomycetes.

Structural features of PIMs, LM and LAM

PIMs and the amphipathic lipoglycans, LM and LAM, are widely found in *Corynebacterineae.* Basically, two forms of acylated PIMs; tri- and tetra-acylated (Ac_1- and Ac_2-) phospho-*myo*-inositol-dimannoside (PIM_2) and tri- and tetra-acylated phospho-*myo*-inositol-hexamannoside (Ac_1/Ac_2PIM_6) (we have used

Ac_1/Ac_2PIMx for two different acylated versions of PIMs throughout the text, and PIM as a synonym where the acylation state is not clear); and different acylated versions of LM and LAM accumulate on the cell wall of mycobacteria (Khoo *et al.*, 1995; Nigou *et al.*, 2003). These glycolipids and lipoglycans are thought to be non-covalently attached to the cell membrane via their mannosyl phosphate inositol (MPI) anchor (Hunter and Brennan, 1990), whilst PIMs serve as precursors for higher mannosylated LM, and subsequently for LAM (Besra *et al.*, 1997; Mishra *et al.*, 2008a) (Fig. 6.1).

Structure of PIMs

PIMs are glycolipids with fatty acids attached to a glycerol unit which in turn is attached to *myo*-inositol by a phospho-diester linkage resulting in PI, which is further substituted at the *O*-2 and *O*-6 positions of *myo*-inositol with α-D-Man*p* units, resulting in a MPI anchor (Ballou and Lee, 1964; Lee and Ballou, 1964; Nigou *et al.*, 2003). There are four potential sites of acylation within the MPI anchor; with different fatty acids at the 1-OH and 2-OH of the glycerol unit in the anchor, the 3-OH of *myo*-inositol and the 6-OH of the Man*p* residue linked at the *O*-2 position of *myo*-inositol (Khoo *et al.*, 1995; Nigou *et al.*, 2003). Two different acylated forms of PIMs accumulate in the cell wall of mycobacteria; Ac_1PIM_x has an acyl group at either the 3-OH of *myo*-inositol or the 6-OH of the Man*p* residue linked at the *O*-2 position of *myo*-inositol, and secondly, Ac_2PIM_x has an acyl group at both positions, Ac_2PIM_x (Ballou and Lee, 1964; Lee and Ballou, 1964; Nigou *et al.*, 2003). Acylated forms of PIM_2 serve as substrates for the synthesis of higher-order PIMs, such as Ac_1/Ac_2PIM_6 (Fig. 6.2), which is a pentamannoside attached to the position *O*-6 of the *myo*-inositol of PI of PIM_1, Man*p*-α(1→2)-Man*p*-α(1→2)-Man*p*-α(1→6)-Man*p*-α(1→6)-Man*p*-α(1→.) (Lee and Ballou, 1965).

Structure of LM and LAM

Ac_1/Ac_2PIM_2 is α(1→6) mannosylated at the 6-OH position by approximately 21–34 Man*p* residues and decorated by 5–10 singular units of α(1→2)-Man*p*, resulting in the formation of LM

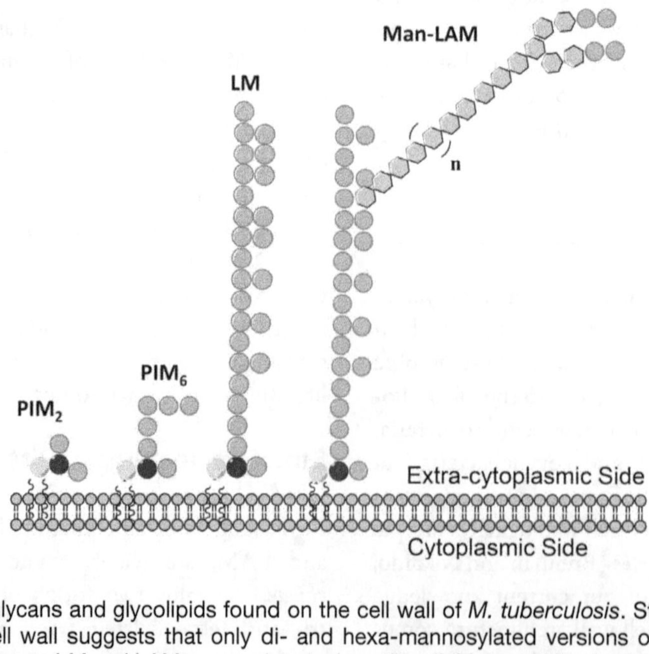

Figure 6.1 Lipoglycans and glycolipids found on the cell wall of *M. tuberculosis*. Structural analysis of the mycobacterial cell wall suggests that only di- and hexa-mannosylated versions of PIMs, and the higher glycosylated polymers LM and LAM accumulate in the cell wall. Mature LM is further linked via an unknown linkage to an arabinan domain made up of 70 arabinose residues which is further capped by mono, -di or -tri α (1→2)-Man*p* units resulting in ManLAM. The 'n' represents different numbers of arabinose which is species specific.

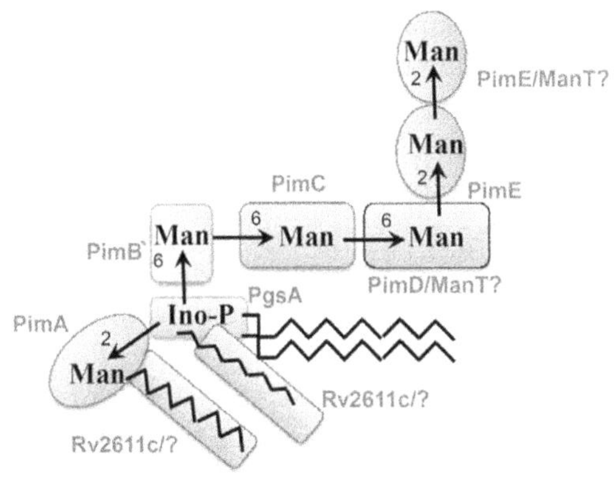

Figure 6.2 PIM biosynthesis in *M. tuberculosis*. On the cytosolic side of the cell, PI is synthesized by PgsA, which is further glycosylated by PimA, PimB′ and an acyltransferase to form Ac_1PIM_2 that is further mannosylated by PimC and PimD? to form Ac_1PIM_4, which is mannosylated by α(1→2) mannopyranosyltransferases, PimE and/or another unidentified enzyme to form Ac_1PIM_6.

(Chatterjee *et al.*, 1992a). The mature LM is further elaborated by an arabinan domain consisting of approximately 55–70 Ara*f* residues in a linear α(1→5)-D-Ara*f* fashion with 3,5-α-D-Ara*f* branches (Chatterjee *et al.*, 1991; Kaur *et al.*, 2008; Birch *et al.*, 2010). However, the exact linkage position of the arabinan domain on LM is yet unknown. The linear α(1→5)-D-Ara*f* chain is highly branched towards its non-reducing end and conserved with two types of chain arrangements (Fig. 6.3): (1) linear tetra-arabinofuranosides (Ara-4) with the structure β-D-Ara*f*(1→2)-α-D-Ara*f*(1→5)-α-D-Ara*f*(1→5)-α-D-Ara*f* and (2) branched (Ara-6) motifs with the structure [β-D-Ara*f*(1→2)-α-D-Ara*f*]$_2$–3,5-α-D-Ara*f*(1→5)-α-D-Ara*f* (Chatterjee *et al.*, 1991, 1993). The penultimate non-reducing end is characterized by the disaccharide unit, Ara*f*-β(1→2)-Ara*f*-α(1→.) (Fig. 6.3) (Chatterjee *et al.*, 1993; McNeil *et al.*, 1994). In several strains of mycobacteria, the tetra-/hexa-arabinofuranoside unit of LAM is further extended by a species specific mono-, di- and tri-α(1→2)-D-Man*p* units (Chatterjee *et al.*, 1992b). A few fast-growing strains of *Mycobacterium* species, such as *Mycobacterium chelonae*, are devoid of any Man*p* caps, and termed AraLAM, (Guerardel *et al.*, 2002) and in some strains, such as *M. smegmatis* strains ATCC 14468 and mc^{2}155, a novel inositol phosphate capping motif, PILAM, has been identified (Chatterjee *et al.*, 1992b).

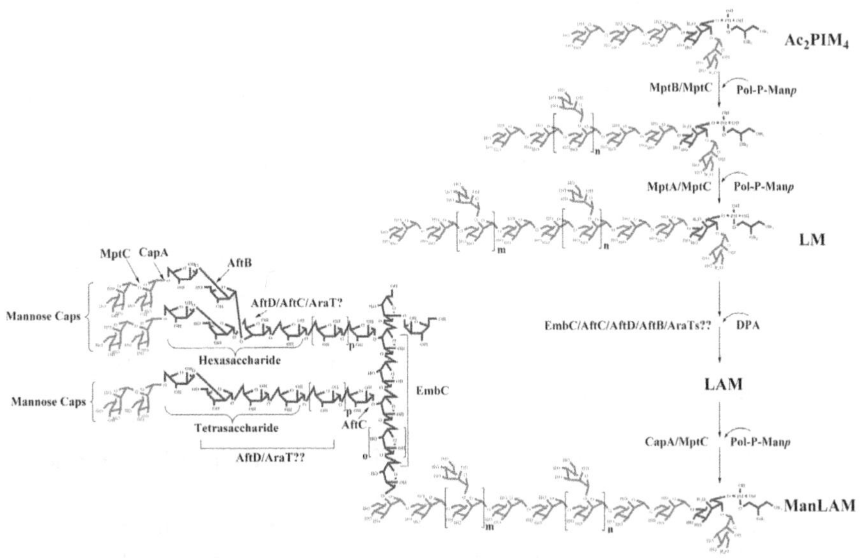

Figure 6.3 Structural features and biogenesis of LM and ManLAM from *M. tuberculosis*. Ac_1/Ac_2PIM_4 is a precursor of LM which is polymerized by MptB and MptA in α(1→6) direction and decorated by singular α(1→2)-Man*p* by MptC. Mature LM is subsequently primed by a singular D-Ara*f* at an unknown position which is extended by EmbC and other arabinosyltransferases such as, AftC, AftD and AftB. The arabinan domain is further capped with Man*p* residues by CapA and MptC to form ManLAM. The o, m, n, and p values are species specific and represent different numbers of sugar residues.

Non-mycobacterial lipoglycans

Apart from the *Mycobacteriaceae* family, most other bacteria from the sub-order *Corynebacterineae* also possess LAM and related glycoconjugates (Table 6.1). *Amycolatopsis orientalis* and *Amycolatopsis sulphurea* possesses a lipoglycan containing an MPI anchor with 14-methyl-pentadecanoate and a short mannan chain. The mannan domain is further linked to a branched arabinan domain composed of α(1→5)-Ara*f* chains substituted at the O-2 position by singular β-Ara*f* residues which are further capped by α-Man*p* residues (Gibson *et al.*, 2003b). Similarly, LAM from *Corynebacterium diphtheriae* is characterized by an MPI anchor with a linear α(1→6)-Man*p* substituted at O-2 by α-Man*p* and at O-4 by α-Ara*f* (Moreira *et al.*, 2008). LAM like lipoglycans have also been reported from the opportunistic pathogen, *Dietzia maris* (Sutcliffe, 2000), *Corynebacterium matruchotii* (Sutcliffe, 1995) and *Gordonia bronchialis* (Garton and Sutcliffe, 2006).

Rhodococcus equi is responsible for approximately 3% of global foal mortality (Prescott, 1991) and *Rhodococcus ruber* has been isolated and cultured from infected human sputa (Osoagbaka, 1989). LAM from *R. equi* has a lipoglycan with a linear α(1→6)-Man*p* with a singular α-Man*p* residue capped with α-Ara*f* (Garton *et al.*, 2002). In contrast, *R. ruber* LAM is a linear α(1→6)-Man*p* chain primed with singular t-Ara*f* units (Gibson *et al.*, 2003c). Similarly, another opportunistic pathogen *Tsukamurella paurometabola* also possesses a LAM variant with pro-inflammatory activity (Gibson *et al.*, 2004). LAM from *T. paurometabola* is a linear α(1→6)-Man*p* chain, which is further linked to an α(1→5)-Ara*f* chain substituted at O-2 by a Man*p*-α(1→2)-Man*p*→ dimannoside motif (Table 6.1) (Gibson *et al.*, 2004). Apart from these, there are a few species which have lost the ability to synthesize mycolic acids, but are still able to synthesize LAM; such as *Turicella otitidis*, an opportunistic pathogen of the human ear (Dana *et al.*, 2001), which has a LAM-like species where the α(1→6)-Man*p* core is substituted with t-Ara*f* units (Gilleron *et al.*, 2005). To date, a number of LAM variants have been identified and characterized from non-mycobacterial strains. These exhibit a common macrostructure and members of closely related taxa display similarities in lipoglycan composition and distribution.

Biogenesis of PIMs, LM and LAM

The present model of lipoglycan and related glycoconjugates biogenesis is reported from biochemical and genetic studies done with mycobacteria and corynebacteria and follows a linear pathway from PI→PIM_2→PIM_4→PIM_6 for PIMs and PI→PIM_2→PIM_4→LM→LAM for LAM biogenesis (Besra and Brennan, 1997; Besra *et al.*, 1997; Morita *et al.*, 2004). Basically, glycosylation of PI by different α-mannopyranosyltransferases, PimA, PimB′, PimC, PimE, and acylation by acyltransferase(s), result in the synthesis of Ac_1/Ac_2PIMs (Fig. 6.2) (Kordulakova *et al.*, 2002, 2003; Kremer *et al.*, 2002; Morita *et al.*, 2006; Lea-Smith *et al.*, 2008; Mishra *et al.*, 2008b, 2009), of which Ac_1/Ac_2PIM_2 and Ac_1/Ac_2PIM_6 accumulate on the mycobacterial cell wall. Furthermore, Ac_1/Ac_2PIM_4 acts as an acceptor for MptB (Mishra *et al.*, 2008a) and are further polymerized by MptA and MptC (Mishra *et al.*, 2011) resulting in a mature LM (Kaur *et al.*, 2006, 2007, 2008; Mishra *et al.*, 2007; Sena *et al.*, 2010). LM is further extended by an array of arabinofuranosyltransferases, AftC, EmbC, AftD, and AftB; and capped by CapA and MptC yielding ManLAM (Fig. 6.3) (Zhang *et al.*, 2003; Dinadayala *et al.*, 2006; Shi *et al.*, 2006; Seidel *et al.*, 2007; Appelmelk *et al.*, 2008; Kaur *et al.*, 2008; Skovierova *et al.*, 2009; Birch *et al.*, 2010). Recently the role of an acyltransferase, Rv1565c was also suggested in the synthesis of mannose caps in *Mycobacterium marinum* (Driessen *et al.*, 2010). These glycosyltransferases (GTs) and their mutual regulation and interaction determines the size and distribution of these glycoconjugates on the bacterial cell wall (Sena *et al.*, 2010).

GTs involved in LAM biogenesis in mycobacteria

GTs catalyse the specific transfer of a sugar moiety from an activated sugar donor to a particular substrate. According to CAZy (Cantarel *et al.*, 2009), GTs are classified into 65 families, based on sequence similarities and catalytic mechanisms and there are currently 42 defined GTs in the genome of *M. tuberculosis* H_{37}Rv, accounting

Table 6.1 Non-mycobacterial lipoglycans

Organism	MW (kDa)	Fatty acids composition	Structure	Reference
Amycolatopsis sulphurea	~10 (Ara*f*/Man*p*)	$C_{16:0}$: 38%, $iC_{16:0}$: 24%, $C_{15:0}$: 7%, $C_{17:0}$: 15%, $C_{18:0}$: 5%	Short α(1→6)-Man*p* domain, occasionally substituted by singular Man*p* residues at *O*-2 or *O*-3	Gibson *et al*. (2003b)
Rhodococcus equi	~8 (Ara*f*/Man*p*)	$C_{16:0}$: 56%, $C_{19:0}$: 19%, $C_{22:0}$: 11%, $C_{18:0}$: 6%	Mannan domain is similar to mycobacterial LM, which is further decorated by single Ara*f* residues	Garton *et al*. (2002)
Rhodococcus ruber	~7 (Ara*f*/Man*p*)	$C_{16:0}$: 56%, $C_{19:0}$: 33%, $C_{18:0}$: 6%, $C_{16:1}$: 3%, $C_{18:1}$: 3%	Linear α(1→6)-Man*p* backbone is substituted at some *O*-2 positions by Ara*f* and Man*p* units	Gibson *et al*. (2003c)
Tsukamurella paurometabola	~12.5 (Ara*f*/Man*p*)	$C_{16:0}$: 50%, $C_{18:1}$, 20%, $C_{18:0}$: 15%, $C_{19:0}$: 15%	Very similar to mycobacterial LAM with α(1→5)-Ara*f* side-chain which is further substituted at the *O*-2 position by a Man*p*-α(1→2)-Man*p*-α(→.) dimannoside motif	Gibson *et al*. (2004)
Turicella otitidis	~9 (Ara*f*/Man*p*)	$C_{16:0}$: 41%, $C_{18:1}$: 32%, $C_{18:0}$: 21%	Linear α(1→6)-Man*p* core substituted with terminal Ara*f* branches.	Gilleron *et al*. (2005)
Saccharothrix aerocolonigenes	~7 (Man*p*)	$iC_{16:0}$: 33%, $C_{18:1}$: 33%, $C_{16:0}$: 20%, $C_{18:0}$: 6%, $C_{17:0}$: 8%	Conserved α(1→6)-Man*p* backbone, substituted at >80% of the *O*-2 position by side-chains composed of Man*p*-α(1→2)-Man*p*-α(→.) motif	Gibson *et al*. (2005)
Dietzia maris	~20 (Ara*f*/Man*p*)	$C_{16:0}$, $C_{17:0}$, $C_{18:0}$, $C_{19:0}$	Not fully characterized	Sutcliffe (2000)
Propionibacterium acnes	LG I: 24–29 LGII: 14–20 Man*p*, Glu, Gal, diaminohexuronic acid	$iC_{15:0}$, $C_{16:0}$, $C_{18:0}$	Incomplete characterization	Whale *et al*. (2004)
Corynebacterium matruchotii	~15–50 (Ara*f*/Man*p*)	$C_{16:0}$, $C_{18:1}$, $C_{18:0}$	Not fully characterized	Sutcliffe (1995)
Micrococcus spp.	Man*p*	$aC_{15:0}$: 55%, $iC_{15:0}$: 14%, $C_{14:0}$: 11%	Dimannosyl diacylglycerol anchor is suggested	Owen and Salton (1975); Pless *et al*. (1975); Powell *et al*. (1975)

Ara*f*, arabinose; Gal, galactose; Glu, glucose, Man*p*, mannose.

for 1% of the genome (Wimmerova *et al.*, 2003). A more recent classification system, utilizing structural similarities, divides GTs between three main superfamilies: GT-A, GT-B and GT-C (Fig. 6.4) (Liu and Mushegian, 2003). Most of these are involved in cell wall biosynthesis, specifically PIMs/LM/LAM and the mAGP complex (Berg *et al.*, 2007). So far, thirteen of these have been shown to be involved in PIMs/LAM biogenesis including four GT-A/Bs (Ppm1, PimA, PimB' and PimC) and nine GT-Cs (PimE, MptB, MptA, MptC, EmbC, AftC, AftD, AftB and CapA).

GT-A superfamily

GT-A proteins are typically soluble proteins that utilize NDP-sugars as their donor substrate. The family is characterized by a single β/α/β 'Rossman-like fold', consisting of a central twisted parallel β-sheet surrounded by α-helices (Rao and Rossmann, 1973). Although comprising a single fold, GT-As do have separate NDP-sugar and acceptor binding domains (Charnock and Davies, 1999). Many also possess a variation on the DxD motif, which is involved in co-ordinating a divalent cation; usually Mn^{2+} or Mg^{2+} (Wiggins and Munro, 1998).

Members of the GT-A superfamily are involved indirectly in the LAM pathway, producing the polyprenyl-phosphate-sugar donor substrates required by the pathway's GT-C transmembrane proteins. For example, Ppm1 (Rv2051c) transfers mannose from GDP-Man*p* to produce C_{50}-P-Man*p* (Gurcha *et al.*, 2002; Gibson *et al.*, 2003a). Although no structural examples from the LAM pathway exist to date, the structure of a *mycobacterial* GT-A has been solved (Fig. 6.4A). MAP2569c (homologue of Rv1208) utilizes UDP-glucose and is involved in 6-*O*-methylglucose lipopolysaccharide (MGLP) synthesis (Kaur *et al.*, 2009). The catalytic domain of MAP2569c comprises a typical αβα sandwich or GT-A-fold, which consists of a nucleotide binding 'Rossman-like fold' and an acceptor-binding domain. The DxD motif is between these two domains, co-ordinating a Mn^{2+} ion, which in turn contacts the phosphate of UDP. Contacts with the sugar donor from both the catalytic and the C-terminal domain ensure specificity (Fig. 6.4A). Interestingly, a flexible loop linking these two domains has been shown to change conformation on ligation of the donor substrate, enabling it to bind firmly (Fulton

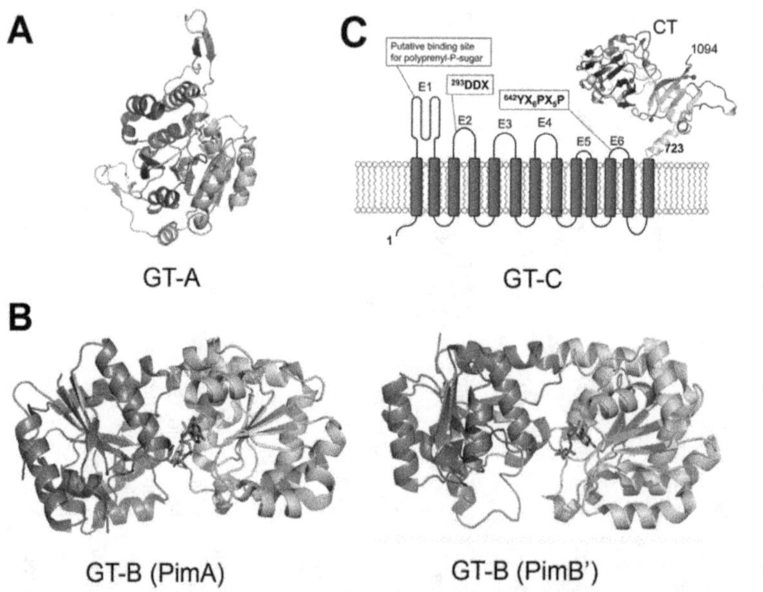

Figure 6.4 Three-dimensional structure of GT-A (MAP2569c), GT-B (PimA and PimB') and GT-C ($EmbC^{CT}$) class of proteins. (A) Ribbon diagram of MAP2569c in complex with UDP. (B) Ribbon diagram of PimA in complex with GDP-Man*p* and PimB' in complex with GDP. (C) Topology diagram of EmbC.

et al., 2008). Although expected to resemble MAP2569c, with a single Rossman-like fold, Ppm1 from *M. tuberculosis* is an unusual GT-A, possessing several predicted transmembrane regions in its N-terminal domain (domain 1), while the catalytic domain (domain 2) is at the C-terminus (Gurcha *et al.*, 2002). Interestingly, other mycobacterial species, such as *M. smegmatis*, *M. avium* and *M. leprae*, express the domains as two separate proteins which probably interact *in vivo* (Baulard *et al.*, 2003).

GT-B superfamily

Similarly to GT-A proteins, proteins of the GT-B superfamily are mostly soluble and transfer sugar from an NDP-sugar substrate. Examples from the PIM pathway include PimA, PimB' and PimC. In contrast to GT-As, GT-Bs consist of two β/α/β Rossman fold-like domains, with a central catalytic cleft and a hinge region between the two domains that enables flexibility (Liu and Mushegian, 2003). The C-terminal domain, responsible for donor substrate binding, is much more conserved; whilst the N-terminus, which binds the acceptor substrate, is more diverse. Even though not all GT-B proteins are metal dependent, they still possess conserved acidic residues required for substrate binding and catalysis (Cid *et al*, 2000). Most have a GPGTF motif, which is also important for activity (Wrabl and Grishin, 2001). There are two structural examples of GT-B proteins involved in the LAM biosynthetic pathway, PimA (Guerin *et al.*, 2007) and PimB' (Batt *et al.*, 2010), enzymes that catalyse the transfer of the first two mannose moieties to PI (Kordulakova *et al.*, 2002; Lea-Smith *et al.*, 2008; Mishra *et al.*, 2009).

PimA from *M. smegmatis* shows a typical GT-B structure (Guerin *et al.*, 2007); each domain consists of a Rossman-like fold of seven parallel β-strands, which alternate with seven α-helices. PimA crystallized in a closed complex, where the two domains are closed around GDP-Man*p* (Fig. 6.4B) (Guerin *et al.*, 2009b). This conformational change has since been shown to be dependent on donor substrate binding and is thought to stabilize the enzyme and create a site for acceptor substrate binding (PI); the apo-form of the enzyme is in an open conformation (Guerin *et al.*, 2009b). The GDP-Man*p* is held in place by several hydrogen bond contacts made along the length of the substrate. Glu-274 from the GPGTF motif and His-118 are both highly conserved in the superfamily and are thought to be required for catalysis; Glu-274 makes hydrogen bond contacts with the 4-OH of the mannose and His-118 is positioned close to the reacting centre (C-1) of the mannose (Fig. 6.4B). As the acceptor substrate would not co-crystallize, docking calculations were performed; these inferred that Tyr-9 is important for substrate recognition, with the phenol ring predicted to make stacking interactions with the inositiol ring of PI positioning reacting oxygen group (2-OH) close to the C-1 position of the donor mannosyl group. Arg-201 is also predicted to be important in stabilizing the substrate, as it is close enough to contact both the phosphate oxygen groups and the glycerol backbone of the PI. Near to the active site there is a hydrophobic tunnel, which is hypothesized to interact with the lipids of the acceptor substrate. A 'putative interfacial binding surface' near to the acceptor substrate binding region of the N-terminal domain, consists of a hydrophobic patch close to several basic residues and a nearby hydrophobic loop (β3–α2). This is thought to direct membrane attachment, which in turn activates the enzyme, possibly by bringing about a conformational change and/or enabling interaction with the substrate which extends out from the lipid bilayer. The β3–α2 loop may also contact the acyl chains of the PI acceptor substrate and have a role in acceptor recognition.

Similarly, the overall structure of PimB', from *Corynebacterium glutamicum*, also consists of two Rossman-like fold domains (Fig. 6.4B) (Batt *et al.*, 2010); the variable N-terminal domain contains seven parallel β-strands, alternating with six α-helices and the GDP-Man*p* binding C-terminal domain has six parallel β-strands, which alternative with four α-helices. Interestingly, PimB' only crystallized in an open complex, with the GDP-Man*p* not contacting any residues from the N-terminal domain; although direct evidence of flexibility was demonstrated by different crystal forms and mutational analysis, where the addition of a bulky side-chain into the hinge region abrogated activity. Whilst the GDP-Man*p* binding site is similar to that of PimA, the variable N-terminal domain is different, in order to accommodate an altered acceptor substrate PIM_1. The residue Tyr-9, which is important for acceptor recognition in PimA, corresponds to Asp13 in PimB′; a residue that is predicted to form hydrogen bonds with the 3- and 4-hydroxyl groups of the inositol ring. This should orientate the inositol group so that the 6-OH, rather than the 2-OH group as in PimA, is near to the reacting C-1 carbon centre of the mannose (Fig. 6.4B) (Batt *et al.*, 2010). However, there must be several more residues involved in such substrate specificity, since changing Asp-13 to a tyrosine did not alter PimB's substrate preferences, but instead abolished all activity. Amongst these differences is the more open active site in PimB′ and the absence of an extended β3–α3 loop, which may confer substrate specificity in PimA (β3–α2 loop). An additional role of Asp-13 in PimB′ could be that of a nucleophile or general acid, since replacing it with an asparagine, which is also capable of hydrogen bonding but has no charge, abolishes activity. Glu1-99 from PimA could have a similar role (Batt *et al.*, 2010).

GT-C superfamily

Members of this superfamily are integral membrane GTs, with multiple transmembrane (8–13) helices and an extracellular C-terminal domain. A conserved DxD motif (or ExD, DxE, DDx, DEx) is present on a long extracellular loop at the N-terminal end (Liu and Mushegian, 2003); although it is not known if this is involved in cation binding as are the DxD motifs in GT-As. The donor substrates of GT-Cs are lipid-linked sugars, a feature that enables the sugar to transverse the phospholipid bilayer (Liu and Mushegian, 2003). There are several examples of GT-Cs in the LAM biosynthetic pathway, such as PimE, which adds mannose to Ac_1/Ac_2PIM_4. (Morita *et al.*, 2006). However, owing to the constraints of generating a detergent-solubilized functional protein, there are no complete structures as yet.

The extracellular C-terminus of EmbC ($EmbC^{CT}$) from *M. tuberculosis* is one of only two structures available from the Emb family of proteins (Fig. 6.4C) (Alderwick *et al.*, 2011). This protein is an arabinosyl transferase, which extends arabinan of LAM in an α(1→5) direction (Zhang *et al.*, 2003; Shi *et al.*, 2006). $EmbC^{CT}$ is organized into two subdomains, each involved in carbohydrate binding. Subdomain I has some similarities with the soluble C-terminal domain of an archeal GT-C (STT3) from *Pyrococcus furiosus* (Igura *et al.*, 2008), consisting of a mixed α/β structure, which includes a semi-barrel formed by five β-strands. Subdomain II resembles a lectin- or 'carbohydrate binding module' (CBM)-like domain, with an anti-parallel β-sandwich fold. A bound Ca^{2+} ion within subdomain II, contacted by a single side-chain, aspartic acid (Asp-949), was shown to be required to stabilize $EmbC^{CT}$. Stability is also thought to be enhanced by the presence of a disulphide bond Cys-749–Cys-993, a structural element that may be conserved for EmbC-terminal domains. Mutational analysis of tryptophans from each subdomain (985 and 868, from I and II, respectively) has shown that both subdomains are involved in binding to different acceptor substrates, based on their chain length; subdomain I binds to short polysaccharides, whilst subdomain II prefers longer chains. The catalytic activity of EmbC is associated with several of its extracellular loops/domains, such as loop E2, which has the conserved DDx motif (Liu and Mushegian, 2003), and loop E6, which possesses a conserved proline motif (Fig. 6.4C) (Berg *et al.*, 2005). In addition, the crystal structure of domain E1 of a related protein shows it to contain a β-sandwich fold which could be a third carbohydrate binding domain, perhaps for the sugar donor DPA (Tan *et al.*, unpublished) (Fig. 6.4C).

Catalytic mechanism of GTs

GTs can be classified as either 'inverting' or 'retaining', depending on whether the anomeric bond from the sugar (C-1) is 'inverted' or 'retained' (Sinnott, 1991). Invertion reactions alter the configuration of the anomeric bond, for example an α-linked donor forming a β-glycosidic bond. Retaining reactions do not alter the bond, for instance an α-linked donor forming an α-glycosidic bond. Inverting enzymes, such as Ppm1, use an S_N2-like direct displacement mechanism; a residue in the enzyme's active site is thought to deprotonate the acceptor which then carries out a nucleophilic attack of the donor sugar's C-1, displacing the nucleoside diphosphate (NDP). In proteins with a GT-A fold, a co-ordinated divalent cation usually aids the displaced NDP's departure, whereas in GT-B fold proteins, positively charged residues and/or hydroxyls and helix dipoles perform this role (Lairson *et al.*, 2008). The mechanism utilized by retaining enzymes, such as PimA and PimB, is more ambiguous; there are two proposed, each with its own drawbacks. The S_N2-like double displacement reaction is two-step: in the first step, a glycosyl-enzyme intermediate is formed, releasing the NDP; this is followed by a nucleophilic attack by the acceptor (Zechel and Withers, 2001). However, the existence of a glycosyl-enzyme intermediate has never been proven and an appropriately positioned nucleophilic residue has not been found in the active sites of enzymes crystallized to date (Withers *et al.*, 2002; Sheng *et al.*, 2009; Batt *et al.*, 2010; Frantom *et al.* 2010; Steiner *et al.*, 2010). The other proposed mechanism is substrate mediated S_Ni (internal return) (Persson *et al.*, 2001). In this mechanism, the enzyme orientates the two substrates in order to facilitate the formation of an oxocarbenium intermediate, whilst shielding the substrates from a nucleophilic attack on the opposite side. The reaction has three steps: in the first, the sugar–phosphate bond is decayed, resulting in formation of the intermediate; in the second, the hydroxyl group of the acceptor is deprotonated by the leaving phosphate; thirdly, nucleophilic attack of the oxocarbenium ion by the acceptor results in formation of the glycosidic bond (Lairson *et al.*, 2008).

Concluding remarks

An important rationale for studying LAM and related glycoconjugates is that they are immunomodulatory in nature and can be utilized to develop diagnostic tools for tuberculosis. Furthermore, many of the genes involved in the biosynthesis of these molecules, especially those encoding GTs are essential (Jackson *et al.*, 2000; Kordulakova *et al.*, 2002; Sassetti *et al.*, 2003; Guerin *et al.*, 2009a; Skovierova *et al.*, 2009); and since LAM is not present in eukaryotic cells, represent excellent drug targets. However, both GT-A and GT-B superfamilies are fairly ubiquitous and are found in all three domains of life (Cantarel *et al.*, 2009); nevertheless, despite their structural similarities, GT-A and GT-B proteins have very different amino acid sequences and are thought not to be evolutionarily related (Liu and Mushegian, 2003). GTs transfer sugars from a specific substrate to a diverse set of acceptor substrates, from proteins in eukaryotes, to the cell wall in prokaryotes. This results in an assorted set of enzymes, where even those that take subsequent substrates in a pathway cannot be engineered to accept the other's substrate, reinforcing the fact that GTs are certainly ideal antibiotic candidates against pathogenic bacteria such as *M. tuberculosis*. The current surge in the number of solved GT structures is providing much insight into the architectures of their catalytic domains and the mechanisms they employ. As a consequence, future antibiotics will be tailor made to bind specifically to a GT's active site, inhibiting activity and, therefore, the fate of the bacilli.

Assessing the role of LAM and related glycoconjugates in mycobacterial pathogenicity requires the availability of mycobacterial mutants defective in their respective biosynthetic pathways; as most studies which use purified molecules do not represent the true *in vivo* condition during infection. The availability of complete genome sequences of several mycobacteria and related actinomycetes, and the development of novel tools for genetic manipulation is leading the way towards new exciting research in the area of mutant construction and infection studies.

Acknowledgements

G.S.B. acknowledges support in the form of a Personal Research Chair from Mr James Bardrick, Royal Society Wolfson Research Merit Award, as a former Lister Institute-Jenner Research Fellow, the Medical Research Council, and The Wellcome Trust (081569/Z/06/Z).

References

Appelmelk, B.J., den Dunnen, J., Driessen, N.N., Ummels, R., Pak, M., Nigou, J., Larrouy-Maumus, G., Gurcha, S.S., Movahedzadeh, F., Geurtsen, J., *et al.* (2008). The mannose cap of mycobacterial lipoarabinomannan does not dominate the Mycobacterium–host interaction. Cell. Microbiol. *10*, 930–944.

Alderwick, L.J., Lloyd, G.S., Ghadbane, H., May, J.W., Bhatt, A., Eggeling, L., Fütterer, K., Besra, G.S. (2011). The C-terminal domain of the Arabinosyltransferase *Mycobacterium tuberculosis* EmbC is a lectin-like carbohydrate binding module. PLoS Pathog. 7(2), e1001299.

Ballou, C.E., and Lee, Y.C. (1964). The structure of a myoinositol mannoside from *Mycobacterium tuberculosis* glycolipid. Biochemistry 3, 682–685.

Batt, S.M., Jabeen, T., Mishra, A.K., Veerapen, N., Krumbach, K., Eggeling, L., Besra, G.S., and Futterer, K. (2010). Acceptor substrate discrimination in phosphatidyl-myo-inositol mannoside synthesis: Structural and mutational analysis of mannosyltransferase *Corynebacterium glutamicum* PimB'. J. Biol. Chem. *285(48)*, 37741–37752.

Baulard, A.R., Gurcha, S.S., Engohang-Ndong, J., Gouffi, K., Locht, C., and Besra, G.S. (2003). *In vivo* interaction between the polyprenol phosphate mannose synthase Ppm1 and the integral membrane protein Ppm2 from *Mycobacterium smegmatis* revealed by a bacterial two-hybrid system. J. Biol. Chem. *278*, 2242–2248.

Berg, S., Starbuck, J., Torrelles, J.B., Vissa, V.D., Crick, D.C., Chatterjee, D., and Brennan, P.J. (2005). Roles of conserved proline and glycosyltransferase motifs of EmbC in biosynthesis of lipoarabinomannan. J. Biol. Chem. *280*, 5651–5663.

Berg, S., Kaur, D., Jackson, M., and Brennan, P.J. (2007). The glycosyltransferases of *Mycobacterium tuberculosis* – roles in the synthesis of arabinogalactan, lipoarabinomannan, and other glycoconjugates. Glycobiology *17*, 35–56R.

Besra, G.S., and Brennan, P.J. (1997). The mycobacterial cell wall: biosynthesis of arabinogalactan and lipoarabinomannan. Biochem. Soc. Trans. *25*, 845–850.

Besra, G.S., Khoo, K.H., McNeil, M.R., Dell, A., Morris, H.R., and Brennan, P.J. (1995). A new interpretation of the structure of the mycolyl–arabinogalactan complex of *Mycobacterium tuberculosis* as revealed through characterization of oligoglycosylalditol fragments by fast-atom bombardment mass spectrometry and 1H nuclear magnetic resonance spectroscopy. Biochemistry *34*, 4257–4266.

Besra, G.S., Morehouse, C.B., Rittner, C.M., Waechter, C.J., and Brennan, P.J. (1997). Biosynthesis of mycobacterial lipoarabinomannan. J. Biol. Chem. *272*, 18460–18466.

Birch, H.L., Alderwick, L.J., Appelmelk, B.J., Maaskant, J., Bhatt, A., Singh, A., Nigou, J., Eggeling, L., Geurtsen, J., and Besra, G.S. (2010). A truncated lipoglycan from mycobacteria with altered immunological properties. Proc. Natl. Acad. Sci. U.S.A. *107*, 2634–2639.

Brennan, P.J. (2003). Structure, function, and biogenesis of the cell wall of *Mycobacterium tuberculosis*. Tuberculosis (Edin.) *83*, 91–97.

Brennan, P.J., and Nikaido, H. (1995). The envelope of mycobacteria. Annu. Rev. Biochem. *64*, 29–63.

Briken, V., Porcelli, S.A., Besra, G.S., and Kremer, L. (2004). Mycobacterial lipoarabinomannan and related lipoglycans: from biogenesis to modulation of the immune response. Mol. Microbiol. *53*, 391–403.

Cantarel, B.L., Coutinho, P.M., Rancurel, C., Bernard, T., Lombard, V., and Henrissat, B. (2009). The carbohydrate-active EnZymes database (CAZy): an expert resource for glycogenomics. Nucleic Acids Res. *37*, D233–238.

Charnock, S.J., and Davies, G.J. (1999). Structure of the nucleotide-diphospho-sugar transferase, SpsA from *Bacillus subtilis*, in native and nucleotide-complexed forms. Biochemistry *38*, 6380–6385.

Chatterjee, D., Bozic, C.M., McNeil, M., and Brennan, P.J. (1991). Structural features of the arabinan component of the lipoarabinomannan of *Mycobacterium tuberculosis*. J. Biol. Chem. *266*, 9652–9660.

Chatterjee, D., Hunter, S.W., McNeil, M., and Brennan, P.J. (1992a). Lipoarabinomannan. Multiglycosylated form of the mycobacterial mannosylphosphatidylinositols. J. Biol. Chem. *267*, 6228–6233.

Chatterjee, D., Lowell, K., Rivoire, B., McNeil, M.R., and Brennan, P.J. (1992b). Lipoarabinomannan of *Mycobacterium tuberculosis*. Capping with mannosyl residues in some strains. J. Biol. Chem. *267*, 6234–6239.

Chatterjee, D., Khoo, K.H., McNeil, M.R., Dell, A., Morris, H.R., and Brennan, P.J. (1993). Structural definition of the non-reducing termini of mannose-capped LAM from *Mycobacterium tuberculosis* through selective enzymatic degradation and fast atom bombardment-mass spectrometry. Glycobiology 3, 497–506.

Chatterjee, D., and Khoo, K.H. (1998). Mycobacterial lipoarabinomannan: an extraordinary lipoheteroglycan with profound physiological effects. Glycobiology *8*, 113–120.

Daffe, M., Brennan, P.J., and McNeil, M. (1990). Predominant structural features of the cell wall arabinogalactan of *Mycobacterium tuberculosis* as revealed through characterization of oligoglycosyl alditol fragments by gas chromatography/mass spectrometry and by 1H and 13C NMR analyses. J. Biol. Chem. *265*, 6734–6743.

Dana, A., Fader, R., and Sterken, D. (2001). *Turicella otitidis* mastoiditis in a healthy child. Pediatr. Infect. Dis. J. *20*, 84–85.

Dinadayala, P., Kaur, D., Berg, S., Amin, A.G., Vissa, V.D., Chatterjee, D., Brennan, P.J., and Crick, D.C. (2006). Genetic basis for the synthesis of the immunomodulatory mannose caps of lipoarabinomannan in *Mycobacterium tuberculosis*. J. Biol. Chem. *281*, 20027–20035.

Dover, L.G., Cerdeno-Tarraga, A.M., Pallen, M.J., Parkhill, J., and Besra, G.S. (2004). Comparative cell wall core biosynthesis in the mycolated pathogens, *Mycobacterium tuberculosis* and *Corynebacterium diphtheriae*. FEMS Microbiol. Rev. *28*, 225–250.

Driessen, N.N., Stoop, E.J., Ummels, R., Gurcha, S.S., Mishra, A.K., Larrouy-Maumus, G., Nigou, J., Gilleron, M., Puzo, G., Maaskant, J.J., *et al.* (2010). *Mycobacterium marinum* MMAR_2380, a predicted transmembrane acyltransferase, is essential for the presence of the mannose cap on lipoarabinomannan. Microbiology *156(Pt 11)*, 3492–3502.

Frantom, P.A., Coward, J.K., and Blanchard, J.S. (2010). UDP-(5F)-GlcNAc acts as a slow-binding inhibitor of MshA, a retaining glycosyltransferase. J. Am. Chem. Soc. *132*, 6626–6627.

Fulton, Z., McAlister, A., Wilce, M.C., Brammananth, R., Zaker-Tabrizi, L., Perugini, M.A., Bottomley, S.P., Coppel, R.L., Crellin, P.K., Rossjohn, J., *et al.* (2008). Crystal structure of a UDP-glucose-specific glycosyltransferase from a *Mycobacterium* species. J. Biol. Chem. *283,* 27881–27890.

Garton, N.J., and Sutcliffe, I.C. (2006). Identification of a lipoarabinomannan-like lipoglycan in the actinomycete *Gordonia bronchialis*. Arch. Microbiol. *184,* 425–427.

Garton, N.J., Gilleron, M., Brando, T., Dan, H.H., Giguere, S., Puzo, G., Prescott, J.F., and Sutcliffe, I.C. (2002). A novel lipoarabinomannan from the equine pathogen *Rhodococcus equi*. Structure and effect on macrophage cytokine production. J. Biol. Chem. *277,* 31722–31733.

Gibson, K.J., Eggeling, L., Maughan, W.N., Krumbach, K., Gurcha, S.S., Nigou, J., Puzo, G., Sahm, H., and Besra, G.S. (2003a). Disruption of Cg-Ppm1, a polyprenyl monophosphomannose synthase, and the generation of lipoglycan-less mutants in *Corynebacterium glutamicum*. J. Biol. Chem. *278,* 40842–40850.

Gibson, K.J., Gilleron, M., Constant, P., Puzo, G., Nigou, J., and Besra, G.S. (2003b). Identification of a novel mannose-capped lipoarabinomannan from Amycolatopsis sulphurea. Biochem. J. *372,* 821–829.

Gibson, K.J., Gilleron, M., Constant, P., Puzo, G., Nigou, J., and Besra, G.S. (2003c). Structural and functional features of *Rhodococcus ruber* lipoarabinomannan. Microbiology *149,* 1437–1445.

Gibson, K.J., Gilleron, M., Constant, P., Brando, T., Puzo, G., Besra, G.S., and Nigou, J. (2004). *Tsukamurella paurometabola* lipoglycan, a new lipoarabinomannan variant with pro-inflammatory activity. J. Biol. Chem. *279,* 22973–22982.

Gibson, K.J., Gilleron, M., Constant, P., Sichi, B., Puzo, G., Besra, G.S., and Nigou, J. (2005). A lipomannan variant with strong TLR-2-dependent pro-inflammatory activity in *Saccharothrix aerocolonigenes*. J. Biol. Chem. *280,* 28347–28356.

Gilleron, M., Garton, N.J., Nigou, J., Brando, T., Puzo, G., and Sutcliffe, I.C. (2005). Characterization of a truncated lipoarabinomannan from the Actinomycete *Turicella otitidis*. J. Bacteriol. *187,* 854–861.

Guerardel, Y., Maes, E., Elass, E., Leroy, Y., Timmerman, P., Besra, G.S., Locht, C., Strecker, G., and Kremer, L. (2002). Structural study of lipomannan and lipoarabinomannan from *Mycobacterium chelonae*. Presence of unusual components with α-1,3-mannopyranose side-chains. J. Biol. Chem. *277,* 30635–30648.

Guerin, M.E., Kordulakova, J., Schaeffer, F., Svetlikova, Z., Buschiazzo, A., Giganti, D., Gicquel, B., Mikusova, K., Jackson, M., and Alzari, P.M. (2007). Molecular recognition and interfacial catalysis by the essential phosphatidylinositol mannosyltransferase PimA from mycobacteria. J. Biol. Chem. *282,* 20705–20714.

Guerin, M.E., Kaur, D., Somashekar, B.S., Gibbs, S., Gest, P., Chatterjee, D., Brennan, P.J., and Jackson, M. (2009a). New insights into the early steps of phosphatidylinositol mannoside biosynthesis in mycobacteria: PimB' is an essential enzyme of *Mycobacterium smegmatis*. J. Biol. Chem. *284,* 25687–25696.

Guerin, M.E., Schaeffer, F., Chaffotte, A., Gest, P., Giganti, D., Kordulakova, J., van der Woerd, M., Jackson, M., and Alzari, P.M. (2009b). Substrate-induced conformational changes in the essential peripheral membrane-associated mannosyltransferase PimA from mycobacteria: implications for catalysis. J. Biol. Chem. *284,* 21613–21625.

Gurcha, S.S., Baulard, A.R., Kremer, L., Locht, C., Moody, D.B., Muhlecker, W., Costello, C.E., Crick, D.C., Brennan, P.J., and Besra, G.S. (2002). Ppm1, a novel polyprenol monophosphomannose synthase from *Mycobacterium tuberculosis*. Biochem. J. *365,* 441–450.

Hill, D.L., and Ballou, C.E. (1966). Biosynthesis of mannophospholipids by *Mycobacterium phlei*. J. Biol. Chem. *241,* 895–902.

Hoffmann, C., Leis, A., Niederweis, M., Plitzko, J.M., and Engelhardt, H. (2008). Disclosure of the mycobacterial outer membrane: cryo-electron tomography and vitreous sections reveal the lipid bilayer structure. Proc. Natl. Acad. Sci. U.S.A. *105,* 3963–3967.

Hunter, S.W., and Brennan, P.J. (1990). Evidence for the presence of a phosphatidylinositol anchor on the lipoarabinomannan and lipomannan of *Mycobacterium tuberculosis*. J. Biol. Chem. *265,* 9272–9279.

Hunter, S.W., Gaylord, H., and Brennan, P.J. (1986). Structure and antigenicity of the phosphorylated lipopolysaccharide antigens from the leprosy and tubercle bacilli. J. Biol. Chem. *261,* 12345–12351.

Igura, M., Maita, N., Kamishikiryo, J., Yamada, M., Obita, T., Maenaka. K., and Kohda, D. (2008). Structure-guided identification of a new catalytic motif of oligosaccharyltransferase. EMBO J. *27,* 234–243.

Jackson, M., Crick, D.C., and Brennan, P.J. (2000). Phosphatidylinositol is an essential phospholipid of mycobacteria. J. Biol. Chem. *275,* 30092–30099.

Kaur, D., Berg, S., Dinadayala, P., Gicquel, B., Chatterjee, D., McNeil, M.R., Vissa, V.D., Crick, D.C., Jackson, M., and Brennan, P.J. (2006). Biosynthesis of mycobacterial lipoarabinomannan: role of a branching mannosyltransferase. Proc. Natl. Acad. Sci. U.S.A. *103,* 13664–13669.

Kaur, D., McNeil, M.R., Khoo, K.H., Chatterjee, D., Crick, D.C., Jackson, M., and Brennan, P.J. (2007). New insights into the biosynthesis of mycobacterial lipomannan arising from deletion of a conserved gene. J. Biol. Chem. *282,* 27133–27140.

Kaur, D., Obregon-Henao, A., Pham, H., Chatterjee, D., Brennan, P.J., and Jackson, M. (2008). Lipoarabinomannan of Mycobacterium: mannose capping by a multifunctional terminal mannosyltransferase. Proc. Natl. Acad. Sci. U.S.A. *105,* 17973–17977.

Kaur, D., Pham, H., Larrouy-Maumus, G., Riviere, M., Vissa, V., Guerin, M.E., Puzo, G., Brennan, P.J., and Jackson, M. (2009). Initiation of methylglucose lipopolysaccharide biosynthesis in mycobacteria. PLoS One 4, e5447.

Khoo, K.H., Dell, A., Morris, H.R., Brennan, P.J., and Chatterjee, D. (1995). Structural definition of acylated phosphatidylinositol mannosides from *Mycobacterium tuberculosis*: definition of a common anchor for lipomannan and lipoarabinomannan. Glycobiology 5, 117–127.

Kordulakova, J., Gilleron, M., Mikusova, K., Puzo, G., Brennan, P.J., Gicquel, B., and Jackson, M. (2002). Definition of the first mannosylation step in phosphatidylinositol mannoside synthesis. PimA is essential for growth of mycobacteria. J. Biol. Chem. *277,* 31335–31344.

Kordulakova, J., Gilleron, M., Puzo, G., Brennan, P.J., Gicquel, B., Mikusova, K., and Jackson, M. (2003). Identification of the required acyltransferase step in the biosynthesis of the phosphatidylinositol mannosides of mycobacterium species. J. Biol. Chem. *278,* 36285–36295.

Kremer, L., Gurcha, S.S., Bifani, P., Hitchen, P.G., Baulard, A., Morris, H.R., Dell, A., Brennan, P.J., and Besra, G.S. (2002). Characterization of a putative α-mannosyltransferase involved in phosphatidylinositol trimannoside biosynthesis in *Mycobacterium tuberculosis.* Biochem. J. *363,* 437–447.

Lairson, L.L., Henrissat, B., Davies, G.J., and Withers, S.G. (2008). Glycosyltransferases: structures, functions, and mechanisms. Annu. Rev. Biochem. *77,* 521–555.

Lea-Smith, D.J., Martin, K.L., Pyke, J.S., Tull, D., McConville, M.J., Coppel, R.L., and Crellin, P.K. (2008). Analysis of a new mannosyltransferase required for the synthesis of phosphatidylinositol mannosides and lipoarbinomannan reveals two lipomannan pools in corynebacterineae. J. Biol. Chem. *283,* 6773–6782.

Lee, Y.C., and Ballou, C.E. (1964). Structural studies on the myo-inositol mannodides from the glycolipids of *Mycobacterium tuberculosis* and *Mycobacterium phlei.* J. Biol. Chem. *239,* 1316–1327.

Lee, Y.C., and Ballou, C.E. (1965). Complete structures of the glycophospholipids of mycobacteria. Biochemistry *4,* 1395–1404.

Liu, J., and Mushegian, A. (2003). Three monophyletic superfamilies account for the majority of the known glycosyltransferases. Protein Sci. *12,* 1418–1431.

McNeil, M., Daffe, M., and Brennan, P.J. (1990). Evidence for the nature of the link between the arabinogalactan and peptidoglycan of mycobacterial cell walls. J. Biol. Chem. *265,* 18200–18206.

McNeil, M., Daffe, M., and Brennan, P.J. (1991). Location of the mycolyl ester substituents in the cell walls of mycobacteria. J. Biol. Chem. *266,* 13217–13223.

McNeil, M.R., Robuck, K.G., Harter, M., and Brennan, P.J. (1994). Enzymatic evidence for the presence of a critical terminal hexa-arabinoside in the cell walls of *Mycobacterium tuberculosis.* Glycobiology *4,* 165–173.

Maeda, N., Nigou, J., Herrmann, J.L., Jackson, M., Amara, A., Lagrange, P.H., Puzo, G., Gicquel, B., and Neyrolles, O. (2003). The cell surface receptor DC-SIGN discriminates between Mycobacterium species through selective recognition of the mannose caps on lipoarabinomannan. J. Biol. Chem. *278,* 5513–5516.

Mishra, A.K., Alderwick, L.J., Rittmann, D., Tatituri, R.V., Nigou, J., Gilleron, M., Eggeling, L., and Besra, G.S. (2007). Identification of an α(1→6) mannopyranosyltransferase (MptA), involved in *Corynebacterium glutamicum* lipomanann biosynthesis, and identification of its orthologue in *Mycobacterium tuberculosis.* Mol. Microbiol. *65,* 1503–1517.

Mishra, A.K., Alderwick, L.J., Rittmann, D., Wang, C., Bhatt, A., Jacobs, W.R., Jr., Takayama, K., Eggeling, L., and Besra, G.S. (2008a). Identification of a novel α(1→6) mannopyranosyltransferase MptB from Corynebacterium glutamicum by deletion of a conserved gene, NCgl*1505,* affords a lipomannan- and lipoarabinomannan-deficient mutant. Mol. Microbiol. *68,* 1595–1613.

Mishra, A.K., Klein, C., Gurcha, S.S., Alderwick, L.J., Babu, P., Hitchen, P.G., Morris, H.R., Dell, A., Besra, G.S., and Eggeling, L. (2008b). Structural characterization and functional properties of a novel lipomannan variant isolated from a *Corynebacterium glutamicum* pimB' mutant. Antonie Van Leeuwenhoek *94,* 277–287.

Mishra, A.K., Batt, S., Krumbach, K., Eggeling, L., and Besra, G.S. (2009). Characterization of the *Corynebacterium glutamicum* ΔpimB' ΔmgtA double deletion mutant and the role of *Mycobacterium tuberculosis* orthologues Rv2188c and Rv0557 in glycolipid biosynthesis. J. Bacteriol. *191,* 4465–4472.

Mishra, A.K., Krumbach, K., Rittmann, D., Appelmelk, B., Pathak, V., Pathak, A.K., Nigou, J., Geurtsen, J., Eggeling, L., and Besra, G.S. (2011). Lipoarabinomannan biosynthesis in Corynebacterineae: the interplay of two α(1→2)-mannopyranosyltransferases MptC and MptD in mannan branching. Mol. Microbiol. *80*(5), 1241–1259.

Moreira, L.O., Mattos-Guaraldi, A.L., and Andrade, A.F. (2008). Novel lipoarabinomannan-like lipoglycan (CdiLAM) contributes to the adherence of *Corynebacterium diphtheriae* to epithelial cells. Arch. Microbiol. *190,* 521–530.

Morita, Y.S., Patterson, J.H., Billman-Jacobe, H., and McConville, M.J. (2004). Biosynthesis of mycobacterial phosphatidylinositol mannosides. Biochem. J. *378,* 589–597.

Morita, Y.S., Sena, C.B., Waller, R.F., Kurokawa, K., Sernee, M.F., Nakatani, F., Haites, R.E., Billman-Jacobe, H., McConville, M.J., Maeda, Y., *et al.* (2006). PimE is a polyprenol-phosphate-mannose-dependent mannosyltransferase that transfers the fifth mannose of phosphatidylinositol mannoside in mycobacteria. J. Biol. Chem. *281,* 25143–25155.

Nigou, J., Gilleron, M., Rojas, M., Garcia, L.F., Thurnher, M., and Puzo, G. (2002). Mycobacterial lipoarabinomannans: modulators of dendritic cell function and the apoptotic response. Microbes Infect. *4,* 945–953.

Nigou, J., Gilleron, M., and Puzo, G. (2003). Lipoarabinomannans: from structure to biosynthesis. Biochimie *85,* 153–166.

Osoagbaka, O.U. (1989). Evidence for the pathogenic role of Rhodococcus species in pulmonary diseases. J. Appl. Bacteriol. *66,* 497–506.

Owen, P., and Salton, M.R. (1975). Isolation and characterization of a mannan from mesosomal membrane vesicles of *Micrococcus lysodeikticus*. Biochim. Biophys. Acta *406,* 214–234.

Persson, K., Ly, H.D., Dieckelmann, M., Wakarchuk, W.W., Withers, S.G., and Strynadka, N.C. (2001). Crystal structure of the retaining galactosyltransferase LgtC from *Neisseria meningitidis* in complex with donor and acceptor sugar analogs. Nat. Struct. Biol. *8,* 166–175.

Pless, D.D., Schmit, A.S., and Lennarz, W.J. (1975). The characterization of mannan of *Micrococcus lysodeikticus* as an acidic lipopolysaccharide. J. Biol. Chem. *250,* 1319–1327.

Powell, D.A., Duckworth, M., and Baddiley, J. (1975). A membrane-associated lipomannan in micrococci. Biochem. J. *151,* 387–397.

Prescott, J.F. (1991). *Rhodococcus equi*: an animal and human pathogen. Clin. Microbiol. Rev. *4,* 20–34.

Rao, S.T., and Rossmann, M.G. (1973). Comparison of super-secondary structures in proteins. J. Mol. Biol. *76,* 241–256.

Sassetti, C.M., Boyd, D.H., and Rubin, E.J. (2003). Genes required for mycobacterial growth defined by high density mutagenesis. Mol. Microbiol. *48,* 77–84.

Schlesinger, L.S., Hull, S.R., and Kaufman, T.M. (1994). Binding of the terminal mannosyl units of lipoarabinomannan from a virulent strain of *Mycobacterium tuberculosis* to human macrophages. J. Immunol. *152,* 4070–4079.

Seidel, M., Alderwick, L.J., Birch, H.L., Sahm, H., Eggeling, L., and Besra, G.S. (2007). Identification of a novel arabinofuranosyltransferase AftB involved in a terminal step of cell wall arabinan biosynthesis in Corynebacterianeae, such as *Corynebacterium glutamicum* and *Mycobacterium tuberculosis*. J. Biol. Chem. *282,* 14729–14740.

Sena, C.B., Fukuda, T., Miyanagi, K., Matsumoto, S., Kobayashi, K., Murakami, Y., Maeda, Y., Kinoshita, T., and Morita, Y.S. (2010). Controlled expression of branch-forming mannosyltransferase is critical for mycobacterial lipoarabinomannan biosynthesis. J. Biol. Chem. *285,* 13326–13336.

Sheng, F., Jia, X., Yep, A., Preiss, J., and Geiger, J.H. (2009). The crystal structures of the open and catalytically competent closed conformation of *Escherichia coli* glycogen synthase. J. Biol. Chem. *284,* 17796–17807.

Shi, L., Berg, S., Lee, A., Spencer, J.S., Zhang, J., Vissa, V., McNeil, M.R., Khoo, K.H., and Chatterjee, D. (2006). The carboxy terminus of EmbC from *Mycobacterium smegmatis* mediates chain length extension of the arabinan in lipoarabinomannan. J. Biol. Chem. *281,* 19512–19526.

Sinnott, M.L. (1991). Catalytic mechanisms of enzymatic glycosyl transfer. Chem. Rev. *90,* 1170–1202.

Skovierova, H., Larrouy-Maumus, G., Zhang, J., Kaur, D., Barilone, N., Kordulakova, J., Gilleron, M., Guadagnini, S., Belanova, M., Prevost, M.C., *et al.* (2009). AftD, a novel essential arabinofuranosyltransferase from mycobacteria. Glycobiology *19,* 1235–1247.

Steiner, K., Hagelueken, G., Messner, P., Schaffer, C., and Naismith, J.H. (2010). Structural basis of substrate binding in WsaF, a rhamnosyltransferase from *Geobacillus stearothermophilus*. J. Mol. Biol. *397,* 436–447.

Sutcliffe, I.C. (1995). Identification of a lipoarabinomannan-like lipoglycan in *Corynebacterium matruchotii*. Arch. Oral Biol. *40,* 1119–1124.

Sutcliffe, I.C. (2000). Characterisation of a lipomannan lipoglycan from the mycolic acid containing actinomycete *Dietzia maris*. Antonie Van Leeuwenhoek *78,* 195–201.

Whale, G.A., Sutcliffe, I.C., Morrisson, A.R., Pretswell, E.L., and Emmison, N. (2004). Purification and characterisation of lipoglycan macroamphiphiles from *Propionibacterium acnes*. Antonie Van Leeuwenhoek *86,* 77–85.

Wiggins, C.A., and Munro, S. (1998). Activity of the yeast MNN1 α-1,3-mannosyltransferase requires a motif conserved in many other families of glycosyltransferases. Proc. Natl. Acad. Sci. U.S.A. *95,* 7945–7950.

Wimmerova, M., Engelsen, S.B., Bettler, E., Breton, C., and Imberty, A. (2003). Combining fold recognition and exploratory data analysis for searching for glycosyltransferases in the genome of *Mycobacterium tuberculosis*. Biochimie *85,* 691–700.

Withers, S.G., Wakarchuk, W.W., and Strynadka, N.C. (2002). One step closer to a sweet conclusion. Chem. Biol. *9,* 1270–1273.

Wrabl, J.O., and Grishin, N.V. (2001). Homology between O-linked GlcNAc transferases and proteins of the glycogen phosphorylase superfamily. J. Mol. Biol. *314,* 365–374.

Zechel, D.L., and Withers, S.G. (2001). Dissection of nucleophilic and acid-base catalysis in glycosidases. Curr. Opin. Chem. Biol. *5,* 643–649.

Zhang, N., Torrelles, J.B., McNeil, M.R., Escuyer, V.E., Khoo, K.H., Brennan, P.J., and Chatterjee, D. (2003). The Emb proteins of mycobacteria direct arabinosylation of lipoarabinomannan and arabinogalactan via an N-terminal recognition region and a C-terminal synthetic region. Mol. Microbiol. *50,* 69–76.

7 Protein Glycosylation in the Third Domain of Life: the Archaea

John F. Kelly and Ken F. Jarrell

Abstract

Archaea are single-celled microorganisms that are sufficiently distinct as to constitute a third domain of life. It has been known for some time that Archaea express glycoproteins. However, research to understand the nature of their glycan modifications, the pathways used to produce them as well as their role in archaeal biology has lagged behind similar efforts in Eukarya and Bacteria, at least until recently. Here we describe in some detail the efforts made to determine the composition and structure of archaeal glycan post-translational modifications and to elucidate their biosynthetic pathways.

Introduction

The Archaea are single-celled microorganisms that constitute the third domain of life alongside Bacteria and Eukarya (Woese and Fox, 1977). The first recognized archaeal species were extremophiles, living in environments harsh enough to prevent colonization by most other forms of life. However, it is now recognized that Archaea are widely distributed in nature and are found in more conventional ecological niches such as soil, seawater and intestinal flora (Chaban *et al.*, 2006a). It is thought that Archaea constitute nearly half of the microbial biomass in the world's oceans and achieve even higher percentages in marine sediments (Karner *et al.*, 2001; Lipp *et al.*, 2008). Thus, Archaea can have a major impact on their immediate environment and on the Earth's ecological balance as a whole. Recognized archaeal species number in the 100s and belong primarily to the Euryarchaeota and Crenarchaeota phyla (Garrity *et al.*, 2007). However, there is evidence of the existence of many more archaeal species that have yet to be isolated and studied.

Archaea were once considered to be an unusual group of bacteria as, generally speaking, they are similar in size and shape to bacteria and have no nucleus or eukaryotic-like organelles. However, it became clear that they were a fundamentally distinct form of life when their 16S ribosomal (r)RNA sequences were compared with that of other prokaryotes (Fox *et al.*, 1977; Woese and Fox, 1977). In addition, they possess other features that distinguish them from bacteria and eukaryotes. For example, the phospholipids that constitute archaeal cell membranes are composed of polyisoprenoid lipids that are ether-linked to *sn*-2,3 positions on phosphoglycerol (Sprott, 1992). These molecules are not found in bacteria and eukaryotes. Furthermore, the archaeal flagellum is different from that found in bacteria (Thomas *et al.*, 2001; Logan, 2006), more closely resembling the type IV pilus in aspects of its composition, structure and probable assembly (Trachtenberg and Cohen-Krausz, 2006; Jarrell *et al.*, 2009). In addition, the composition and organization of the archaeal cell wall differs from that of bacteria (Kandler and Konig, 1998). Archaea make other structural appendages such as cannulae (Nickell *et al.*, 2003) and hami (Moissl *et al.*, 2005) that have no equivalents in Bacteria. In other respects Archaea resemble eukaryotes. For example, transcription (Thomm, 2007) and translation (Londei, 2007) in Archaea more closely resemble these processes in eukaryotes than in bacteria and certain aspects of protein glycosylation in Archaea, the subject of this chapter, resemble those of eukaryotes (Yurist-Doutsch *et al.*, 2008b).

Protein glycosylation in Archaea

Once thought to be an exclusive trait of eukaryotes, it is now known that protein glycosylation is relatively common in prokaryotes. The first reports of protein glycosylation in an archaeon, indeed the first in any prokaryote, appeared in the early 1970s (Koncewicz, 1972; Mescher *et al.*, 1974). Since then protein glycosylation has been inferred and/or confirmed in numerous other archaeal species (Eichler and Adams, 2005; Calo *et al.*, 2010b). Recently, Magidovich and Eichler examined the genomes of 56 archaea and found homologues of AglB, an oligosaccharyltranferase that appends glycans to the side-chain of asparagine (Asn), in all but two of them (Magidovich and Eichler, 2009). This suggests that protein N-glycosylation is indeed widespread in Archaea.

Protein glycosylation in Bacteria and Archaea is mainly associated with the cell surface. S-layer and flagellin proteins, both relatively abundant and easy to isolate, have been the most studied to date. However, glycosylation of other cell surface proteins, such as pilin and abundant membrane proteins, has also been studied in some species. *O*-glycosylation is predominant in Bacteria whereas N-glycosylation appears to be more common in Archaea. However, it has to be said that *O*-glycosylation has not been well studied in Archaea (Calo *et al.*, 2010b).

Generally speaking, protein glycosylation in Eukaryotes is a highly conserved process that uses a handful of monosaccharide species to produce comparatively few glycan structures. In contrast, prokaryotes produce a far greater range of sugars and structures that are often species specific (Abu-Qarn *et al.*, 2008a). Thus, it has proven difficult to characterize prokaryotic glycans using methods that were originally developed for eukaryotic glycans and glycoproteins. This explains, in part, why protein glycosylation has been understudied in Archaea and Bacteria. However, this field of research has benefited greatly from the renaissance that many analytical technologies have experienced in recent years, especially mass spectrometry (MS) and nuclear magnetic resonance spectroscopy (NMR). Numerous examples will be described in this chapter.

Determining glycan structures on archaeal glycoproteins

Halophiles

Prokaryotic protein glycosylation was first reported in the haloarchaeon, *Halobacterium salinarum* (also previously designated Halobacterium halobium). Cell wall extracts were shown to contain an abundant high molecular weight (S-layer) protein at ~200 kDa which, when separated by SDS-PAGE and treated with periodate-Schiff reagent, stained positively for glycosylation (Koncewicz, 1972; Mescher *et al.*, 1974). As it turned out, the glycosylation of this particular protein is extraordinarily complicated and it took almost two decades of determined effort to elucidate the different glycan structures, their linkages to the S-layer protein as well as their biosynthesis. Not long after, it was reported that the S-layer protein is both *N*- and *O*-glycosylated (Mescher and Strominger, 1976b). Trypsin digestion followed by gel filtration yielded two fractions containing large glycopeptides, one containing mainly hexose sugars and the other amino sugars. Gas chromatography (GC) analysis of the hexose-containing glycopeptide revealed the presence of glucose (Glc) and galactose (Gal) in 1:1 ratios while colorimetric analysis suggested the presence of a smaller proportion of uronic acid. Further analysis indicated that even though this glycopeptide was very large (~55 kDa) the glycans were short (2–3 residues long) and were O-linked to many threonine (Thr) residues via the Gal residue (Fig. 7.1A). Analysis of the amino sugar-containing glycopeptide fraction indicated that the glycan was large, composed of multiple sugar species and that it was N-linked.

A few years later it was proven that the large N-linked glycan as well as some of the smaller oligosaccharides are sulphated (Wieland *et al.*, 1980). When *H. salinarum* was grown in medium containing $^{35}SO_4^{2-}$ the S-layer protein became radioactive.

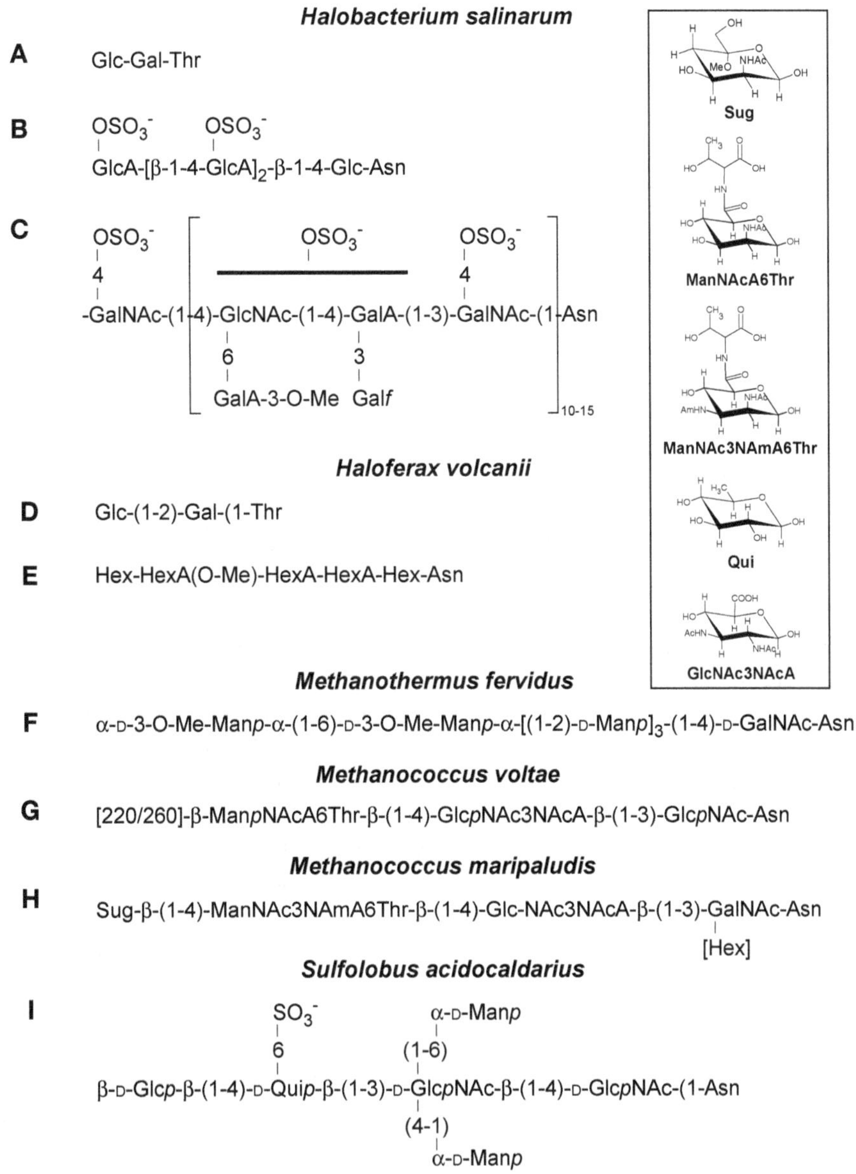

Figure 7.1 The archaeal N- and O-linked glycan post-translational modifications that have been determined to date. Some glycans have been more fully characterized than others. The structures of some of the more unusual sugar residues are presented in the side panel.

Using a variety of analyses, it was determined that there were approximately 40 sulphate residues per glycoprotein molecule and that the sulphates are attached to the large N-linked glycan as well as smaller N-linked oligosaccharides. Furthermore, it was determined that the large N-linked glycan was actually a family of related oligosaccharides differing in the number of sulphated-containing sugar residues. Subsequently, the short sulphated glycans were shown to contain Glc and glucuronic acid (GlcA) and that they were N-linked to Asn via the Glc moiety (Fig. 7.1B) (Wieland *et al.*, 1983). Pronase treatment followed by exposure to HF removed all but the linking sugar and amino acid. Fast atom bombardment mass spectrometry (FAB–MS) analysis of the pronase/HF product yielded an ion at m/z 295 (the expected MH^+ ion for Glc-Asn) and proton NMR confirmed that the product was indeed Glc-Asn. More detailed carbohydrate analysis of the short Glc-Asn-linked oligosaccharides indicated that approximately one third of the GlcA residues were substituted with iduronic acid (Wieland *et al.*, 1986).

The composition and structure of the large N-linked hetero-oligosaccharide was reported by Paul *et al.* in a pair of exceptional publications (Paul *et al.*, 1986; Paul and Wieland, 1987). The carbohydrate composition of this glycan was determined to be Gal, galacturonic acid (GalA), *N*-acetylglucosamine (GlcNAc), *N*-acetylgalactosamine (GalNAc) and 3–0-methylgalacturonic acid, all present in roughly stochiometric amounts (Paul *et al.*, 1986). It was determined that this hetero-oligosaccharide is N-linked through GalNAc. Furthermore, the amino acid sequence surrounding the linking Asn residue was partially determined. The second publication from these investigators reported the complete structure of the large hetero-oligosaccharide which is composed of a pentasaccharide repeating unit, 10–15 units long, and with two sulphates associated with each unit (Fig. 7.1C) (Paul and Wieland, 1987). At this time the amino acid sequence of the S-layer protein was also published and correlated well with the carbohydrate analysis described above (Lechner and Sumper, 1987). The mature protein contains 13 sites of potential N-glycosylation but only the site adjacent to the N-terminus (Asn-2) has the amino acid sequence shown to be associated with the large hetero-oligosaccharide (Paul *et al.*, 1986). The remainder of the N-glycosylation sites, with the exception of Asn-17, are occupied with the smaller sulphated tetrasaccharides (Fig. 7.1B). A cluster of 14 threonine residues located near the hydrophobic C-terminal tail is the probable location for most of the O-linked disaccharides (Fig. 7.1A).

The S-layer protein of a second haloarchaeon, *Haloferax volcanii*, has been the focus of an effort to elucidate the biosynthesis of its glycan modifications. In the first attempt to characterize the *H. volcanii* S-layer protein, Sumper *et al.* reduced the apparent molecular weight of the S-layer protein by treatment with anhydrous HF, a strong indication of the presence of carbohydrate (Sumper *et al.*, 1990). Edman sequencing of tryptic peptides provided sufficient information to generate PCR primers which in turn allowed the S-layer gene to be cloned and sequenced. The amino acid sequences of S-layer proteins from *H. volcanii* and *H. salinarum* are very similar except for the *N*-terminal region. Interestingly, *H. volcanii* S-layer protein also possesses a threonine-rich region near the hydrophobic C-terminus. On the other hand, *H. volcanii* S-layer protein has less N-linkage sequons (7 instead of 13). Furthermore, the investigators determined that the O-linked disaccharide observed on the high molecular weight (HMW) protein from *H. salinarum* was present here also (Fig. 7.1D). Beta-elimination in the presence of $NaBD_4$ liberated a disaccharide composed of Glc and Gal with the deuterium incorporated at the C-1 position of Gal indicating that the disaccharide is linked to the protein via Gal. Here again, Thr was determined to be the site of *O*-glycosylation. Threonine was converted to α–aminobutyric acid during the beta-elimination step to release the O-linked disaccharides. GC–MS detected α-aminobutyric acid in the amino acid pool obtained by acid hydrolysis of the deglycosylated protein.

Sumper and Mengele reported that the N-linked sites were modified with either a β1-4-Glc decamer, or oligosaccharides containing Glc, idose and Gal (Mengele and Sumper, 1992). However, more recent studies appear to contradict these findings although in the two studies cells were grown under very different Nacl concentrations. Abu-Qarn *et al* compared N-glycosylation of S-layer protein from WT and an archaeal glycosylation gene D (AglD) mutant of *H. volcanii* (Abu-Qarn *et al.*, 2007). The S-layer proteins were resolved by SDS-PAGE, in-gel digested and analysed by MS which revealed that the WT N-linked glycans consisted of a family of related oligomers ranging from a single hexose monosaccharide to a linear pentasaccharide with the following mass sequence, Asn–162 Da–176 Da–176 Da–190 Da–162 Da (Fig. 7.1E). Methyl esterification in 2 M HCl/MeOH revealed that the 176-Da sugar residues were hexuronic acids. The same analysis indicated that the 190-Da residue did not contain a free carboxylic acid and may be a dimethylated hexose or a methylated hexuronic acid. The terminal hexose residue is absent in the mutant strain indicating that AglD is a glycosyltransferase. More recently, the 190-Da residue was identified as a methylated hexuronic acid (Magidovich *et al.*, 2010) and the terminal hexose is thought to be mannose (Man). However, the full elucidation of the N-linked glycan has yet to be reported.

Methanogens

S-layer glycoprotein was extracted from the thermophilic methanogen, *Methanothermus fervidus*, and the glycans were split off by hydrazinolysis (Karcher *et al.*, 1993). After re-*N*-acetylation, reduction and multiple rounds of chromatography, a single N-linked oligosaccharide was isolated. Chemical analysis indicated that this glycan was composed of D-Man, D-3-*O*-methylmannose (and/or D-3-*O*-methylglucose) and D-GalNAc in a ratio of 3:2:1. NMR was able to detect the presence of methylmannose while GC analysis confirmed the presence of methylglucose in a 7:1 (methylmannose to methylglucose) ratio. Positive ion plasma desorption mass spectrometry (PD–MS) on a linear time-of-flight mass spectrometer (TOF–MS) gave a sodiated ion ($[M+Na]^+$) at m/z 1084.82 which corresponded to the expected hexasaccharide and minor ion at m/z 1436.03 corresponding to the glycan plus two additional methylhexoses. Extensive NMR analyses (1D and 2D TOCSY, H–H COSY, HMBC, ROESY, HMQC and HMQC-TOCSY) were performed on the purified oligosaccharide and its structure was determined to be α-D-3-*O*-methylmannose-(1–6)-α-D-3-*O*-methylmannose-((–1–2)-α-D-mannose)3-(1–4)-D-GalNAc-Asn (Fig. 7.1F).

The glycosylation on the flagellins of methanogens have been the focus of vigorous study. Flagellin from a variety of methanogens was tested for glycosylation using thymol blue-sulphuric acid which detects underivatized hexoses and pentoses (Kalmokoff *et al.*, 1992). The flagellin from *Methanococcus deltae* and *Methanothermus fervidus* stained positively for this type of glycosylation while flagellin isolated from other methanogens, including *M. maripaludis*, did not. However, the addition of bacitracin, an inhibitor of some forms of glycosylation in archaea, to the culture medium produced lower molecular weight flagellin proteins in *M. deltae* and *M. maripaludis* (Bayley *et al.*, 1993) suggesting that glycosylation was inhibited. The collective evidence suggested that flagellar glycosylation is common among the methanogens but that the nature of glycosylation varies from one species to another.

Some years were to pass before the first structure of a flagellar glycan on a methanogen was determined. Flagellin isolated from *Methanococcus voltae* was analysed by MS and NMR (Voisin *et al.*, 2005). *M. voltae* flagellin is composed of four flagellar proteins and good amino acid coverage was obtained for three of them, FlaB1, B2 and B3, by proteolytic digestion (trypsin and chymotrypsin) and nano-liquid chromatography–tandem mass spectrometry (nLC–MS/MS) analysis. Two peptides (one of them glycosylated) were also detected for FlaA, the least abundant component of the flagellin. Glycopeptides covering 15 of the 16 N-linked sequons in these proteins were detected. Mass spectrometry determined that the glycan has a mass of 779 Da, is composed of three residues and that it is N-linked. MS/MS and exact mass analysis of the glycan oxonium ions indicated that the linking sugar is a HexNAc (203 Da) and that the middle residue is a di-*N*-acetylhexuronic acid (256 Da). However, the terminal sugar (318 Da) was more complex and it was not possible to hypothesize a structure from the mass spectrometry data alone. Therefore, a large amount of the glycan was isolated from the flagellin by pronase digestion and size exclusion chromatography and exhaustively analysed by NMR (^{1}H NMR, TOCSY, NOESY, COSY and ^{1}H-^{13}C-HMBC). The terminal sugar was determined to be a *N*-acetylated mannuronic acid substituted at the C-6 position with threonine (ManNAcA6Thr). Thus, using a combination of MS and NMR, the glycan structure was determined to be β-ManpNAcA6Thr-(1–4)-β-GlcpNAc3NAcA-β-GlcNac-Asn (Fig. 7.1G). Furthermore, it was also shown that at least one of the two N-linkage sites on the S-layer protein from *M. voltae* is modified with the same glycan (Voisin *et al.*, 2005). A peptide or glycopeptide containing the second sequon was not detected. More recently, the flagellar glycan from a closely related strain of *M. voltae* was determined to be a tetrasaccharide composed of the same trisaccharide but with an extra 220- or 262-Da moiety attached to the ManNAcThr residue (Chaban *et al.*, 2009) (Fig. 7.1G). The extra residues are believed to be carbohydrates but their structures have yet to be determined.

The structure of the glycan on the flagellar proteins from *M. maripaludis* was reported recently (Kelly *et al.*, 2009). Here again MS and NMR were used exclusively to elucidate the glycan composition and structure. MS analysis revealed that the flagellar N-linkage sites were modified with a 1036 Da glycan. The glycan MS/MS spectra, though difficult to interpret, demonstrated that the glycan is a tetrasaccharide: a linking HexNAc (203 Da), a di-*N*-acetylhexuronic acid (258 Da), a 358-Da sugar and a terminal 217-Da sugar. The glycan was isolated in the same manner as for *M. voltae* and NMR analysis determined the structure to be Sug-β-ManNAc3NAmA6Thr-4-β-GlcNAc3NAcA-3-β-GalNAc-Asn where Sug is the novel monosaccharide, (5S)-2-acetamido-2,4-dideoxy-5-*O*-methyl-a-L-erythro-hexos-5-ulo-1,5-pyranose (Fig. 7.1H). *M. maripaludis* also express glycosylated pilin structures on the outer surface. Pilin from a non-flagellated mutant of *M maripaludis* was analysed by genetic mutational studies and by MS (Ng *et al.*, 2011). The protein components of this pilin were initially expected to be MMP0233, 0236 and 0237 based on previous *in silico* analysis (Szabo *et al.*, 2007) Indeed, deletion of MMP2036 and 0237 resulted in no pili while deletion of MMP0233 resulted in greatly reduced numbers of pili on the cell surface. However, using a variety of 'top-down' and 'bottom-up' MS analyses, the major component of the pilin was determined to be another protein, MMP1685. Furthermore, this protein is N-glycosylated with a pentasaccharide composed of the same tetrasaccharide observed on the *M. maripaludis* flagellar proteins but with an extra hexose appended as a side-branch to the linking GalNAc (Fig. 7.1H). Unfortunately, NMR analysis was not feasible in this instance as only small quantities of the pilin could be isolated from the mutant strain.

Protein glycosylation in other methanogens has also been reported. For example, *in vivo* biotin tagging and MS were used to identify outer membrane proteins in *Methanosarcina acetivorans* and *Methanosarcina mazei*, two methanogenic archaeons that are widely distributed in the environment and play important roles in the decay and recycling of organic matter (Francoleon *et al.*, 2009). In all, 12 and 17 proteins were confidently identified in *M. acetovirans* and *M. mazei*, respectively, including the putative S-layer proteins. Many of the isolated proteins stained with Pro-Q Emerald, an indicator of glycosylation, and reacted with ConA, a lectin that binds to terminal glucose and/or mannose. The staining patterns were complex. For example, the S-layer proteins from both species were resolved into multiple bands by SDS-PAGE but not all of the bands reacted with Pro-Q Emerald or with ConA, indicating variation in glycosylation.

Thermophiles and acidophiles

The glycoproteins produced by a number of thermophilic and acidophilic archaeons have also been investigated. Cytochrome b588/566 from *Sulfolobus acidocaldarius*, a crenarchaeon that grows optimally at pH 2–3 and temperatures of 75–80°C, was shown to be glycosylated (Hettman *et al.*, 1998). The mass of the gene-derived sequence is 50,526 Da and contains 16 potential N-linkage sites (NXT, NXS) as well as many serines (Ser) and Thr residues suitable for O-linkage. The apparent mass of the purified cytochrome, as determined by SDS-PAGE, was approximately 65 kDa and treatment with alkali reduced the mass by about 12%, a good indication of significant glycosylation. Matrix-assisted laser desorption ionization–time-of-flight mass spectrometry (MALDI–TOF–MS) of the intact glycoprotein gave a MH^+ ion at approximately 64,210 Da while chemical deglycosylation using a commercial kit for *N*- and *O*-glycan removal gave an apparent mass of approximately 55 kDa. Total sugar analysis by GC (as alditol acetates) identified Man, Glc and GlcNAc in a 7:2:2 ratio while the N-linked fraction contained Glc, Man and GlcNAc in roughly stoichiometric amounts. On the other hand, the O-linked sugar fraction yielded only mannose. No attempt was made to characterize the O-linked modifications or to identify their linkage sites.

The structure of the N-linked glycan was determined using classic carbohydrate analysis combined with modern MS and NMR technologies (Zahringer *et al.*, 2000). The N-linked glycan was purified from the cytochrome either by proteolysis and high performance anion exchange chromatography (HPAEC) or cleavage with PNGase F and size exclusion chromatography. GC analysis of the sugars (as alditol acetates) derived from the PNGase F product revealed the presence of Glc, Man and GlcNAc in a 1:2:2 ratio. However, electrospray ionization mass spectrometry (ESI–MS) in negative ion mode (–ve) gave a mass at m/z 1135.9 indicating an additional sugar residue with a mass of 226 Da. NMR confirmed the presence of a hexasaccharide. MALDI–TOF–MS analysis of one of the glycopeptides isolated by proteolysis gave an ion at m/z 1412.8 corresponding to this oligosaccharide plus Asn and tyrosine (Tyr). A full NMR analysis was performed on the proteolytic glycopeptides which confirmed the identity of the 6-sulfo-quinovose and determined the structure of the tribranched, N-linked hexasaccharide (Fig. 7.1I).

It has been demonstrated that *S. acidocaldarius* produces other glycoproteins (Michel *et al.*, 1980; Lin and Tang, 1990) as do other *Sulfolobus* species (Faguy *et al.*, 1996). Recently, the same N-linked hexasaccharide identified on cytochrome b588/566 was also observed on the S-layer glycoprotein (Peyfoon *et al.*, 2010). LC–MS/MS identified tryptic glycopeptides corresponding to 9 of the 11 N-linkage sites in the C-terminal region of this glycoprotein. Interestingly, there is evidence that each site was substituted with a family of glycans corresponding to the hexasaccharide in various stages of depolymerization.

Soon after the first reports of prokaryotic protein glycosylation were published an attempt was made to characterize the glycosylation on membrane proteins from another thermophilic and acidophilic archaeon, *Thermoplasma acidophilum* (Yang and Haug, 1979). The most abundant membrane glycoproteins were isolated and purified using ConA lectin chromatography. After an exhaustive analysis, the investigators concluded that the protein was N-glycosylated with very large glycans composed mostly of mannose, which were linked to Asn residues via a chitobiose core (two GlcNAc residues), in the same manner as in eukaryotes. However, this result has never been confirmed and, indeed, recent data generated in our laboratory contradict the proposed structure.

Biosynthesis and assembly of the glycans

Significant progress has been made in identifying genes encoding glycosyltransferases and oligosaccharyltransferases used in the assembly and transfer of the *N*-glycans in the model organisms, *M. voltae, M. maripaludis* and *H. volcanii* (Yurist-Doutsch *et al.*, 2008b; Calo *et al.*, 2010b; Jarrell *et al.*, 2010a,b; Yurist-Doutsch *et al.*, 2010). To date these are the only Archaea in which links between the glycan structures and specific gene knockouts have been reported. In other archaea, there are N-linked glycan structures determined but no genetic work is available on the genes involved in either assembly or biosynthesis of the glycans. Unfortunately many archaea simply lack any tractable genetic system. The almost universal presence of *agl* (genes involved in archaeal glycosylation as first proposed by Chaban *et al.* (2006b)) loci in archaeal genomes (Magidovich and Eichler, 2009) suggest that N-linked glycosylation probably is a very common post-translational modification in archaea. Among the three model organisms, in addition to the glycosyltransferases and oligosaccharyltransferase that together assemble and transfer the glycan to the target proteins, some progress has been made in identifying other gene products that play a role in the biosynthesis and modification of the glycan component sugars (as discussed in detail below). Some of this work has coupled *in vitro* biochemical assays with genetic knockouts and MS of the mutant glycans. However, this work is in its infancy.

Oligosaccharyltransferase AglB

As in eukaryotic cells and bacteria, the N-linked glycan assembled on a lipid carrier is transferred to Asn residues within a consensus sequon by an oligosaccharyltransferase, designated AglB in Archaea, and homologous to PglB in Bacteria and the STT3 subunit of the oligosaccharyltransferase complex in the Eukarya. Unlike the oligosaccharyltransferase complex in eukaryotes, AglB, like PglB, functions as a single component. Like the STT3 subunit and PglB, AglB contains several membrane-spanning domains at the N-terminus while the conserved WWDYG motif implicated in catalysis is found in the C-terminus. AglB can efficiently transfer both fully assembled glycans as well as partial glycans formed in various mutant strains unable to assemble the complete version of the glycan.

Recent analysis of the oligosaccharyltransferases from the three domains indicates three different possible catalytic domains, a bacterial version (B-type), a eukaryotic version (E-type) and an archaeal A type (Maita *et al.*, 2010). The A type is only found in some Archaea (some Euryarchaeota, such as *H. volcanii*), while other archaea can have either the B type (other Euryarchaeota, such as *Methanococcus* species) or E type (various Crenarchaeota, Korarchaeota, Nanoarchaeota and Thaumarchaeota as well as some Euryarchaeota, like *Pyrococcus*) catalytic centres.

There appears to be significant flexibility in the substrates recognized by the archaeal oligosaccharyltransferase, perhaps best exemplified in *H. salinarum*. Here, two very different N-linked glycans can be attached to the S-layer protein, a repeating unit type attached to a single position at the very N-terminus of the protein (Asn-2) and an oligosaccharide glycan attached to several other sites (Sumper, 1987). The completely sequenced genome reveals only a single oligosaccharyltransferase gene, suggesting this one enzyme is capable of attaching either glycan. Additionally, it has been shown that this enzyme is also flexible in the recognition of the N-linked glycan, Asn-X-Thr/Ser since site-directed mutagenesis of the repeating unit glycan attachment site from Asn-X-Ser to Asn-X-Val or Asn-X-Leu or Asn-X-Asn did not prevent proper glycan attachment (Zeitler *et al.*, 1998). In addition, the truncated glycans found in many *agl* gene mutants, even glycans composed of a single sugar, are efficiently transferred to the target proteins in both *Methanococcus* (Chaban *et al.*, 2006b; Vandyke *et al.*, 2009) and *Haloferax* (Abu-Qarn and Eichler, 2006; Abu-Qarn *et al.*, 2007).

Knockouts of *aglB* in *M. voltae* (Chaban *et al.*, 2006b), *M. maripaludis* (Vandyke *et al.*, 2009) and *H. volcanii* (Abu-Qarn and Eichler, 2006; Abu-Qarn *et al.*, 2007) have all shown evidence that this gene product is responsible for transfer of the glycan to the target proteins. In such knockouts, the target proteins, flagellins and S-layer migrate as much smaller proteins than those of wild-type cells. However, it is only so far in *Pyrococcus furiosus* that an *in vitro* assay has shown directly that AglB can transfer glycan to a target peptide containing the N-linked sequon (Igura *et al.*, 2008). The ability to generate *aglB* mutants in different archaea shows that the N-linked system is not essential for cell viability although disruption of the system can have deleterious effects on the cells.

Below the assembly and biosynthesis of N-linked glycans in the two best-studied model systems, *Methanococcus* and *Haloferax*, are reviewed.

Methanococcus

Glycosyltransferases

A combination of genetic, biochemical and MS analyses have led to the identification of most of the glycosyltransferases used to assemble the three sugar glycan N-linked to flagellins and the S-layer protein of *M. voltae* and the four sugar glycan N-linked to flagellins of *M. maripaludis* (Voisin *et al.*, 2005; Chaban *et al.*, 2006b, 2009; Shams-Eldin *et al.*, 2008; VanDyke *et al.*, 2008, 2009; Kelly *et al.*, 2009). The first archaeon in which genes involved in N-linked glycosylation were identified was *M. voltae*. Besides the oligosaccharyltransferase, the initial publication identified a glycosyltransferase involved in attachment of the terminal sugar (Chaban *et al.*, 2006b). Inactivation of this gene, *aglA*, resulted in flagellins and S-layer protein that migrated faster in SDS-PAGE and western blots. MS analysis of isolated flagella showed that this mobility shift was caused by the loss of the terminal sugar. A subsequent publication described the inactivation of a pair of contiguous genes (*aglC* and *aglK*), both of which were annotated as glycosyltransferases and both of which were involved in the biosynthesis/transfer of the second sugar (Chaban *et al.*, 2009). Unlike the *aglA* mutants, these mutants were unable to assemble flagella. The glycosyltransferase responsible for the attachment of the first sugar, GlcNAc, to the dolichol lipid carrier in the first step on the N-glycosylation pathway was identified in a different way (Shams-Eldin *et al.*, 2008). The candidate gene, designated *aglH*, was the only one in the *M. voltae* genome that belonged to Pfam family PF00953 (glycosyltransferase family 4), the family that includes the eukaryotic gene *alg7*. Alg7 is the GlcNAc-1-phosphate transferase that attaches GlcNAc to the dolichol lipid carrier in eukaryotic cells. Repeated attempts to inactivate *aglH* were unsuccessful but evidence that *aglH* did encode the first glycosyltransferase in the N-glycosylation pathway in *M. voltae* was obtained when it was shown that the archaeal gene could complement a conditionally lethal mutation in *alg7* in the yeast, *Saccharomyces cerevisiae*.

In the related methanogen, *M. maripaludis*, genetic tools were introduced for in-frame deletion and complementation studies that were not available in *M. voltae* (Moore and Leigh, 2005) and studies on N-glycosylation continued in this organism. In *M. maripaludis*, glycosyltransferases required for the attachment of the final three sugars of the four sugar glycan were found clustered in one area of the chromosome, although on separate transcripts (Kelly *et al.*, 2009; Vandyke *et al.*, 2009). These genes, designated *aglO*, *aglA* and *aglL*, were separated by other genes, some of which were later shown to be involved in the biosynthesis and modification of the component sugars. Deletion of any of *aglO*, *aglA* and *aglL* led to a decrease in the molecular mass of the flagellins compared with wild-type cells and all deletion mutants could be successfully complemented back to wild type with plasmid-borne versions of the deleted genes. In order, the flagellin masses decreased in size from wild type to the *aglL*, *aglA* and finally to the *aglO* mutant version. The flagellins of the *aglO* mutants were slightly larger than those of an *aglB* mutant, which would represent completely non-glycosylated flagellins. While all the mutants in *agl* genes still make flagellin subunits, the extent of truncation of the attached glycan determines whether the flagellins are incorporated into flagella structures. The *aglL* and *aglA* mutants were still able to assemble flagella while mutants with further truncation of the glycan (*aglO* and *aglB*) were non-flagellated. MS analysis of flagella from the *aglL* mutant identified two defects in its glycan; it lacked the terminal sugar but also, surprisingly, was lacking the Thr attached to the third sugar. This suggested that AglL may be responsible for either (or both) of these differences. However, since AglL shows strong sequence homology to known glycosyltransferases, it was reasoned that it was responsible for the attachment of the unusual terminal sugar. If this is true, it further suggested that the Thr is added to the third sugar after the transfer of the terminal sugar and if this final sugar is not present, then the third sugar cannot be modified with the threonine. Genes involved in the Thr attachment have not been identified. In the case of the *aglA* mutant, MS analysis of isolated flagella indicated that only the first two sugars were left of the glycan, indicating AglA was the glycosyltransferase responsible for the transfer of the third sugar. This enzyme has 55% similarity to the *M. voltae* AglA responsible for catalysing an almost identical reaction. Since *aglO* mutants do not make flagella, these could not be analysed for glycan defects. However, from western blot analysis, flagellins from this mutant migrate at a molecular mass intermediate between flagellins with no glycan attachment (in the *aglB* mutant) and flagellins of the *aglA* mutant (with a two sugar glycan). Thus, it is believed that the *aglO* mutants possess only a single sugar, GalNAc, attached to their flagellins and AglO is the glycosyltransferase responsible for transfer of the second sugar. In *M. maripaludis*, the linking sugar is GalNAc and not GlcNAc (Kelly *et al.*, 2009) as found in *M. voltae* (Voisin *et al.*, 2005). The glycosyltransferase responsible for this initial step in the N-glycosylation pathway in *M. maripaludis* is currently unknown.

Recently, the main pilus structural protein, encoded by MMP1685, was identified and shown to possess an N-linked glycan attached at multiple sites. Unexpectedly, this glycan is not identical to the one attached to flagellins. It is modified in having a hexose branch attachment to the GalNAc. To date the glycosyltransferase responsible for this attachment has not been identified. How the cell differentiates between the flagellins and pilins in the attachment of the slightly different glycan and whether this extra sugar is added after the transfer of the four-sugar 'flagellin' glycan to the pilin has already occurred is also unknown.

Biosynthesis genes

In addition to the glycosyltransferases and the oligosaccharyltransferase used to assemble and attach the glycan to target proteins, progress is being made in the identification of genes and proteins involved in the biosynthesis of the component sugars of the N-linked glycan in the methanococci. Some genes have been identified, as with the glycosyltransferases, by inframe deletion analysis and MS to confirm the exact nature of the altered glycan. The first gene so identified was *MMP0350*, annotated as an acetyltransferase and targeted because both the first and second sugars in the glycan are acetylated sugars (Vandyke *et al.*, 2008). Deletion of *MMP0350* led to flagellins which migrated in western blots as much smaller proteins than the wild-type flagellins, initially suggesting either completely non-glycosylated forms or ones that had a very truncated glycan attached. Subsequent comparisons to the flagellins of the various glycosyltransferase and oligosaccharyltransferase mutants showed that the flagellins of the *MMP0350* mutant were the size of the flagellins in the *aglO* mutant (Vandyke *et al.*, 2009). This would indicate that the glycan on the flagellins of the *MMP0350* mutant was composed of a single sugar, suggesting that the *MMP0350*-encoded acetyltransferase was probably responsible for one of the acetyl groups found on the di-acetylated second sugar. MMP0350 appears to transfer the acetyl group to the C-3 position of the di-acetyl glucuronic acid as it has been shown in *in vitro* assays that the purified gene product of *MMP1076* (a glucosamine-1-phosphate uridyltransferase/acetylase), acetylates the C-2 position of glucosamine-1-phosphate to generate *N*-acetylglucosamine (Namboori and Graham, 2008). The absence of the MMP0350-derived modification appears to prevent the subsequent addition of the second sugar to the glycan chain (Vandyke *et al.*, 2009).

Other genes in the same locus as the three glycosyltransferases identified have recently been shown to modify the third and terminal sugars. A deletion in *MMP1081* led to smaller molecular mass flagellins (as detected by western blotting) and MS analysis of flagellins from this mutant showed that the glycan was missing the terminal sugar and the acetamidino group of the third sugar. Since *MMP1081* has homology to bacterial proteins involved in acetamidino transfer, it seems that the function of MMP1081 is in the transfer of this group and, although this undermodified sugar is added to the glycan, the subsequent addition of the final sugar is prevented. Finally, deletion of another gene in this region, *MMP1085* which is annotated as a methyltransferase, leads to a very small decrease in flagellin molecular mass. MS analysis of this mutant's flagellins indicated the sole defect to be a lack of the methyl group normally found on the terminal sugar.

Genes predicted to encode enzymes involved in acetamido sugar biosynthesis in the Methanococcales were identified by Namboori and Graham (2008). The genes were expressed in *E. coli*, the enzymes then purified and used in *in vitro* assays designed to test for their predicted functions. Identified in that key study as enzymes involved in the biosynthesis of acetamido sugars were MMP1680, MMP1076, MMP1077, MMP0705 and MMP0706. MMP1680 catalysed the transamidation and epimerization reactions that converted fructose-6-phosphate and L-glutamine to glucosamine-6-phosphate and glutamate. MMP1077 is a phosphomutase which converts glucosamine-6-phosphate to glucosamine-1-phosphate while its neighbouring gene *MMP1076* encodes a protein with two functions. It acts first to acetylate the glucosamine-1-phosphate and then it transfers the sugar to UTP to generate UDP-GlcNAc. Two other neighbouring genes, *MMP0705* and *MMP0706*, serve to convert UDP-GlcNAc to UDP-*N*-acetylmannosaminuronate (UDP-ManNAcA). In this pathway, the direct precursors to sugars found in the N-linked glycan are produced, although genetic knockout studies specifically linking these enzymes to a role in N-linked glycosylation is lacking. Given the homology of MMP1680 to bacterial GlmS, MMP1076 to GlmU and MMP1077 to GlmM, it appears that methanogens use the bacterial pathways to generate UDP-N-GlcNAc and UDP-ManNAcA, the precursors to sugars found in the N-linked glycan of several surface proteins. Interestingly, outside of their presence in N-linked glycans, these sugars may be part of the structures of glycolipids and the methanogenic cofactor, coenzyme B. Coenzyme

B is an essential cofactor in methanogenesis and was reported to contain a ManNAcA β-1,4-UDP-GlcNAc disaccharide attachment (Sauer *et al.*, 1990). If that glycan portion is necessary for function, then enzymes involved in the biosynthesis of the disaccharide components would be essential. Interestingly, attempts in our laboratory to delete either *MMP0706* or *MMP0705* have been unsuccessful.

Impact of disruption in the N-linked system

The importance of N-glycosylation has been studied in only a few Archaea for which reports of *N*-glycosylated proteins and a suitable genetic system to generate mutants in the pathway are available. To date, this has occurred only in *M. voltae, M. maripaludis* and *H. volcanii*. In all three of these model organisms it has been possible to generate mutants in *aglB*, showing that the N-glycosylation system is not essential for cell survival. However this does not mean that interference in the pathway does not have significant measurable effects on the cell.

In both *M. voltae* and *M. maripaludis*, interesting observations have been made concerning the effect of attachment of partial glycans or no glycan at all to the flagellins. Mutants in *M. voltae* that have insertions in *aglA*, responsible for attachment of the last of the three sugars in the glycan are still able to assemble flagella (Chaban *et al.*, 2006b) while insertional inactivation of *aglC* or *aglK*, both required for attachment/biosynthesis of the second sugar, are non-flagellated (Chaban *et al.*, 2009). Insertional inactivation of the *aglB* gene, leaving the flagellins completely non-glycosylated, also resulted in non-flagellated cells (Chaban *et al.*, 2006b). Seemingly, attachment of a glycan composed of at least the first two sugars is necessary for the flagellins to assemble stably into a flagellar filament. This observation was confirmed and extended in subsequent studies on the related organism, *M. maripaludis*. This organism has a four sugar glycan and mutants carrying deletions in *aglA* or *aglL* encoding glycosyltransferases which attach the third and fourth sugar, respectively, are still able to assemble flagella. On the other hand, deletions in the glycosyltransferase that attaches the second sugar (*aglO*) or in the oligosaccharyltransferase (*aglB*) cannot assemble flagellar filaments (Vandyke *et al.*, 2009). Among the glycosylation-defective mutants that can assemble flagella, those mutants with a more complete glycan can swim farther in swarm agar (0.25%) (Vandyke *et al.*, 2009). A deletion in *MMP0350*, whose product is responsible for attachment of an acetyl group to the diacetylated second sugar of the glycan, leads to a glycan of apparently only one sugar and the resulting strain is non-flagellated (Vandyke *et al.*, 2008). Why a minimum of two sugars in the glycan is needed for assembly of flagella is unknown. Two possibilities are that the assembly mechanism at the base of the flagellum recognizes only flagellins with this basal glycan modification or that all flagellins, fully modified or not, are initially sent through the assembly apparatus but only flagellins with at least the minimum two-sugar glycan remain assembled in a filament.

In the case of *M. voltae*, defects in glycosylation also result in easily detected-changes in the migration of the S-layer protein which can be easily seen in Coomassie stained gels of total cell lysates (Chaban *et al.*, 2006b). The S-layer protein here has two potential modification sites with at least one carrying the same glycan as the flagellins. No such change in S-layer migration was observed in the case of the *M. maripaludis* glycosylation mutants. In this organism there is only one potential *N*-glycan attachment site and it is currently not known whether this lone site is occupied.

Other earlier reports had indicated an important role for glycosylated flagellins in flagella structure and attachment in Archaea, although in these early reports there was no genetic or structural data to complement the observations. In *H. salinarum*, a mutant that overproduced flagella and shed them into the media was found to be producing flagellins of lower molecular weight compared with wild-type cells, suggestive of underglycosylation. Treatment of cultures of *M. deltae* and *M. maripaludis* with bacitracin led to flagellins shifted to a much lower apparent molecular weight, suggestive of underglycosylation or non-glycosylation, and the cells became non-flagellated (Bayley *et al.*, 1993).

In addition to effects on flagellation, the

deletion of *MMP0350* also had effects on the type IV-like pili of *M. maripaludis* (Vandyke *et al.*, 2008). *MMP0350* mutant cells typically lacked attached pili, although apparently fully formed pilus structures were found in the culture supernatants. This suggested that the glycan defect had not interfered with pilus formation *per se* but had instead prevented proper attachment of the formed pili to the cell surface. This study suggested a more diverse role for the N-linked glycan in expression of the surface structures of Archaea than previously assumed.

A summary of the effects of the deletions of various genes involved in the synthesis and assembly of the N-linked flagellin glycan in *M. maripaludis* is presented in Table 7.1.

Haloferax volcanii

Glycosyltransferases

As mentioned earlier, the S-layer protein of *H. volcanii* is modified at least at two positions, Asn-13 and Asn-83 (out of seven potential sites), with a pentasaccharide linked via a hexose followed by two hexuronic acids, a 190-Da sugar (subsequently identified as a methyl ester of a hexuronic acid) and, finally, another hexose, which has recently been identified as mannose (Abu-Qarn *et al.*, 2007; Calo *et al.*, 2010b; Kaminski *et al.*, 2010; Magidovich *et al.*, 2010). With the exception of *aglD,* all genes known to be involved in the biosynthesis and assembly of the glycan, including the olgosaccharyltransferase *aglB,* are located in a single large locus (Yurist-Doutsch and Eichler, 2009). A number of the gene products have been identified by deletion and complementation analyses, coupled with specific glycan staining and MS of S-layer protein, as the specific glycosyltransferases involved in assembling the pentasaccharide. AglD is responsible for attachment of the terminal mannose, AglE for the addition of the 190-Da fourth sugar and AglI and AglG for addition of the hexuronic acids at positions 3 and 2, respectively (Abu-Qarn *et al.*, 2007, 2008b; Yurist-Doutsch *et al.*, 2008a). Recently, evidence has emerged that AglJ is the glycosyltransferase responsible for transfer of the first, linking, hexose (Kaminski *et al.*, 2010). The exact nature of this linking hexose remains unknown (Kaminski *et al.*, 2010). As with the methanogens, deletions in each of the glycosyltransferase genes, except for *aglE,* led to increased migration of the S-layer protein in SDS-PAGE (Jarrell *et al.*, 2010a). However, unlike the case in *Methanococcus* flagellins, a logical stepwise decrease in molecular mass of the *H. volcanii* S-layer protein correlating to the decreased size of the attached glycan has not been observed (Abu-Qarn *et al.*, 2007, 2008b). Indeed, the greatest shifts in mass of the S-layer protein were observed in *aglD* mutants where only the terminal sugar is missing (Abu-Qarn *et al.*, 2007).

More detailed work on the terminal glycosyltransferase, AglD, has recently been reported. Deletions of the *aglD* gene lead to a faster migrating S-layer protein upon SDS-PAGE analysis as well as one that does not stain with the periodic acid–Schiff reagent. Complementation of the mutant with *aglD in trans* leads to the restoration of the wild-type S-layer protein size in SDS-PAGE as well as its wild-type glycostaining properties. Complementation of the *H. volcanii aglD* mutant to wild-type status could also be attained when the complementing gene was a homologue of *aglD* from several different archaeal extreme halophiles, such as *H. salinarum, Haloarcula marismortui* and *Haloquadratum walsbyi* (Calo *et al.*, 2010a). The presence of wild-type glycan was confirmed by MS of the S-layer protein in these heterologously complemented mutants.

Using this same glycosyltransferase, Eichler's group examined, for the first time in Archaea, key residues important for its activity (Kaminski and Eichler, 2010). Utilizing the complementation protocol outlined above they complemented an *aglD* mutant with both a wild-type version of the gene as well as various site-directed mutant versions looking for restoration of wild-type S-layer migration in SDS-PAGE as well as glycostaining. AglD is a member of the GT-A fold containing glycosyltransferases and, as such, has a signature DXD divalent cation interaction motif. Asp-110 and Asp-112 were shown to be the elements of the DXD motif in AglD and to be crucial for glycosyltransferase activity. Furthermore, Asp-201 was identified as the probable catalytic base for the enzyme as complementation of *aglD* mutants with D201A or D201N versions of the

Table 7.1 Effects of *agl* gene deletions on flagella glycosylation in *M. maripaludis*

Gene deletion[1]	Function	Glycan structure	MW shift of flagellin	Flagella	Effect on motility
aglL	4th GTase	Missing 4th sugar and Thr on 3rd sugar	Yes	Yes	Decreased compared with WT
aglA	3rd GTase	Missing 3rd and 4th sugars	Yes	Yes	Decreased compared with *aglL* mutant
aglO	2nd GTase	Missing 2nd, 3rd and 4th sugars	Yes	No	Non-motile
aglB	OTase	No glycan	Yes	No	Non-motile
MMP0350	Acetyl transferase	Missing 2nd, 3rd and 4th sugars	Yes	No	Non-motile
MMP1081	Acetamidino transfer	Missing 4th sugar and acetamidino group of 3rd sugar	Yes	Yes	Decreased compared with WT
MMP1085	Methyl transferase	Missing methyl group of 4th sugar	Yes	Yes	To be determined

[1]References for each of these gene experiments are cited in the text.

GTase, glycosyltransferase; OTase, oligosaccharyltransferase; MW, molecular weight; WT, wild-type.

enzyme did not restore the S-layer SDS-PAGE migration or glycostaining to the wild-type phenotype. Also crucial for activity of AglD is the conserved Arg-139, although other conserved amino acids tested (e.g. Asp-133 and Gly-177) were not.

Biosynthesis genes

Currently, a few of the gene products predicted to be involved in biosynthesis of the S-layer glycan in *H. volcanii* have been analysed by *in vitro* assays in support of genetic studies aimed at deleting the genes as well as MS analysis of the glycans produced in these mutants. The gene products involved in these studies are AglF, AglM and AglP. In the case of AglP, deletion of the gene led to a modified S-layer glycan lacking the methyl group on the fourth sugar of the glycan (Yurist-Doutsch and Eichler, 2009; Magidovich *et al.*, 2010). The *aglP* gene product was purified and an assay developed to demonstrate it was acting as an S-adenosyl-L-methionine-dependent methyltransferase, as had been initially predicted using bioinformatics analysis, and was responsible for adding the methyl group to the hexuronic acid at position 4 in the glycan. Interestingly, this addition of a methyl group to the fourth sugar appears to be a critical step needed for attachment of the terminal sugar since cells deleted for *aglP* are not only missing the methyl group on sugar 4 of the glycan but are also missing sugar 5 as well. In a similar fashion, purification of AglM and AglF was followed by *in vitro* assays that showed AglM was a UDP-glucose dehydrogenase, generating UDP-GlcA from UDP-Glc while AglF was able to generate UDP-Glc from Glc-1-phosphate and UTP in a NAD-dependent manner, functioning as a UDP-glucose dehydrogenase (Yurist-Doutsch *et al.*, 2010). In the presence of both proteins with the addition of Glc-1-phosphate, UTP and

NAD, UDP-GlcA was made. The *in vitro* analysis of halobacterial gene products involved in the N-linked glycosylation process go hand in hand with similar studies in *Methanococcus* by Namboori and Graham as important first steps towards development of a complete *in vitro* generation of the archaeal N-glycosylation process. In both *Methanococcus* and *Haloferax*, truncated glycans are still transferred to target proteins with high efficiency, indicating that both the proposed but still unidentified flippase (responsible for transferring the glycan-dolichol phosphate lipid carrier from the cytoplasmic side of the cytoplasmic membrane to the external side) as well as the oligosaccharyltransferase have a relaxed specificity and can recognize lipid carriers loaded with even a single sugar (Chaban *et al.*, 2006b; Abu-Qarn *et al.*, 2007; Vandyke *et al.*, 2009; Calo *et al.*, 2010b; Kaminski *et al.*, 2010).

Recently, important information has been reported on the lipid carriers involved in assembly of the *H. volcanii* glycan (Guan *et al.*, 2010). Here, dolichol phosphate linked to sugars one, two, three and four of the glycan were identified but no dolichol phosphate modified with the entire pentasaccharide was found. Instead, the terminal mannose was bound to its own dolichol phosphate carrier, indicating distinct dolichol phosphate carriers modified with different parts of the final glycan are involved in the assembly of the glycan, a situation also seen in eukaryotic cells. Interestingly, the glycosyltransferase responsible for transfer of the mannose subunit is encoded by *aglD*, thus far the only gene identified in the *H. volcanii* N-linked system that is not contained within a large *agl* gene cluster (Yurist-Doutsch and Eichler, 2009). This finding may indicate that the final sugar is transferred after the tetrasaccharide version of the glycan has already been transferred to the target protein, a hypothesis supported by the previous finding of tetrasaccharide-modified S-layer peptides.

Impact of disruption of the N-linked system

Early experiments in *H. salinarum* had already suggested an important role for glycosylation in cellular structure. Growth in the presence of bacitracin led to the normally rod-shaped cells instead growing as spheres, suggesting a role of the glycan attached to the S-layer protein in structural support of the cell shape (Mescher and Strominger, 1976a). These kinds of studies on the effects of glycosylation in extreme halophiles have been greatly expanded in the study of *H. volcanii* and its many *agl* gene mutants. In *H. volcanii*, the main reporter protein studied in regards to N-glycosylation has been the S-layer protein and interference in its normal glycosylation has significant effects on a number of properties related to this protein. It has been suggested that the glycosylation of the S-layer protein in this organism may be an adaptive response as transcription of various *agl* genes differs under varying growth conditions (Yurist-Doutsch *et al.*, 2008a, 2010). Unlike the case in *M. voltae*, mutants carrying deletions in genes involved in N-linked glycosylation did not always demonstrate S-layers which migrated faster on SDS-PAGE. The S-layer protein of *H. volcanii* has seven potential sites for attachment of N-linked glycan and it also has O-linked glycans. At least two N-linked sites are occupied, Asn-13 and Asn-83. In some *agl* gene deletion mutants, such as *aglD* and *aglB*, growth defects in media containing higher salt concentrations were observed (Abu-Qarn *et al.*, 2007) but these were not seen in other *agl* mutants (such as *aglG, F,E* and *I*) (Abu-Qarn *et al.*, 2008b; Jarrell *et al.*, 2010a). Defective N-glycosylation was shown to lead to an unstructured S-layer on the cell surface while in *aglB* mutants that lacked N-glycan attachment completely, S-layer stability was compromised (Abu-Qarn *et al.*, 2007). Finally, the resistance of the S-layer to proteolysis was lost in many of the *agl* gene mutants isolated in *H. volcanii* (Abu-Qarn *et al.*, 2007; Jarrell *et al.*, 2010a; Yurist-Doutsch *et al.*, 2010).

A summary of the effects of the deletions of various genes involved in the synthesis and assembly of the N-linked S-layer glycan in *H. volcanii* is presented in Table 7.2.

Conclusion and future directions

We are learning more about the Archaea every day. They are far more numerous than was once thought and they are not solely restricted to

Table 7.2 Effects of *agl* gene deletions on S-layer glycosylation in *H. volcanii*

Gene deletion[1]	Function	Glycan structure	Growth in high salt	MW shift of S-layer	Susceptibility of S-layer to proteases	Shedding of S-layer	S-layer symmetry
aglB	OTase	No glycan	Deficient	Yes	Same as WT	Increased	No effect
aglD	5th GTase	Missing 5th sugar	Deficient	Yes	Less than WT	Decreased	Symmetry defect
aglE	4th GTase	Missing 4th and 5th sugars	No effect	No	Same as WT	Same as WT	No effect
aglF	Glucose-1-P uridyltransferase (sugar 3)	Mix of mono- and disaccharides	No effect	Yes	More than WT	Same as WT	No effect
aglG	2nd GTase	Missing last four sugars	No effect	Yes	More than WT	Same as WT	No effect
aglI	3rd GTase	Mix of mono and di-saccharides	No effect	Yes	More than WT	Same as WT	No effect
aglM	UDP-glucose dehydrogenase (sugars 2 and 3)	Missing last four sugars	NR	Yes	More than WT	Not reported	Not reported
aglP	Methyltransferase (sugar 4)	Missing 5th sugar and methyl group of sugar 4	Not reported	Not reported	More than WT	Not reported	Not reported

[1]References for each of these gene experiments are cited in the text.

GTase, glycosyltransferase; OTase, oligosaccharyltransferase; MW, molecular weight; WT, wild type.

extreme environments. The first archaeal glycoprotein was discovered over 30 years and now it is believed that protein glycosylation is an almost universal trait in this domain of life. Initially, slow progress was made in determining the structures of the archaeal glycan modifications and understanding the biological pathways by which they are synthesized. However, recent advances in analytical and biomolecular technologies combined with the development of a couple of archaeal glycoprotein model systems have improved this situation significantly. A new archaeal glycan can now be structurally determined in a fraction of the time it once took by employing modern MS and NMR technologies. To date, only a handful of glycan structures have been fully determined but this situation is likely to improve significantly in the coming years. Bioinformatics combined with the availability of numerous sequenced archaeal genomes has enabled the identification of many putative glycosylation genes whose function can be determined, at least in those archaea with manipulatable genetic systems, using gene knockouts. Archaea invest a great deal of energy and resources in producing glycoproteins which suggests they are important for survival in the often harsh conditions these organisms inhabit. Our understanding of archaeal protein glycosylation and its relevance in Archaea is improving but there is still a great deal to learn.

References

Abu-Qarn, M., and Eichler, J. (2006). Protein N-glycosylation in Archaea: defining *Haloferax volcanii* genes involved in S-layer glycoprotein glycosylation. Mol. Microbiol. *61,* 511–525.

Abu-Qarn, M., Yurist-Doutsch, S., Giordano, A., Trauner, A., Morris, H.R., Hitchen, P., Medalia, O., Dell, A., and Eichler, J. (2007). *Haloferax volcanii* AglB and AglD are involved in N-glycosylation of the S-layer glycoprotein and proper assembly of the surface layer. J. Mol. Biol. *374,* 1224–1236.

Abu-Qarn, M., Eichler, J., and Sharon, N. (2008a). Not just for Eukarya anymore: protein glycosylation in Bacteria and Archaea. Curr. Opin. Struct. Biol. *18,* 544–550.

Abu-Qarn, M., Giordano, A., Battaglia, F., Trauner, A., Hitchen, P.G., Morris, H.R., Dell, A., and Eichler, J. (2008b). Identification of AglE, a second glycosyltransferase involved in N-glycosylation of the *Haloferax volcanii* S-layer glycoprotein. J. Bacteriol. *190,* 3140–3146.

Bayley, D.P., Kalmokoff, M.L., and Jarrell, K.F. (1993). Effect of bacitracin on flagellar assembly and presumed glycosylation of the flagellins of *Methanococcus deltae.* Arch. Microbiol. *160,* 179–185.

Calo, D., Eilam, Y., Lichtenstein, R.G., and Eichler, J. (2010a). Towards glycoengineering in archaea:replacement of *Haloferax volcanii* algD with homologous glycosyltransferases from other halophilic archaea. Appl. Environ. Microbiol. *76,* 5684–5692.

Calo, D., Kaminski, L., and Eichler, J. (2010b). Protein glycosylation in Archaea: sweet and extreme. Glycobiology *20,* 1065–1076

Chaban, B., Ng, S.Y., and Jarrell, K.F. (2006a). Archaeal habitats – from the extreme to the ordinary. Can. J. Microbiol. *52,* 73–116.

Chaban, B., Voisin, S., Kelly, J., Logan, S.M., and Jarrell, K.F. (2006b). Identification of genes involved in the biosynthesis and attachment of *Methanococcus voltae* N-linked glycans: insight into N-linked glycosylation pathways in Archaea. Mol. Microbiol. *61,* 259–268.

Chaban, B., Logan, S.M., Kelly, J.F., and Jarrell, K.F. (2009). AglC and AglK are involved in biosynthesis and attachment of diacetylated glucuronic acid to the N-glycan in *Methanococcus voltae.* J. Bacteriol. *191,* 187–195.

Eichler, J., and Adams, M.W. (2005). Post-translational protein modification in Archaea. Microbiol. Mol. Biol. Rev. *69,* 393–425.

Faguy, D.M., Bayley, D.P., Kostyukova, A.S., Thomas, N.A., and Jarrell, K.F. (1996). Isolation and characterization of flagella and flagellin proteins from the thermoacidophilic archaea *Thermoplasma volcanium* and *Sulfolobus shibatae.* J. Bacteriol. *178,* 902–905.

Fox, G.E., Magrum, L.J., Balch, W.E., Wolfe, R.S., and Woese, C.R. (1977). Classification of methanogenic bacteria by 16S ribosomal RNA characterization. PNAS *74,* 4537–4541.

Francoleon, D.R., Boontheung, P., Yang, Y., Kin, U., Ytterberg, A.J., Denny, P.A., Denny, P.C., Loo, J.A., Gunsalus, R.P., and Loo, R.R. (2009). S-layer, surface-accessible, and concanavalin A binding proteins of *Methanosarcina acetivorans* and *Methanosarcina mazei.* J. Proteome Res. *8,* 1972–1982.

Garrity, G.M., Lilburn, T.G., Cole, J.R., Harrison, S.H., Euzeby, J., and Tindall, B.J. (2007). Taxonomic outline of the Bacteria and Archaea release 7.7. (Michigan State University Board of Trustees). Available from: http://www.taxonomicoutline.org.

Guan, Z., Naparstek, S., Kaminski, L., Konrad, Z., and Eichler, J. (2010). Distinct glycan-charged phosphodolichol carriers are rquired for the assembly of the pentasaccharide N-linked to the *Haloferax volcanii* S-layer. Mol. Microbiol. *78,* 1293–1303.

Hettman, T., Schmidt, C.L., Anemüller, S., Zähringer, U., Moll, H., Petersen, A., and Schäfer, G. (1998). Cytochrome $b_{558/566}$ from the archaeon *Sulfolobus acidocaldarius*: a novel highly glycosylated, membrane-bound B-type hemoprotein. J. Biol. Chem. *273,* 12032–12040.

Igura, M., Maita, N., Kamishikiryo, J., Yamada, M., Obita, T., Maenaka, K., and Kohda, D. (2008). Structure-guided identification of a new catalytic motif of oligosaccharyltransferase. EMBO J. *27,* 234–243.

Jarrell, K.F., VanDyke, D.J., and Wu, J. (2009). Archaeal flagella and pili. In Pili and Flagella: Current Research and Future Trends, Jarrell, K.F., ed., (Norfolk, UK: Caister Academic Press), pp. 215–234.

Jarrell, K.F., Jones, G.M., Kandiba, L., Nair, D.B., and Eichler, J. (2010a). S-layer glycoproteins and flagellins: reporters of archaeal post-translational modifications. Archaea doi: 10.1155/2010/612948.

Jarrell, K.F., Jones, G.M., and Nair, D.B. (2010b). Role of N-linked glycosylation in cell surface structures of Archaea with a focus on flagella and S-layers. Int. J. Microbiol. doi: 10.1155/2010/470138.

Kalmokoff, M.L., Koval, S.F., and Jarrell, K.F. (1992). Relatedness of the flagellins from methanogens. Arch. Microbiol. *157,* 481–487.

Kaminski, L., and Eichler, J. (2010). Identification of residues important for the activity of *Haloferax volcanii* AglD, a component of the archaeal N-glycosylation pathway. Archaea doi: 10.1155/2010/315108.

Kaminski, L., Abu-Qarn, M., Guan, Z., Naparstek, S., Ventura, V.V., Raetz, C.R.H., Hitchen, P.G., Dell, A., and Eichler, J. (2010). AglJ adds the first sugar of the N-linked pentasaccharide decorating the *Haloferax volcanii* S-layer glycoprotein. J. Bacteriol. *192,* 5572–5579.

Kandler, O., and Konig, H. (1998). Cell wall polymers in Archaea (Archaebacteria). Cell Mol. Life Sci. *54,* 305–308.

Karcher, U., Schroder, H., Haslinger, E., Allmaier, G., Schreiner, R., Wieland, F., Haselbeck, A., and Konig, H. (1993). Primary structure of the heterosaccharide of the surface glycoprotein of *Methanothermus fervidus.* J. Biol. Chem. *268,* 26821–26826.

Karner, M.B., DeLong, E.F., and Karl, D.M. (2001). Archaeal dominance in the mesopelagic zone of the Pacific Ocean. Nature *409,* 507–510.

Kelly, J., Logan, S.M., Jarrell, K.F., Vandyke, D.J., and Vinogradov, E. (2009). A novel N-linked flagellar glycan from *Methanococcus maripaludis.* Carbohydr. Res. *344,* 648–653.

Koncewicz, M.A. (1972). Glycoproteins in the cell envelope of *Halobacterium halobium*. Biochem. J. *128*, 124P.

Lechner, J., and Sumper, M. (1987). The primary structure of a procaryotic glycoprotein. Cloning and sequencing of the cell surface glycoprotein gene of halobacteria. J. Biol. Chem. *262*, 9724–9729.

Lin, X.L., and Tang, J. (1990). Purification, characterization, and gene cloning of thermopsin, a thermostable acid protease from *Sulfolobus acidocaldarius*. J. Biol. Chem. *265*, 1490–1495.

Lipp, J.S., Morono, Y., Inagaki, F., and Hinrichs, K.-. (2008). Significant contribution of Archaea to extant biomass in marine subsurface sediments. Nature *454*, 991–994.

Logan, S.M. (2006). Flagellar glycosylation – a new component of the motility repertoire? Microbiology *152*, 1249–1262.

Londei, P. (2007). Translation. In Archaea Molecular and Cellular Biology, Cavicchioli, R., ed. (Washington, D.C.: ASM Press), pp. 175–197.

Magidovich, H., and Eichler, J. (2009). Glycosyltransferases and oligosaccharyltransferases in Archaea: putative components of the N-glycosylation pathway in the third domain of life. FEMS Microbiol. Lett. *300*, 122–130.

Magidovich, H., Yurist-Doutsch, S., Konrad, Z., Ventura, V.V., Dell, A., Hitchen, P.G., and Eichler, J. (2010). AglP is an S-adenosyl-L-methionine-dependent methyltransferase that participates in the N-glycosylation pathway in *Haloferax volcanii*. Mol. Microbiol. *76*, 190–199.

Maita, N., Nyirenda, J., Igura, M., Kamishikiryo, J., and Kohda, D. (2010). Comparative structural biology of eubacterial and archaeal oligosaccharyltransferases. J. Biol. Chem. *285*, 4941–4950.

Mengele, R., and Sumper, M. (1992). Drastic differences in glycosylation of related S-layer glycoproteins from moderate and extreme halophiles. J. Biol. Chem. *267*, 8182–8185.

Mescher, M.F., and Strominger, J.L. (1976a). Structural (shape-maintaining) role of the cell surface glycoprotein of *Halobacterium salinarium*. Proc. Natl. Acad. Sci. U.S.A. *73*, 2687–2691.

Mescher, M.F., and Strominger, J.L. (1976b). Purification and characterization of a prokaryotic glucoprotein from the cell envelope of *Halobacterium salinarium*. J. Biol. Chem. *251*, 2005–2014.

Mescher, M.F., Strominger, J.L., and Watson, S.W. (1974). Protein and carbohydrate composition of the cell envelope of *Halobacterium salinarium*. J. Bacteriol. *120*, 945–954.

Michel, H., Neugebauer, D.-., and Oesterhelt, D. (1980). The 2D crystalline cell wall of *Sulfolobus acidocaldarius*: structure, solubilization and reassembly. In Electron Microscopy at Molecular Dimension, Baumeister, W., and Vogell, W., eds. (Berlin: Springer-Verlag), pp. 25–37.

Moissl, C., Rachel, R., Briegel, A., Engelhardt, H., and Huber, R. (2005). The unique structure of archaeal 'hami', highly complex cell appendages with nano-grappling hooks. Mol. Microbiol. *56*, 361–370.

Moore, B.C., and Leigh, J.A. (2005). Markerless mutagenesis in *Methanococcus maripaludis* demonstrates roles for alanine dehydrogenase, alanine racemase, and alanine permease. J. Bacteriol. *187*, 972–979.

Namboori, S.C., and Graham, D.E. (2008). Acetamido sugar biosynthesis in the Euryarchaea. J. Bacteriol. *190*, 2987–2996.

Ng, S.Y.M., Wu, J., Nair, D.B., Logan, S.M., Robotham, A., Tessier, L., Kelly, J.F., Uchida, K., Aizawa, S.-., and Jarrell, K.F. (2011). Genetic and mass spectrometry analysis of the unusual type IV-like pili of the archaeon *Methanococcus maripaludis*.. J. Bacteriol. *193*, 804–814.

Nickell, S., Hegerl, R., Baumeister, W., and Rachel, R. (2003). *Pyrodictium* cannulae enter the periplasmic space but do not enter the cytoplasm, as revealed by cryo-electron tomography. J. Struct. Biol. *141*, 34–42.

Paul, G., and Wieland, F. (1987). Sequence of the halobacterial glycosaminoglycan. J. Biol. Chem. *262*, 9587–9593.

Paul, G., Lottspeich, F., and Wieland, F. (1986). Asparaginyl-*N*-acetylglucosamine: linkage unit of halobacterial glycosaminoglycan. J. Biol. Chem. *261*, 1020–1024.

Peyfoon, E., Meyer, B., Hitchen, P.G., Panico, M., Morris, H.R., Haslam, S.M., Albers, S.-., and Dell, A. (2010). The S-layer glycoprotein of the crenarchaeote *Sulfolobus acidocaldarius* is glycosylated at multiple sites with the chitobiose-linked N-glycans. Archaea doi: 10.1155/2010/754101.

Sauer, F.D., Blackwell, B.A., Kramer, J.K., and Marsden, B.J. (1990). Structure of a novel cofactor containing N-(7-mercaptoheptanoyl)-O-3-phosphothreonine. Biochemistry *29*, 7593–7600.

Shams-Eldin, H., Chaban, B., Niehus, S., Schwarz, R.T., and Jarrell, K.F. (2008). Identification of the archaeal *alg7* gene homolog (encoding N-acetylglucosamine-1-phosphate transferase) of the N-linked glycosylation system by cross-domain complementation in *Saccharomyces cerevisiae*. J. Bacteriol. *190*, 2217–2220.

Sprott, G.D. (1992). Structures of archaebacterial membrane lipids. J. Bioenerg. Biomembr. *24*, 555–566.

Sumper, M. (1987). Halobacterial glycoprotein biosynthesis. Biochim. Biophys. Acta *906*, 69–79.

Sumper, M., Berg, E., Mengele, R., and Strobel, I. (1990). Primary structure and glycosylation of the S-layer protein of *Haloferax volcanii*. J. Bacteriol. *172*, 7111–7118.

Szabo, Z., Stahl, A.O., Albers, S.V., Kissinger, J.C., Driessen, A.J.M., and Pohlschroder, M. (2007). Identification of diverse archaeal proteins with class III signal peptides cleaved by distinct archaeal prepilin peptidases. J. Bacteriol. *189*, 772–778.

Thomas, N.A., Bardy, S.L., and Jarrell, K.F. (2001). The archaeal flagellum: a different kind of prokaryotic motility structure. FEMS Microbiol. Rev. *25*, 147–174.

Thomm, M. (2007). Transcription: mechanism and regulation. In Archaea Molecular and Cellular Biology,

Cavicchioli, R., ed. (Washington, D.C.: ASM Press), pp. 139–157.

Trachtenberg, S., and Cohen-Krausz, S. (2006). The archaeabacterial flagellar filament: a bacterial propeller with a pilus-like structure. J. Mol. Microbiol. Biotechnol. *11,* 208–220.

Vandyke, D.J., Wu, J., Ng, S.Y., Kanbe, M., Chaban, B., Aizawa, S.I., and Jarrell, K.F. (2008). Identification of putative acetyltransferase gene, MMP0350, which affects proper assembly of both flagella and pili in the archaeon *Methanococcus maripaludis.* J. Bacteriol. *190,* 5300–5307.

Vandyke, D.J., Wu, J., Logan, S.M., Kelly, J.F., Mizuno, S., Aizawa, S.I., and Jarrell, K.F. (2009). Identification of genes involved in the assembly and attachment of a novel flagellin N-linked tetrasaccharide important for motility in the archaeon *Methanococcus maripaludis.* Mol. Microbiol. *72,* 633–644.

Voisin, S., Houliston, R.S., Kelly, J., Brisson, J.R., Watson, D., Bardy, S.L., Jarrell, K.F., and Logan, S.M. (2005). Identification and characterization of the unique N-linked glycan common to the flagellins and S-layer glycoprotein of *Methanococcus voltae.* J. Biol. Chem. *280,* 16586–16593.

Wieland, F., Dompert, W., Bernhardt, G., and Sumper, M. (1980). Halobacterial glycoprotein saccharides contain covalently linked sulphate. FEBS Lett. *120,* 110–114.

Wieland, F., Heitzer, R., and Schaefer, W. (1983). Asparaginylglucose: Novel type of carbohydrate linkage. Proc. Natl. Acad. Sci. U.S.A. *80,* 5470–5474.

Wieland, F., Lechner, J., and Sumper, M. (1986). Iduronic acid: constituent of sulphated dolichyl phosphate oligosaccharides in halobacteria. FEBS Lett. *195,* 77–81.

Woese, C.R., and Fox, G.E. (1977). Phylogenetic structure of the prokaryotic domain: the primary kingdoms. Proc. Natl. Acad. Sci. U.S.A. *74,* 5088–5090.

Yang, L.L., and Haug, A. (1979). Purification and partial characterization of a procaryotic glycoprotein from the plasma membrane of *Thermoplasma acidophilum.* Biochim. Biophys. Acta *556,* 265–277.

Yurist-Doutsch, S., and Eichler, J. (2009). Manual annotation, transcriptional analysis, and protein expression studies reveal novel genes in the agl cluster responsible for N glycosylation in the halophilic archaeon *Haloferax volcanii.* J. Bacteriol. *191,* 3068–3075.

Yurist-Doutsch, S., Abu-Qarn, M., Battaglia, F., Morris, H.R., Hitchen, P.G., Dell, A., and Eichler, J. (2008a). AglF, aglG and aglI, novel members of a gene island involved in the N-glycosylation of the *Haloferax volcanii* S-layer glycoprotein. Mol. Microbiol. *69,* 1234–1245.

Yurist-Doutsch, S., Chaban, B., VanDyke, D.J., Jarrell, K.F., and Eichler, J. (2008b). Sweet to the extreme: protein glycosylation in Archaea. Mol. Microbiol. *68,* 1079–1084.

Yurist-Doutsch, S., Magidovich, H., Ventura, V.V., Hitchen, P.G., Dell, A., and Eichler, J. (2010). N-glycosylation in Archaea: On the coordinated actions of *Haloferax volcanii* AglF and AglM. Mol. Microbiol. *75,* 1047–1058.

Zahringer, U., Moll, H., Hettmann, T., Knirel, Y.A., and Schafer, G. (2000). Cytochrome b558/566 from the archaeon *Sulfolobus acidocaldarius* has a unique Asn-linked highly branched hexasaccharide chain containing 6-sulfoquinovose. Eur. J. Biochem. *267,* 4144–4149.

Zeitler, R., Hochmuth, E., Deutzmann, R., and Sumper, M. (1998). Exchange of Ser-4 for Val, Leu or Asn in the sequon Asn-Ala-Ser does not prevent N-glycosylation of the cell surface glycoprotein from *Halobacterium halobium.* Glycobiology *8,* 1157–1164.

Identification and Characterization of Glycosylation in Bacterial Proteins

8

Nichollas E. Scott, Stuart J. Cordwell, John F. Kelly and Susan M. Twine

Abstract

There are increasing numbers of reports of bacterial glycosylation in pathogenic bacteria, with well-characterized bacterial glycoproteins including pilins, flagellin and other surface-associated proteins. However, the discovery of bacterial glycoproteins can be challenging owing to the diversity of glycans bacteria use to modify proteins. At the protein level, so-called 'top-down' mass spectrometry studies of intact protein can rapidly characterize bacterial glycan ions. At the peptide level, interpretation of individual bacterial glycopeptide tandem mass spectra can be challenging, owing to the diverse range of bacterial glycans produced. Reports of methods to specifically isolate bacterial glycopeptides are advancing knowledge of bacterial glycoproteomes. Herein, we provide an overview of protein and peptide centric mass spectrometry and related analytical techniques for the enrichment and analysis of bacterial glycoproteins.

Introduction

For many years glycoproteins were thought to be expressed only by eukaryotes and archaea (Mescher and Strominger, 1976a,b). It has become evident in recent years however, that glycoproteins are found in all domains of life. In eukarytotes, comparatively few monosaccharide species are used to build N- and O-linked glycan structures using highly conserved processes. By contrast, prokaryotes, including medically relevant pathogens such as *Campylobacter jejuni* (Thibault *et al.*, 2001), *Helicobacter pylori* (Schirm *et al.*, 2005), *Pseudomonas aeruginosa* (Castric *et al.*, 2001; Voisin *et al.*, 2007) and *Clostridium difficile* (Twine *et al.*, 2009), are increasingly being shown to synthesize a vast array of unusual monosaccharides and glycan structures that are often difficult to characterize using the analytical technologies developed for the more 'predictable' eukaryotic glycans and glycoproteins. Thus, while bacterial glycoprotein analysis shares many of the same approaches as those used in studies of eukaryotic glycoproteins, it also poses some unique challenges.

Here, we describe a variety of analytical approaches that can be used for bacterial glycoprotein discovery and characterization. The chapter is organized into three major sections: (1) gel-based methods for detecting and analysing bacterial glycoproteins; (2) glycopeptide preparation and enrichment strategies suitable for bacterial glycoproteomics studies; and (3) mass spectrometry methods for glycoprotein and glycopeptide analysis. Examples of the application of these techniques to bacterial glycoprotein analysis will be given throughout.

Protein level bacterial glycoprotein analysis

Gel-based detection of bacterial glycoproteins

Polyacrylamide gel electrophoresis has played an important role in the discovery of new glycoproteins, as often the first indication that a protein may be modified is its aberrant migration on a 1D or 2D PAGE gel. The visualization of glycosylated proteins by 1D/2D gels or by Western blotting can be aided by lectin binding, or glycostaining reagents (Packer *et al.*, 1997; Taverna *et al.*, 1998; Koketsu and Linhardt, 2000). These methods are favoured by many researchers because they utilize equipment readily available in most laboratories, are simple to conduct, and are perceived as being relatively cheap compared with mass spectrometry based approaches. When using these detection methods, the putative glycoprotein can be identified from the corresponding stained gel spot or band by in-gel proteolytic digestion and MS analysis.

Glycoprotein detection via periodic acid chemistry

In the last decade, advances in the probes used for protein and glycoprotein detection have increased the ease of use and sensitivity of these dyes, so that they are readily incorporated into most standard gel-based proteomics workflows. The most widely used approaches exploit the periodic acid–Schiff (PAS) chemistry. Specifically, during the staining process, periodic acid converts the vicinal diol of susceptible carbohydrate moieties into two aldehyde functionalities (Fig. 8.1). The dye reagent, for example a fluorescent probe, can then be coupled to the chemically reactive aldehyde on the glycan moiety to generate a detectable conjugate that can be visualized using a suitable gel scanner. Commercially available reagents, such as the Pro-Q Emerald 300 dye reagent from Molecular Probes (Invitrogen) are widely used and it is reported that as little as 300 pg of α_1-acid glycoprotein (40% carbohydrate) and 1 ng of glucose oxidase (12% carbohydrate) are detectable in gels after staining with Pro-Q Emerald 300 dye (Steinberg *et al.*, 2001). The sensitivity of this approach is directly proportional to the density of periodic acid-susceptible carbohydrates within the protein. Gels may then also be post-stained with a protein specific dye, such as SYPRO Ruby, thus allowing sequential two-colour detection of both glycosylated and non-glycosylated proteins, as illustrated by the example in Fig. 8.2. This figure shows 2D-PAGE separation of a crude membrane extract of the pathogen *Francisella novicida* in the pH range 3–6. Staining with the Pro-Q Emerald 300 dye reagent revealed two reactive spots, indicated in Fig. 8.2A, which were then aligned with protein staining areas of the same gel that was subsequently stained with Sypro Ruby (Fig. 8.2B). The protein spots were identified by tandem mass spectrometry of their tryptic digests as a protein disulphide isomerase (PDI) homologue. Interestingly, the protein stained gel revealed two protein spots, not reactive with glycostain, which migrated at a slightly lower molecular mass compared with the glycoreactive spots. These spots, indicated in Fig. 8.2B, were also identified as the PDI homologue. The lack of reactivity with glycostain, combined with a slightly reduced molecular mass, suggest that these spots correspond to unmodified forms of the protein. Although 2D-PAGE gels have well-documented limitations (Rabilloud *et al.*, 2010), this example highlights their utility in the first identification of novel glycoproteins, in a relatively straightforward manner. The major drawback of this approach for

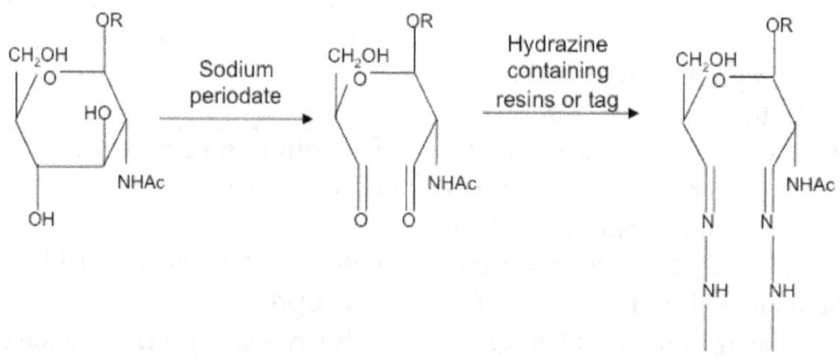

Figure 8.1 Periodate oxidation and hydrazide capture of glycopeptides. Vicinal cis-diol groups are converted using periodic acid to reactive aldehyde groups that can be further reacted with hydrazide resin and non-glycosylated components removed. R represents the glycan chains attached to the peptide.

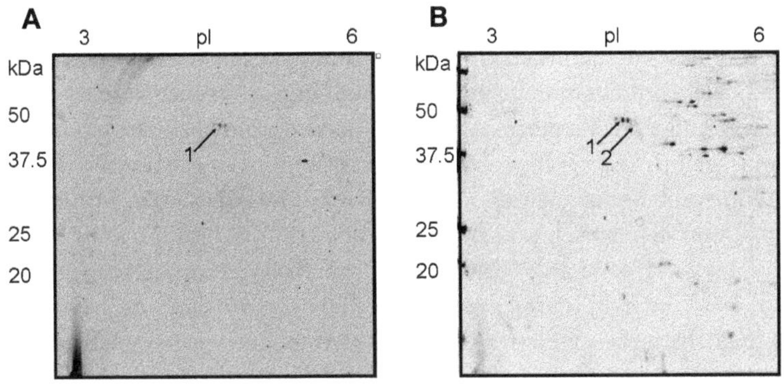

Figure 8.2 Two-dimensional PAGE of membrane extracts of *Francisella novicida* U112. (A) Stained with Pro-Emerald Q glycostain, a periodic acid based stain with a fluorescent conjugate for sensitive visualisation. (B) The same gel subsequently stained with Sypro Ruby protein stain. Gel (A) showed two glyco-reactive protein spots, denoted 1 in the figure. Alignment of the glyco stained gel with gel (B) determined the glyco reactive spots to align to an areas of protein staining (denoted 1). These protein spots were excised from the gel and protein identified from MS/MS of their proteolytic digests as shown in (A) and (B). The protein was identified as protein-disulfide isomerase (YP_898416). The spots denoted 2 were also identified as the same protein-disulfide isomerase. Their lack of reactivity with glyco stain indicates they are probably unmodified forms of the protein.

bacterial glycoprotein studies is the unusual and often unknown nature of bacterial glycans, which may lack the vicinal diol required for successful PAS-based glycostaining.

Glycoprotein detection using lectin affinity approaches

Lectins, receptor proteins that bind to glycans and antibodies, are regularly employed in Western blotting for sensitive detection and characterization of eukaryotic glycoproteins. Lectins recognize specific carbohydrates and or carbohydrate combinations within a defined configuration and are independent of the linkage used for glycosylation (N versus O). When combined with endo- and exo-glycosidase treatments, lectin blotting can provide additional information regarding *N*- versus O-linkage and glycan composition of the glycoproteins in a given sample. In eukaryotic glycoproteomics studies, panels of lectins with well-defined carbohydrate binding specificities are used to probe for specific sugar epitopes (Cummings, 1994). The binding of lectins to bacterial sugars, however, is less predictable.

An elegant bacterial glycoproteomics study of the pathogen *C. jejuni* highlights the utility of lectins. Specifically, a GalNAc-specific lectin was used in an affinity chromatography approach to enrich for N-linked glycoproteins from the periplasmic subproteome of *C. jejuni* (Young *et al.*, 2002). The enriched proteins were separated by 2D-PAGE, and proteins identified by tandem mass spectrometry of their tryptic digests. Sequence alignments of the identified protein revealed that the usual sequon for N-glycosylation (N-X-S/T) is extended in *Campylobacter* to D/E-Z-N-X-S/T (where Z, X≠P) (Kowarik *et al.*, 2006). This, a relatively simple glycoproteomics experiment led to the discovery of a novel glycosylation phenomenon that appears to be found only in *C. jejuni* and closely related species; an N-glycosylation system that modifies membrane and periplasmic proteins with a heptamer composed primarily of an α1,4-linked GalNAc oligomer linked to the asparagine side-chain via bacillosamine (2,4-diacetamido-2,4,6-trideoxyglucopyranose) (Young *et al.*, 2002).

Glycan removal

Removing the glycan modifications from a glycoprotein is a requirement of many assays. For example, confirmation of protein glycosylation by the gel-shift assay is based on comparing sample migration on a gel before and after deglycosylation. Similarly, *N*- or *O*-glycans can be distinguished from one another by their reactivity to certain endoglycosidases. In addition, removal of the glycan can make a protein more amenable to proteolytic cleavage, allowing for greater amino acid sequence coverage by mass spectrometry. In fact, many eukaryotic proteomics analyses incorporate a deglycosylation step in their preparation protocols for this very reason. Lastly, selective deglycosylation of a glycoprotein is used to study the biological role of the glycan moieties themselves. Glycans can be liberated from proteins by chemical or enzymatic means, both of which are briefly described here.

Glycan removal by enzymatic deglycosylation

Endoglycosidases cleave carbohydrate moieties *en bloc* from proteins. There are several endoglycosidases that specifically remove a wide range of glycans from the side-chains of asparagine residues (Maley *et al.*, 1989), although similar options for removal of O-linked glycans are somewhat limited. The most commonly used endoglycosidase employed in glycoproteomics studies is peptide-N4-(acetyl-β-glucosaminyl)-asparagine amidase (EC 3.5.1.52) otherwise known as PNGase F (peptide-*N*-glycosidase F), which is typically isolated from culture supernatants of *Elizabethkingia* (*Chryseobacterium* or *Flavobacterium*) *meningosepticum*. This enzyme can be used to investigate whether a glycan is attached via N-linkage (gel shift assay). In the process of glycan cleavage, asparagine is deamidated to aspartic acid; therefore if the site of modification is within an N-linked sequon, as described above, the exact site of glycan modification can often be mapped. The oligosaccharide itself remains intact and can be isolated for further analysis. Glycan cleavage occurs more efficiently if the protein is denatured, and the enzyme is active in the presence of detergents or 2–5 M urea (Maley *et al.*, 1989). The enzyme itself has broad substrate specificity but is not universally applicable. Oligosaccharides containing -(1,3)-linked core fucose and also some bacterial glycans (e.g. the *C. jejuni* N-linked glycan) are resistant to PNGase F removal. The enzyme has been used in a limited number of studies of bacterial glycoproteins, to aid in determining whether a putative glycan linkage is N-linked. For example, a study of polymorphic outer membrane proteins of *Chlamydophila abortus* showed a loss of lectin binding upon incubation with PNGase F, and a decrease in apparent protein mass (Vretou *et al.*, 2001). Although this can be taken as a good first indication that a glycan may be N-linked, mass spectrometry studies must be carried out to confirm the type and sites of glycan linkage.

For enzymatic deglycosylation of O-linked glycans, options are much more limited, with only the *O*-glycanase from *Diplococcus pneumoniae* being widely available. This enzyme is specific for removal of Galβ1–3GalNAc units linked to Ser/Thr residues and does not release more complex *O*-glycans (Umemoto *et al.*, 1978). Such enzymes have been used in studies of bacterial glycoproteins (Stephenson *et al.*, 2002), although once again the diversity and unknown nature of many bacterial glycans can make negative results difficult to interpret.

Glycan removal by chemical methods

Chemical deglycosylation is independent of glycan structure or linkage, by contrast with enzymatic approaches, but can result in destruction of the glycan [e.g. trifluoromethanesulphonic acid (TFMS) treatment] and/or the protein (e.g. hydrazinolysis). Several approaches have been described and we highlight two in this section – glycan removal with TFMS and base catalysed beta elimination (β-elimination).

Deglycosylation with TFMS has been used extensively to remove carbohydrates from glycoproteins, while leaving the protein backbone intact (Edge *et al.*, 1981). N-linked and O-linked sugars, and glycosaminoglycan chains, can all be removed by this procedure. The method exploits the sensitivity of linkages between sugars to TFMS and the relative stability of the peptide bond to the same conditions. The susceptibility of individual sugars in glycosidic linkage varies with the substituents at C-2 and the presence of amido and acetyl groups, but even the most stable sugars can be removed with extended TFMS incubation (Edge, 2003). This approach has the advantage over enzymatic removal of sugars in that all carbohydrate groups tested to date can be removed in a single step, with no prior information needed about the glycans or their attachment sites. In addition, the removal of glycan allows the mass of the deglycosylated protein to be determined. By comparing the mass of the intact and deglycosylated glycoproteins the relative contribution of the carbohydrate to the original glycoprotein can be estimated and this molecular mass shift can be readily visualized by SDS-PAGE. It should be noted, however, that the size difference between glycosylated and deglycosylated proteins can be greater than expected in some cases due to the anomalous migration of glycoproteins on SDS gels.

This approach has been widely used, in both eukaryotes and prokaryotes. For example, TFMS treatment of S-layer glycoproteins from *Clostridium thermohydrosulfuricum* aided in the identification of a novel glucose-tyrosine linkage (Christian *et al.*, 1993). Other examples of the successful application of this method include the characterization of S-layer protein from *C. thermosaccharolyticum* (Altman *et al.*, 1995), as well as the *M. tuberculosis* cell wall-associated antigens (Dhiman *et al.*, 1997). In the latter case, deglycosylation yielded core proteins of 71 kDa, 60 kDa, and 45 kDa respectively, which were used for the generation of a vaccine that was protective in mice (Dhiman *et al.*, 1997). Of note, the deglycosylated 45 kDa antigen was less immunogenic compared with the glycosylated form (Romain *et al.*, 1999).

Owing to the paucity of enzymes that specifically cleave the sugar-Ser/Thr linkage, β-elimination is often used for removal of O-linked glycan moieties and for mapping of the sites of glycan modification. This approach exploits the alkali-labile nature of the linkage between the sugar and the Ser/Thr residue, which can be cleaved under relatively mild basic conditions. Several variations of the β-elimination method have been developed and used for this purpose. For example, the Carlson degradation method heats the glycoprotein in sodium hydroxide for several hours in the presence of excess sodium borohydride (Carlson, 1968). The excess sodium borohydride ensures fast reduction and conversion of the released oligosaccharides into base-stable alditols, thus protecting them from further degradation under the alkaline conditions. This method, however, results in high levels of residual salts in the sample that can interfere with downstream mass spectrometry analyses.

A related method uses aqueous ammonia to β-eliminate glycans from proteins and consequently to identify the sites of O-linkage on the protein/peptide using tandem mass spectrometry (Rademaker *et al.*, 1998). Elimination of the glycan (Fig. 8.3) and subsequent Michael addition results in the loss of 1 Da (due to the exchange of H_2O for NH_3) from the Ser/Thr residue and thus tagging the site of modification enabling identification by mass spectrometry. This can be a key step in confirming that a protein is modified by

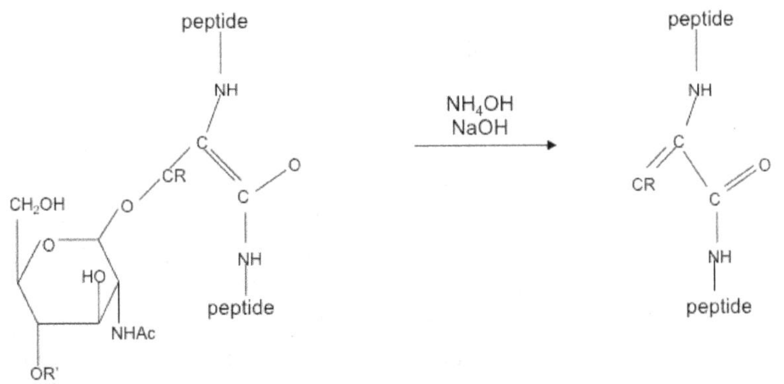

Figure 8.3 β-Elimination of O-linked carbohydrates from peptides using alkali conditions. Glycopeptides are subjected to high pH resulting in the deglycosylation and conversion of the initial glycosylation site to dehydroalanine and dehydroamino butyric acid from serine and threonine, respectively. R represents H_2 for serine and H_2CH_3 for threonine. The dehydro products of this reaction can then be further reacted to produce stabile products enabling to localisation of glycosylation.

glycan. More recently, this method was modified to use borane-ammonia. Both the ammonia and borane ammonia complex can be readily removed by evaporation, which significantly reduces the sample handling complexity and also allows the released glycans to be maintained intact for further analysis. Recently, it was shown that β-elimination could be adapted to eliminate glycan from glycoproteins separated by SDS-PAGE (Taylor *et al.*, 2006). Although mass spectrometry approaches, such as electron transfer dissociation and electron capture dissociation are increasingly being used to map sites of modification on glycopeptides, β-elimination can still prove useful in many situations.

Peptide-centric bacterial glycoproteomics

Owing to the heteropolymeric nature of proteins, a multitude of chemically unique polypeptides exist within bacterial cells. This diversity leads to a range of proteins that, owing to their intrinsic chemical properties, can be resistant to isolation, enrichment and analysis techniques typically employed to characterize proteins in their intact forms. In the field of bacterial glycoproteomics this represents a unique problem as to confirm the glycosylated status of a suspected glycoprotein both the detection of the glycosylation event, defined as the confirmation of the chemical linkage of a carbohydrate to the protein, and the protein identity are required. To overcome the challenges associated with the characterization of these proteins, both eukaryotic (Chalkley *et al.*, 2009; Nilsson *et al.*, 2009; Zielinska *et al.*, 2010) and prokaryotic (Ding *et al.*, 2009; Scott *et al.*, 2010) glycoproteomic studies have begun to adopt the use of peptide-centric approaches. This progression in experimental design mirrors the widespread application of peptide-based approaches for the analysis of recalcitrant subproteomes and protein classes by providing an alternative workflow to overcome limitations such as insolubility (Cordwell, 2006; Speers and Wu, 2007), incompatible mass (Klein *et al.*, 2007), charge (Bae *et al.*, 2003) or ionization (Kelly *et al.*, 2009). These approaches, which are also known as 'shotgun' or 'bottom-up' proteomics (Bogdanov and Smith, 2005; Han *et al.*, 2008; Gundry *et al.*, 2009), enable the identification and confirmation of the glycosylation site by the analysis of smaller peptide components, which owing to their size and composition may be more amenable to specific enrichment techniques and downstream analysis compared with the intact parent protein.

The conversion of a protein into smaller peptide units affects multiple aspects of the downstream glycosylation elucidation workflow, offering both strengths and weaknesses in the ultimate analysis. The strength of peptide-based approaches lies largely in simplification, predominantly associated with the reduction in size and composition complexity, of the glycosylated units being isolated. This advantage has led to the use of peptide-based glycosylation analysis in bacterial systems and has been shown to enable the confirmation of the site of glycan attachment (Thibault *et al.*, 2001; Young *et al.*, 2002; Chaban *et al.*, 2006; Kowarik *et al.*, 2006; Gross *et al.*, 2008; Sartain and Belisle, 2009; Scott *et al.*, 2009; Twine *et al.*, 2009f; Vik *et al.*, 2009), the assessment of glycoform diversity on specific glycopeptides (Thibault *et al.*, 2001; Aas *et al.*, 2006; Vik *et al.*, 2009) and the identification of both the attached glycan and the peptide sequence using mass spectrometry (Thibault *et al.*, 2001; Young *et al.*, 2002; Gross *et al.*, 2008; Chaban *et al.*, 2009; Scott *et al.*, 2009; Twine *et al.*, 2009; Vik *et al.*, 2009). These advantages however, come at the expense of obtaining information that requires the context of the intact protein and its isoforms, such as the determination of stoichiometric ratios of specific glycosylation or other modification events (Fig. 8.4). Although this stoichiometric information is of increasing interest owing to the apparent heuristic nature of some bacterial modification systems, such as the *C. jejuni* N-linked glycosylation system (Scott *et al.*, 2009) and the glycosylation/phospho-form modification system of *Neisseria gonorrhoeae* (Aas *et al.*, 2006), the generation of peptides does enable parallel detection of modifications from multiple proteins, which ultimately enhances analysis of events found on isoforms (Fig. 8.4).

Within most bottom-up proteomic approaches the first step is the generation of peptides that can be subjected to further enrichment or analysis (Bogdanov and Smith, 2005; Han *et al.*, 2008;

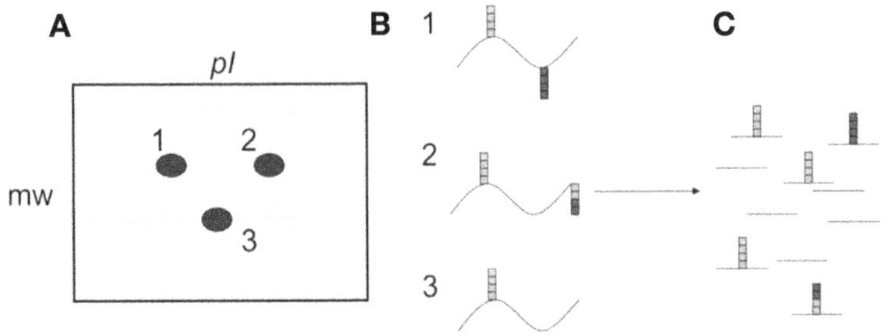

Figure 8.4 The loss of glycosylation context due to the generation of glycopeptides. (A) Proteins can undergo different modifications at several sites leading to the generation of multiple protein isoforms with distinct properties such as molecular weight and isoelectric point. (B) When examined at the protein level, differences in the combination of modification are maintained. (C) Upon generation of a complex peptide mixture the context of these modification events are lost. The resulting glycopeptides from multiple isoforms become a single peptide population.

Gundry *et al.*, 2009). Peptide generation can be accomplished using enzymatic (Swaney *et al.*, 2010b), chemical (Han *et al.*, 1986; Samyn *et al.*, 2006) or radiation-induced (Swatkoski *et al.*, 2008; Hauser and Basile, 2008) methodologies that aim to induce cleavage at a specific motif or amino acids, providing controlled, predictable and reproducible production. The most widely used methodology for the generation of peptides within proteomic-based studies is the enzymatic treatment of protein samples with the endoproteinase trypsin. Although trypsinization has been used in multiple bacterial glycosylation studies (Kowarik *et al.*, 2006; Ding *et al.*, 2009; Sartain and Belisle, 2009; Scott *et al.*, 2010), it is unlikely to provide total coverage of a complete protein amino acid sequence, as most will lack the adequate spacing of lysine and arginine to ensure the generation of peptides that can be detected or analysed for all regions of the protein. To overcome this shortcoming, multiple cleavage approaches such as using enzymes with different substrate specificity (Chen *et al.*, 2009; Gauci *et al.*, 2009; Swaney *et al.*, 2010) can enhance protein coverage and thus increase modification site identifications. Within the context of glycoproteomics, several groups have used this strategy to both enhance coverage and identification of glycosylation events (Scott *et al.*, 2009, 2010; Sartain and Belisle, 2009; Vik *et al.*, 2009; Voisin *et al.*, 2007). These studies have also utilized non-specific proteases such as chymotrypsin, thermolysin, pronase E or proteinase K that also provide unique advantages (Thibault *et al.*, 2001; Young *et al.*, 2002; Sartain and Belisle, 2009; Scott *et al.*, 2009; Vik *et al.*, 2009). These proteinases result in the generation of short peptides that further simplify the confirmation of glycosylation sites, as only a limited number of amino acids can be occupied. Furthermore, the addition of a glycan moiety limits the access of proteinase to the peptide backbone via steric hindrance, thus limiting the digestion of amino acids immediately surrounding the site of glycosylation (Larsen *et al.*, 2005). This results in the generation of peptides containing the modification, while non-modified peptides are completely digested. These reactions are also tuneable by the alteration of the length of time used for digestion (Clowers *et al.*, 2007), providing further control over the generation of glycopeptides for downstream analysis.

Enrichment of bacterial glycopeptides

Upon the generation of peptides from glycoproteins, two chemically discrete populations exist – glycosylated and non-glycosylated peptides. The addition of a glycan(s) affects the chemical properties of the peptide(s), such as charge, hydrophobicity or chemical reactivity, of these peptides compared with their unglycosylated counterparts. This change in chemistry of the glycopeptides in turn affects the enrichment and isolation strategies that are suitable for their analysis. Within bacterial glycoproteomics, the isolation/enrichment strategies that have been used to date are largely borrowed from the study of eukaryotic glycosylation but with one key difference, the need to consider the unique glycan type that may be present in the bacterial system under investigation. This consideration results from the fact that unlike the largely ubiquitous and conserved glycosylation systems within eukaryotes (Helenius and Aebi, 2004; Weerapana and Imperiali, 2006), bacterial systems display a range of unique glycosylation pathways that can incorporate bacterial-specific carbohydrates, such as 2,4-diacetamido-2,4,6-trideoxy-glucopyranore (commonly referred to as bacillosamine). (Young *et al.*, 2002), 2,4-diacetamido-2,4,6-trideoxyhexose (DATDH). (Stimson *et al.*, 1995) and sialic acid analogues including pseudaminic acid (Thibault *et al.*, 2001) and legionaminic acid (McNally *et al.*, 2007), into a diverse array of glycan structures (Szymanski and Wren, 2005; bu-Qarn *et al.*, 2008). This diversity requires strategies that take into account the unique properties of these glycans as not all approaches will be compatible with all glycan structures.

The enrichment of glycopeptides differs from that of glycoproteins since the glycan confers a much more dramatic relative influence upon the physico-chemical properties of small versus large polypeptides, resulting in a significant effect on the behaviour/properties of peptides during methodologies such as separation by liquid chromatography (Wuhrer *et al.*, 2005). This alteration in behaviour enables the enrichment and isolation of glycopeptides by exploiting the change in affinity of these molecules. Within the field of glycoproteomics these enrichment regimes can be grouped into four general approaches; approaches which utilize the change in charge associated with the addition of a non-neutral glycan; approaches which utilize the unique chemical reactivity of the glycans or carbohydrate moiety to enable enrichment; approaches which use biological components such as lectins to specifically recognize/bind modified peptides and finally, approaches which utilize the change in hydrophobicity owing to the addition of a glycan. It should be noted that not all approaches have been applied to peptide-centric bacterial glycoproteomics as this field is still in its infancy, but the proof of concept has been demonstrated within eukaryotic studies, proving the effectiveness and potential of these approaches to investigate bacterial glycosylation.

Enrichment of charged glycopeptides

Multiple carbohydrates can be utilized by biological systems to construct glycans. In both eukaryotic and prokaryotic systems, many of these carbohydrates are non-neutral in nature, such as sialic acids in eukaryotes (Schauer, 2009; Varki, 1992) or similar analogues found to decorate the surface of prokaryotes (Szymanski and Wren, 2005; Severi *et al.*, 2007). The addition of charged glycans such as pseudaminic acid and legionaminic acid found on the flagellin of *C. jejuni* (Thibault *et al.*, 2001; McNally *et al.*, 2007), *Clostridium botulinum* (Twine *et al.*, 2008), *Aeromonas caviae* (Tabei *et al.*, 2009), the sulfated glycans of *Halobacterium salinarium* (Sumper, 1987) and the phosphorylated glycan of *Pseudomonas aeruginosa* b-type flagellin (Verma *et al.*, 2006), result in glycosylated peptides with a higher net negative charge compared with their non-glycosylated counterparts. Although the analysis of these glycosylation events has typically been accomplished using purified protein (Thibault *et al.*, 2001; Verma *et al.*, 2006; McNally *et al.*, 2007), the presence of the charged modification enables separation by anion exchange chromatography utilizing ion–ion interaction where the presence of the negatively charged glycan enables the retention, or selective removal if cation exchange chromatography is used, of the analyte of interest and therefore simplifies the mixture for downstream analysis (Sandra *et al.*, 2009). Anion exchange has been applied to the analysis of eukaryotic glycopeptides (Townsend *et al.*, 1996; Larsen *et al.*, 2007), but few examples of its application to bacterial glycosylation currently exist, with the notable exception of studies on the N-linked glycan of *Methanococcus maripaludis* (Kelly *et al.*, 2009).

An extension of anion exchange chromatography has become widespread in proteomic studies, using resins such as titanium dioxide to enable the enrichment of electronegative components, including phosphopeptides (Larsen *et al.*, 2005) and sialylated glycopeptides (Larsen *et al.*, 2007). This metal oxide, and others such as zirconium oxide (Kweon and Hakansson, 2006), that are essentially anion exchange resins, enable the enrichment of these compounds by the formation of multidentate ion interactions (Larsen *et al.*, 2007). The applicability of such resins to bacterial glycoproteomes has also been established, for example pseudaminic acid modified peptides from the *C. jejuni* flagellin could be effectively enriched using titanium dioxide (Fig. 8.5). Anion exchange approaches do however, require the use of a high-pH mobile phase that results in a lack of direct compatibility to many reverse phase chromatography workflows, the most common peptide infusion methodology for positive ion mass spectrometry. As such, further sampling handling, such as acidification and solid phase extraction are required, which will lead to a potential loss of material.

Enrichment of glycopeptides using carbohydrate chemistry

The glycan-peptide modification can be targeted by using chemical reactions that specifically alter carbohydrate moieties. Such approaches are being increasingly implemented in eukaryotic glycoproteomics (Zhang *et al.*, 2003; Sun *et al.*, 2007; Nilsson *et al.*, 2009), but have not yet been readily applied to peptide-centric bacterial glycoproteomics, although the compatibility with bacterial glycosylation has been shown (Gerwig *et al.*, 1993; Doig *et al.*, 1996; Thibault *et al.*, 2001; Schirm *et al.*, 2004). The most widely used of these chemical approaches is periodate oxidation coupled to immobilization of the glycosylated substrate to hydrazide resins. In this process, treatment with periodic acid leads to the opening of saccharide rings at vicinal cis-diol groups and the generation of aldehyde functional groups, which can then be reacted with hydrazide to form a covalent linkage that enables extensive washing to remove non-specific binders (Fig. 8.1) (Zhang *et al.*, 2003; Sun *et al.*, 2007). The binding of glycosylated peptides enables selective enrichment and concentration, followed by liberation from the resin using either enzymatic cleavage or a further chemical cleavage of the glycan-peptide.

Cleavage of the glycan–peptide bond represents the most challenging step for bacterial glycosylation as unlike N-linked eukaryotic glycosylation, for which PNGase F can cleave the glycan–peptide bond (Gerwig *et al.*, 1993; Doig *et al.*, 1996), no endoglycosidase exists that recognizes or effectively cleaves the majority of bacteria-specific carbohydrates. Developments within the study of eukaryotic O-linked glycosylation suggest however, that chemical cleavage can provide a robust alterative. It should also be noted that the type of carbohydrate and glycosylation linkage will govern the effectiveness of chemical liberation techniques, and not all glycopeptides will be compatible, for example previous work has shown that the N-linked glycan of *C. jejuni* cannot be released by β-elimination (Young *et al.*, 2002).

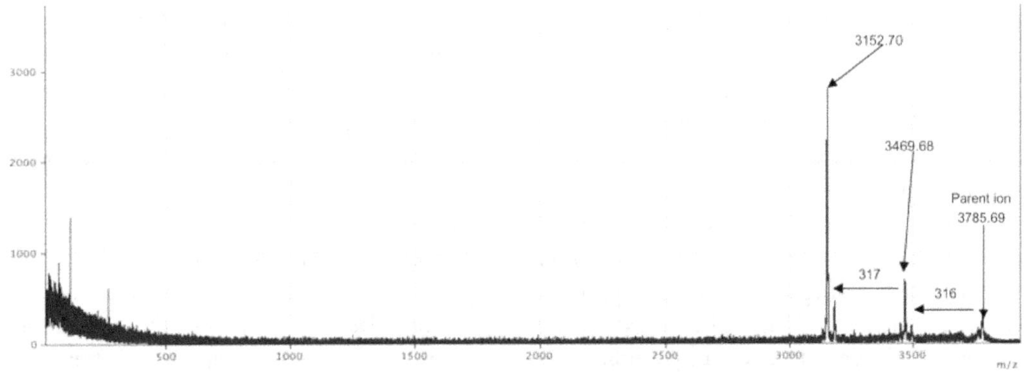

Figure 8.5 Pseudaminic acid-containing peptide from *C. jejuni* flagellin enriched with titanium dioxide. CID fragmentation results in a spectrum dominated by the loss of two carbohydrate moieties but no assignment of the peptide sequence.

Furthermore, periodate oxidation and immobilization is not suitable for all peptide substrates as it is known to have adverse effects on *N*-terminal Ser- and Thr-containing peptides as the presence of the vicinal amino alcohol undergoes rapid oxidation to generate reactive aldehyde groups that are then captured by the hydrazide resin (Geoghegan and Stroh, 1992).

Chemical strategies that lead to the release of glycopeptides can be grouped into two regimes; the liberation of completely deglycosylated peptides from hydrazide by β-elimination (Klement *et al.*, 2010); and the liberation of partial glycan-containing peptides by hydroxylamine cleavage (Klement *et al.*, 2010) or acid hydrolysis (Nilsson *et al.*, 2009). β-elimination of glycopeptides, in conjunction with Michael addition, has been utilized for mapping bacterial glycosylation sites (Thibault *et al.*, 2001; Schirm *et al.*, 2004; Verma *et al.*, 2006), and results in the loss of the hydroxyl side-chain of glycan-attached Ser and Thr residues leading to the formation of dehydroalanine and dehydroamino butyric acid, respectively (Fig. 8.3). Conversion of glycosylated residues to their dehydrated forms tags the site of modification and facilitates assignment of the glycosylation event. The disadvantages of this procedure are the use of a strong base, such as sodium hydroxide, causing peptide samples to undergo degradation or secondary chemical modification at other hydroxyl residues and thus potentially leading to mis-assignment of glycosylation sites or peptide sequence (Rademaker *et al.*, 1998; Klement *et al.*, 2010). As an alternative to β-elimination, chemically milder liberation approaches can be utilized such as the release of glycosylated or partially glycosylated peptides by hydroxylamine cleavage and acid hydrolysis. Hydroxylamine cleaves the hydrazone bond between the open saccharide ring and the support resin (Fig. 8.6) and has recently been demonstrated as an effective means for enrichment of eukaryotic O-linked glycopeptides (Klement *et al.*, 2010). Another more carbohydrate-dependent cleavage approach is the use of acid hydrolysis (Nilsson *et al.*, 2009) to cleave the labile ketosidic linkage utilized by some carbohydrates, such as sialic acids (Yu and Ledeen, 1969). Both these mild chemical cleavage approaches leave glycans or open ring saccharide structures associated with the site of glycosylation, which when subjected to tandem mass spectrometry (MS/MS) fragmentation behave similarly to the initial attached carbohydrates (Nilsson *et al.*, 2009; Klement *et al.*, 2010e).

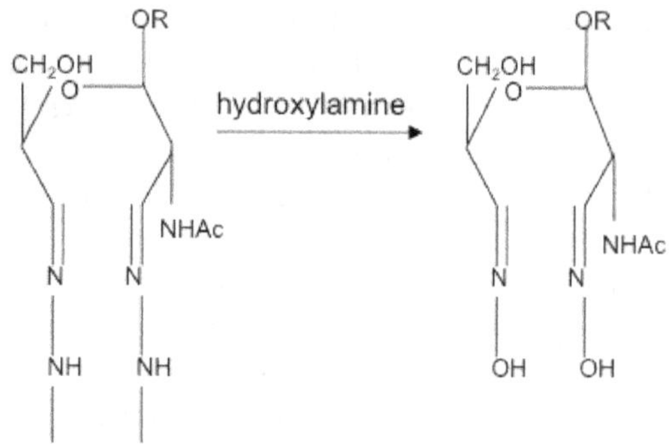

Figure 8.6 Liberation of bound glycopeptides using hydroxylamine cleavage. The treatment of glycopeptides with hydroxylamine enables their release from the hydrazide support and generates a modified glycan consisting of the open ring initially attached to the support.

Enrichment of glycopeptides using biological recognition

One of the key requirements for biological systems is the ability to recognize specific compounds and biomolecules. This prerequisite has resulted in the generation of proteins, such as lectins, which have specificity and selectivity for carbohydrates in all domains of life (Sharon, 2008). The use of lectins and others sugar binding proteins, such as carbohydrate-specific antibodies, have enabled the enrichment of glycoproteins from complex mixtures in multiple protein-centric bacterial glycosylation studies (Doig *et al.*, 1996; Linton *et al.*, 2002; Young *et al.*, 2002; Gonzalez-Zamorano *et al.*, 2009; Vik *et al.*, 2009); however, no examples of this type of enrichment to peptide-centric bacterial glycosylation have been performed. Lectin affinity enrichment has been applied to peptide-centric studies of both *N*- (Kaji *et al.*, 2003; Zielinska *et al.*, 2010) and O-linked (Chalkley *et al.*, 2009; Darula and Medzihradszky, 2009) glycopeptides, demonstrating the compatibility of this approach with peptide substrates. The strength of lectin-based enrichment strategies lies in the selectivity of chosen lectins for specific carbohydrates and/or glycan configurations (Sharon,

2008). This specificity and the commercial availability of multiple lectins enabling enrichment of many different specific glycans allows identification of a broad range of glycan-containing structures (Yang and Hancock, 2005; Zielinska *et al.*, 2010). The interaction between carbohydrate and lectin, however, is generally of low affinity, particularly if only a single lectin binding site exists in the complex glycan structure (Duverger *et al.*, 2003). Low-affinity interactions are not ideal for enrichment and thus large amounts of sample, stringent washing procedures or the use of long affinity columns may be required to gain the resolving power necessary to separate glycosylated components from their non-glycosylated counterparts. Lectin affinity is also tempered by the requirement for essential co-factors, such as divalent metal ions for binding activity, and lectins can also be inactivated, owing to denaturation in the presence of chaotropic agents and detergents (Lotan *et al.*, 1977).

Another approach that can be utilized for the enrichment of glycopeptides is metabolic labelling coupled to chemical enrichment (Koenigs *et al.*, 2009; Breidenbach *et al.*, 2010). In this approach, unnatural carbohydrates containing functional groups that can be subjected to selective chemistry, such as CLICK chemistry (Rostovtsev *et al.*, 2002), are incorporated into the bacterial glycan by the endogenous glycosylation machinery. These modified glycans can then be subjected to the selective chemistry and coupled to a high-affinity tag or immobilized, enabling the enrichment of the glycopeptide (Fig. 8.7). This approach has only recently been demonstrated but may offer the ability to probe glycosylation pathways with great specificity. For example, the contribution of specific enzyme reactions within a given glycosylation pathway can be monitored in conjunction with genetic manipulation to observe the effects of the modified glycan at specific points in glycan biogenesis (Rostovtsev *et al.*, 2002). Bacterial glycosylation systems may also be uniquely amenable to this approach as studies on the oligosaccharyltransferase of both *N. meningitidis* and *C. jejuni* have shown these enzymes are able to utilize multiple glycans for glycosylation (Wacker *et al.*, 2006; Faridmoayer *et al.*, 2008). An extension of the use of metabolic labelling was also recently demonstrated in the elucidation of O-linked *N*-acetylglucosamine glycosylation (Khidekel *et al.*, 2007; Wang *et al.*, 2010). In this approach, incorporation of the chemically reactive carbohydrate occurs after generation of the glycan by an engineered galactosyltransferase, which is able to ligate the unnatural carbohydrate to *N*-acetylglucosamine and thus enable the selective enrichment of these glycopeptides (Khidekel *et al.*, 2007; Wang *et al.*, 2010). Although the method is still under development, it may represent one of the best strategies to enrich and identify modifications based on a single carbohydrate residue from complex mixtures – for example, the O-linked mannosylation of *M. tuberculosis* and Actinomycetes (Espitia *et al.*,

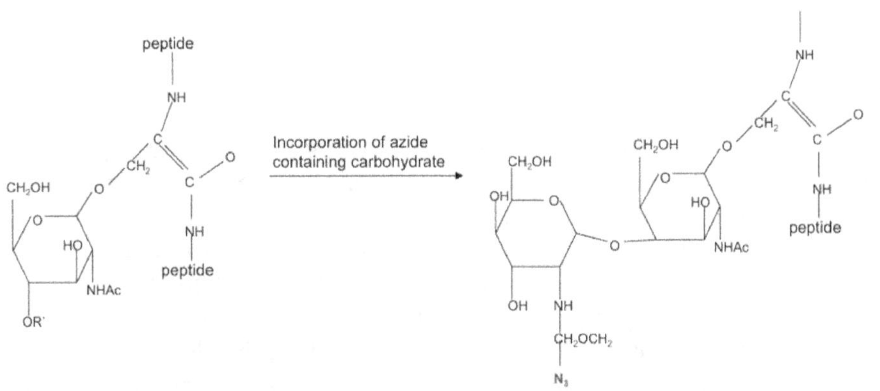

Figure 8.7 Chemical/enzymatic tagging of glycopeptides. By using *in vitro* metabolic labelling, carbohydrates containing reactive functional groups, in this case an azide, N_3, can be reacted with alkynes linked to a resin or affinity tag.

2010), if a suitable mannose-specific transferase could be identified.

Enrichment/isolation of glycopeptides using hydrophilic and hydrophobic chromatography

Although the diverse nature of bacterial glycans can result in differences in charge, susceptibility to chemical reactions and biological recognition, one property that is similarly altered in all glycopeptides compared with their unmodified counterparts is an increase in hydrophilicity. Hydrophilicity increases with glycan size and is the result of increased dipole–dipole interactions, ionic interactions and hydrogen bonding (Wuhrer *et al.*, 2009). Glycosylation therefore affects the behaviour of peptides in hydrophilic and hydrophobic chromatography. During the separation of glycopeptides using hydrophobic stationary phases such as reverse phase chromatography, glycopeptides generally exhibit poorer retention leading to the earlier elution of these peptides compared with their non-glycosylated forms (Wuhrer *et al.*, 2005). This phenomenon can be utilized within less complex samples to enable the isolation and identification of glycopeptides and has been largely used for single protein studies (Young *et al.*, 2002; Kowarik *et al.*, 2006). For samples of increasing complexity, this approach becomes less effective as the presence of hydrophilic, yet not glycosylated, peptides with similar retention profiles may mask the presence of the glycopeptides of interest.

The enrichment of glycopeptides using hydrophilic stationary phase strategies is increasingly being applied for bacterial glycoproteomics (Liu *et al.*, 2006; Ding *et al.*, 2009; Scott *et al.*, 2009, 2010) and stems from interest in bacterial glycosylation systems responsible for the modification of multiple proteins found in *C. jejuni* (Young *et al.*, 2002), *N. gonorrhoeae* (Vik *et al.*, 2009) and *Bacteroides fragilis* (Fletcher *et al.*, 2009), where the targets of glycan modification are not completely known. Enrichment approaches have included normal phase chromatography (Ding *et al.*, 2009), hydrophilic interaction liquid chromatography (HILIC) (Scott *et al.*, 2010) and graphite/carbon enrichment (Liu *et al.*, 2006; Scott *et al.*, 2009), with *C. jejuni* being the system most intensely examined. Liu *et al.* (2006) demonstrated the ability of graphite enrichment to purify protein-bound glycan in *C. jejuni* following pronase digestion to generate the glycan attached to a single amino acid. The binding of glycopeptides during graphite enrichment is mediated by both the carbohydrate and the peptide (Fan *et al.*, 1994), and it has been suggested that larger peptides and those containing aromatic groups may adsorb strongly to carbon (Yamaki *et al.*, 1996), preventing elution using conditions suitable for smaller glycopeptides. As such, this limits the size of the peptide component of the glycopeptide that can be eluted from the stationary phase and limits its utility with most peptide mixtures (e.g. trypsin derived).

Normal-phase chromatography (Ding *et al.*, 2009) and HILIC (Scott *et al.*, 2010) enrich glycopeptides in a similar manner, which is believed to be via the partitioning of glycopeptides into the aqueous-layer formed on the hydrophilic stationary phase and enabling hydrophobic components to be eluted in the mobile phase (Alpert, 1990). The presence of mobile components such as acids and ion pairing agents augment the effectiveness of glycopeptide binding (Ding *et al.*, 2009). Work on eukaryotic glycopeptides has shown that the addition of these agents increases hydrophilicity differences between glycopeptides and non-glycosylated peptides and enhances the removal of previously co-purifying non-glycosylated peptides (Mysling *et al.*, 2010). These approaches have both been applied to *C. jejuni* and have enabled the most in-depth analysis of bacterial glycosylation with over 80 modification sites identified (Ding *et al.*, 2009; Scott *et al.*, 2010).

Mass spectrometry analysis of bacterial proteins and peptides

Mass spectrometric analysis of bacterial glycopeptides

The confirmation of glycan attachment is essential in determining the glycosylation status of proteins. Owing to this requirement, the use of mass spectrometric analysis has become the methodology of choice for the confirmation of glycosylation from isolated proteins and during discovery-based studies coupled to the enrichment approaches discussed above (Harvey, 2001; Wuhrer *et al.*, 2005). Although all mass spectrometric analysis involves the measurement of mass and charge, these approaches are not a uniform methodology but rather a collection of related techniques that can utilize different ionization and fragmentation methods as well as instrumental configurations. These differences influence the information that can be generated and the ability to identify both glycan and peptide components, as well as decipher the location of the glycosylation site. This section therefore outlines the information that can be generated using multiple ionization, fragmentation and instrumental configurations and will focus on the development and applications of these to bacterial glycosylation.

Ionisation methodologies and their application to bacterial glycosylation

Mass spectrometric analyses of biopolymers are typically undertaken using two ionization methods: matrix-assisted laser desorption-ionization (MALDI) (Hillenkamp *et al.*, 1991) and electrospray ionization (ESI) (Fenn *et al.*, 1989). For protein identification and characterization, MALDI is typically used for 'peptide mass fingerprinting' (Hardouin, 2007), enabling the detection of singly charged ions of peptides derived from cleavage of proteins with a protease of known specificity and via comparison of the experimentally derived peptide masses (the 'fingerprint') with theoretically derived fingerprints available through publicly available databases (Gevaert and Vandekerckhove, 2000). Although this method is robust and sensitive for the identification of most biopolymers, it typically does not allow the effective analysis of glycopeptides in the presence of non-glycosylated peptides due to the suppression of ion signal under these conditions (Harvey, 2001; Larsen *et al.*, 2005). This is largely due to the harsh ionization nature of MALDI which can also lead to the loss of labile carbohydrates (Harvey, 2001). Although this limits the utility of MALDI in most bacterial glycosylation studies, such phenomena can be exploited to probe aspects of glycosylation, such as occupancy. This has been accomplished in the study of the *C. jejuni* protein JlpA, where it was shown that the loss of specific peptides from the MALDI mass spectrum corresponded with the occupation of these peptides by the *C. jejuni* glycan (Scott *et al.*, 2009).

ESI has in recent years become the method of choice for the study of glycopeptides and indeed it has been used in multiple bacterial glycosylation studies (Thibault *et al.*, 2001; Kowarik *et al.*, 2006; Verma *et al.*, 2006; Young *et al.*, 2002; McNally *et al.*, 2007; Ding *et al.*, 2009; Scott *et al.*, 2009, 2010). ESI results in less in-source fragmentation of carbohydrate structures compared with other ionization methods (Dell and Morris, 2001; Harvey, 2001). Unlike MALDI, ESI results in the generation of multiply charged ions that are the result of peptide composition, size, elution solvents and instrumental configuration (Kebarle and Verkerk, 2009). ESI has also been suggested to suffer from ion suppression of glycopeptide signals due to the competition with co-ionizing compounds (Annesley, 2003; Wohlgemuth *et al.*, 2010). This suppression can be largely overcome by coupling ESI to reversed-phase liquid chromatography (RP–LC), which provides enhanced separation and simplification of samples. Although this provides a clear advantage over MALDI, ESI requires samples to be largely free of salt and chargeable contaminants, while MALDI can typically operate with trace amounts present. Many contaminants result in significant signal suppression due to adduct formation in ESI (Kebarle and Verkerk, 2009) and this may require the use of additional preparative handling to remove them prior to analysis.

Fragmentation of glycopeptides

Confirmation of glycosylation sites requires the detection of the chemical linkage of a carbohydrate to the peptide and can be achieved by the use of MS-based fragmentation via MS^2 or MS^n if multiple rounds of fragmentation are utilized within specific mass spectrometers such as linear ion trap (Douglas *et al.*, 2005) and ion cyclotron resonance (ICR) mass spectrometers (Marshall *et al.*, 1998). Within all MS instruments the fragmentation of glycopeptides is governed by three factors, the peptide backbone, the glycan and the fragmentation approach (Scott *et al.*, 2010). All three of these parameters can potentially be manipulated to enhance the resulting fragmentation to produce ions that will provide informative spectra. The manipulation of fragmentation has only recently become widely possible with the advent of electron-based dissociation methods such as electron capture dissociation (ECD) (Zubarev *et al.*, 1998) and electron transfer dissociation (ETD) (Syka *et al.*, 2004). Prior to this, glycopeptide fragmentation was typically accomplished by the use of vibrational fragmentation approaches such as collision-induced dissociation (CID) and infrared multiphoton dissociation (IRMPD). Both electron-based and vibrational dissociation methods induce glycopeptide fragmentation differently and can lead to the generation of complementary MS information that can be essential to confirm the identity of the glycan, peptide and site of occupation.

Vibrational-induced fragmentation

Vibrational fragmentation approaches occur by the energetic bombardment of a selected ion species with non-reactive particles such as photons during IRMPD or the inert gases, nitrogen, helium or argon in the case of CID, the most common tandem fragmentation technique used. Both CID and IRMPD involve multiple low energy collisions, where the resulting kinetic energy is allowed to distribute through the selected ion leading to an increase in the internal energy and fragmentation predominantly at sites with the lowest energy barrier, or sites able to undergo unique chemical processes, such as residues located on the N-terminal side of prolines (Paizs and Suhai, 2005). In the case of multiply charged glycopeptides, the use of vibrational fragmentation typically yields spectra dominated by ions derived from cleavage within the glycan structure at the glycosidic bonds owing to the labile, low energy barrier of these bonds compared with other bond types, particularly peptide bonds (Figs. 8.5 and 8.8; Seipert *et al.*, 2008, 2009). Glycan fragmentation, which is denoted using the carbohydrate nomenclature of Domon and Costello (Domon and Costello, 1988) enables the identification of bacterial glycan structure and confirmation of the glycosylated status of a peptide. It should be noted that peptide fragmentation does still occur but is typically an order of magnitude less intense than that of carbohydrate fragmentation (Fig. 8.8). Although not

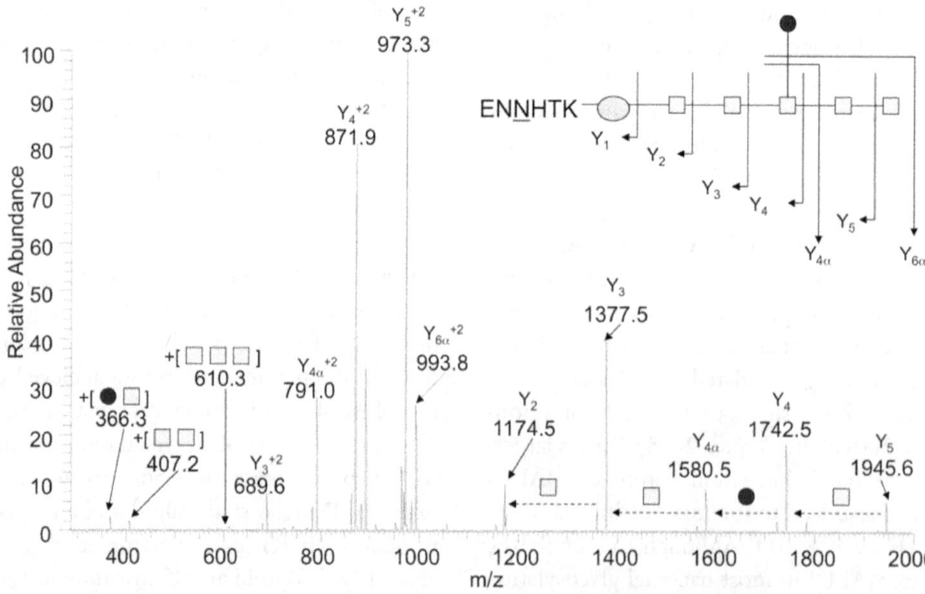

Figure 8.8 CID fragmentation of the *C. jejuni* N-linked glycopeptide ENNHTK from phosphoethanolamine transferase (Cj0256c). The complete composition of the N-linked glycan can be determined, including the position of the branching glucose residue (blue circle) but no information on the peptide composition can be obtained.

ideal, this has still enabled the identification of peptide components in a number of glycopeptide studies (Kelly *et al.*, 2006; Scott *et al.*, 2009, 2010). Vibrational fragmentation also has varying effects on different types of peptide–carbohydrate linkages with N-linked glycans being largely less labile than O-linked (Seipert *et al.*, 2008, 2009; Chalkley *et al.*, 2009). This property enables the localization of N-linked glycan attachment and has been used in both bacterial (Young *et al.*, 2002a) and eukaryotic studies (Chalkley *et al.*, 2009) to confirm the site of glycan attachment.

The dominant fragmentation of glycans during vibrational fragmentation generally results in difficulty in the determination of the underlying peptide identity. To overcome this, alterative types of vibrational fragmentation approaches are available via different instrumental configurations and can be performed to generate peptide sequence-related ions. These alterative types of vibrational fragmentation overcome the low energy barrier of glycans in two ways; (i) using higher energy collision events (Scott *et al.*, 2010); or (ii) by undergoing multiple rounds of fragmentation (Seipert *et al.*, 2008, 2009; Nilsson *et al.*, 2009). The first of these approaches can be accomplished on the higher-energy C-trap dissociation (HCD) cell of the Thermo Orbitrap mass spectrometer (Olsen *et al.*, 2007), and has been applied to the analysis of *C. jejuni* glycopeptides enabling the almost simultaneous identification of the glycan and peptide component of a glycopeptide due to the generation of peptide ions and detection with ultra-high mass accuracy (Fig. 8.9) (Scott *et al.*, 2010). Although the exact mechanism of HCD glycopeptide fragmentation is still poorly understood (Segu and Mechref, 2010), it is suggested that the high efficiency of the HCD collision results in higher internal energy, which enables the generation of amide bond cleavage events. The second group of alternative vibrational methods is to enable the generation of peptide ions by the fragmentation of deglycosylated or reduced glycan-containing ion products. This can be done by either the isolation of ions to perform MS^n in trapping instruments such as ion traps (Douglas *et al.*, 2005) and ICR mass spectrometers (Marshall *et al.*, 1998), or with fragmentation methods that act in a non-resonant manner, affecting both daughter and parent ions simultaneously (such as

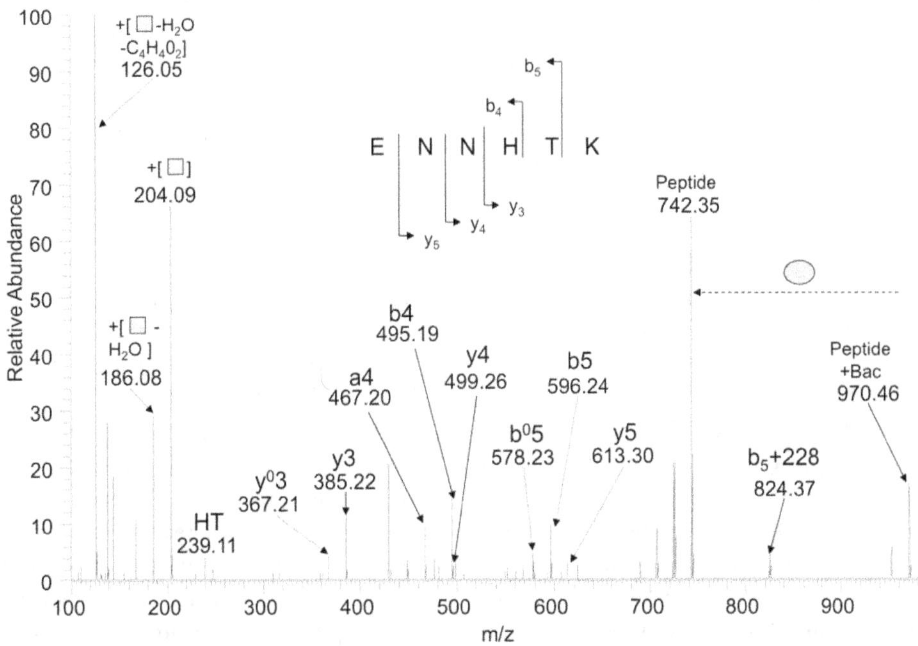

Figure 8.9 HCD fragmentation of the *C. jejuni* N-linked glycopeptide ENNHTK from phosphoethanolamine transferase (Cj0256c). Using HCD, peptide-related ions can be detected as well as the intense carbohydrate-related oxonium ion 204 (Gal*N*Ac) and associated cross ring cleavage products.

IRMPD; Seipert *et al.*, 2008, 2009). In respects to bacterial glycosylation, MS^n has been used in multiple studies (Young *et al.*, 2002; Ding *et al.*, 2009; Twine *et al.*, 2009), while IRMPD fragmentation has not been as widely used.

Electron-induced fragmentation

The alterative to vibrational fragmentation approaches is the use of electron driven, (ECD and ETD) fragmentation methodologies. The use of electrons leads to the selective cleavage of the peptide backbone and the generation of *c* and *z* ions compared with predominantly *b* and *y* ion generation produced by vibrational fragmentation. Importantly, electron-driven fragmentation achieves cleavage of the peptide backbone while maintaining labile modifications such as carbohydrate structures in their complete form (Wiesner *et al.*, 2008). This selectivity for peptide cleavage has been demonstrated using eukaryotic glycopeptides (Catalina *et al.*, 2007; Alley Jr., *et al.*, 2009) and more recently, glycopeptides isolated from *N. gonorrhoeae* (Vik *et al.*, 2009), *C. jejuni* (Scott *et al.*, 2010) and *C. difficile* (Twine *et al.*, 2009). Both ETD and ECD require specialized instrumentation, with ECD requiring high magnetic field and thus is best utilized in ICR mass spectrometers (Zubarev *et al.*, 1998; Wiesner *et al.*, 2008), while ETD is performed using ion–ion chemistry within ion traps (Syka *et al.*, 2004) or time-of-flight instruments (Han *et al.*, 2008). The resulting fragmentation patterns are similar with minor differences in the ability to fragment +2 ions, where ECD leads to greater dissociation (Zhang, 2010). ECD/ETD fragmentations are believed to occur by a non-ergodic process leading to cleavage in a sequence-independent manner with the exception of proline where no observable cleavage is seen (Wiesner *et al.*, 2008) and are performed more effectively at higher charge states (≥ +3; Good *et al.*, 2007) with observed masses under 850 m/z) (Darula and Medzihradszky, 2009). These optimal conditions are not always met with glycopeptides owing to the large neutral or acidic nature of some glycans, such as the N-linked heptasaccharide of *C. jejuni* (Scott *et al.*, 2010). In these situations, the resulting fragmentation will result in the generation of largely charge-reduced species, such as $[M+2H]^{+2}$ being converted to $[M+2H]^{+1}$ with the addition of an electron and poor *c* and *z* product ion series.

The ability of ETD/ECD to maintain labile modifications in their native state has resulted in considerable interest in the development of methods to enhance the efficiency of fragmentation for characterization of post-translationally modified proteins. These methods augment the basicity, charge and observed m/z of ions to those that favour generation of *c* and *z* ions and can be accomplished by either altering the peptide composition; by using enzymes which lead to larger peptides that typically exhibit higher charge states (Syka *et al.*, 2004), or by using derivatization strategies such as the alteration of N-termini with imidazole (Hennrich *et al.*, 2009), carboxylic acid groups with quaternary ammonium groups (Krusemark *et al.*, 2009), or the addition of (3-acrylamidopropyl)-trimethyl ammonium to the reactive side-chain of cysteine residues (Vasicek and Brodbelt, 2009). These aim to increase the charge or basicity of the analyte leading to enhanced fragmentation. Alternatively, the conditions used during the infusion of peptides into the mass spectrometer can be altered with agents known to increase the charge state of ions, such as *m*-nitrobenzyl alcohol (Kjeldsen *et al.*, 2007) and glycerol (Iavarone and Williams, 2003). These additives increase the charge state by affecting the amount of charge that can be tolerated on droplets formed in the ESI process (Iavarone and Williams, 2003; Kjeldsen *et al.*, 2007), but can have a detrimental effect on reverse phase chromatography (Kjeldsen *et al.*, 2007).

Multiple fragmentation approaches and high mass accuracy instrumental configurations for glycopeptide analysis

Although only recently implemented, the benefits of using both electron and vibrational fragmentation approaches to further characterize glycopeptides has already been demonstrated (Vik *et al.*, 2009; Scott *et al.*, 2010). Not all peptides will respond ideally to a single fragmentation approach, with fragmentation efficiency of vibrational fragmentation being observed for +2 ions while electron fragmentation appears more efficient when ions exhibit charge states ≥ +3

(Good *et al.*, 2007; Kalli and Hakansson, 2008). These observations have led to the development of approaches that enhance the number of identifications within a mixture and improve the generation of sequence-related ions by subjecting the selected precursor to both electron and vibrational fragmentation. Decision tree-driven tandem mass spectrometry (Swaney *et al.*, 2008) is one such approach that involves the determination of charge state during the MS experiment and is coupled to the use of statistical models to decide which fragmentation approach is most appropriate based on the specific charge states or m/z (Swaney *et al.*, 2008). The advantage of this approach is that within a single experiment both electron and vibrational fragmentation can be implemented and under ideal situations leads to a greater number of glycopeptide identifications. A second approach utilizes both fragmentation methods to enhance generation of sequence-specific ions and can be accomplished using supplemental collision activation to enhance ETD or simultaneous electron fragmentation and CID. In supplemental collision activation, vibrational energy is used to enhance the fragmentation of +2 ions subjected to ETD by enabling the dissociation of *c* and *z* ion products that owing to the mild nature of ETD fail to separate within the mass spectrometer (Swaney *et al.*, 2007). This approach uses lower levels of vibrational energy than standard vibrational fragmentation and leads to an increase in the detection of ETD generated *c* and *z* product ions from +2 precursors, as well as the absence of CID-related ions, however labile modifications can still be subject to vibrational-induced loss (Swaney *et al.*, 2007). In simultaneous electron fragmentation and CID, both electron and vibrational activation occur leading to the generation of *b*, *c*, *y* and *z* ions, as well as enhancing electron fragmentation in a similar manner to supplemental activation (Bushey *et al.*, 2009; Campbell *et al.*, 2009). This enables the generation of multiple ion series from two different fragmentation mechanisms, which if applied to glycopeptides may enable the identification of both the peptide and glycan within a single scan.

Another emerging mass spectrometric workflow is the use of mass spectrometers that can generate high resolution and high mass accuracy MS–MS spectra (Mann and Kelleher, 2008). This is a departure from previous developments that aimed to increase the number of MS–MS events and demonstrates the value of resolution and mass accuracy over scan events. There is already strong evidence that resolution and high mass accuracy improve the number of confident peptide identifications in comparison to increasing the sheer number of scans (McAlister *et al.*, 2010). High mass accuracy ensures much improved specificity in database searching (improved data *quality* over increased data *quantity*) and furthermore, improves differentiation of similar peptides (Mann and Kelleher, 2008; McAlister *et al.*, 2010). Increases in specificity are best demonstrated by the ability to accurately differentiate almost identical amino acid masses such as oxidized methionine (147.0354 Da) compared with phenylalanine (147.06841 Da), which on lower resolution instruments is not possible. Hybrid instruments such as the Orbitrap mass spectrometer afford users greater flexibility as two mass analysers are present allowing measurements to be performed quickly in the low resolution linear ion trap or slower on the high resolution orbital trap. A further advantage of this instrument is the ability to perform MS^n (Macek *et al.*, 2006) and ETD (McAlister *et al.*, 2008) within the linear ion trap and analyse the product ions in the high mass accuracy orbital trap. A consequence of the two mass analysers is the ability to measure MS–MS product ions concurrently in both the linear ion trap and the orbital trap, and this feature was utilized during analysis of the *C. jejuni* glycoproteome providing the confirmation of glycosylation, using the linear ion trap, and confirmation of the glycopeptide sequence, using HCD fragmentation in the orbital trap (Scott *et al.*, 2010).

The analysis of mass spectra from bacterial glycopeptides

Analysis of mass spectral data is unfortunately often viewed as a computational 'black box' and many studies do not use manual interpretation or validation of data. While this is sometimes unavoidable for large peptide identification data sets, the analysis of glycoproteomic mass spectral data is a highly specialized endeavour requiring informed consideration by the end user. The difficulties in data analysis arise from the non-uniform nature of bacterial glycosylation, which leads to a complete lack of standard approaches (or 'rules') and software capable of accurately interpreting the resulting fragmentation. As such the analysis of glycopeptide data warrants detailed discussion and the description of strategies that can be employed to generate meaningful results.

As discussed above, glycopeptide ionization is typically suppressed within complex mixtures. On the removal of non-glycosylated peptides by enrichment or isolation, glycopeptides become detectible and can typically form adducts with alkali metals, such as sodium or potassium (Fig. 8.10) (Harvey, 2001). The formation of these adducts are the result of multidentate interactions with the glycan and peptide (Seipert *et al.*, 2008) and can be used as diagnostic for selection of ions for further characterization by MS–MS. As these adducts are the result of metal ion interaction with mainly the hydroxyl oxygen of carbohydrates (Seipert *et al.*, 2008), the composition and size of the glycan can affect adduct formation, but generally for large neutral or acidic glycans an adduct can be formed. It is important to note the formation of metal adducts are not proof of glycosylation and confirmation of the glycan is required. As previously described, this confirmation can be acquired by fragmentation of glycopeptides with vibrational fragmentation methodologies. The dominant ions within the resulting MS–MS spectra typically correspond to cleavage events within the glycan, but these glycosidic fragments also present another set of diagnostic ions corresponding to the monomeric carbohydrate fragments known as carbohydrate oxonium ions (Figs. 8.8 and 8.9) (Carr *et al.*, 1993). As multiple carbohydrates of the same composition can exist in a glycan structure, these ions and the associated cross-ring cleavage products, can be orders of intensity above peptide-related product ions (Fig. 8.9).

The resulting sequential fragmentation of carbohydrates and the presence of oxonium ions, although useful for glycan determination, adversely affect the ability of peptide search algorithms to match glycopeptides even if complete peptide product ions are present. This is due to the negative effect intense unassigned glycan-related ions have on search algorithms such as MASCOT. To overcome this, manual validation and assignment of glycan and peptide ions must be undertaken. Although no strict guidelines exist for manual validation it is generally considered

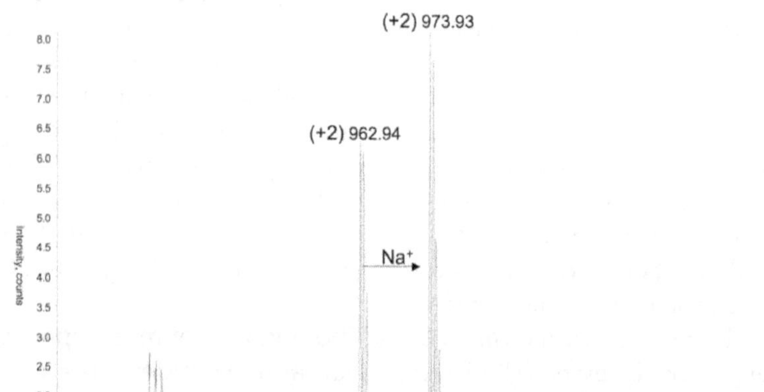

Figure 8.10 Formation of sodium adduct on the *C. jejuni* N-linked glycopeptide ^{144}DINAS^{148} from lipoprotein JlpA. The formation of these adducts are a by-product of the presence of the glycan and can be used as a preliminary scan to identify possible glycopeptides.

that the following criteria must be met to assign a spectrum manually:

- All major peaks must be assigned.
- The resulting product ions can only be assigned to a specific peptide if no other possible amino acid combination can be determined.
- The error, as defined by the difference in the observed mass to the expected, must be consistent between all ions observed in the spectrum.
- Peptide sequence variations (e.g. errors in the genome sequence) and unusual modifications are extremely rare and should only be assigned if no other explanation can be found and if other supporting evidence confirms the assignment.

Manual validation, although highly useful and accurate, typically lacks statistical validation, which can result in a glycopeptide being mis-assigned owing to unforeseen redundancy or modification within proteomes. To address this, methodology to enable the implementation of existing search algorithms are desirable. A form of post-acquisition manipulation known as spectral processing can be used to remove carbohydrate-related ions that influence scoring, thus enabling assignment of the peptide identity. This processing approach mirrors techniques recently demonstrated to increase the accuracy scoring of ETD data (Good *et al.*, 2009, 2010) and can be accomplished by first assessing the ions associated with glycosylation and removing these from the peptide search query (Fig. 8.11). This approach has been shown to improve scoring of glycopeptides from *C. jejuni*. It should be noted that prior knowledge of the glycan structure aids in the design and implementation of spectral processing workflows, as well as the use of fragmentation methods where the glycan is maintained, such as ETD or ECD, as without this data algorithms are unable to successfully assign modified peptides.

Mass spectrometric analysis of intact bacterial glycoproteins

The characterization of protein glycosylation can be a complex procedure, especially for prokaryotic

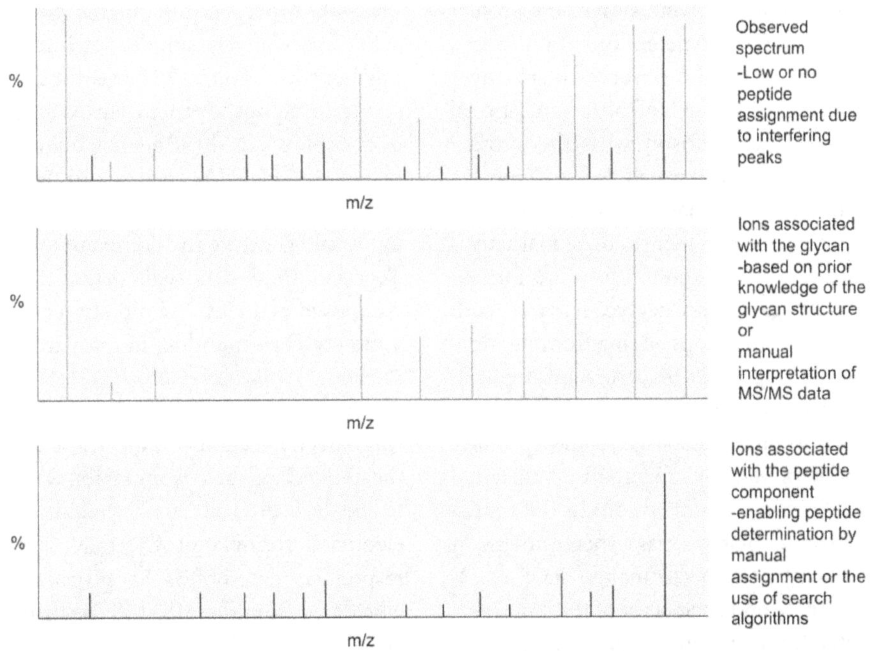

Figure 8.11 Spectral processing of glycopeptides. Within vibrational fragmentation spectra the overlapping peptide and glycan fragmentation prevents the determination of peptide composition. By identifying the glycan-related ions and subtracting these the determination of the peptide can be simplified.

O-linked glycoproteins whose glycans are often unique and for which no glycosylation consensus motif is known. Complementary to the peptide centric, 'bottom-up' approaches described earlier, a 'top-down' approach, described for bacterial glycoproteins by Schirm *et al.* (2005) rapidly provides information on the extent of glycosylation, the molecular masses of putative glycan moieties, and the composition of oligosaccharide residues on bacterial glycoproteins. Typically, top-down protein studies are conducted using high-resolution Fourier transform mass spectrometers that can yield isotopically resolved spectra even for 112 kDa proteins (Kelleher, 2004). The approach outlined by Schirm *et al.*, however, can be carried out using lower resolution and less costly Q-TOF instrumentation (Schirm *et al.*, 2005). Minimal sample preparation is required and the mass spectrometry analysis can be carried out in minutes.

Typically, this technique has been applied to purified protein samples, although it has been demonstrated to be effective in protein mixtures (Schirm *et al.*, 2005) or culture supernatants (Twine *et al.*, 2008). The observed mass of the protein can be determined from spectral deconvolution of the envelope of multiply charged protein ions observed upon direct infusion of the protein into an ESI mass spectrometer. This is illustrated in Fig. 8.12A and B. The observed mass gives an indication to the extent of post-translational modification, by comparison with the protein mass predicted from the translated gene sequence.

Next, multiply protonated protein precursor ions are subjected to low-energy collisional activation. Typically, glycan oxonium ions are formed as a result of gas-phase deglycosylation with relatively few fragment ions arising from peptidic bond cleavages (Fig. 8.12). Structural information on putative carbohydrate residues can be obtained either by performing a second-generation product ion scan of oxonium ions formed by collisional activation of the intact protein ions in the source region, or by multi-stage mass spectrometry, if using an ion trap mass spectrometer (Fig. 8.12C).

The selectivity and specificity of this approach was exploited for the identification of the unusual glycan modifications on the flagellin from *H. pylori* (Schirm *et al.*, 2005). In some instances, labile peptide ions are also liberated from the glycoprotein and these can be used to identify/confirm the protein sequence. This top-down approach was used to characterize flagellin glycosylation from a wide range of *C. botulinum* isolates (Twine *et al.*, 2008b; Carter *et al.*, 2009). All strains produced the same peptide fragment ions but the glycan ions varied significantly (Twine *et al.*, 2008; Carter *et al.*, 2009). Thus there is potential in utilizing flagellin glycan-associated fragment ions for strain differentiation.

Characteristic bacterial glycan oxonium ions

Bacterial glycans are so diverse that mass determination alone is not sufficient to identify even those that have been previously characterized. Further information regarding the glycan composition is required and can often be obtained by performing multi-stage mass spectrometry analysis on the glycan oxonium ions observed in the bottom-up and top-down studies described above. For example, an intense glycan oxonium ion at *m/z* 399.1 is observed in the infusion–MS spectrum of the intact *C. difficile* flagellin protein as well as in the MS–MS spectrum of the intact protein ion at *m/z* 1119.7 (Fig. 8.12A and B, respectively). The MS–MS spectrum of this glycan ion, Fig. 8.12C, is relatively simple but informative. The fragment ion at *m/z* 284 was thought to correspond to a phoshorylated Hex*N*Ac. Further loss of phosphoric acid (98 amu) would give rise to a dehydrated Hex*N*Ac at *m/z* 186. Neutral loss of a dehydrated Hex*N*Ac from the precursor glycan ion would generate the dominant ion at *m/z* 214. Together, these data suggest that the glycan was composed of a Hex*N*Ac moiety, a phosphate and a moiety corresponding in mass to aspartic acid (115 amu) (Twine *et al.*, 2009). Accurate mass measurement provided further evidence that this was indeed the composition. The accurate mass of the *C. difficile* glycan oxonium ion was determined to be 399.1164 ± 0.0012, yielding a plausible elemental formula of $C_{13}H_{24}N_2O_{10}P$ that corresponded exactly to the glycan composition elucidated from the MS–MS spectrum (Twine *et al.*, 2009).

A thorough analysis of a novel bacterial glycan usually requires NMR analysis. Powerful though NMR is for these studies, it is very insensitive

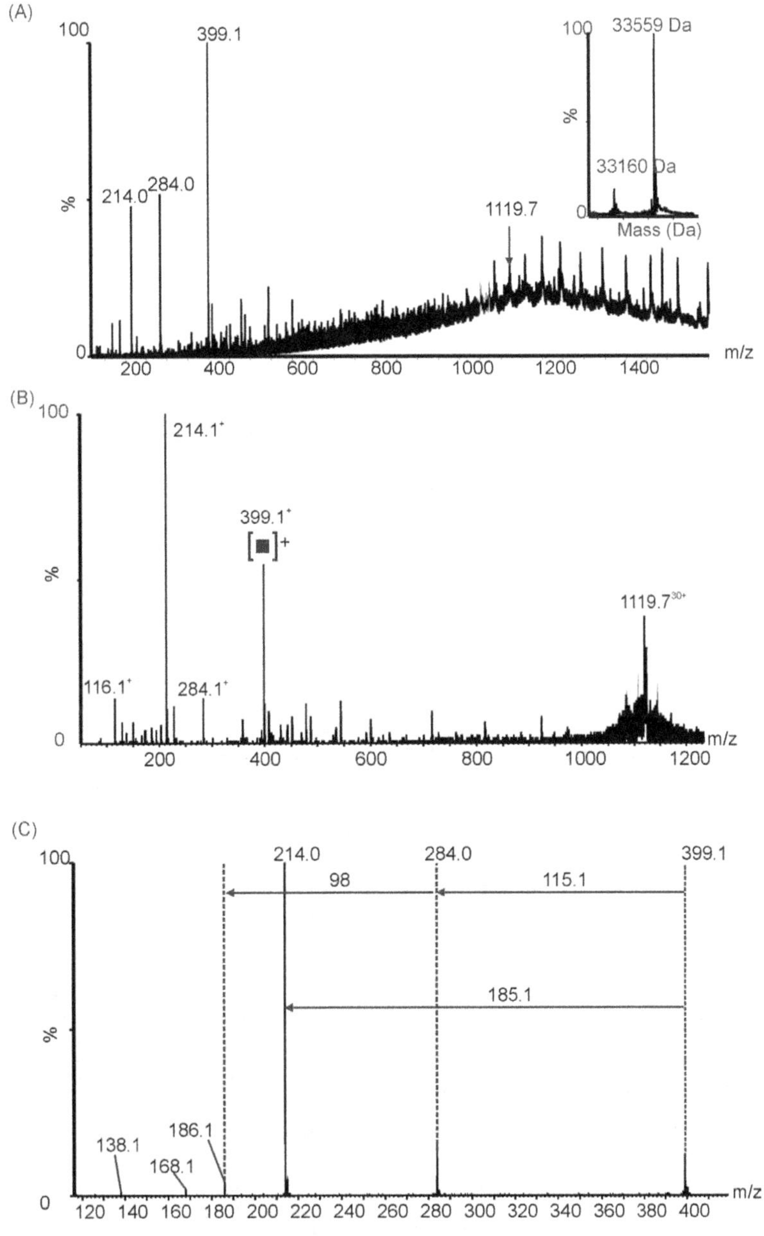

Figure 8.12 Electrospray mass spectrometry and tandem mass spectrometry analyses of intact flagellin protein from *C. difficile* strain 630 (A). Electrospray mass spectrum of intact flagellin from *Clostridium difficile* strain 630. Note that the presence of glycan oxonium ion at *m/z* 399.1 and glycan related fragment ion at *m/z* 284.0 and 214.0. Inset is the reconstructed molecular mass profile showing two major peaks at 33,559 and, 33,160 Da, respectively. Note that the separation between these glycoprotein isoforms is equivalent to one glycan (398 Da). (B) Tandem mass spectrometry analyses of the multiply charged glycoprotein ion at *m/z* 1119.7^{30+} showing fragment ions at *m/z* 399.1, 284.1, 214. 1 and 116.1, indicating that these ions arise from the glycoprotein. (C) Second-generation product ion spectra of the glycan oxonium ion at *m/z* 399.1. The glycan precursor ion was generated by in-source fragmentation of the multiply charged flagellin ions. Loss of putative dehydrated HexNAc (185 amu) from the precursor ion gave rise to the intense fragment ion at *m/z* 284.1. Loss of 115 amu from the precursor gave rise to the fragment ion at *m/z* 284, thought to correspond to a phoshorylated HexNAc. Loss of phosphoric acid (98 amu) from this ion gave rise to a dehydrated HexNAc at *m/z* 186. Together, these data suggested that the unknown glycan was a HexNAc moiety, a phosphate and a moiety corresponding in mass to aspartic acid. These data were reported in Twine *et al.* (2009) and have been modified from the original format.

compared with mass spectrometry, requiring at least 20–50 μg of purified glycan or glycopeptide. In many studies of bacterial glycans, this can prove challenging to obtain. Larger glycans such as the heptasaccharide found attached to the antigen PEB3 from *C. jejuni* can be isolated by size exclusion chromatography after pronase/proteinase K digestion (Young *et al.*, 2002). Smaller glycans are harder to purify in quantities sufficient for NMR analysis. They can be isolated as short glycopeptides by HPLC (Twine *et al.*, 2008) or by releasing the glycan from the glycoprotein by β-elimination. In addition, purification of intracellular sugar metabolites has proved useful in the structural elucidation of bacterial glycans in a few cases (Soo *et al.*, 2004; McNally *et al.*, 2006, 2007; Howard *et al.*, 2009; Logan *et al.*, 2009; Soo and Hui, 2010). These metabolites are typically analysed on Q-TOF instruments which proves only a single round of MS fragmentation but high mass accuracy (<100 ppm). To gain further insight into the composition of the novel glycan it is often advantageous to perform further MS^n (MS^3 or greater) analyses (Schirm *et al.*, 2005). The caveat of these MS^n studies is the requirement of intense glycan ions to start with as the intensity of the fragment ions generated will decrease approximately 10-fold with each sequential MS step in the analysis.

Conclusion

The rapid evolution of mass spectrometry is allowing more rapid and detailed characterization of bacterial glycoproteins. Accompanied by continued increasingly sophisticated analytical approaches for enriching glycoprotein and peptides, this opens the door to a more rapid and extensive inventory of the bacterial glycoproteome. Techniques not described in this chapter also provide additional glycan structural information, such as NMR. Alternative approaches, metabolomics studies offer an alternative approach to study N- or O-linked protein glycosylation biosynthetic pathways. In the future, the focus will be higher throughput and encompassing studies of the entire bacterial glycoproteme, requiring the continued development of analytical approaches. The largest challenge, however, remains to determine the biological role of bacterial glycoproteins. This is of particular interest in pathogenic organisms, where there is a potential role in host–pathogen interactions.

Acknowledgement

The authors thank Kelly Fulton for glycostaining analyses of *Francisella novicida* and critical reading of this manuscript.

References

Aas, F.E., Egge-Jacobsen, W., Winther-Larsen, H.C., Lovold, C., Hitchen, P.G., Dell, A., and Koomey, M. (2006). *Neisseria gonorrhoeae* type IV pili undergo multisite, hierarchical modifications with phosphoethanolamine and phosphocholine requiring an enzyme structurally related to lipopolysaccharide phosphoethanolamine transferases. J. Biol. Chem. *281*, 27712–27723.

Alley, W.R., Jr., Mechref, Y., and Novotny, M.V. (2009). Characterization of glycopeptides by combining collision-induced dissociation and electron-transfer dissociation mass spectrometry data. Rapid Commun. Mass Spectrom. *23*, 161–170.

Alpert, A.J. (1990). Hydrophilic-interaction chromatography for the separation of peptides, nucleic acids and other polar compounds. J. Chromatogr. *499*, 177–196.

Altman, E., Schaffer, C., Brisson, J.R., and Messner, P. (1995). Characterization of the glycan structure of a major glycopeptide from the surface layer glycoprotein of *Clostridium thermosaccharolyticum* E207–71. Eur. J. Biochem. *229*, 308–315.

Annesley, T.M. (2003). Ion suppression in mass spectrometry. Clin. Chem. *49*, 1041–1044

Bae, S.H., Harris, A.G., Hains, P.G., Chen, H., Garfin, D.E., Hazell, S.L., Paik, Y.K., Walsh, B.J., and Cordwell, S.J. (2003). Strategies for the enrichment and identification of basic proteins in proteome projects. Proteomics 3, 569–579.

Bogdanov, B., and Smith, R.D. (2005). Proteomics by FTICR mass spectrometry, top down and bottom up. Mass Spectrom. Rev. *24*, 168–200.

Breidenbach, M.A., Gallagher, J.E., King, D.S., Smart, B.P., Wu, P., and Bertozzi, C.R. (2010). Targeted metabolic labelling of yeast N-glycans with unnatural sugars. Proc. Natl. Acad. Sci. U.S.A. *107*, 3988–3993.

bu-Qarn, M., Eichler, J., and Sharon, N. (2008). Not just for Eukarya anymore, protein glycosylation in Bacteria and Archaea. Curr. Opin. Struct. Biol. *18*, 544–550.

Bushey, J.M., Baba, T., Glish, G.L. (2009). Simultaneous collision induced dissociation of the charge reduced parent ion during electron capture dissociation. Anal. Chem. *81*, 6156–6164.

Campbell, J.L., Hager, J.W., and Le Blanc, J.C. (2009). On performing simultaneous electron transfer dissociation and collision-induced dissociation on multiply protonated peptides in a linear ion trap. J. Am. Soc. Mass Spectrom. *20*, 1672–1683.

Carlson, D.M. (1968). Structures and immunochemical properties of oligosaccharides isolated from pig submaxillary mucins. J. Biol. Chem. *243,* 616–626.

Carr, S.A., Huddleston, M.J., and Bean, M.F. (1993). Selective identification and differentiation of N- and O-linked oligosaccharides in glycoproteins by liquid chromatography-mass spectrometry. Protein Sci. *2,* 183–196.

Carter, A.T., Paul, C.J., Mason, D.R., Twine, S.M., Alston, M.J., Logan, S.M., Austin, J.W., and Peck, M.W. (2009). Independent evolution of neurotoxin and flagellar genetic loci in proteolytic Clostridium botulinum. BMC Genomics *10,* 115.

Castric, P., Cassels, F.J., and Carlson, R.W. (2001). Structural characterization of the *Pseudomonas aeruginosa* 1244 pilin glycan. J. Biol. Chem. *276,* 26479–26485.

Catalina, M.I., Koeleman, C.A., Deelder, A.M., and Wuhrer, M. (2007). Electron transfer dissociation of N-glycopeptides, loss of the entire N-glycosylated asparagine side-chain. Rapid Commun. Mass Spectrom. *21,* 1053–1061.

Chaban, B., Voisin, S., Kelly, J., Logan, S.M., and Jarrell, K.F. (2006). Identification of genes involved in the biosynthesis and attachment of *Methanococcus voltae* N-linked glycans, insight into N-linked glycosylation pathways in Archaea. Mol. Microbiol. *61,* 259–268.

Chaban, B., Logan, S.M., Kelly, J.F., and Jarrell, K.F. (2009). AglC and AglK are involved in biosynthesis and attachment of diacetylated glucuronic acid to the N-glycan in Methanococcus voltae. J. Bacteriol. *191,* 187–195.

Chalkley, R.J., Thalhammer, A., Schoepfer, R., and Burlingame, A.L. (2009). Identification of protein O-GlcNAcylation sites using electron transfer dissociation mass spectrometry on native peptides. Proc. Natl. Acad. Sci. U.S.A. *106,* 8894–8899.

Chen, R., Jiang, X., Sun, D., Han, G., Wang, F., Ye, M., Wang, L., and Zou, H. (2009). Glycoproteomics analysis of human liver tissue by combination of multiple enzyme digestion and hydrazide chemistry. J. Proteome Res. *8,* 651–661.

Christian, R., Schulz, G., Schuster-Kolbe, J., Allmaier, G., Schmid, E.R., Sleytr, U.B., and Messner, P. (1993). Complete structure of the tyrosine-linked saccharide moiety from the surface layer glycoprotein of Clostridium thermohydrosulfuricum S102–70. J. Bacteriol. *175,* 1250–1256.

Clowers, B.H., Dodds, E.D., Seipert, R.R., and Lebrilla, C.B. (2007). Site determination of protein glycosylation based on digestion with immobilized nonspecific proteases and Fourier transform ion cyclotron resonance mass spectrometry. J. Proteome Res. *6,* 4032–4040.

Cordwell, S.J. (2006). Technologies for bacterial surface proteomics. Curr. Opin. Microbiol. *9,* 320–329.

Cummings, R.D. (1994). Use of lectins in analysis of glycoconjugates. Methods Enzymol. *230,* 66–86.

Darula, Z., and Medzihradszky, K.F. (2009). Affinity enrichment and characterization of mucin core-1 type glycopeptides from bovine serum. Mol. Cell Proteomics *8,* 2515–2526.

Dell, A., and Morris, H.R. (2001). Glycoprotein structure determination by mass spectrometry. Science *291,* 2351–2356

Dhiman, N., Verma, I., and Khuller, G.K. (1997). Cellular immune responses to cell wall peptidoglycan associated protein antigens in tuberculosis patients and healthy subjects. Microbiol. Immunol. *41,* 495–502.

Ding, W., Nothaft, H., Szymanski, C.M., and Kelly, J. (2009). Identification and quantification of glycoproteins using ion-pairing normal-phase LC and MS. Mol. Cell Proteomics. *8,* 2170–2185.

Doig, P., Kinsella, N., Guerry, P., and Trust, T.J. (1996). Characterization of a post-translational modification of *Campylobacter flagellin,* identification of a sero-specific glycosyl moiety. Mol. Microbiol. *19,* 379–387.

Domon, B., and Costello, C.E. (1988). A systematic nomenclature for carbohydrate fragmentation in FAB–MS/MS spectra of glycoconjugate. Glycoconjugate J. *5,* 397–409.

Douglas, D.J., Frank, A.J., and Mao, D. (2005). Linear ion traps in mass spectrometry. Mass Spectrom. Rev. *24,* 1–29.

Duverger, E., Frison, N., Roche, A.C., and Monsigny, M. (2003). Carbohydrate–lectin interactions assessed by surface plasmon resonance. Biochimie *85,* 167–179.

Edge, A.S. (2003). Deglycosylation of glycoproteins with trifluoromethanesulphonic acid, elucidation of molecular structure and function. Biochem. J. *376,* 339–350.

Edge, A.S., Faltynek, C.R., Hof, L., Reichert, L.E., Jr., and Weber, P. (1981). Deglycosylation of glycoproteins by trifluoromethanesulfonic acid. Anal. Biochem. *118,* 131–137.

Espitia, C., Servin-Gonzalez, L., and Mancilla, R. (2010). New insights into protein O-mannosylation in actinomycetes. Mol. Biosyst. *6,* 775–781.

Fan, J.Q., Kondo, A., Kato, I., and Lee, Y.C. (1994). High-performance liquid chromatography of glycopeptides and oligosaccharides on graphitized carbon columns. Anal. Biochem. *219,* 224–229.

Faridmoayer, A., Fentabil, M.A., Haurat, M.F., Yi, W., Woodward, R., Wang, P.G., and Feldman, M.F. (2008). Extreme substrate promiscuity of the *Neisseria* oligosaccharyl transferase involved in protein O-glycosylation. J. Biol. Chem. *283,* 34596–34604.

Fenn, J.B., Mann, M., Meng, C.K., Wong, S.F., and Whitehouse, C.M. (1989). Electrospray ionization for mass spectrometry of large biomolecules. Science *246,* 64–71.

Fletcher, C.M., Coyne, M.J., Villa, O.F., Chatzidaki-Livanis, M., and Comstock, L.E. (2009). A general O-glycosylation system important to the physiology of a major human intestinal symbiont. Cell *137,* 321–331.

Gauci, S., Helbig, A.O., Slijper, M., Krijgsveld, J., Heck, A.J., and Mohammed, S. (2009). Lys-N and trypsin cover complementary parts of the phosphoproteome in a refined SCX-based approach. Anal. Chem. *81,* 4493–4501.

Geoghegan, K.F., and Stroh, J.G. (1992). Site-directed conjugation of nonpeptide groups to peptides and proteins via periodate oxidation of a 2-amino alcohol. Application to modification at N-terminal serine. Bioconjug. Chem. *3,* 138–146.

Gerwig, G.J., Kamerling, J.P., Vliegenthart, J.F., Morag, E., Lamed, R., and Bayer, E.A. (1993). The nature of the carbohydrate-peptide linkage region in glycoproteins from the cellulosomes of *Clostridium thermocellum* and *Bacteroides cellulosolvens.* J. Biol. Chem. *268,* 26956–26960.

Gevaert, K., and Vandekerckhove, J. (2000). Protein identification methods in proteomics. Electrophoresis *21,* 1145–1154.

Gonzalez-Zamorano, M., Mendoza-Hernandez, G., Xolalpa, W., Parada, C., Vallecillo, A.J., Bigi, F., and Espitia, C. (2009). *Mycobacterium tuberculosis* glycoproteomics based on ConA-lectin affinity capture of mannosylated proteins. J. Proteome Res. *8,* 721–733.

Good, D.M., Wenger, C.D., McAlister, G.C., Bai, D.L., Hunt, D.F., and Coon, J.J. (2009). Post-acquisition ETD spectral processing for increased peptide identifications. J. Am. Soc. Mass Spectrom. *20,* 1435–1440.

Good, D.M., Wirtala, M., McAlister, G.C., and Coon, J.J. (2007). Performance characteristics of electron transfer dissociation mass spectrometry. Mol. Cell. Proteomics *6,* 1942–1951.

Good, D.M., Wenger, C.D., and Coon, J.J. (2010). The effect of interfering ions on search algorithm performance for electron-transfer dissociation data. Proteomics *10,* 164–167

Gross, J., Grass, S., Davis, A.E., Gilmore-Erdmann, P., Townsend, R.R., and St, G.J., III. (2008). The *Haemophilus influenzae* HMW1 adhesin is a glycoprotein with an unusual N-linked carbohydrate modification. J. Biol. Chem. *283,* 26010–26015.

Gundry, R.L., White, M.Y., Murray, C.I., Kane, L.A., Fu, Q., Stanley, B.A., and Van Eyk, J.E. (2009). Preparation of proteins and peptides for mass spectrometry analysis in a bottom-up proteomics workflow. Curr. Protoc. Mol. Biol. Unit 10.

Han, H., Xia, Y., Yang, M., and McLuckey, S.A. (2008). Rapidly alternating transmission mode electron-transfer dissociation and collisional activation for the characterization of polypeptide ions. Anal. Chem. *80,* 3492–3497.

Han, K.K., Richard, C., Zhang, G.Y., and Delacourte, A. (1986). Sequence homology analysis of proteins by chemical cleavages, using a mono and two dimensional sodium dodecyl sulfate-polyacrylamide gel electrophoresis. Int. J. Biochem. *18,* 1073–1082.

Han, X., Aslanian, A., and Yates, J.R., III. (2008). Mass spectrometry for proteomics. Curr. Opin. Chem. Biol. *12,* 483–490.

Hardouin, J. (2007). Protein sequence information by matrix-assisted laser desorption/ionization in-source decay mass spectrometry. Mass Spectrom. Rev. *26,* 672–682.

Harvey, D.J. (2001). Identification of protein-bound carbohydrates by mass spectrometry. Proteomics *1,* 311–328.

Hauser, N.J., and Basile, F. (2008). Online microwave D-cleavage LC-ESI–MS/MS of intact proteins, site-specific cleavages at aspartic acid residues and disulfide bonds. J. Proteome Res. *7,* 1012–1026.

Helenius, A., and Aebi, M. (2004). Roles of N-linked glycans in the endoplasmic reticulum. Annu. Rev. Biochem. *73,* 1019–1049.

Hennrich, M.L., Boersema, P.J., van den, T.H., Mischerikow, N., Heck, A.J., and Mohammed, S. (2009). Effect of chemical modifications on peptide fragmentation behavior upon electron transfer induced dissociation. Anal. Chem. *81,* 7814–7822.

Hillenkamp, F., Karas, M., Beavis, R.C., and Chait, B.T. (1991). Matrix-assisted laser desorption/ionization mass spectrometry of biopolymers. Anal. Chem. *63,* 1193A–1203A.

Howard, S.L., Jagannathan, A., Soo, E.C., Hui, J.P., Aubry, A.J., Ahmed, I., Karlyshev, A., Kelly, J.F., Jones, M.A., Stevens, M.P., *et al.* (2009). *Campylobacter jejuni* glycosylation island important in cell charge, legionaminic acid biosynthesis, and colonization of chickens. Infect. Immun. *77,* 2544–2556.

Iavarone, A.T., and Williams, E.R. (2003). Mechanism of charging and supercharging molecules in electrospray ionization. J. Am. Chem. Soc. *125,* 2319–2327.

Kaji, H., Saito, H., Yamauchi, Y., Shinkawa, T., Taoka, M., Hirabayashi, J., Kasai, K., Takahashi, N., and Isobe, T. (2003). Lectin affinity capture, isotope-coded tagging and mass spectrometry to identify N-linked glycoproteins. Nat. Biotechnol. *21,* 667–672.

Kalli, A., and Hakansson, K. (2008). Comparison of the electron capture dissociation fragmentation behavior of doubly and triply protonated peptides from trypsin, Glu-C, and chymotrypsin digestion. J. Proteome Res. *7,* 2834–2844.

Kebarle, P., and Verkerk, U.H. (2009). Electrospray, from ions in solution to ions in the gas phase, what we know now. Mass Spectrom. Rev. *28,* 898–917.

Kelleher, N.L. (2004). Top-down proteomics. Anal. Chem. *76,* 197A–203A

Kelly, J., Jarrell, H., Millar, L., Tessier, L., Fiori, L.M., Lau, P.C., Allan, B., and Szymanski, C.M. (2006). Biosynthesis of the N-linked glycan in *Campylobacter jejuni* and addition onto protein through block transfer. J. Bacteriol. *188,* 2427–2434.

Kelly, J., Logan, S.M., Jarrell, K.F., Vandyke, D.J., and Vinogradov, E. (2009). A novel N-linked flagellar glycan from *Methanococcus maripaludis.* Carbohydr. Res. *344,* 648–653.

Khidekel, N., Ficarro, S.B., Clark, P.M., Bryan, M.C., Swaney, D.L., Rexach, J.E., Sun, Y.E., Coon, J.J., Peters, E.C., and Hsieh-Wilson, L.C. (2007). Probing the dynamics of O-GlcNAc glycosylation in the brain using quantitative proteomics. Nat. Chem. Biol. *3,* 339–348

Kjeldsen, F., Giessing, A.M., Ingrell, C.R., and Jensen, O.N. (2007). Peptide sequencing and characterization of post-translational modifications by enhanced ion-charging and liquid chromatography electron-transfer dissociation tandem mass spectrometry. Anal. Chem. *79,* 9243–9252.

Klein, C., Aivaliotis, M., Olsen, J.V., Falb, M., Besir, H., Scheffer, B., Bisle, B., Tebbe, A., Konstantinidis, K., Siedler, F., *et al.* (2007). The low molecular weight proteome of *Halobacterium salinarum*. J. Proteome Res. *6,* 1510–1518.

Klement, E., Lipinszki, Z., Kupihar, Z., Udvardy, A., and Medzihradszky, K.F. (2010). Enrichment of O-GlcNAc modified proteins by the periodate oxidation-hydrazide resin capture approach. J. Proteome Res. *9,* 2200–2206.

Koenigs, M.B., Richardson, E.A., and Dube, D.H. (2009). Metabolic profiling of *Helicobacter pylori* glycosylation. Mol. Biosyst. *5,* 909–912.

Koketsu, M., and Linhardt, R.J. (2000). Electrophoresis for the analysis of acidic oligosaccharides. Anal. Biochem. *283,* 136–145.

Kowarik, M., Young, N.M., Numao, S., Schulz, B.L., Hug, I., Callewaert, N., Mills, D.C., Watson, D.C., Hernandez, M., Kelly, J.F., *et al.* (2006). Definition of the bacterial N-glycosylation site consensus sequence. EMBO J. *25,* 1957–1966.

Krusemark, C.J., Frey, B.L., Belshaw, P.J., and Smith, L.M. (2009). Modifying the charge state distribution of proteins in electrospray ionization mass spectrometry by chemical derivatization. J. Am. Soc. Mass Spectrom. *20,* 1617–1625.

Kweon, H.K., and Hakansson, K. (2006). Selective zirconium dioxide-based enrichment of phosphorylated peptides for mass spectrometric analysis. Anal. Chem. *78,* 1743–1749.

Larsen, M.R., Hojrup, P., and Roepstorff, P. (2005a). Characterization of gel-separated glycoproteins using two-step proteolytic digestion combined with sequential microcolumns and mass spectrometry. Mol. Cell Proteomics *4,* 107–119.

Larsen, M.R., Thingholm, T.E., Jensen, O.N., Roepstorff, P., and Jorgensen, T.J. (2005b). Highly selective enrichment of phosphorylated peptides from peptide mixtures using titanium dioxide microcolumns. Mol. Cell Proteomics *4,* 873–886.

Larsen, M.R., Jensen, S.S., Jakobsen, L.A., and Heegaard, N.H. (2007). Exploring the sialiome using titanium dioxide chromatography and mass spectrometry. Mol. Cell Proteomics *6,* 1778–1787.

Linton, D., Allan, E., Karlyshev, A.V., Cronshaw, A.D., and Wren, B.W. (2002). Identification of *N*-acetylgalactosamine-containing glycoproteins PEB3 and CgpA in *Campylobacter jejuni*. Mol. Microbiol. *43,* 497–508.

Liu, X., McNally, D.J., Nothaft, H., Szymanski, C.M., Brisson, J.R., and Li, J. (2006). Mass spectrometry-based glycomics strategy for exploring N-linked glycosylation in eukaryotes and bacteria. Anal. Chem. *78,* 6081–6087.

Logan, S.M., Hui, J.P., Vinogradov, E., Aubry, A.J., Melanson, J.E., Kelly, J.F., Nothaft, H., and Soo, E.C. (2009). Identification of novel carbohydrate modifications on *Campylobacter jejuni* 11168 flagellin using metabolomics-based approaches. FEBS J. *276,* 1014–1023.

Lotan, R., Beattie, G., Hubbell, W., and Nicolson, G.L. (1977). Activities of lectins and their immobilized derivatives in detergent solutions. Implications on the use of lectin affinity chromatography for the purification of membrane glycoproteins. Biochemistry *16,* 1787–1794.

McAlister, G.C., Berggren, W.T., Griep-Raming, J., Horning, S., Makarov, A., Phanstiel, D., Stafford, G., Swaney, D.L., Syka, J.E., Zabrouskov, V., Coon, J.J. (2008). A proteomics grade electron transfer dissociation-enabled hybrid linear ion trap-orbitrap mass spectrometer. J Proteome Res 7, 3127–3136.

McAlister, G.C., Phanstiel, D., Wenger, C.D., Lee, M.V., Coon, J.J. (2010). Analysis of tandem mass spectra by FTMS for improved large-scale proteomics with superior protein quantification. Anal. Chem. *82,* 316–322.

Macek, B., Waanders, L.F., Olsen, J.V., and Mann, M. (2006). Top-down protein sequencing and MS3 on a hybrid linear quadrupole ion trap-orbitrap mass spectrometer. Mol. Cell. Proteomics 5, 949–958.

McNally, D.J., Hui, J.P., Aubry, A.J., Mui, K.K., Guerry, P., Brisson, J.R., Logan, S.M., and Soo, E.C. (2006). Functional characterization of the flagellar glycosylation locus in *Campylobacter jejuni* 81-176 using a focused metabolomics approach. J. Biol. Chem. *281,* 18489–18498.

McNally, D.J., Aubry, A.J., Hui, J.P., Khieu, N.H., Whitfield, D., Ewing, C.P., Guerry, P., Brisson, J.R., Logan, S.M., and Soo, E.C. (2007a). Targeted metabolomics analysis of *Campylobacter coli* VC167 reveals legionaminic acid derivatives as novel flagellar glycans. J. Biol. Chem. *282,* 14463–14475.

McNally, D.J., Aubry, A.J., Hui, J.P., Khieu, N.H., Whitfield, D., Ewing, C.P., Guerry, P., Brisson, J.R., Logan, S.M., and Soo, E.C. (2007b). Targeted metabolomics analysis of *Campylobacter coli* VC167 reveals legionaminic acid derivatives as novel flagellar glycans. J. Biol. Chem. *282,* 14463–14475.

Maley, F., Trimble, R.B., Tarentino, A.L., and Plummer, T.H., Jr. (1989). Characterization of glycoproteins and their associated oligosaccharides through the use of endoglycosidases. Anal. Biochem. *180,* 195–204.

Mann, M., and Kelleher, N.L. (2008). Precision proteomics, the case for high resolution and high mass accuracy. Proc. Natl. Acad. Sci. U.S.A. *105,* 18132–18138.

Marshall, A.G., Hendrickson, C.L., and Jackson, G.S. (1998). Fourier transform ion cyclotron resonance mass spectrometry, a primer. Mass Spectrom. Rev. *17,* 1–35.

Mescher, M.F., and Strominger, J.L. (1976a). Purification and characterization of a prokaryotic glucoprotein from the cell envelope of *Halobacterium salinarium*. J. Biol. Chem. *251,* 2005–2014.

Mescher, M.F., and Strominger, J.L. (1976b). Structural (shape-maintaining) role of the cell surface glycoprotein of *Halobacterium salinarium*. Proc. Natl. Acad. Sci. U.S.A. *73,* 2687–2691.

Mysling, S., Palmisano, G., Hojrup, P., and Thaysen-Andersen, M. (2010). Utilizing ion-pairing hydrophilic interaction chromatography solid phase extraction for efficient glycopeptide enrichment in glycoproteomics. Anal. Chem. *82,* 5598–5609.

Nilsson, J., Ruetschi, U., Halim, A., Hesse, C., Carlsohn, E., Brinkmalm, G., and Larson, G. (2009). Enrichment of glycopeptides for glycan structure and attachment site identification. Nat. Methods *6,* 809–811.

Olsen, J.V., Macek, B., Lange, O., Makarov, A., Horning, S., and Mann, M. (2007). Higher-energy C-trap dissociation for peptide modification analysis. Nat. Methods *4,* 709–712.

Packer, N.H., Pawlak, A., Kett, W.C., Gooley, A.A., Redmond, J.W., and Williams, K.L. (1997). Proteome analysis of glycoforms, a review of strategies for the microcharacterisation of glycoproteins separated by two-dimensional polyacrylamide gel electrophoresis. Electrophoresis *18,* 452–460.

Paizs, B., and Suhai, S. (2005). Fragmentation pathways of protonated peptides. Mass Spectrom. Rev. *24,* 508–548.

Rabilloud, T., Chevallet, M., Luche, S., and Lelong, C. (2010). Two-dimensional gel electrophoresis in proteomics, past, present and future. J. Proteomics *73,* 2064–2077.

Rademaker, G.J., Pergantis, S.A., Blok-Tip, L., Langridge, J.I., Kleen, A., and Thomas-Oates, J.E. (1998). Mass spectrometric determination of the sites of O-glycan attachment with low picomolar sensitivity. Anal. Biochem. *257,* 149–160.

Romain, F., Horn, C., Pescher, P., Namane, A., Riviere, M., Puzo, G., Barzu, O., and Marchal, G. (1999). Deglycosylation of the 45/47-kilodalton antigen complex of *Mycobacterium tuberculosis* decreases its capacity to elicit *in vivo* or *in vitro* cellular immune responses. Infect. Immun. *67,* 5567–5572.

Rostovtsev, V.V., Green, L.G., Fokin, V.V., and Sharpless, K.B. (2002). A stepwise huisgen cycloaddition process, copper(I)-catalysed regioselective 'ligation' of azides and terminal alkynes. Angew Chem. Int. Ed. Engl. *41,* 2596–2599.

Samyn, B., Sergeant, K., and Van, B.J. (2006). A method for C-terminal sequence analysis in the proteomic era (proteins cleaved with cyanogen bromide). Nat. Protoc. *1,* 318–323.

Sandra, K., Moshir, M., D'hondt, F., Tuytten, R., Verleysen, K., Kas, K., Francois, I., and Sandra, P. (2009). Highly efficient peptide separations in proteomics. Part 2, bi- and multidimensional liquid-based separation techniques. J. Chromatogr. B. Anal. Technol. Biomed. Life Sci. *877,* 1019–1039.

Sartain, M.J., and Belisle, J.T. (2009). N-terminal clustering of the O-glycosylation sites in the *Mycobacterium tuberculosis* lipoprotein SodC. Glycobiology *19,* 38–51.

Schauer, R. (2009). Sialic acids as regulators of molecular and cellular interactions. Curr. Opin. Struct. Biol. *19,* 507–514.

Schirm, M., Kalmokoff, M., Aubry, A., Thibault, P., Sandoz, M., and Logan, S.M. (2004). Flagellin from Listeria monocytogenes is glycosylated with beta-O-linked N-acetylglucosamine. J. Bacteriol. *186,* 6721–6727.

Schirm, M., Schoenhofen, I.C., Logan, S.M., Waldron, K.C., and Thibault, P. (2005). Identification of unusual bacterial glycosylation by tandem mass spectrometry analyses of intact proteins. Anal. Chem. *77,* 7774–7782.

Scott, N.E., Bogema, D.R., Connolly, A.M., Falconer, L., Djordjevic, S.P., and Cordwell, S.J. (2009). Mass spectrometric characterization of the surface-associated 42 kDa lipoprotein JlpA as a glycosylated antigen in strains of *Campylobacter jejuni.* J. Proteome Res. *8,* 4654–4664.

Scott, N.E., Parker, B.L., Connolly, A.M., Paulech, J., Edwards, A.V., Crossett, B., Falconer, L., Kolarich, D., Djordjevic, S.P., Hojrup, P., *et al.* (2010). Simultaneous glycan-peptide characterization using hydrophilic interaction chromatography and parallel fragmentation by CID, HCD and ETD-MS applied to the N-linked glycoproteome of *Campylobacter jejuni.* Mol. Cell Proteomics. *10,* M00003–MCP201.

Segu, Z.M., and Mechref, Y. (2010). Characterizing protein glycosylation sites through higher-energy C-trap dissociation. Rapid Commun. Mass Spectrom. *24,* 1217–1225

Seipert, R.R., Dodds, E.D., Clowers, B.H., Beecroft, S.M., German, J.B., and Lebrilla, C.B. (2008). Factors that influence fragmentation behavior of N-linked glycopeptide ions. Anal. Chem. *80,* 3684–3692.

Seipert, R.R., Dodds, E.D., and Lebrilla, C.B. (2009). Exploiting differential dissociation chemistries of O-linked glycopeptide ions for the localization of mucin-type protein glycosylation. J. Proteome Res. *8,* 493–501.

Severi, E., Hood, D.W., and Thomas, G.H. (2007). Sialic acid utilization by bacterial pathogens. Microbiology *153,* 2817–2822.

Sharon, N. (2008). Lectins, past, present and future. Biochem. Soc. Trans. *36,* 1457–1460.

Soo, E.C., Aubry, A.J., Logan, S.M., Guerry, P., Kelly, J.F., Young, N.M., and Thibault, P. (2004). Selective detection and identification of sugar nucleotides by CE-electrospray-MS and its application to bacterial metabolomics. Anal. Chem. *76,* 619–626.

Soo, E.C., and Hui, J.P. (2010). Metabolomics in glycomics. Methods Mol. Biol. *600,* 175–186

Speers, A.E., and Wu, C.C. (2007). Proteomics of integral membrane proteins – theory and application. Chem. Rev. *107,* 3687–3714.

Steinberg, T.H., Berggren, K.N., Kemper, C., Jones, L., Diwu, Z., Haugland, R.P., and Patton, W.F. (2001). Rapid and simple single nanogram detection of glycoproteins in polyacrylamide gels and on electroblots. Proteomics *1,* 841–855.

Stephenson, A.E., Wu, H., Novak, J., Tomana, M., Mintz, K., and Fives-Taylor, P. (2002). The Fap1 fimbrial adhesin is a glycoprotein, antibodies specific for the glycan moiety block the adhesion of Streptococcus parasanguis in an *in vitro* tooth model. Mol. Microbiol. *43,* 147–157.

Stimson, E., Virji, M., Makepeace, K., Dell, A., Morris, H.R., Payne, G., Saunders, J.R., Jennings, M.P., Barker, S., and Panico, M. (1995). Meningococcal pilin, a glycoprotein substituted with digalactosyl 2,4-diacetamido-2,4,6-trideoxyhexose. Mol. Microbiol. *17,* 1201–1214

Sumper, M. (1987). Halobacterial glycoprotein biosynthesis. Biochim. Biophys. Acta *906,* 69–79.

Sun, B., Ranish, J.A., Utleg, A.G., White, J.T., Yan, X., Lin, B., and Hood, L. (2007). Shotgun glycopeptide capture approach coupled with mass spectrometry for comprehensive glycoproteomics. Mol. Cel.l Proteomics *6,* 141–149.

Swaney, D.L., McAlister, G.C., Wirtala, M., Schwartz, J.C., Syka, J.E., and Coon, J.J. (2007). Supplemental activation method for high-efficiency electron-transfer dissociation of doubly protonated peptide precursors. Anal. Chem. *79,* 477–485.

Swaney, D.L., McAlister, G.C., and Coon, J.J. (2008). Decision tree-driven tandem mass spectrometry for shotgun proteomics. Nat. Methods *5,* 959–964.

Swaney, D.L., Wenger, C.D., and Coon, J.J. (2010). Value of using multiple proteases for large-scale mass spectrometry-based proteomics. J. Proteome Res. *9,* 1323–1329.

Swatkoski, S., Gutierrez, P., Wynne, C., Petrov, A., Dinman, J.D., Edwards, N., and Fenselau, C. (2008). Evaluation of microwave-accelerated residue-specific acid cleavage for proteomic applications. J. Proteome Res. *7,* 579–586.

Syka, J.E., Coon, J.J., Schroeder, M.J., Shabanowitz, J., and Hunt, D.F. (2004). Peptide and protein sequence analysis by electron transfer dissociation mass spectrometry. Proc. Natl. Acad. Sci. U.S.A. *101,* 9528–9533.

Szymanski, C.M., and Wren, B.W. (2005). Protein glycosylation in bacterial mucosal pathogens. Nat. Rev. Microbiol. 3, 225–237.

Tabei, S.M., Hitchen, P.G., Day-Williams, M.J., Merino, S., Vart, R., Pang, P.C., Horsburgh, G.J., Viches, S., Wilhelms, M., Tomas, J.M., *et al.* (2009). An *Aeromonas caviae* genomic island is required for both O-antigen lipopolysaccharide biosynthesis and flagellin glycosylation. J. Bacteriol. *191,* 2851–2863.

Taverna, M., Tran, N.T., Merry, T., Horvath, E., and Ferrier, D. (1998). Electrophoretic methods for process monitoring and the quality assessment of recombinant glycoproteins. Electrophoresis *19,* 2572–2594.

Taylor, A.M., Holst, O., and Thomas-Oates, J. (2006). Mass spectrometric profiling of O-linked glycans released directly from glycoproteins in gels using in-gel reductive beta-elimination. Proteomics *6,* 2936–2946

Thibault, P., Logan, S.M., Kelly, J.F., Brisson, J.R., Ewing, C.P., Trust, T.J., and Guerry, P. (2001). Identification of the carbohydrate moieties and glycosylation motifs in *Campylobacter jejuni* flagellin. J. Biol. Chem. *276,* 34862–34870.

Townsend, R.R., Lipniunas, P.H., Tulk, B.M., and Verkman, A.S. (1996). Identification of protein kinase A phosphorylation sites on NBD1 and R domains of CFTR using electrospray mass spectrometry with selective phosphate ion monitoring. Protein Sci. *5,* 1865–1873

Twine, S.M., Paul, C.J., Vinogradov, E., McNally, D.J., Brisson, J.R., Mullen, J.A., McMullin, D.R., Jarrell, H.C., Austin, J.W., Kelly, J.F., *et al.* (2008). Flagellar glycosylation in *Clostridium botulinum.* FEBS J. *275,* 4428–4444.

Twine, S.M., Reid, C.W., Aubry, A., McMullin, D.R., Fulton, K.M., Austin, J., and Logan, S.M. (2009). Motility and flagellar glycosylation in *Clostridium difficile.* J. Bacteriol. 191, 7050–7062.

Umemoto, J., Matta, K.L., Barlow, J.J., and Bhavanandan, V.P. (1978). Action of endo-alpha-N-acetyl-D-galactosaminidase on synthetic glycosides including chromogenic substrates. Anal. Biochem. *91,* 186–193.

Varki, A. (1992). Diversity in the sialic acids. Glycobiology *2,* 25–40.

Vasicek, L., and Brodbelt, J.S. (2009). Enhanced electron transfer dissociation through fixed charge derivatization of cysteines. Anal. Chem. *81,* 7876–7884.

Verma, A., Schirm, M., Arora, S.K., Thibault, P., Logan, S.M., and Ramphal, R. (2006b). Glycosylation of b-Type flagellin of Pseudomonas aeruginosa, structural and genetic basis. J. Bacteriol. *188,* 4395–4403.

Vik, A., Aas, F.E., Anonsen, J.H., Bilsborough, S., Schneider, A., Egge-Jacobsen, W., and Koomey, M. (2009). Broad spectrum O-linked protein glycosylation in the human pathogen Neisseria gonorrhoeae. Proc. Natl. Acad. Sci. U.S.A. *106,* 4447–4452.

Voisin, S., Kus, J.V., Houliston, S., St-Michael, F., Watson, D., Cvitkovitch, D.G., Kelly, J., Brisson, J.R., and Burrows, L.L. (2007). Glycosylation of Pseudomonas aeruginosa strain Pa5196 type IV pilins with mycobacterium-like alpha-1,5-linked D-Araf oligosaccharides. J. Bacteriol. *189,* 151–159.

Vretou, E., Giannikopoulou, P., and Psarrou, E. (2001). Polymorphic outer-membrane proteins of Chlamydophila abortus are glycosylated. Microbiology *147,* 3303–3310.

Wacker, M., Feldman, M.F., Callewaert, N., Kowarik, M., Clarke, B.R., Pohl, N.L., Hernandez, M., Vines, E.D., Valvano, M.A., Whitfield, C., *et al.* (2006). Substrate specificity of bacterial oligosaccharyltransferase suggests a common transfer mechanism for the bacterial and eukaryotic systems. Proc. Natl. Acad. Sci. U.S.A. *103,* 7088–7093.

Wang, Z., Udeshi, N.D., O'Malley, M., Shabanowitz, J., Hunt, D.F., and Hart, G.W. (2010). Enrichment and site mapping of O-linked N-acetylglucosamine by a combination of chemical/enzymatic tagging, photochemical cleavage, and electron transfer dissociation mass spectrometry. Mol. Cell. Proteomics *9,* 153–160.

Weerapana, E., and Imperiali, B. (2006). Asparagine-linked protein glycosylation, from eukaryotic to prokaryotic systems. Glycobiology *16,* 91R–101R.

Wiesner, J., Premsler, T., and Sickmann, A. (2008). Application of electron transfer dissociation (ETD) for the analysis of post-translational modifications. Proteomics *8,* 4466–4483.

Wohlgemuth, J., Karas, M., Jiang, W., Hendriks, R., and Andrecht, S. (2010). Enhanced glyco-profiling by specific glycopeptide enrichment and complementary monolithic nano-LC (ZIC-HILIC/RP18e)/ESI–MS analysis. J. Sep. Sci. *33,* 880–890.

Wuhrer, M., Koeleman, C.A., Hokke, C.H., and Deelder, A.M. (2005). Protein glycosylation analyzed by normal-phase nano-liquid chromatography–mass spectrometry of glycopeptides. Anal. Chem. *77,* 886–894.

Wuhrer, M., de Boer, A.R., and Deelder, A.M. (2009). Structural glycomics using hydrophilic interaction chromatography (HILIC) with mass spectrometry. Mass. Spectrom Rev. *28,* 192–206.

Yamaki, S., Isobe, T., Okuyama, T., and Shinoda, T. (1996). High-performance liquid chromatography of peptides on a microspherical carbon column. J. Chromatogr. A. *729,* 143–153.

Yang, Z., and Hancock, W.S. (2005). Monitoring glycosylation pattern changes of glycoproteins using multi-lectin affinity chromatography. J. Chromatogr. A. *1070,* 57–64.

Young, N.M., Brisson, J.R., Kelly, J., Watson, D.C., Tessier, L., Lanthier, P.H., Jarrell, H.C., Cadotte, N., St, M.F., Aberg, E., *et al.* (2002). Structure of the N-linked glycan present on multiple glycoproteins in the Gram-negative bacterium, *Campylobacter jejuni*. J. Biol. Chem. *277,* 42530–42539.

Yu, R.K., and Ledeen, R. (1969). Configuration of the ketosidic bond of sialic acid. J. Biol. Chem. *244,* 1306–1313

Zhang, H., Li, X.J., Martin, D.B., and Aebersold, R. (2003). Identification and quantification of N-linked glycoproteins using hydrazide chemistry, stable isotope labelling and mass spectrometry. Nat. Biotechnol. *21,* 660–666.

Zhang, Z. (2010). Prediction of electron-transfer/capture dissociation spectra of peptides. Anal. Chem. *82,* 1990–2005.

Zielinska, D.F., Gnad, F., Wisniewski, J.R., and Mann, M. (2010). Precision mapping of an *in vivo N*-glycoproteome reveals rigid topological and sequence constraints. Cell *141,* 897–907.

Zubarev, R.A., Kelleher, N.L., and McLafferty, F.W. (1998). Electron capture dissociation of multiply charged protein cations. A nonergodic process. J. Am. Chem. Soc. *120,* 3265–3266.

Solid-state NMR of the Bacterial Cell Wall

9

Catherine Bougault, Sabine Hediger and Jean-Pierre Simorre

Abstract

Liquid-state nuclear magnetic resonance (NMR) is traditionally used to provide fine chemical and structural information on soluble fragments, but is limited to biomolecules in the fast rotational tumbling regime. On the other hand, solid-state NMR, unlimited by the molecular size, is extensively applied to polymers, but provides a spectral resolution that is hampered by orientational heterogeneity. This chapter analyses the potential of high-resolution solid-state NMR in providing chemical, structural and dynamical information on whole bacteria or bacterial cell envelopes in lyophilized as well as in fully hydrated samples. The first section addresses bacterial strain typing issues and discusses the choice of adequate nuclear probes, the concomitant requirements for efficient isotope labelling schemes and the NMR methods used to characterize the chemical composition of peptidoglycan, teichoic acids, lipopolysaccharides or mycolic acids in bacteria. The second section describes the principles used to characterize molecular interactions of proteins, ions, antibiotic/antimicrobial molecules with the bacterial cell wall by NMR. The third and last section gives an overview of the contribution of solid-state NMR to characterize the cell wall structure and dynamics, and covers the different techniques used to extract specific structural information as well as global mobility.

Introduction

The first study on the bacterial cell wall by solid-state NMR was presented in 1983 by the group of Jacob Schaefer (Jacob *et al.*, 1983), but solid-state NMR remained for a long time an isolated technique in this application field, on the side compared with other characterization tools. Nevertheless a renewed interest in this technique has emerged from the tremendous progress made during the last 10 years in biomolecular solid-state NMR, and this technique is becoming promising for the study of the complex bacterial cell wall.

Traditionally, solid-state NMR is applied to materials in the solid-phase, and for a long time cell wall studies have been performed almost exclusively on lyophilized samples (extracted cell walls or entire cells). However, it can be applied to any sample whose molecules are re-orientating on a time scale which is much slower than the one of the NMR experiment, which is in the ms range. Thus, solid-state NMR can be applied to hydrated insoluble cell wall samples or hydrated intact cells as well, with the advantage of an improved spectral resolution resulting from the local dynamics present in the hydrated state (Tong *et al.*, 1997; Kern *et al.*, 2008, 2010).

As opposed to the spectra obtained in solution NMR, the intrinsic NMR signal for static solid material is very broad, masking any site resolution. The reason for this situation is the dependence of the NMR signal to the orientation of the molecule with respect to the external applied magnetic field. The very rapid molecular tumbling in the liquid state removes naturally the broadening of the NMR signal. For solid samples, an elegant technical solution called Magic Angle Spinning (MAS) was proposed in the late 1950s (Andrew *et al.*, 1958, 1959; Lowe, 1959). It consists of rapidly spinning the sample in a particular orientation with respect to the external magnetic field, called the Magic Angle (54.74°). It is important to note that the distance of any particle to the rotation axis is very small (usually 1 to 2.5 mm, depending on the size of the sample container, called the rotor) and therefore the centrifugal force applied to cells within the sample is limited despite the quite high spinning speeds commonly used (5000 to 20,000 Hz). Bacterial survival after NMR experiments under MAS has been checked by optical microscopy and tests on agar plates, which showed that cells can survive and reproduce after an NMR experiment in such conditions (Kern *et al.*, 2010). The comparison of liquid- and solid-state NMR with respect to resolution is illustrated in Fig. 9.1. Together with the use of specific pulse sequences, solid-state MAS NMR leads to high-resolution spectra allowing the study of various insoluble materials. Among the NMR-active nuclei present in biomolecules, protons are widely detected in liquid-state NMR because of the high sensitivity of their nuclear spin. However the resolution in proton solid-state NMR spectra using MAS is very often not good enough to allow site resolution. Thus other nuclei are usually preferred in the solid-state, as for example ^{13}C, ^{15}N or ^{31}P, the most commonly used ones in biomolecular applications. It is important to note that MAS is sometimes used combined with liquid-state NMR techniques in the presence of heterogeneous samples (technique called HR-MAS, High Resolution MAS). In this case, MAS eliminates the magnetic susceptibility broadening caused by the heterogeneity of the sample, whose sites of interest themselves are mobile enough for the application of liquid-state NMR techniques.

Despite high-resolution conditions provided by MAS, signals in the solid-state NMR spectrum of a biomolecule are often too numerous to afford site-specific resolution. In the context of bacterial cell wall studies, two main approaches have been described to overcome this problem. The first compensates the lack of resolution by the introduction of site-specific isotope labels, like ^{13}C and/or ^{15}N, which greatly enhances the sensitivity and simplicity of the corresponding 1D-NMR experiments (Forrest *et al.*, 1991). This has been mainly performed on lyophilized cell wall samples or lyophilized whole cells. In this case, the quantity of material needed for good quality 1D NMR amounts to 150–300 mg of cell wall samples and to 600 mg of whole cells. The second approach consists of increasing the dimensionality of the

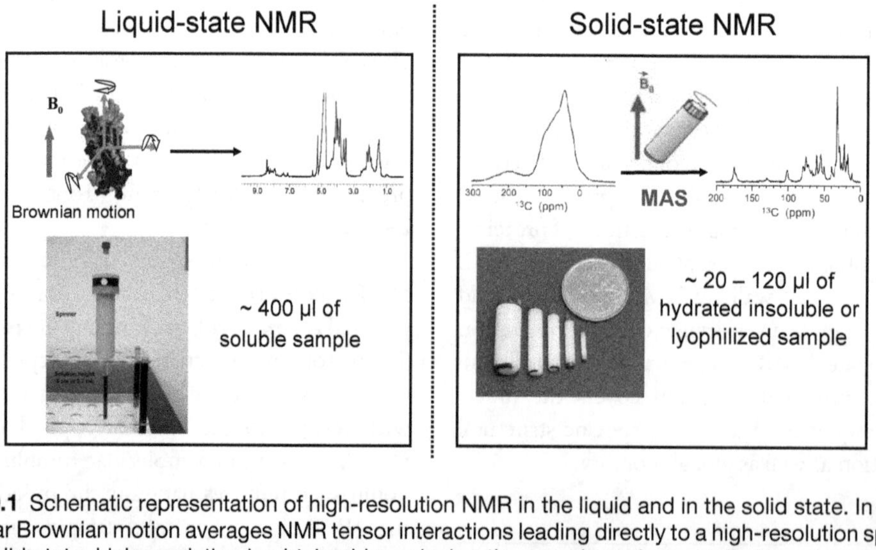

Figure 9.1 Schematic representation of high-resolution NMR in the liquid and in the solid state. In solution, molecular Brownian motion averages NMR tensor interactions leading directly to a high-resolution spectrum. In the solid state, high-resolution is obtained by spinning the sample at the so-called magic angle (magic angle spinning, MAS) with respect to the external static magnetic field B_0, using a small sample container called the rotor (see picture). In both application fields, the amount of sample necessary for the study is given.

NMR experiment by correlating nuclei that are covalently bound or spatially close to each other. It is generally coupled to uniform labelling of the biological sample to preserve enough sensitivity. This approach has been mainly applied to hydrated samples of bacterial cell walls or whole cells, from which about 10–30 mg of hydrated material is necessary.

Because the NMR signal is sensitive to the local environment and dynamics of each detected nuclei, it is a method of choice to investigate the chemical and conformational structure, the three-dimensional arrangement in complex systems, as well as local dynamics which may be present. In the case of the bacterial cell wall, solid-state MAS NMR has been successfully applied to answer a variety of questions. Its application to the characterization of the chemical composition and modifications, which may occur during cell growth or under antibiotic stress, is described in the first part of this chapter. Another crucial question for a better understanding of the bacterial cell wall is its mode of interaction with proteins, with antibiotics or antimicrobial peptides, and how metal ions are trapped inside the cell envelope. Characterization of these different interactions involving the cell wall is discussed in the second section. Finally, advanced solid-state NMR techniques have been applied to get information about the polymer network itself, from the degree of peptide cross-linking to attempts in solving the peptidoglycan architecture itself or describing the cell wall flexibility. This is reviewed in the third part of this contribution.

Bacterial strain typing

Introduction

As pathogenic bacteria pose ever greater challenges to human health, including increased virulence and resistance to antibiotics, identifying bacteria at the strain level has become increasingly important. In this context, different approaches have been considered: phenotyping and genotyping (Li *et al.*, 2009). The question raised here is whether NMR can help in the differentiation of bacteria through the characterization of the gene products, and more particularly, of the bacterial cell-envelope components of Gram-negative, Gram-positive and Mycobacteria (Silhavy *et al.*, 2010).

The NMR signal arising from a given nucleus within the sample will show two major characteristics: (i) its position (chemical shift δ, in ppm) which depends on the chemical environment of the considered nucleus; (ii) its intensity, which in certain conditions is proportional to the number of equivalent sites in the biological sample. Teichoic acids in Gram-positive bacteria (which can account for up to 60% of the cell wall) and lipopolysaccharides in Gram-negative bacteria are abundant. Peptidoglycan is a large repeat (3×10^6 kDa) of identical or highly similar fragments. The signals from the bacterial cell-envelope (Fig. 9.2) will thus dominate the NMR spectrum. The question then comes down to the ability of NMR to discriminate chemical entities or chemical modifications based on their differences in chemical shifts.

In solution, only fragments can be detected owing to their rather low molecular weight and resulting fast molecular tumbling. In contrast, solid-state NMR is not limited by the molecular weight and gives the possibility to target intact cell-wall isolates or even whole cells. However, even with magic angle spinning, which reduces the line width of the NMR resonance drastically, the resolution remains lower than in solution. As stated in the introduction, this can be overcome either by the introduction of specific isotope labels, like ^{13}C and/or ^{15}N, or by increasing the dimensionality of the NMR experiment in uniformly labelled samples, which is obtained by correlating nuclei that are covalently bound or spatially close to each other. The information content of both approaches is examined next for the different bacterial cell-envelope main components as isolated entities or within the whole cell.

Characterization of the peptidoglycan

Peptidoglycan is the common cell wall component of most Gram-positive and Gram-negative bacteria and is essentially composed of glycan strands (containing a repeat of β-1,4-linked *N*-acetylglucosamine and *N*-acetylmuramic acid disaccharide units) cross-linked by short peptides (Vollmer *et*

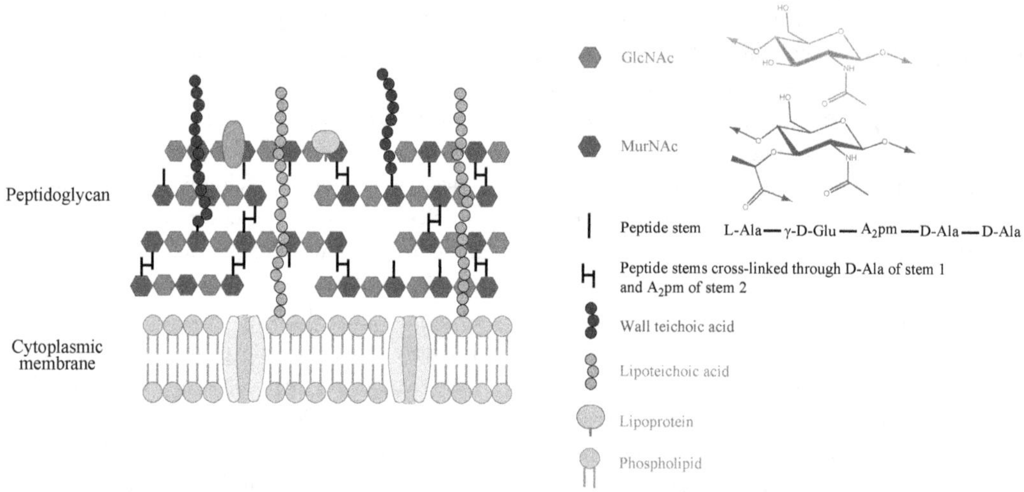

Figure 9.2 Schematic representation of the cell wall of *Bacillus subtilis* strain 168. Representative units of the peptidoglycan are described: β-*N*-acetylglucosamine (GlcNAc) is covalently linked (β-1,4-linkage) to β-*N*-acetylmuramic acid (MurNAc) to constitute the glycan chains, while peptide stems linked at position 3 on the MurNAc can be reticulated or not. Wall teichoic acids and lipoteichoic acids consist of a poly(glycerolphosphate) chain as a major component, but additional components can be detected as emphasized in Fig. 9.4.

al., 2008), as sketched in Fig. 9.2. Owing to their marked implications in bacterial taxonomy, physiology and pathogenesis but also in the field of innate immunity, pathogenicity and anti-bacterial therapy, inter-species and inter-strain variations in the peptidoglycan chemical structure have been widely investigated.

Originally, typical chemical methods to analyse peptidoglycan involved enzymatic fragmentation of purified sacculi, state-of-the-art HPLC separation of the mixture of components and mass spectrometry analysis (Glauner, 1988). Solution-state NMR has then provided useful complementary information on purified fragments at natural abundance (see for example Atrih *et al.*, 1998). Solid-state NMR approaches have the advantage of directly allowing the analysis of complete sacculi or even whole cells, thus providing a significant gain in the time required for the analysis and potentially avoiding fragmentation artefacts.

What can be learned from 1D experiments on specifically labelled samples?

Four main regions can be distinguished in the ^{13}C NMR spectra of peptidoglycan substrates (see Fig. 9.3A): (i) the carbonyl region (170–185 ppm) that contains the CO groups of peptide bonds as well as free carboxylic ends and the *N*-acetyl substituents of *N*-acetylglucosamine and *N*-acetyl muramic acids, all at very similar frequencies; (ii) the sugar C-1 region (90–110 ppm) for which the chemical shift is indicative of the α- or β-glycosidic configuration; (iii) the 50–80 ppm region containing signals from all other sugar-type carbons and from the Cα of amino acids involved in the peptide stems or bridges with the exception of glycines; (iv) the 20–48 ppm region containing signals from the amino acid side-chains (this region is a signature of the nature of the amino acids involved in the peptide stems and bridges). Alternatively two main regions only can be identified in the ^{15}N NMR spectra of peptidoglycan: (i) the amide region (90–110 ppm); (ii) the amine region (0–50 ppm).

Peptidoglycan-specific labelling

In lyophilized sacculi or whole cell samples, different signals from the same region are highly overlapping in 1D spectra and site specific information can only be retrieved if specific labels are introduced. For this purpose, bacteria are grown

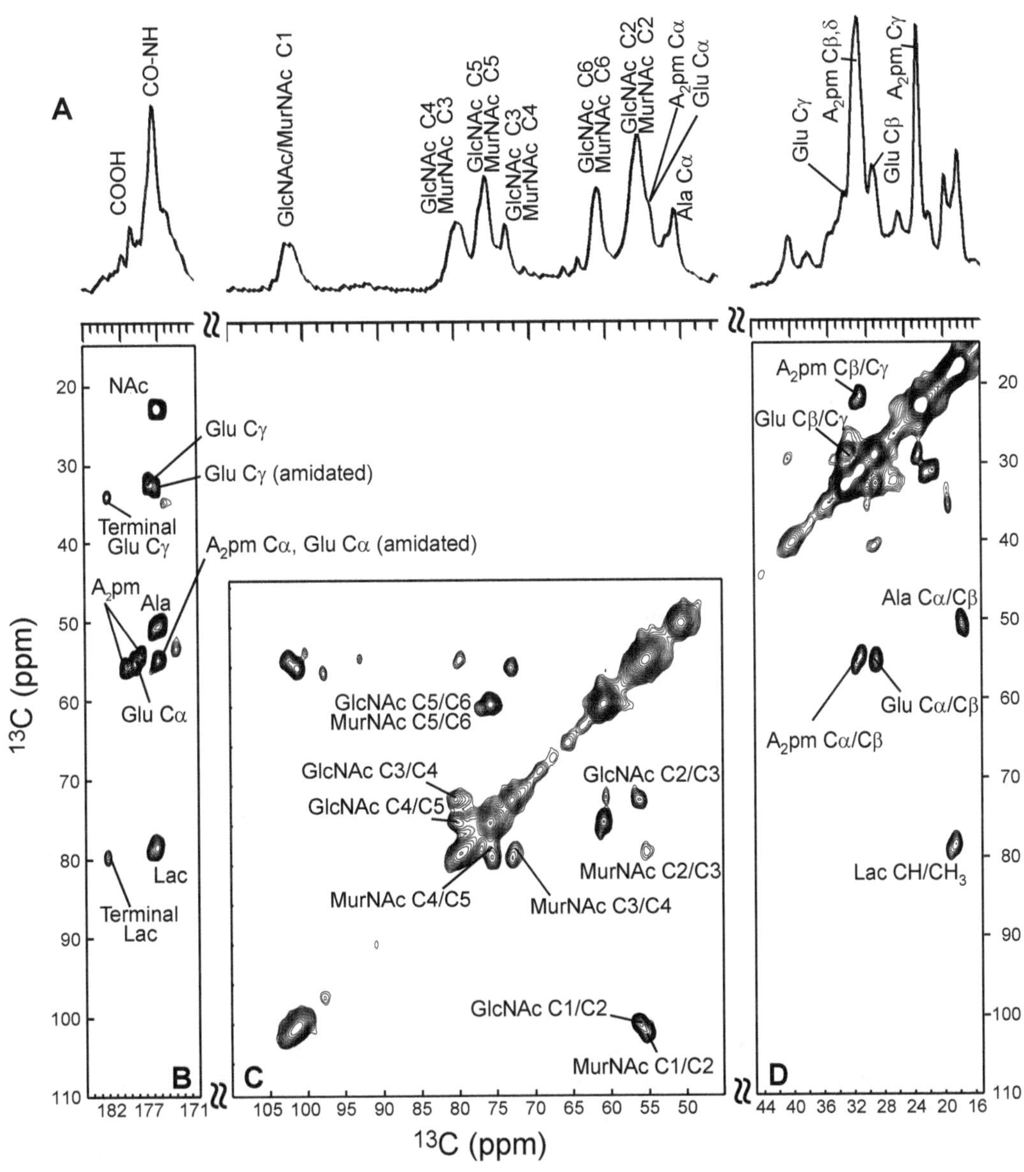

Figure 9.3 Solid-state NMR spectra of a fully hydrated ^{13}C-,^{15}N-uniformly labelled peptidoglycan sample (sacculi) of *Bacillus subtilis* strain 168 at 9.4T and 283K. (A) 1D ^{13}C-CPMAS spectrum and (B–D) 2D ^{13}C–^{13}C correlation spectrum [see Kern *et al.* (2010) for a description of the pulse sequences]. Regions B, C and D of the 2D spectrum respectively show the correlations of the carbonyl to the Cα of individual amino acids, the correlations between the carbons of the sugar rings, and the fingerprint region of the amino acid side-chains. Each region is used to unambiguously identify the nature of the amino acids in the peptide stems and the configuration of the sugars, as well as eventual chemical modifications.

on defined synthetic media supplemented with all 20 common amino acids unlabelled with the exception of the amino acid under study which is ^{13}C and/or ^{15}N-labelled (Forrest *et al.*, 1991). The background of 1.11% and 0.37% natural abundance of respectively ^{13}C and ^{15}N remains beyond detection in typical 1D experiments and only the signals arising from the introduced labelled amino acid or leakage due to metabolic pathways are detected.

Chemical modifications

If different configurations of the covalent network around one particular nucleus are statistically well represented in the cell wall, the chemical shift value of the corresponding signal will be affected and peaks corresponding to each individual situation will be detected in the spectrum. As an example, the ε-nitrogen of lysine in the peptidoglycan of vancomycin-sensitive *Enterococcus faecium* is expected to exist under two configurations: (i) the amide configuration when the lysine is engaged in a bridge of pentaglycine and (ii) the amine configuration when the lysine is free. Nevertheless, the ^{15}N-CP-MAS spectrum on sacculi specifically labelled with L-[^{15}N]lysine shows three signals, one of which at 151 ppm in an unexpected position. This signature has been assigned, in consistency with mass spectrometry data on peptidoglycan fragments, to a lysyl succinimide (Patti *et al.*, 2008a). Solid-state NMR thus appears as an elegant method to report on peptidoglycan chemical modifications.

Differential repartition of D-Ala

Alternatively, ^{13}C-MAS spectra [^{13}C-CP-MAS (Cross-Polarized Magic-Angle Spinning); Cegelski *et al.*, 2006] or full echo ^{13}C{^{15}N}-REDOR (Rotational-Echo Double-Resonance) spectra (Kim *et al.*, 2008) can be used to follow the distribution of D-Ala in D-[1-^{13}C]-Ala and [^{15}N]-Gly-labelled *Staphylococcus aureus* whole cell preparations. In fact, the broad ^{13}C signal resonating around 175 ppm can be deconvoluted into four types of D-Ala carbonyl carbons: (i) the peptide carbonyl carbon in cross-linked stems ending in D-Ala at position 4 (centred at 175 ppm), (ii) the peptide carbonyl carbon of D-Ala in position 4 of the D-Ala-D-Ala acyl-acceptor stem (also centred at 175 ppm), (iii) the peptide carbonyl carbon of D-Ala in position 4 or 5 respectively of a D-Ala-D-Ala stem (centred at 178 ppm) and (iv) various teichoic acid D-Ala esters. The first component (intensity and width) can be obtained from ^{13}C{^{15}N}-REDOR difference spectra (see 'Structure and flexibility of the cell wall' in this chapter and Fig. 9.5B) that quantitatively show only D-Ala carbonyl carbons directly bound to the nitrogen of Gly residues (D-Ala involved in cross-links). Supposing that the line width of the resonance from the carbonyl group does not depend on the chemical environment, the slight chemical shift differences between the other carbonyl groups and an appropriate line-fitting procedure allow the quantification of the other components. This approach was used to quantify the percentage of cross-linked peptide stems in D-[1-^{13}C]-Ala and D,L-[^{15}N]-Asp *E. faecium* cells as well (47%) (Patti *et al.*, 2008b) and gave consistent results with a LC–MS/MS mass spectrometry analysis (40%) (Patti *et al.*, 2008a).

Metabolism

MAS solid-state NMR spectra can additionally be used to follow the metabolism of amino acids involved in peptidoglycan synthesis and serve as a reporter for modifications in the cell wall synthesis as a function of cell ageing or antibiotic stress. When *S. aureus* whole cells were grown with [1-^{13}C]-Gly and L-[ε-^{15}N]-Lys in the presence of vancomycin (Cegelski *et al.*, 2002) or oritavancin (Kim *et al.*, 2008), ^{15}N-MAS spectra showed an intensity decrease of the amide component at 95 ppm, arising from peptidoglycan and from the Lipid II precursor that contained a (L-[ε-^{15}N]-Lys)NH–([L-^{13}C]-Gly)CO bond, and a correlated intensity increase of the amine component at 10 ppm arising from Park's nucleotide which contains a free L-[ε-^{15}N]-Lys NH_2 side-chain. Corresponding ^{13}C-MAS spectra showed no reduction of the [1-^{13}C]-Gly resonance at 171 ppm, but suppression of purine metabolism (signature at 149 ppm) as a consequence of glycine rerouting into cell wall synthesis. The combination of these NMR signatures was unambiguously assigned to the inhibition of the transglycosylation step by the antibiotics.

What can be learned from 2D experiments on uniformly labelled samples?

As emphasized previously specific labelling can offer detailed qualitative and quantitative chemical information, but on the other hand can be rather time-consuming (preparation of multiple samples) and requires a priori knowledge of the peptidoglycan composition. Another alternative is to raise the degeneracy of the problem (overlap of resonance lines in uniformly labelled samples) by increasing the number of dimensions and establishing correlations between covalently bonded nuclei. In this approach, it is advantageous to work on fully hydrated cell wall samples to recover a sharper NMR signal (Tong *et al.*, 1997).

Sample preparation

In this approach, bacteria are grown in minimal M9 media supplemented with ^{13}C-glucose and $^{15}NH_4Cl$ as sole carbon and nitrogen sources respectively or, alternatively, with rich media containing ^{13}C- and ^{15}N-labelled metabolites from other bacterial growth. NMR experiments required approximately 10 mg of hydrated cell paste, which is directly centrifuged into the MAS rotor.

Data acquisition

As a proof of concept, 2D ^{13}C-correlation experiments were collected in our group (Kern *et al.*, 2008) on uniformly ^{13}C,^{15}N-labelled *Escherichia coli* BL21 (DE3) sacculi. Starting from the 170–180 ppm region of the ^{13}C carbonyl groups (Fig. 9.3B), individual cross-peaks to the directly bound carbon atom (Cα for all amino acids and side-chain carbons in some amino acids like glutamic acid) provide unique identification for all amino acids included in peptide stems after analysis of the 20–48 ppm region (Fig. 9.3D).

Analysis of a few different species (Kern *et al.*, 2010) showed that the following peptidoglycan characteristics can be clearly identified in these spectra: the presence of *m*-diaminopimelic acid *versus* lysine in position 3 of the peptide stem (cross-peak correlating 28.5 to 39.6 ppm is unique to Lys), the nature of amino acids such as glycine or aspartic acid in the inter-peptidic bridge (unique resonance of Gly Cα at 43.2 ppm). Modifications of the peptidoglycan components can also be detected with this approach as soon as they represent more than 10% of the representative units. As an example, amidation of γ-D-Glu in position 2 of the peptide stem was detected in *Bacillus subtilis* 168 (Fig. 9.3B). Amidation of the aspartic acid of the bridge in *E. faecium* was also evidenced (unpublished results). Some of these characteristic features are illustrated in Fig. 9.3 that depicts the ^{13}C–^{13}C correlation spectrum obtained for *Bacillus subtilis* strain 168 sacculi before HF treatment, i.e. for samples containing peptidoglycan and wall teichoic acids.

Additional structural variations in the glycan strands such as N-glycosylation of muramic acid in *Actinomycetales* or O-acetylation at position 6 of MurNAc (Vollmer, 2008) could also potentially be identified by this approach. Peak intensities should in principal allow determining the proportion of the different forms, but dynamics differences can prevent a correct quantification. These data, as well as information on the cross-link index, could be more easily obtained after digestion of the peptidoglycan by mutanolysin and subsequent analysis of the fragment mixture by 2D heteronuclear liquid-state NMR methods (unpublished results).

1D and 2D solid-state NMR methods can thus provide detailed information at atomic resolution on the peptidoglycan. If the requirement for isotopic enrichment makes these methods eventually less attractive for taxonomial purposes, they can be very useful to characterize peptidoglycan modifications resulting from a change in the environment or from the presence of external factors such as antibiotics.

Characterization of teichoic acids in Gram-positive bacteria

Teichoic acids, major cell surface anionic glycopolymers, can account for over 60% of the mass of the cell wall in a wide range of Gram-positive organisms. They fall into two categories: the wall teichoic acids that are coupled to the peptidoglycan (WTAs) and the lipoteichoic acids (LTAs) that are anchored to the cell membrane (Neuhaus and Baddiley, 2003) (see Fig. 9.2 for a schematic representation).

In the past ten years, solution NMR has largely contributed, along with mass spectrometry and chemical analysis, to elucidate the chemical structures of LTAs (Theilacker *et al.*, 2006; Rahman *et al.*, 2009; Carballo *et al.*, 2010) and WTAs (Sadovskaya *et al.*, 2004; Vinogradov *et al.*, 2006; Potekhina *et al.*, 2009; Tomita *et al.*, 2009) isolated respectively from butanol extracts or from 5% TCA extracts (obtained after 4% SDS boiling of the cells to eliminate the LTAs) of bacterial cells. As a consequence of these studies, the analysis of 1D ^{13}C-NMR spectra of isolated teichoic acids was proposed as a basis for the identification of different strains within the genus *Nocardiopsis* (Naumova *et al.*, 2001).

While these methods have revealed the large variability in the WTA structures and LTA modifications, they mostly rely on acidic processing of the cell walls which are susceptible to cleave some of the glycerol or ribitol substituents such as D-Ala. Because D-alanylation and glycosylation (Xia *et al.*, 2010b) have been proposed to impact WTA and LTA functions (Xia *et al.*, 2010a), there is a need to extract chemical structural information directly from whole cells. Owing to the embedment of teichoic acid in the cell wall, solid-state NMR appears as the method of choice for this purpose.

What can be learned from 1D experiments on unlabelled samples?

Since teichoic acids are generally formed by repeats of polyol phosphate (in general polyglycerol- or polyribitol-phosphate) or glycosylpolyol phosphate residues linked by phosphodiesters, the performance of ^{31}P-NMR must be examined. ^{31}P nuclei can be detected using 1D experiments with good sensitivity and a correct resolution without the need for specific labelling schemes (100% natural abundance). The chemical nature of the repetitive unit of WTA and LTA can be tentatively assigned in whole cells from ^{31}P solid-state NMR spectra recorded with direct ^{31}P excitation (Kern *et al.*, 2010). In fact, when cells (or bacterial cell wall containing only peptidoglycan and WTA) are resuspended in HEPES buffer at pH 7.5, the phosphorus chemical shift of polyglycerol is between 0 and 1 ppm, while polyribitol chains tend to resonate between 1 and 2 ppm. Although substitutions at position 2 of glycerol and 2, 3 or 4 of ribitol may shift these values to upfield regions, these effects are generally smaller than the large chemical shift difference due to the nature of the alditol. In addition to the previous groups, the presence of phosphocholine (in *S. aureus*) and phosphoglycosyl between 0 and –1 ppm can be readily identified in these conditions.

The chemical shift dispersion and resolution in solid-state ^{31}P-NMR spectra remain insufficient to identify the degree of substitution at different positions of the polyol phosphate groups, which is essential for taxonomy purposes. ^{13}C nuclei appear as better probes for this purpose (Sadovskaya *et al.*, 2004). Yet the polyol signals as well as the acyl or glycosyl substitutions overlap with the carbohydrate resonances arising from the peptidoglycan. The potential of 2D experiments to raise this degeneracy must thus be examined.

What can be learned from 2D experiments on uniformly labelled samples?

Fig. 9.4 illustrates the potential of 2D ^{13}C–^{13}C correlation solid-state NMR techniques in the identification of the chemical components of WTAs within crude cell wall isolates, which contain peptidoglycan as well as WTAs (Kern *et al.*, 2010). This figure emphasizes the clear identification of the resolved resonances of unsubstituted and C-2(α-D-Glc)-substituted phosphoglycerol fragments in the major components of the cell wall teichoic acid of *B. subtilis* strain 168 in the presence of peptidoglycan. The nature of the glycosyl substituent is also identified in a straightforward manner based on the pattern of the carbon chemical shifts. In this sample, minor components arising from α-D-GalNAc and β-D-Glc, that have been identified in liquid-state NMR studies on other strains of *B. subtilis* (Shashkov *et al.*, 2009), are also detected. It must be noted that the chemical shifts measured in these solid-state NMR experiments match the chemical shifts reported in liquid-state NMR studies for WTAs and LTAs containing a polyglycerol phosphate polymer.

Once modifications have been identified, they can be quantified using specific labelling as proposed for D-alanylation of teichoic acids in *S. aureus* (Cegelski *et al.*, 2006) and *E. faecium* (Patti *et al.*, 2009).

In the future, an approach that would combine solution NMR analysis of extracted LTAs and solid-state characterization of the total peptidoglycan–WTA matrix would allow a detailed characterization of different bacterial strains within the same Gram-positive species. Although it may not meet the requirements for high-throughput analysis necessary for taxonomy identification, this methodology shows promising features for the comparison of bacterial strains as a function of external factors.

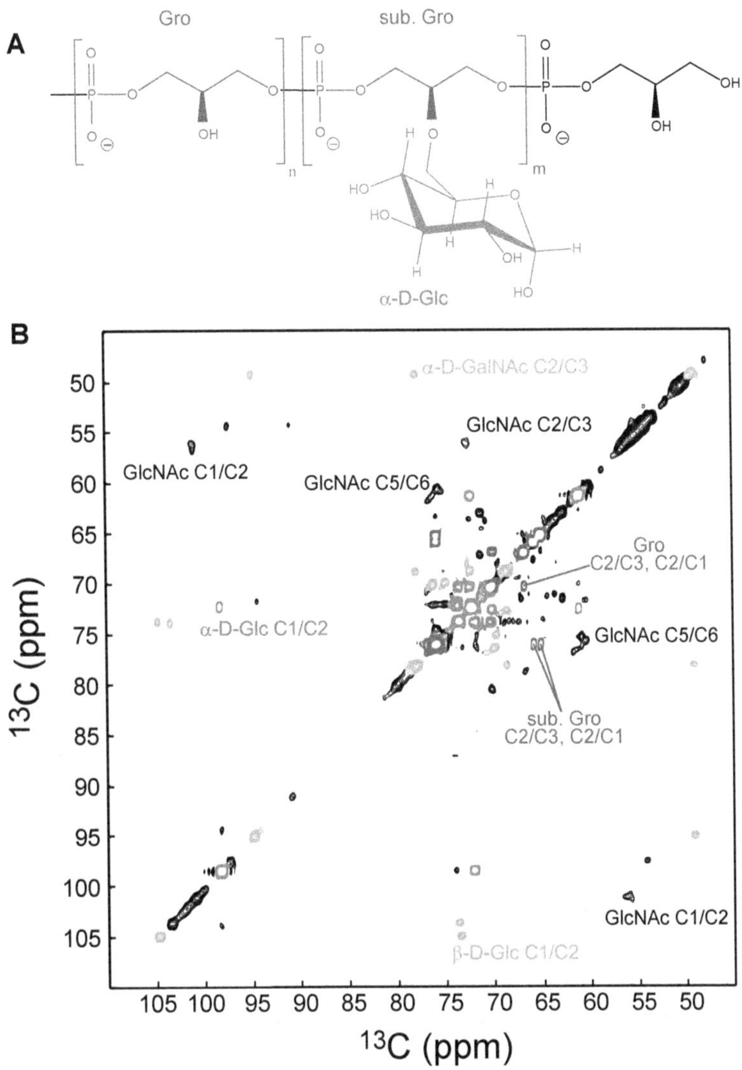

Figure 9.4 Cell wall teichoic acids of *Bacillus subtilis* strain 168. (A) Chemical structure of the major component and (B) fingerprint sugar region of the 2D ^{13}C–^{13}C correlation spectrum recorded at 14.1 T and 285 K on a fully hydrated ^{13}C-,^{15}N-uniformly labelled cell wall sample. Units of the major poly(glycerolphosphate) component of WTAs are detected and labelled with different shades of grey: unsubstituted glyceroldiphosphate (Gro, in dark grey) and C-2-(α-D-glucose, in medium grey)-substituted glyceroldiphosphate (sub. Gro, in dark grey). Minor components include α-D-*N*-acetylgalactose (α-D-GalNAc, in light grey) and β-D-glucose (β-D-Glc, in light grey) which were identified on the basis of their chemical shift and their comparison with teichoic acids identified in *Bacillus subtilis* strain VKM B-501[T] (Shashkov *et al.*, 2009). Peptidoglycan cross peaks (GlcNAc) are shown in black.

Characterization of lipopolysaccharides in Gram-negative bacteria

Lipopolysaccharides are glycolipids exposed on the outer-membrane of Gram-negative bacteria and are responsible for the innate immune system response of host cells. They are composed of a glucosamine disaccharide with six or seven acyl chains, a polysaccharide core and an extended polysaccharide chain called the O-antigen. Solution NMR has extensively contributed to the determination of the chemical structure of the O-antigen carbohydrate moieties (Mulloy, 1996) using identical approaches to the ones described for teichoic acid analysis (Vinogradov *et al.*, 2004). In particular, NMR contributed to show that the serological properties of different strains were in part correlated to different degrees of O-acetylation in the glysosidic components of the polysaccharides (Jones and Lemercinier, 2002; Dag *et al.*, 2004). In 1999, Jachymek *et al.* demonstrated that a ^{1}H-NMR spectrum of a lipopolysaccharide suspension in water or of intact *Yokenella regensburgei* bacterial cells could be obtained using magic angle spinning techniques at low spinning rates (HR-MAS). With this technique, ^{1}H nuclei of the molecules that exhibit a relatively high flexibility can be directly observed without the use of any ^{13}C or ^{15}N labelling. In this case, the large background components from the bacterial cells were filtered out based on the fact that the O-antigen section of the polysaccharide is highly dynamic owing to the free rotation between the individual saccharide units.

This approach can be extended to the detection and analysis of other surface glycan structures in Gram-negative bacteria, like capsular polysaccharides (see Chapter 2 in this book) or N-linked and O-linked glycans (Szymanski *et al.*, 2003), N-linked protein-glycans (see Chapter 4 in this book) (Nothaft *et al.*, 2010) and other carbohydrate signals (see Chapter 6) (Lee *et al.*, 2005). As an example, comparison of ^{1}H HR-MAS spectra of whole cell samples prepared from two specific strains of *Campylobacter jejuni* allowed to prove the structural similarity of the main surface polysaccharide excepted for the presence of free aspartic acid. It could also be used to detect phenotypic variations within the same strain, revealed by various small chemical modifications. In ^{1}H spectra recorded on whole cells, comparison of the anomeric protons also offers a powerful tool to identify genes involved in the biosynthesis of lipo-oligosaccharides, capsular polysaccharides or other glycans (Bernatchez *et al.*, 2005), for example the *mynC* gene coding for a O-acetyltransferase in the serogroup A of *Neisseira meningitidis* (Gudlavalleti *et al.*, 2006).

When resonances are narrow enough, individual components of the repetitive units can be identified and linked by using 1D and 2D ^{1}H-detected experiments recorded under slow sample spinning at the magic angle (HR-MAS) (Wieruszeski *et al.*, 2005). In this context, 1D-NOESY and 2D ^{1}H,^{31}P-HSQC correlation spectra were used to establish the non-stoechiometric fructofuranose and O-methylphosphoramidate substitutions in the HS:1 serostrain capsular polysaccharide of *C. jejuni* (McNally *et al.*, 2005).

Characterization of mycolic acids

Like lipopolysaccharides, mycolic acids are involved in diverse host immune responses to the *Corynebacteria–Mycobacteria–Nocardia* group and there is a need to fully characterize mycolic acids to establish structure-biological activity relationship. Liquid-state NMR studies, in combination with mass spectrometry, contributed to the identification of the long-chain α-alkyl, β-hydroxy fatty acid (Schroeder and Barry, 2001; Kremer *et al.*, 2005; Uenishi *et al.*, 2008) or arabinogalactan or trehalose (Gilleron *et al.*, 2005; Niescher *et al.*, 2006) components.

Due to the exposure of mycolic acids to the surface of bacterial cells and their intrinsic mobility, ^{1}H-MAS at low spinning rate (HR-MAS) can again be used to follow mycolic acids in whole cells. In fact, *Mycobacterium bovis* BCG showed a well-isolated ^{1}H signal at –0.33 ppm that was specifically assigned to the *cis*-cyclopropyl ring of α- and oxygenated-mycolates. This signal was used to quantify *cis*-cyclopropane rings in cells and to follow its reduction by 75% under the administration of 1 μg/ml thiacetazone, an inexpensive antitubercular, bacteriostatic drug and its derivatives. This conclusion contributed to identify the genes inactivated by thiacetazones (Alahari *et al.*, 2007). Similarly, the modulation of the cyclopropanation activity of α-mycolic acids could be measured as a function of temperature in *Mycobacterium smegmatis* (Alibaud *et al.*, 2010).

Conclusions and future prospects of the applications of NMR in bacterial strain typing

A combination of solution NMR on isolated materials or fragments, in-cell solution NMR (Azurmendi *et al.*, 2007) and whole cell HR-MAS or solid-state NMR techniques (Metzger *et al.*, 2009) can provide unique information on the chemical structure of the glycopolymers constituting the bacterial cell wall. Whole-cell NMR techniques are particularly suited to follow the details of *in vivo* chemical modifications as a function of developmental environment of the organisms as they do not depend on the reliability of degradation procedures. This is particularly important when quantification comes into play.

^{1}H-, ^{31}P- and ^{13}C-MAS NMR perform well in the comparison of different strains or even species, especially when teichoic acids, lipopolysaccharides, or other cell-wall glycopolymers are considered. Using these methods for taxonomy purposes would require the set-up of extended databases for individual cell wall components. In this respect, other techniques like mass-spectrometry (see Chapter 10) are more advanced (Sauer and Cooke, 2010).

Cell wall interactions

Introduction

In Gram-positive bacteria, peptidoglycan and teichoic acids provide a scaffold for the anchoring of proteins to the cell surface, in addition to lipoproteins and other proteins that are directly inserted into the cytoplasmic membrane. Despite the presence of the outer membrane in Gram-negative bacteria, peptidoglycan plays a similar anchoring role. Some of these anchored proteins are covalently attached to the peptidoglycan, but some of them are known to be non-covalently bound to the peptidoglycan. Among them, there are a large number of enzymes involved in the final stages of peptidoglycan biosynthesis which determine the shape of the bacteria and regulate processes such as cell elongation and division (Mattei *et al.*, 2010). Furthermore, microbial recognition motifs used by the immune system are mainly localized in the bacterial cell wall (polysaccharide and peptidoglycan) (Guan and Mariuzza, 2007). In mammalians, soluble peptidoglycan recognition proteins (PGRPs) also target the peptidoglycan. They directly act as a bactericide through binding to the peptidoglycan and, in certain cases, even hydrolysing the peptidoglycan (Carneiro *et al.*, 2008). These are only a few examples of molecules that interact with the cell wall.

Regarding the importance of the peptidoglycan biosynthesis in a context of increasing resistance to antibiotic treatments but also its implication in the response of the innate immune system, it is crucial to get more information about cell–wall interaction with proteins, antibiotics and ions. This section addresses the question whether solid-state NMR can contribute to this area.

NMR probes to study molecular interaction

Solid-state NMR methods have been successfully used to study protein–protein or protein–ligand interactions (e.g. Baldus, 2006). For the study of peptidoglycan–protein, peptidoglycan–antibiotic or peptidoglycan–ion interactions, different approaches can be considered and chosen depending on the affinity between the two interacting partners. In fact, molecular interaction can induce:

i A modification of the chemical shift observed in an NMR experiment (chemical shift mapping). Each nucleus of a molecule gives rise to a NMR signal at a specific resonance frequency that depends on its local environment. This isotropic chemical shift is a very sensitive probe for binding events. Slight conformational changes due to the chemical interaction can modify the local electronic environment of the nucleus and therefore its chemical shift. Under MAS, only the isotropic component of the chemical shift tensor remains and can be recorded in a multi-dimensional correlation experiment. Comparison of the spectra recorded before and after complex formation allows the localization of the binding interface, and the study of possible structural changes due to the molecular interaction. Successful use of chemical shift mapping as a probe and proof of interaction has been demonstrated by Zech *et al.* (2004).

ii A change in the dynamic behaviour. Molecular dynamics is an important factor for complex formation and stability. It implies the measurement of either tensor interactions (as Chemical Shift Anisotropy (CSA) or dipolar coupling), or relaxation parameters such as the spin-lattice relaxation time constant T_1. Comparison of molecular dynamics before and after complex formation can reveal the localization of the interface and allow the analysis of molecular motion involved in the binding event. More details concerning dynamic studies by solid-state NMR are given in the last section of this contribution.

iii A proximity between particular nuclei of interacting partners lying in the detectable range of solid-state NMR (< 7 Å). Dipolar couplings, which are directly linked to the distance between the coupled spins, are measured. The obtained inter-molecular distances are then used as structural restraints between the individual molecular constituents in structure calculation protocols. This procedure leads to a more precise characterization of the complex than the previously described sources of information. In this approach, long-range distance measurements require a selective labelling strategy based on a pre-knowledge of the interface position in both interacting partners. Indeed, in uniformly labelled samples, intermolecular interactions are truncated by the usually stronger intramolecular couplings. Even though the complete structure determination is difficult to obtain, site-specific atomic-level information can be extracted for heterogeneous systems by a combination of selective labelling and adapted NMR dipolar recoupling experiments such as for example REDOR (Rotational-Echo Double-Resonance NMR) (Gullion and Schaefer, 1989).

The choice of the appropriate method will strongly depend on the molecular system under study and on the possibility of introducing isotopic labels. In fact, except for ^{31}P, all described methods will take advantage of efficient ^{13}C and/or ^{15}N labelling strategies involving either complete labelling of only one of the interacting molecules, or a combination of selective labelling of the two molecular components.

Most published NMR studies have been recorded on isolated cell wall constituents (peptidoglycan, teichoic acids) to simplify the spectrum analysis. Fewer of them have been described on intact lyophilized or fully hydrated cell wall samples. However, modulation of the interaction in the presence of different physiological conditions can be assessed by recording data on intact bacteria. This motivates the extension of NMR techniques to the study of intact bacteria in physiological environments, which would further open the field for a large number of potential applications.

Interaction of the cell wall with proteins

Some of the proteins interacting with the peptidoglycan are able to bind fragments with an affinity that can sustain a structural study. In this case, structural and functional information on protein–peptidoglycan interactions can be retrieved from liquid-state NMR (Lehotzky *et al.*, 2010), but requires, for efficiency purposes, $^{13}C/^{15}N$-labelling of peptidoglycan fragments. The latter can be prepared from bacteria grown on minimal media including labelled carbon and nitrogen sources, purification of the peptidoglycan, and digestion with specific hydrolases. Using a combination of different enzymes and chemical treatment (Glauner, 1988; Arbeloa *et al.*, 2004), a variety of fragments can be generated and further screened for interactions with target proteins by liquid-state NMR or crystallography. This strategy is foreseen to lead to the identification of the chemical structure required for interaction. Initial models of interaction derived from these studies will then serve as a working hypothesis and a starting point to computationally build a more elaborate model involving the complete bacterial cell-wall entity.

However, interaction studies involving large fragments or the entire peptidoglycan polymer are not accessible to liquid-state NMR. Solid-state NMR is in this case a promising method to analyse protein recognition of the peptidoglycan. The effect of the interaction on the NMR parameters described above can be observed by detecting the nuclear signal emanating from the cell wall or/and from the protein. Since the proton is difficult to observe by solid-state NMR (except for very flexible regions), other nuclei such as carbon, nitrogen and phosphorus can be used. ^{31}P has the advantage to be very sensitive without specific labelling and can be chosen in the case of a specific interaction with the teichoic acids (Rice and Wickham, 2005; Wickham and Rice, 2008). To observe protein interaction with the peptidoglycan, ^{13}C- and/or ^{15}N-isotopic labelling of the peptidoglycan or of the protein has to be used. In general, ^{13}C-labelling of the peptidoglycan is easier and ensures good quality of the spectra even in cases of interaction with very large proteins. For the preparation of the NMR sample, a protocol similar to the one used in pull-down experiments can be designed (Kern *et al.*, 2008). In that case, a few milligrams of purified peptidoglycan or cell wall preparation are incubated with the protein. After incubation, the sample is centrifuged and the precipitate is transferred into the solid-state NMR rotor. The recorded spectra can then be compared with reference spectra previously obtained on the labelled peptidoglycan or cell-wall preparation alone (before any interaction with the protein). Chemical shift changes or variations in cross-peak intensities or relaxation rates upon interaction can permit to localize the interaction on the cell wall. For example, on the basis of through-bond ^{13}C-correlation spectra and spin-lattice relaxation measurements, it was shown that glycan and lactic acid sites are clearly and specifically affected by the presence of the YajG protein, while peptide cross-peaks remain nearly unchanged in the same conditions (Kern *et al.*, 2008), thus clearly demonstrating the directional interaction of YajG with the GlcNAc-MurNAc glycan chains of the peptidoglycan.

The precedent strategy is valid for quite strong interactions between the peptidoglycan and the protein partner. To extend this application to protein-cell wall interactions of lower affinity, a paramagnetic centre such as for example nitroxide (Nadaud *et al.*, 2007) can be covalently linked to the protein by using a site-directed spin-labelling approach. The presence of a paramagnetic centre leads to enhanced relaxation rates, as well as in some cases to electron-nucleus distance-dependent changes of the NMR chemical shifts. Paramagnetic relaxation effects induced by the paramagnetic centre can be measured with a good sensitivity on the cell wall resonances and can be transcribed in terms of intermolecular distances.

Interaction of the cell wall with antibiotics or antimicrobial peptides

Some glycopeptide antibiotics inhibit the synthesis of the cell wall by a specific interaction with the peptidoglycan precursor and the peptidoglycan itself. For example, vancomycin inhibits the transglycosylation step by binding the D-Ala-D-Ala carboxy termini of the peptidoglycan pentapeptide. A careful analysis of the modifications induced on the peptidoglycan polymer (degree of cross-linking, length of the glycosidic and peptidic chains) when the cell is growing in presence of the antibiotics can confirm its specific activity on the transpeptidation and transglycosylation enzymatic steps (Cegelski *et al.*, 2002; Patti *et al.*, 2008a,b; Kim *et al.*, 2009). As a complementary strategy, structural information on the glycopeptide/peptidoglycan complex allowed the rationalization of the mode of action of this class of antibiotics. Specific nuclei distances were measured by using an appropriate strategy of specific labelling (^{15}N- and/or ^{13}C-labelling of Gly, Ala and Lys) (Kim *et al.*, 2002) on lyophilized samples. To obtain a direct measurement of inter-distances between one specific residue of the peptidoglycan and the unlabelled antibiotic, a ^{19}F-labelled substituent was attached to different glycopeptide compounds (Kim *et al.*, 2006, 2008). A model of interaction was finally proposed by energy-minimization of a peptidoglycan peptidic chain and the use of few measured internuclear distances.

Interaction of the cell wall with metal ions and surfaces

Presence of counter-ions interacting with different charged polymers as liposaccharides or teichoic acids has been discussed in the literature (Neuhaus and Baddiley, 2003). The presence of ions can modulate the electrostatic interactions of the cell wall and play a role in the regulation of ion homeostasis and trafficking of ions in and out of the cell, as well as in the virulence and cell adhesion capability of the bacteria.

Interaction of isolate phophodiester polymers with different surfaces coated with titanium dioxide, cellulose and peptidoglycan have been studied by ^{31}P solid-state NMR by Wickham and Rice. For this purpose, the authors used the ^{31}P chemical shift anisotropy measured on lyophilized samples as a probe for structural and dynamic changes induced by surface adhesion (Rice and Wickham, 2005; Wickham and Rice, 2008).

To analyse the interaction of cell-wall polymers with important biological metal ions such as Ca^{2+} and Mg^{2+}, the influence of the metal ion on the isotopic or/and anisotropic chemical shift of the ^{13}C and ^{31}P nuclei of the polymer can be used. Since ^{31}P is the natural isotope present without the need for specific labelling, this nucleus has been used in many different studies. In particular, CSA (chemical shift anisotropy) profiles have been measured on dried or partially dried samples. Distances between the ion and the proximate phosphorus nuclei could be extrapolated by comparing profiles calculated from molecular models built on different types of ion/phosphate interaction (Wickham *et al.*, 2009). A model based on a bidentate interaction between one ion and two adjacent phosphate of the teichoic chain has thus been proposed. To validate the model, direct determination of ion–phosphate distances were obtained by using a NMR visible cadmium ion and ^{31}P-Cd^{2+} REDOR experiments (Halye and Rice, 2010).

In the case of fully hydrated samples, the isotropic chemical shift is directly measured in 1D MAS ^{31}P solid-state NMR spectra. Isotropic chemical shift variations upon ion titration were used to extract binding constants on intact sacculi (Kern *et al.*, 2010).

Since the effect induced by the ion on the ^{31}P chemical shift is small and strongly decrease with the interatomic distance, an alternative is the use of paramagnetic ions that quench the NMR signal of proximate nuclei up to a distance of ~15 Å. By using paramagnetic Mn^{2+}, it has been demonstrated that this ion interacts with the wall teichoic acid phosphate groups but that its affinity for the cell wall is reinforced by an extra interaction with the carboxyl groups of the peptidoglycan peptide stems (Kern *et al.*, 2010).

Interestingly, a possible modulation of the cell wall repulsive charge induced by the presence of a phosphate-D-alanine zwitterrion on teichoic acids has been investigated by NMR as well. D-Ala can be specifically labelled in the cell wall by integrating commercial labelled D-Ala in the bacterial culture with alaphosphin, a racemase inhibitor to limit the D-Ala to L-Ala conversion (Kim *et al.*, 2006; Garimella *et al.*, 2009). Distance measurements between the phosphate 31p and the Ala ^{15}N sites were performed using the $^{15}N\{^{31}P\}$ REDOR technique in presence and in absence of Mg^{2+} ions on lyophilized sacculi samples (Garimella *et al.*, 2009). The authors show that magnesium chelation is not prevented by the D-Ala group but only pushes the D-Ala further away from the phosphate group.

Structure and flexibility of the cell wall

Structure of the peptidoglycan

Introduction

As described above, several techniques, amongst them solution or solid-state NMR, are suited to investigate the chemical structure of the peptidoglycan polymer. A much more difficult task consists of resolving the architecture of the cell wall, the knowledge of which is needed to understand the growing and dividing mechanisms of the cell. Indeed, the large size of the peptidoglycan sacculus and its flexibility prevent crystal formation or solubilization. Typical structure determination techniques as X-ray or solution NMR can therefore not be applied. Current models of the peptidoglycan architecture are based on chemical and biophysical data, postulating glycan chains which run either parallel or perpendicular to the cell membrane plane. Recent results by atomic force microscopy on *B. subtilis* (Hayhurst *et al.*, 2008) suggest that even more complicated architecture may have to be envisaged, as for example, a rope-like structure. A short overview of the present knowledge on peptidoglycan architecture can be found in reference (Vollmer and Seligman, 2010).

With respect to the peptidoglycan architecture itself, solid-state NMR seems promising, as it can be applied to cell-wall samples with an intact architecture. However, as NMR is a very local probe of molecular structure, a whole set of local structural information will be needed to solve the problem of architecture on this very flexible and loose polymer. It is important to note here, that the NMR signal itself, although being very sensitive to the local environment, arises from the entire sample. It can therefore not detect sample heterogeneity as could for example an imaging-type technique. Although the characteristics of the peptide cross-linking does not directly come under peptidoglycan architecture, we decided to treat it in this context, as it is at the base of the polymer-network formation and has in this sense an impact on the final architecture of the cell wall.

Architecture

A first contribution to solve the peptidoglycan architecture has been recently presented by Schaefer *et al.* on *S. aureus* (Sharif *et al.*, 2009b). Using a carbon-13 spin diffusion technique, they were able to estimate the distance between the pentaglycyl bridge and the neighbouring glycan chain to be within 5 Å. This short distance is then shown to be only compatible with a lattice where the glycan main chains are parallel to each other. This result has been obtained with a modified version of the CODEX (centreband-only detection of exchange) technique (deAzevedo *et al.*, 2000). In order to limit the data acquisition to one-dimensional spectra, the glycyl residues of the bridge were selectively labelled on the carbonyl position by adding labelled glycine to the culture medium. This selective labelling scheme has the additional advantage of avoiding relayed spin-diffusion mechanisms that are present in fully labelled samples. The distance between the labelled ^{13}C spins located on the bridge and other carbon sites in natural abundance is estimated from the speed of the carbon spin-diffusion between both sites. It is to note that distance measurements based on spin-diffusion are quite imprecise and can only be indicative. Notably the following points need to be considered: (i) the relation between spin-diffusion speed and inter-nuclear distance is not trivial; it has been estimated for specific samples (Vanderhart, 1987; Manolikas *et al.*, 2008), where it is used for distance estimation but depends on several parameters, as the mobility, proton density, chemical shift difference between the two involved ^{13}C sites, magic-angle spinning speed, or strength of the external static magnetic field; (ii) the experiment has to be performed on a lyophilized sample as mobility of the peptidoglycan will impact on spin-diffusion; (iii) the resolution in lyophilized samples is often not good enough to resolve carbon sites with similar chemical shift; (iv) in a polymer like the peptidoglycan, it can in principle not be distinguished whether spin-diffusion happens between carbon sites belonging to the same monomer or not. In order to properly interpret spin-diffusion experiments, pre-knowledge of precise chemical shifts and distances in the monomer are required. Thus, in the case of *S. aureus*, the intra-monomer distance between the glycyl bridge and the glycan chain was known to be about 10 Å, which is too far away to be detected by spin-diffusion. The detection of spin-diffusion characteristic for a shorter distance of about 5 Å between these sites could therefore be assigned to an intermolecular interaction.

Peptide cross-linking

Using again the possibility of selectively labelling specific sites in the amino acid residues, the solid-state NMR technique REDOR (rotational-echo double resonance NMR) was shown to be successful in characterizing the cross-linking between peptide stems in different peptidoglycans (Tong *et al.*, 1997; Cegelski *et al.*, 2002; Patti *et al.*, 2008b; Sharif *et al.*, 2009a). Thus for example in *S. aureus*, the degree of cross-linking (number of peptide stem D-Ala covalently bound to a pentaglycyl bridge) and bridge linking (number of peptide-stem L-Lys covalently bound to a pentaglycyl bridge) could be determined in the presence or absence of vancomycin, suggesting that the antibiotic interferes with transglycosylation (Cegelski *et al.*, 2002). The number of glycyl residues in the peptide bridge of different Fem-deletion mutants of methicillin-resistant *S. aureus* and their impact on cross-linking has been investigated using REDOR as well (Sharif *et al.*, 2009a), showing that the *fem*B mutant reduces the pentaglycyl bridge to 25% of monoglycyl and 75% of triglycyl bridges, with no significant changes in cross- or bridge-linking. In the *fem*A mutant, the bridge is reduced as expected to a single glycyl residue, inducing, however, a decrease in the cross-linking level by more than 20%. The REDOR technique used to obtain these different results is applied on a set of samples with different pairs of specific labels located as close as possible on each side of the peptide connection. Thus, in *S. aureus*, D-[1-^{13}C]alanine and [^{15}N]glycine labelling scheme was used to investigate the degree of cross-linking, [1-^{13}C]glycine and L-[ε-^{15}N]lysine for the degree of bridge-linking, and [1-^{13}C,^{15}N]glycine and L-[ε-^{15}N]lysine for the number of glycyl residues in the inter-peptide bridge. These different labelling schemes are highlighted in Fig. 9.5A. The REDOR technique is applied in two steps (see Fig. 9.5B): in a first experiment, the signal of one of the spin labels, e.g. A (^{13}C or ^{15}N in this case), is detected after an evolution period where all possible interactions are removed by the mean of the pulse sequence, such that nothing happens during this time (except relaxation). All spin labels A contribute to the signal, which is used as reference and is dubbed S_0. The same experiment is then repeated, but this time the heteronuclear dipolar interaction

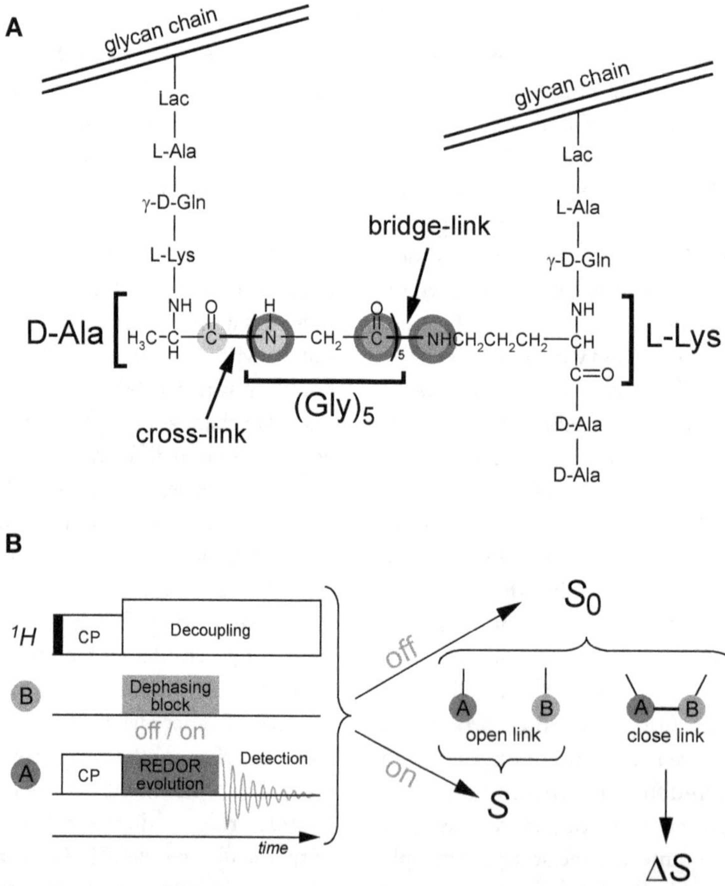

Figure 9.5 (A) Schematic representation of the interpeptide cross-linking in *Staphylococcus aureus*. The selective spin label pair introduced to measure with the REDOR technique the degree of cross-linking, the degree of bridge-linking, and the number of glycyl residues in the inter-peptide bridge are highlighted in light, medium, and dark grey, respectively. (B) Schematic representation of the pulse sequence used in the REDOR technique (each line represents the successive radio-frequency pulses performed on each type of nuclei). Two spectra of nuclei A are acquired, the first one with the REDOR dephasing block off (no pulse) leading to the reference signal S_0, the second one in the presence of REDOR dephasing pulses (REDOR dephasing block on) giving rise to the signal *S*. The contribution to each signal of the spin labels present in open and close links is given. Δ*S* represents $S_0 - S$. The exact composition in radio-frequency pulses of the REDOR dephasing and evolution block can be found in, for example, Gullion and Schaefer (1989) and Toke *et al.* (2006).

between spins A and B is re-introduced by the use of 180° pulses (REDOR dephasing block in Fig. 9.5B). For those spin labels A which are in close contact to a spin label B, this leads to a dephasing of their signal during the evolution period. The detected signal intensity S, which now only arises from non-linked spin pairs (open links), is therefore decreased with respect to S_0. To remove relaxation effects, the REDOR signal S or REDOR difference $\Delta S = S_0 - S$ are scaled by the reference signal S_0. Under certain conditions the relative REDOR difference $\Delta S/S_0$ directly gives the ratio of covalently bound spin labels with respect to the total number of detected spins, corresponding therefore to the degree of cross-linking. As stated in the references cited above, the following aspects have to be taken into account: (i) the sample has to be lyophilized to avoid a reduction of dephasing due to motion-averaged dipolar couplings; (ii) the REDOR difference needs to be properly corrected for the degree of isotopic enrichment obtained in the peptidoglycan sample for the dephasing label B (an incomplete isotope labelling leads to an underestimation of $\Delta S/S_0$); (iii) the presence of unwanted and disturbing labels introduced at other sites by isotope scrambling or amino acid conversion during the biosynthesis of the peptidoglycan may interfere with the REDOR difference (they contribute to S_0, although they are not at a link position); (iv) a proper REDOR evolution time has to be chosen such that the dephasing of spin A is complete and only due to the presence of the spin label(s) next to it; (v) the REDOR difference cannot be quantitatively interpreted in whole cells as some of the labels are incorporated into cytoplasmic proteins (see point iii). In this case the TEDOR (Transferred Echo DOuble Resonance) experiment is preferred (Cegelski *et al.*, 2006). Differences in chemical shift for labels located in a bridge or in an open stem can greatly help the interpretation of data. Interestingly, in cases where the REDOR experiment can be quantitatively interpreted, the REDOR dephasing curve (REDOR difference as a function of the evolution time) contains distance and conformational information. Thus, for example, in Tong *et al.* (1997), the pentaglycyl inter-peptide bridge was shown to be essentially in a compact conformation, and not extended.

Structure in hydrated fully labelled peptidoglycan

An alternative to these techniques based on selective labelling of lyophilized samples is the use of multidimensional solid-state NMR applied to fully ^{13}C-enriched hydrated cell walls. As stated above, the extension of the NMR technique to two (or more) dimensions allows raising the ambiguities coming from resonance overlap in fully labelled samples. Similar to the case of fully labelled proteins, the hydrated state of the solid cell-wall sample has been shown to increase the resolution of NMR spectra (Tong *et al.*, 1997). Thus, high-resolution 2D $^{13}C/^{13}C$ correlation spectra acquired on different Gram-negative (*E. coli*) and Gram-positive (*B. subtilis* and *S. aureus*) peptidoglycans, were recently compared in our group (Kern *et al.*, 2008, 2010). Despite the various structural characteristics of the studied peptidoglycans and the sensitivity of the carbon chemical shift for its surrounding local structure, the only differences observed in the spectra were arising from chemical modifications. This can be explained by the fact that the polymer structure is quite loose and differences in structure could then only appear on a much larger length scale. Considering the approximate length of peptide stems, the distance between glycan strands can be greater than 30 Å. These considerations could call into question the capacity of solid-state NMR alone in solving the three-dimensional structure of the peptidoglycan network. Additionally, the possible high dynamics of the peptidoglycan in the hydrated, biologically relevant state would interfere with any structural solid-state NMR technique based on distance information. Therefore, information about the dynamic properties of the cell wall in the hydrated state needs to be obtained for a complete understanding of the cell wall, and of its interactions with other biological or chemical entities. This is the subject of the next section.

Peptidoglycan dynamics

Impact of dynamics on the NMR spectra

Presence of different types of flexibility inside the cell wall can directly impact the recorded solid-state NMR spectra, depending on the technique used. This point is illustrated in Fig. 9.6. The first spectrum (A) is a direct acquisition of carbon-13 spins in a hydrated ^{13}C-uniformly labelled sample of *B. subtilis* cell wall. It contains signals from all the carbons in the sample. We can already note that some resonances are much sharper than others; this reflects a higher dynamics of the corresponding nuclei. This is confirmed by spectra B and C, which select either only the more rigid part of the sample (spectrum B) by using a filter based on dipolar interactions (cross-polarization, CP), or the very mobile part (spectrum C) which is selected by a sequence working only on 'liquid'-like samples (INEPT, Insensitive Nucleus Enhanced by Polarization Transfer). Spectrum B contains essentially resonances from the peptidoglycan, whereas carbons from the teichoic acids are present in spectrum C, demonstrating the much higher mobility of the long extended teichoic acid chains compared with the cross-linked peptidoglycan network. This kind of filter can be combined with multidimensional techniques in order to select some parts of the sample according to their mobility. The higher mobility of hydrated cell wall samples when compared with lyophilized or frozen samples has been demonstrated using similar techniques (Tong *et al.*, 1997). On the base of 2D spin-diffusion experiments, glycan chains were shown to be more rigid than the peptidic chains (Kern *et al.*, 2008).

Detection of mobility via relaxation

Using solid-state NMR, quantification of the flexibility can essentially be obtained by two methods. The first one consists in measuring the spin-lattice relaxation time constant called T_1 (the time needed for the spins to get back to equilibrium magnetization after a perturbation). The impact of dynamics on this parameter has been checked

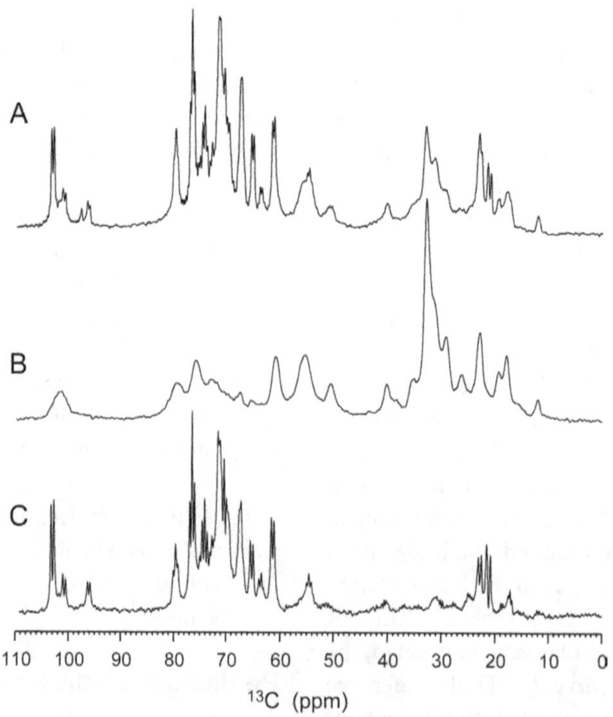

Figure 9.6 ^{13}C solid-state NMR spectra acquired on a hydrated uniformly ^{13}C-labelled cell wall sample of *Bacillus subtilis* using (A) direct carbon excitation, (B) ^{1}H- to-^{13}C cross-polarization (CP) and (C) INEPT transfer. In spectrum A, all carbon labels contribute to the signal. In spectrum B, only the more rigid part of the sample, essentially the peptidoglycan carbons are visible, whereas in spectrum C the very mobile carbon atoms of the teichoic acid chains are selected.

as a function of temperature for *B. subtilis* peptidoglycan (Kern *et al.*, 2010). For higher temperature and therefore higher dynamics, decreasing of T_1 relaxation time constants is observed. Spin-lattice relaxation is sensitive to very fast local motion in the MHz range. Although a quantitative characterization of the dynamics could in principle be obtained with a carbon relaxation study, this would be only possible on a very insensitive non-^{13}C-labelled sample (sample in natural abundance or with diluted labels). In fully labelled samples, spin-lattice relaxation can be interpreted qualitatively only, keeping in mind that the measured T_1 can be averaged over different sites in close proximity by spin-diffusion mechanisms. This way of measuring dynamics has recently been used in our group to compare the dynamics of different peptidoglycans (*E. coli*, *B. subtilis*, and *S. aureus*) (Kern *et al.*, 2010). We found that *S. aureus* has the most rigid peptidoglycan, followed by *B. subilis* whose peptidoglycan is more mobile. The least rigid peptidoglycan was found for *E. coli*. Comparing this information with known structural cell-wall parameters, as the peptidoglycan thickness, the averaged glycan chain length, and the peptide cross-linking, the flexibility of the peptidoglycan was shown to be inversely correlated to the degree of peptide cross-linking in the cell wall. As already described above, a protein (YajG) interaction with the peptidoglycan of *E. coli* could be brought to light by a change in T_1-values, and therefore of dynamics, of the glycan chains (Kern *et al.*, 2008).

Detection of mobility through reduction of an interaction tensor

The second way to measure dynamics is to determine the impact of motion on orientation–dependent interactions, like the chemical shift anisotropy (CSA), the dipolar, or the quadrupolar (for nuclei like ^{2}H) interaction. These interactions depend on the orientation of each molecule with respect to the external magnetic field leading to very broad lines. To obtain high-resolution solid-state NMR spectra, the anisotropy of the interactions is usually removed by spinning the sample at the magic angle (MAS). The different interactions are then reduced to their isotropic values, the isotropic chemical shift for the CSA and 0 for the dipolar interaction. The orientation–dependent interactions can, however, be selectively reintroduced by use of adequate pulse sequences. An eventual motional averaging due to sample dynamics can then be observed on the re-introduced tensor interaction. It is interesting to note that CSA and dipolar interactions are also sensitive to motion in the kHz range, and therefore to network fluctuations in the cell wall.

All dynamic studies on the cell wall presented so far in the literature were only qualitative. The next step would be to describe quantitatively the dynamics for different sites of the cell wall. This is possible for example by comparing motion-averaged dipolar couplings or CSAs with the corresponding full interaction. Studies in this direction are currently in progress in our group.

Future prospects and challenges

With the availability of higher magnetic fields, improvement in sophisticated instrumentation such as fast spinning of the rotor, progress in uniform and/or specific labelling strategies, high-resolution solid-state NMR provides the required sensitivity and resolution for the acquisition of site-specific chemical and structural information on bacterial cell walls from a variety of species and strains. Fully hydrated cell wall sacculi or whole cell samples can be manipulated at temperatures close to the physiological conditions, and attempts to follow *in vivo* modifications of the cell wall under antibiotic-stress or incubation with antimicrobial peptides or simply along the cell cycle can be foreseen in the near future. As regards peptidoglycan or teichoic acids, new quantitative NMR tools need to be developed in order to further characterize the dynamics of these components within the cell, and monitor changes along cell developments. Future developments will focus on the determination of motional time-scale and amplitude, as well as their restriction or amplification by external factors such as bacterial glycomics products or homeostasis-relevant metals. To date, none of the available biophysical methods has been able to provide structural data, at atomic resolution, on the geometry of the interaction of peptidoglycan or cell wall components with wall-manufacturing or host recognition proteins. Solid-state NMR in combination with efficient labelling strategies can address this important biological question in order to provide new routes for pathogen inhibition. In case of interactions with favourable thermodynamical and exchange regimes, intermolecular distances could be retrieved using the latest NMR sequences developed for biomolecular solid-state NMR applications. Experimental restraints thus obtained can then be considered for the exploration of the energy surface in molecular modelling calculations. Exciting new macromolecular complex structures are expected as an outcome. Liquid- and solid-state NMR approaches developed for the study of bacterial cell walls can be extended to other challenging biological systems such as human epithelial cells or viruses.

References

Alahari, A., Trivelli, X., Guerardel, Y., Dover, L.G., Besra, G.S., Sacchettini, J.C., Reynolds, R.C., Coxon, G.D., and Kremer, L. (2007). Thiacetazone, an antitubercular drug that inhibits cyclopropanation of cell wall mycolic acids in Mycobacteria. PLoS One *2*, e1343.

Alibaud, L., Alahari, A., Trivelli, X., Ojha, A.K., Hatfull, G.F., Guerardel, Y., and Kremer, L. (2010). Temperature-dependent regulation of mycolic acid cyclopropanation in saprophytic Mycobacteria: Role of the *Mycobacterium smegmatis* 1351 gene (MSMEG_1351) in *cis*-cyclopropanation of alpha-mycolates. J. Biol. Chem. *285*, 21698–21707.

Andrew, E.R., Bradbury, A., and Eades, R.G. (1958). Nuclear magnetic resonance spectra from a crystal rotated at high speed. Nature *182*, 1659–1659.

Andrew, E.R., Bradbury, A., and Eades, R.G. (1959). Removal of dipolar broadening of nuclear magnetic resonance spectra of solids by specimen rotation. Nature *183*, 1802–1803.

Arbeloa, A., Hugonnet, J.E., Sentilhes, A.C., Josseaume, N., Dubost, L., Monsempes, C., Blanot, D., Brouard, J.P., and Arthur, M. (2004). Synthesis of mosaic peptidoglycan cross-bridges by hybrid peptidoglycan assembly pathways in Gram-positive bacteria. J. Biol. Chem. *279*, 41546–41556.

Atrih, A., Zollner, P., Allmaier, G., Williamson, M.P., and Foster, S.J. (1998). Peptidoglycan structural dynamics during germination of *Bacillus subtilis* 168 endospores. J. Bacteriol. *180*, 4603–4612.

Azurmendi, H.F., Vionnet, J., Wrightson, L., Trinh, L.B., Shiloach, J., and Freedberg, D.I. (2007). Extracellular structure of polysialic acid explored by on cell solution NMR. Proc. Natl. Acad. Sci. U.S.A. *104*, 11557–11561.

de Azevedo, E.R., Hu, W.G., Bonagamba, T.J., and Schmidt-Rohr, K. (2000). Principles of centerband-only detection of exchange in solid-state nuclear magnetic resonance, and extension to four-time centerband-only detection of exchange. J. Chem. Phys. *112*, 8988–9001.

Baldus, M. (2006). Molecular interactions investigated by multi-dimensional solid-state NMR. Curr. Opin. Struct. Biol. *16*, 618–623.

Bernatchez, S., Szymanski, C.M., Ishiyama, N., Li, J.J., Jarrell, H.C., Lau, P.C., Berghuis, A.M., Young, N.M., and Wakarchuk, W.W. (2005). Single bifunctional UDP-GlcNAc/Glc 4-epimerase supports the synthesis of three cell surface glycoconjugates in *Campylobacter jejuni*. J. Biol. Chem. *280*, 4792–4802.

Carballo, P.M.S., Vilen, H., Palva, A., and Holst, O. (2010). Structural characterization of teichoic acids from *Lactobacillus brevis*. Carbohydr. Res. *345*, 538–542.

Carneiro, L.A.M., Magalhaes, J.G., Tattoli, I., Philpott, D.J., and Travassos, L.H. (2008). Nod-like proteins in inflammation and disease. J. Pathol. *214*, 136–148.

Cegelski, L., Kim, S.J., Hing, A.W., Studelska, D.R., O'Connor, R.D., Mehta, A.K., and Schaefer, J. (2002). Rotational-echo double resonance characterization of the effects of vancomycin on cell wall synthesis in *Staphylococcus aureus*. Biochemistry *41*, 13053–13058.

Cegelski, L., Steuber, D., Mehta, A.K., Kulp, D.W., Axelsen, P.H., and Schaefer, J. (2006). Conformational and quantitative characterization of oritavancin–peptidoglycan complexes in whole cells of *Staphylococcus aureus* by *in vivo* C-13 and N-15 labelling. J. Mol. Biol. *357*, 1253–1262.

Dag, S., Niedziela, T., Dzieciatkowska, M., Lukasiewicz, J., Jachymek, W., Lugowski, C., and Kenne, L. (2004). The O-acetylation patterns in the O-antigens of *Hafnia alvei* strains PCM 1200 and 1203, serologically closely related to PCM 1205. Carbohydr. Res. *339*, 2521–2527.

Forrest, T.M., Wilson, G.E., Pan, Y., and Schaefer, J. (1991). Characterization of cross-linking of cell-walls of *Bacillus subtilis* by a combination of magic-angle spinning NMR and gas-chromatography mass-spectrometry of both intact and hydrolyzed C-13-labelled and N-15-labelled cell-wall peptidoglycan. J. Biol. Chem. *266*, 24485–24491.

Garimella, R., Halye, J.L., Harrison, W., Klebba, P.E., and Rice, C.V. (2009). Conformation of the phosphate D-alanine zwitterion in bacterial teichoic acid from nuclear magnetic resonance spectroscopy. Biochemistry *48*, 9242–9249.

Gilleron, M., Garton, N.J., Nigou, J., Brando, T., Puzo, G., and Sutcliffe, I.C. (2005). Characterization of a truncated lipoarabinomannan from the actinomycete *Turicella otitidis*. J. Bacteriol. *187*, 854–861.

Glauner, B. (1988). Separation and quantification of muropeptides with high-performance liquid-chromatography. Anal. Biochem. *172*, 451–464.

Guan, R.J., and Mariuzza, R.A. (2007). Peptidoglycan recognition proteins of the innate immune system. Trends Microbiol. *15*, 127–134.

Gudlavalleti, S.K., Szymanski, C.M., Jarrell, H.C., and Stephens, D.S. (2006). *In vivo* determination of *Neisseria meningitidis* serogroup A capsular polysaccharide by whole cell high-resolution magic angle spinning NMR spectroscopy. Carbohydr. Res. *341*, 557–562.

Gullion, T., and Schaefer, J. (1989). Rotational-echo double-resonance NMR. J. Magn. Reson. *81*, 196–200.

Halye, J.L., and Rice, C.V. (2010). Cadmium chelation by bacterial teichoic acid from solid-state nuclear magnetic resonance spectroscopy. Biomacromolecules *11*, 333–340.

Hayhurst, E.J., Kailas, L., Hobbs, J.K., and Foster, S.J. (2008). Cell wall peptidoglycan architecture in *Bacillus subtilis*. Proc. Natl. Acad. Sci. U.S.A. *105*, 14603–14608.

Jachymek, W., Niedziela, T., Petersson, C., Lugowski, C., Czaja, J., and Kenne, L. (1999). Structures of the O-specific polysaccharides from *Yokenella regensburgei* (*Koserella trabulsii*) strains PCM 2476, 2477, 2478, and 2494: High-resolution magic-angle spinning NMR investigation of the O-specific polysaccharides in native lipopolysaccharides and directly on the surface of living bacteria. Biochemistry *38*, 11788–11795.

Jacob, G.S., Schaefer, J., and Wilson, G.E. (1983). Direct measurement of peptidoglycan cross-linking in bacteria by N-15 nuclear magnetic-resonance. J. Biol. Chem. *258*, 824–826.

Jones, C., and Lemercinier, X. (2002). Use and validation of NMR assays for the identity and O-acetyl content of capsular polysaccharides from *Neisseria meningitidis* used in vaccine manufacture. J. Pharm. Biomed. Anal. *30*, 1233–1247.

Kern, T., Hediger, S., Müller, P., Giustini, C., Joris, B., Bougault, C., Vollmer, W., and Simorre, J.P. (2008). Toward the characterization of peptidoglycan structure and protein – Peptidoglycan interactions by solid-state NMR spectroscopy. J. Am. Chem. Soc. *130*, 5618–5619.

Kern, T., Giffard, M., Hediger, S., Amoroso, A., Giustini, C., Bui, N.K., Joris, B., Bougault, C., Vollmer, W., and Simorre, J.P. (2010). Dynamics characterization of fully hydrated bacterial cell walls by Solid-State NMR: Evidence for cooperative binding of metal ions. J. Am. Chem. Soc. *132*, 10911–10919.

Kim, S.J., Cegelski, L., Studelska, D.R., O'Connor, R.D., Mehta, A.K., and Schaefer, J. (2002). Rotational-echo double resonance characterization of vancomycin binding sites in *Staphylococcus aureus*. Biochemistry *41*, 6967–6977.

Kim, S.J., Cegelski, L., Preobrazhenskaya, M., and Schaefer, J. (2006). Structures of *Staphylococcus aureus* cell–wall complexes with vancomycin, eremomycin, and chloroeremomycin derivatives by C-13{F-19} and N-15{F-19} rotational-echo double resonance. Biochemistry *45*, 5235–5250.

Kim, S.J., Cegelski, L., Stueber, D., Singh, M., Dietrich, E., Tanaka, K.S.E., Parr, T.R., Far, A.R., and Schaefer, J. (2008). Oritavancin exhibits dual mode of action to inhibit cell-wall biosynthesis in *Staphylococcus aureus*. J. Mol. Biol. *377*, 281–293.

Kim, S.J., Singh, M., and Schaefer, J. (2009). Oritavancin binds to isolated protoplast membranes but not intact protoplasts of *Staphylococcus aureus*. J. Mol. Biol. *391*, 414–425.

Kremer, L., de Chastellier, C., Dobson, G., Gibson, K.J.C., Bifani, P., Balor, S., Gorvel, J.P., Locht, C., Minnikin, D.E., and Besra, G.S. (2005). Identification and structural characterization of an unusual mycobacterial monomeromycolyl-diacylglycerol. Mol. Microbiol. *57*, 1113–1126.

Lee, R.E.B., Li, W., Chatterjee, D., and Lee, R.E. (2005). Rapid structural characterization of the arabinogalactan and lipoarabinomannan in live mycobacterial cells using 2D and 3D HR-MAS NMR: structural changes in the arabinan due to ethambutol treatment and gene mutation are observed. Glycobiology *15*, 139–151.

Lehotzky, R.E., Partch, C.L., Mukherjee, S., Cash, H.L., Goldman, W.E., Gardner, K.H., and Hooper, L.V. (2010). Molecular basis for peptidoglycan recognition by a bactericidal lectin. Proc. Natl. Acad. Sci. U.S.A. *107*, 7722–7727.

Li, W.J., Raoult, D., and Fournier, P.E. (2009). Bacterial strain typing in the genomic era. FEMS Microbiol. Rev. *33*, 892–916.

Lowe, I.J. (1959). Free induction decays of rotating solids. Phys. Rev. Lett. 2, 285–287.

McNally, D.J., Jarrell, H.C., Li, J., Khieu, N.H., Vinogradov, E., Szymanski, C.M., and Brisson, J.R. (2005). The HS: 1 serostrain of *Campylobacter jejuni* has a complex teichoic acid-like capsular polysaccharide with nonstoichiometric fructofuranose branches and O-methyl phosphoramidate groups. FEBS J. *272*, 4407–4422.

Manolikas, T., Herrmann, T., and Meier, B.H. (2008). Protein structure determination from C-13 spin-diffusion solid-state NMR spectroscopy. J. Am. Chem. Soc. *130*, 3959–3966.

Mattei, P.J., Neves, D., and Dessen, A. (2010). Bridging cell wall biosynthesis and bacterial morphogenesis. Curr. Opin. Struct. Biol. *278*, 414–426.

Metzger, U., Lankes, U., Fischpera, K., and Frimmel, F.H. (2009). The concentration of polysaccharides and proteins in EPS of *Pseudomonas putida* and *Aureobasidum pullulans* as revealed by C-13 CPMAS NMR spectroscopy. Appl. Microbiol. Biotechnol. *85*, 197–206.

Mulloy, B. (1996). High-field NMR as a technique for the determination of polysaccharide structures. Mol. Biotechnol. *6*, 241–265.

Nadaud, P.S., Helmus, J.J., Hofer, N., and Jaroniec, C.P. (2007). Long-range structural restraints in spin-labelled proteins probed by solid-state nuclear magnetic resonance spectroscopy. J. Am. Chem. Soc. *129*, 7502–7503.

Naumova, I.B., Shashkov, A.S., Tul'skaya, E.M., Streshinskaya, G.M., Kozlova, Y.I., Potekhina, N.V., Evtushenko, L.I., and Stackebrandt, E. (2001). Cell wall teichoic acids: structural diversity, species specificity in the genus *Nocardiopsis*, and chemotaxonomic perspective. FEMS Microbiol. Rev. *25*, 269–283.

Neuhaus, F.C., and Baddiley, J. (2003). A continuum of anionic charge: structures and functions of D-alanyl-teichoic acids in Gram-positive bacteria. Microbiol. Mol. Biol. Rev. *67*, 686–723.

Niescher, S., Wray, V., Lang, S., Kaschabek, S.R., and Schlomann, M. (2006). Identification and structural characterisation of novel trehalose dinocardiomycolates from *n*-alkane-grown *Rhodococcus opacus* 1CP. Appl. Microbiol. Biotechnol. *70,* 605–611.

Nothaft, H., Liu, X., McNally, D.J., and Szymanski, C.M. (2010). N-linked protein glycosylation in a bacterial system. In Functional Glycomics: Methods and Protocols, Li, J., ed. (Hertfordshire, Humana Press), pp. 227–243.

Patti, G.J., Chen, J.W., Schaefer, J., and Gross, M.L. (2008a). Characterization of structural variations in the peptidoglycan of vancomycin-susceptible *Enterococcus faecium*: understanding glycopeptide-antibiotic binding sites using mass spectrometry. J. Am. Chem. Soc. Mass Spectrom. *19,* 1467–1475.

Patti, G.J., Kim, S.J., and Schaefer, J. (2008b). Characterization of the peptidoglycan of vancomycin-susceptible *Enterococcus faecium*. Biochemistry *47,* 8378–8385.

Patti, G.J., Kim, S.J., Yu, T.Y., Dietrich, E., Tanaka, K.S.E., Parr, T.R., Far, A.R., and Schaefer, J. (2009). Vancomycin and oritavancin have different modes of action in *Enterococcus faecium*. J. Mol. Biol. *392,* 1178–1191.

Potekhina, N.V., Streshinskaya, G.M., Kozlova, Y.I., Kudryashova, E.B., Senchenkova, S.N., Shashkov, A.S., and Anan'ina, L.N. (2009). Heterogeneous set of cell wall teichoic acids in strains of *Bacillus subtilis* VKM B-760 and VKM B-764. Biochemistry (Mosc.) *74,* 1368–1374.

Rahman, O., Pfitzenmaier, M., Pester, O., Morath, S., Cummings, S.P., Hartung, T., and Sutcliffe, I.C. (2009). Macroamphiphilic components of thermophilic *Actinomycetes*: identification of lipoteichoic acid in *Thermobifida fusca*. J. Bacteriol. *191,* 152–160.

Rice, C.V., and Wickham, J.R. (2005). Heterogeneous binding of lipoteichoic acid to the surface of titanium dioxide as determined with ^{31}P solid-state NMR spectroscopy. J. Am. Chem. Soc. *127,* 856–857.

Sadovskaya, I., Vinogradov, E., Li, J.J., and Jabbouri, S. (2004). Structural elucidation of the extracellular and cell-wall teichoic acids of *Staphylococcus epidermidis* RP62A, a reference biofilm-positive strain. Carbohydr. Res. *339,* 1467–1473.

Sauer, K., and Cooke, M.P. (2010). Regulation of immune cell development through soluble inositol-1,3,4,5-*tetrakis*phosphate. Nat. Rev. Immunol. *10,* 257–271.

Schroeder, B.G., and Barry, C.E. (2001). The specificity of methyl transferases involved in trans mycolic acid biosynthesis in *Mycobacterium tuberculosis* and *Mycobacterium smegmatis*. Bioorg. Chem. *29,* 164–177.

Sharif, S., Kim, S.J., Labischinski, H., and Schaefer, J. (2009a). Characterization of peptidoglycan in Fem-deletion mutants of methicillin-resistant *Staphylococcus aureus* by solid-state NMR. Biochemistry *48,* 3100–3108.

Sharif, S., Singh, M., Kim, S.J., and Schaefer, J. (2009b). *Staphylococcus aureus* peptidoglycan tertiary structure from Carbon-13 spin diffusion. J. Am. Chem. Soc. *131,* 7023–7030.

Shashkov, A.S., Potekhina, N.V., Senchenkova, S.N., and Kudryashova, E.B. (2009). Anionic polymers of the cell wall of *Bacillus subtilis* subsp subtilis VKM B-501(T). Biochemistry (Mosc.) *74,* 543–548.

Silhavy, T.J., Kahne, D., and Walker, S. (2010). The bacterial cell envelope. Cold Spring Harb. Persp. Biol. *2,* a000414.

Szymanski, C.M., St Michael, F., Jarrell, H.C., Li, J.J., Gilbert, M., Larocque, S., Vinogradov, E., and Brisson, J.R. (2003). Detection of conserved N-linked glycans and phase-variable lipooligosaccharides and capsules from *Campylobacter* cells by mass spectrometry and high resolution magic angle spinning NMR spectroscopy. J. Biol. Chem. *278,* 24509–24520.

Theilacker, C., Kaczynski, Z., Kropec, A., Fabretti, F., Sange, T., Holst, O., and Huebner, J. (2006). Opsonic antibodies to *Enterococcus faecalis* strain 12030 are directed against lipoteichoic acid. Infect. Immun. *74,* 5703–5712.

Toke, O., Cegelski, L., and Schaefer, J. (2006). Peptide antibiotics in action: Investigation of polypeptide chains in insoluble environments by rotational-echo double resonance. Biochim. Biophys. Acta (Biomembranes) *1758,* 1314–1329.

Tomita, S., Furihata, K., Nukada, T., Satoh, E., Uchimura, T., and Okada, S. (2009). Structures of two monomeric units of teichoic acid prepared from the cell wall of *Lactobacillus plantarum* NRIC 1068. Biosci. Biotechnol. Biochem. *73,* 530–535.

Tong, G., Pan, Y., Dong, H., Pryor, R., Wilson, G.E., and Schaefer, J. (1997). Structure and dynamics of pentaglycyl bridges in the cell walls of *Staphylococcus aureus* by ^{13}C,^{15}N REDOR NMR. Biochemistry *36,* 9859–9866.

Uenishi, Y., Fujita, Y., Kusunose, N., Yano, I., and Sunagawa, M. (2008). Comprehensive analysis of mycolic acid subclass and molecular species composition of *Mycobacterium bovis* BCG Tokyo 172 cell wall skeleton (SMP-105). J. Microbiol. Methods *72,* 149–156.

Vanderhart, D.L. (1987). Natural-abundance C-13-C-13 spin exchange in rigid crystalline organic solids. J. Magn. Reson. *72,* 13–47.

Vinogradov, E., Li, J.J., Sadovskaya, I., Jabbouri, S., and Helander, I.M. (2004). The structure of the carbohydrate backbone of the lipopolysaccharide of *Pectinatus frisingensis* strain VTT E-79104. Carbohydr. Res. *339,* 1637–1642.

Vinogradov, E., Sadovskaya, I., Li, J.J., and Jabbouri, S. (2006). Structural elucidation of the extracellular and cell-wall teichoic acids of *Staphylococcus aureus* MN8m, a biofilm forming strain. Carbohydr. Res. *341,* 738–743.

Vollmer, W. (2008). Structural variation in the glycan strands of bacterial peptidoglycan. FEMS Microbiol. Rev. *32,* 287–306.

Vollmer, W., and Seligman, S.J. (2010). Architecture of peptidoglycan: more data and more models. Trends Microbiol. *18,* 59–66.

Vollmer, W., Blanot, D., and de Pedro, M.A. (2008). Peptidoglycan structure and architecture. FEMS Microbiol. Rev. *32,* 149–167.

Wickham, J.R., and Rice, C.V. (2008). Solid-state NMR studies of bacterial lipoteichoic acid adsorption on different surfaces. Solid State Nucl. Magn. Reson. *34,* 154–161.

Wickham, J.R., Halye, J.L., Kashtanov, S., Khandogin, J., and Rice, C.V. (2009). Revisiting magnesium chelation by teichoic acid with phosphorus solid-state NMR and theoretical calculations. J. Phys. Chem. B. *113,* 2177–2183.

Wieruszeski, J.M., Talaga, P., and Lippens, G. (2005). Development of a high-resolution magic-angle spinning nuclear magnetic resonance identity assay of the capsular polysaccharide from *Haemophilus influenzae* type b present in cetavlon precipitate. Anal. Biochem. *338,* 20–25.

Xia, G.Q., Kohler, T., and Peschel, A. (2010a). The wall teichoic acid and lipoteichoic acid polymers of *Staphylococcus aureus.* Int. J. Med. Microbiol. *300,* 148–154.

Xia, G.Q., Maier, L., Sanchez-Carballo, P., Li, M., Otto, M., Holst, O., and Peschel, A. (2010b). Glycosylation of wall teichoic acid in *Staphylococcus aureus* by TarM. J. Biol. Chem. *285,* 13405–13415.

Zech, S.G., Olejniczak, E., Hajduk, P., Mack, J., and McDermot, A.E. (2004). Characterization of protein–ligand interactions by high-resolution solid-state NMR spectroscopy. J. Am. Chem. Soc. *126,* 13948–13953.

Methods for the Analysis of Microbial Lipid-Linked Oligosaccharides by Mass Spectrometry

10

Christopher W. Reid

Abstract

Bacteria and Archaea produce a variety of glycoconjugates such as capsular polysaccharides, lipopolysaccharides, and glycoproteins that are assembled on a polyisoprenyl-phosphate lipid in the cytoplasmic membrane. Traditional methods to analyse the membrane-associated steps of glycan biosynthesis involved the use of metabolic radiolabelling or the indirect detection of lipid-associated glycans. Recent advances in analytical biochemistry now provide the microbial glycobiologist with a number of tools for the direct detection and characterization of low abundance lipid-linked oligosaccharides. Approaches include targeted glycolipidomics strategies, such as affinity-capture capillary electrophoresis–mass spectrometry and global approaches such as separation on porous graphite carbon and normal phase liquid chromatography–mass spectrometry (LC–MS). These techniques provide opportunities to probe the lipid-associated steps in glycan biosynthesis in greater detail as well as provide an enabling technology for the exploitation of these pathways in glycoengineering.

Introduction

Lipids are essential to the proper functioning of the microbial membrane. Lipids exist in hundreds of different forms and are involved in myriad of processes including metabolism, protein transport across membranes, and the synthesis of biosynthetic intermediates of microbial polysaccharides such as peptidoglycan (PG), lipopolysaccharides (LPS), capsular polysaccharides (CPS), and *N*- and O-linked glycans. As microbiologists strive to understand the complex interactions between the components of biological systems, a variety of approaches are utilized from the 'omics cascade' in order to obtain a whole system understanding of a particular biological process. One area of the 'omics cascade' that is gaining increased attention is lipidomics, the study of the pathways and networks of cellular lipids. A subset of the microbial lipidome is the glycolipidome, lipids which possess a polar carbohydrate modification.

Technological advancements in genomics, transcriptomics and proteomics have allowed scientists to probe biological systems with increased precision and sensitivity. While the analysis of lipids from microorganisms has been around since the 1950s, the tools for the examination of the lipidome and in particular the glycolipidome are still emerging (Bino *et al.*, 2004; Mashego *et al.*, 2004, 2007). Traditionally, analysis of microbial lipidomes has been centred on fatty acid analysis using tools such as gas chromatography–mass spectrometry (GC–MS) (Lodowska *et al.*, 1999; Andreotti *et al.*, 2006; Byss *et al.*, 2007; Eliyahu, 2009), fast atom bombardment mass spectrometry (FAB–MS) (Ballou and Dell, 1985; Takayama *et al.*, 1986), matrix assisted laser desorption ionization time-of-flight mass spectrometry (MALDI–TOF–MS) (Schiller *et al.*, 2007) and LC–MS (Nakanishi *et al.*, 2009; Bird *et al.*, 2011; Nygren *et al.*, 2011). While these studies have provided valuable information regarding the fatty acid composition of microbial membranes under a variety of conditions, the study of glycolipids has not advanced to the same degree. The role of glycolipids in host–pathogen interactions is an emerging field with the bulk of the scientific literature revolving around the analysis of lipid A (Kussak and Weintraub, 2002; Murphy *et al.*, 2005). In the realm of bacterial glycolipidomics, there is an increasing body of work on the membrane-associated steps involved in microbial polysaccharides biosynthesis (Szymanski and Wren, 2005; Aas *et al.*, 2007; Faridmoayer *et al.*, 2007, 2008; Guerry and Szymanski, 2008; Nothaft and Szymanski, 2010). For many microbial polysaccharides, the oligosaccharides or oligosaccharide repeat units such as those found in LPS, glycoprotein, CPS, and PG biosynthesis are assembled on a polyisoprene lipid. Unlike eukaryotes and Archaea which use dolichol as the lipid carrier, bacteria utilize undecaprenol (Fig. 10.1). While there is a substantial body of work on the cytoplasmic steps in these pathways, a comprehensive understanding of the membrane-associated steps has remained elusive. In many cases this has been due to the complexity of the systems and a lack of amenable techniques for studying bacterial glycolipids. A summary of lipid-linked oligosaccharides (LLOs) isolated from a variety of bacterial and archaeal sources is presented in Table 10.1.

The currently available methods for the detection of LLOs derive from studies of eukaryotic systems, where these molecules are of greater interest (Lehrman, 2007). Traditionally, the detection of LLOs has been achieved using a variety of techniques including metabolic radiolabelling (Lehrman, 2007), solid-state analysis of LLOs with conjugated lectins (Kelleher *et al.*, 2001), oligosaccharide transferase (OST) assays (Kohda *et al.*, 2007), or fluorophore-assisted carbohydrate electrophoresis (FACE) (Gao and Lehrman, 2006). While these techniques have proven effective, there are several notable disadvantages. For instance, metabolic radiolabelling of LLOs is hampered by the variable rates of LLO turnover, isotope dilution effects, and difficulties associated with certain cells or tissues not being amenable to metabolic labelling. The solid-state assay employing horseradish peroxidase-conjugated Concanavalin A (ConA) lectin, while selective and semi-quantitative, requires prior knowledge of the glycan under study for the appropriate selection of lectins. Both the OST assay and FACE analysis of LLOs rely on detection of the LLO-associated glycans through the use of fluorescent labelling of glycans or acceptor peptides. With the exception of the solid-state assay, all current methods require chromatographic or electrophoretic separation and are not readily amenable to structural characterization. In microbial glycolipidomics, the analysis of LLOs has traditionally relied on the use of metabolic radiolabelling coupled to chromatographic separations such as thin-layer chromatography (TLC) or ion-exchange chromatography (Troy *et al.*, 1971; Masson and Holbein, 1985).

This chapter will focus on the recent advances in the direct analysis of polyisoprenyl LLOs from both Bacteria and Archaea utilizing capillary electrophoresis (CE) or LC–MS. An overview of the various extraction and separation techniques available to the microbial glycobiologist will be covered. As this is a young and emerging field of bacterial glycomics, the techniques discussed are focused on elucidating the membrane-associated steps of microbial protein glycosylation.

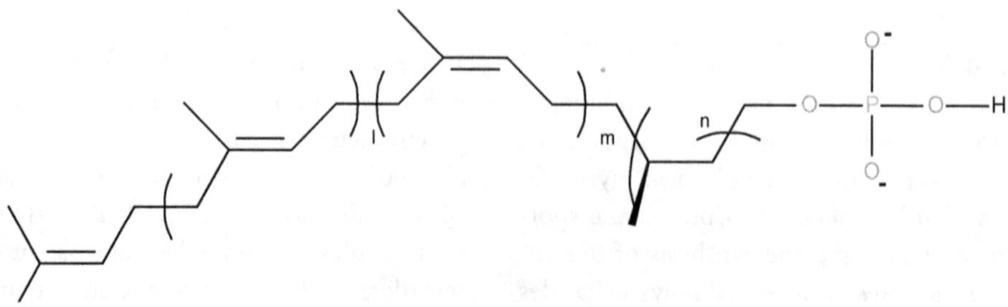

Figure 10.1 General structure of polyisoprenyl phosphates found in Bacteria and Archaea. Undecaprenyl phosphate, l=3, m=7, *n*=0; Dolicyl 17–19 phosphate l=2, m=13–15, *n*=1.

Table 10.1 Summary of reported analyses of LLOs from Bacteria and Archaea

Organism	LLO[1]	Reference
Campylobacter jejuni	Und-PP-Bac-(GalNAc)$_5$Glc	Reid *et al.* (2008, 2009)
Salmonella typhimurium	Und-PP-Ara4N	Trent *et al.* (2001)
Escherichia coli	Und-PP-Ara4N; Lipid-II	Trent *et al.* (2001), Guan *et al.* (2005)
Francisella novicida	Und-PP-GalNAc, Und-PP-Glc	Wang *et al.* (2009)
Haloferax volcanii	Dol-P-monosaccharide; Dol-P-oligosaccharide	Kuntz *et al.* (1997), Guan *et al.* (2010)

[1]Und-PP, undecaprenyl pyrophosphate.

Dol-P, dolichyl phosphate.

Analysis of LLOs

Extraction procedures

While many of the extraction protocols used in modern glycolipidomics can be traced back to the 1950s (Bligh and Dyer, 1959), there have been some modifications and advancements to meet the needs of modern high-throughput glycolipidomics and the application of MS to the analysis and characterization of glycolipids. A summary of the common extraction techniques and their application to the isolation of microbial glycolipids is provided in Table 10.2.

By far the most widely used glycolipid extraction techniques are single phase chloroform/methanol and chloroform/methanol/water methods (Kuntz *et al.*, 1997; Kelleher *et al.*, 2001; Guan *et al.*, 2005, 2010; Reid *et al.*, 2008, 2009). While simple and highly effective, this approach results in the generation of crude extracts of lipids and other soluble biomolecules, many of which may interfere with the glycolipid analysis. Two-phase systems such as the classical Bligh–Dyer extraction, have found extensive use in lipidomics (Bligh and Dyer, 1959; Trent *et al.*, 2001; Murphy *et al.*, 2005; Oursel *et al.*, 2007; Matyash *et al.*, 2008). The major drawback to the Bligh–Dyer extraction comes from the fact that the glycolipids are typically found in the chloroform phase. This poses a problem in high-throughput lipidomics strategies that utilize robotic liquid handling systems, as sample contamination occurs when the sampling needle must pass through the aqueous-layer to reach the lipid-enriched chloroform phase. This can be overcome through the use of solvents like methyl-*t*-butyl ether (MTBE), which is less dense than water thereby allowing easier sampling with robotic systems. This method has been successfully applied to the analysis of the *Escherichia coli* lipidome (Matyash *et al.*, 2008). It should be noted that, in the case of glycolipids containing large carbohydrate moieties, the glycolipid can in some instances partition into the aqueous phase potentially complicating lipidomic profiling.

LLO separation strategies

In microbial glycolipidomics, the most widely used separation strategy to isolate polyisoprenyl-linked LLOs is ion exchange chromatography

Table 10.2 Summary of lipid and glycolipid extraction protocols employed in bacterial glycolipidomics

Extraction protocol	Applications
Bligh-Dyer and modified Bligh-Dyer	Mycobacterial phenolic glycolipid (Young and Buchanan, 1983); Lipid II (Guan *et al.*, 2005); Archaeal dolichyl-oligosaccharides (Kuntz *et al.*, 1997)
Folch, Lees, Sloane Stanely	*Spirochaeta aurantia* glycolipid (Paul *et al.*, 2009)
Single phase $CHCl_3$:MeOH:H_2O	*Campylobacter jejuni N*-glycan LLO (Reid *et al.*, 2008, 2009)
Methyl-*tert*-butyl ether	*Escherichia coli* lipidomics (Matyash *et al.*, 2008)

on resins such as diethylaminoethyl (DEAE), an anion exchange resin (Troy *et al.*, 1971). TLC is also a widely used, fast and inexpensive method for the separation of complex mixtures that has found wide-spread use in bacterial lipidomics. The drawback to this separation strategy lies in the unambiguous assignment of the obtained spots to defined compounds. Identification of spots is often not possible by common staining methods and the lack of appropriate reference standards for bacterial glycolipids makes this a less than ideal strategy (Fuchs *et al.*, 2009). These challenges can be overcome through coupling TLC with mass spectrometry, in particular MALDI–MS. The combination of these two techniques has found growing use in lipidomics (Vieler *et al.*, 2007; Roberts *et al.*, 2008). Techniques developed for the analysis of eukaryotic LLOs such as FACE have found limited use in bacterial glycolipidomics (Gopal *et al.*, 2001), most probably owing to a lack of appropriate standards. The lack of available microbial glycolipid standards is due in large part to the vast diversity of microbial glycans synthesized, the larger pool of monosaccharides used, and the potential for a wide array of stoichiometric and non-stoichiometric substitutions. In light of this there is a need for the development of targeted and non-targeted glycolipidomics strategies that can be applied to bacterial systems. With the increased interest in the exploitation of bacterial polysaccharide biosynthetic pathways for glycoengineering (Calo *et al.*, 2010; Fisher *et al.*, 2011), there is a need for tools to characterize the lipid-mediated steps in the biosynthesis of these polysaccharides. The next section will focus on the description of a targeted (affinity capture CE–MS) and non-targeted glycolipidomics strategies and their application to microbial glycomics (Table 10.3).

Affinity capture CE–MS

CE is a high-resolution technique for the separation of complex biological mixtures that has been widely applied to biological analyses (Hermentin *et al.*, 1992; Desiderio *et al.*, 2010; Suntornsuk, 2010). The coupling of CE with MS provides a powerful approach for rapid identification of target analytes present at trace levels in biological matrices, and for structural characterization of complex biomolecules (Moini, 2002; Li and Richards, 2007; Li *et al.*, 2007). The application of CE–MS combined with an initial affinity capture to enrich for LLOs can be a powerful and sensitive targeted glycolipidomics tool for the analysis of glycan biosynthetic pathways (Reid *et al.*, 2008). Affinity-capture technologies have found numerous applications in bacterial enrichment (Bundy, *et al.*, 1999; Desai, *et al.*, 2008), but limited examples exist for the analysis of bacterial glycolipids. Affinity-capture CE–MS is a suitable approach if the structure of the glycan of interest is known and there is a suitable lectin available. Alternatively, an immobilized antibody which reacts to the desired glycan could be used as a lectin replacement. The effectiveness of affinity capture CE–MS was demonstrated in the analysis of the *Campylobacter jejuni* N-linked protein glycosylation (Pgl) pathway (Reid 2008). The *N*-glycan of *C. jejuni* is a conserved heptasaccharide with the structure GalNAc-α-(1,4)-GalNAc-α-(1,4)-[Glc-β-(1,3)]-GalNAc-α-(1,4)-GalNAc-α-(1,4)-GalNAc-α-(1,3)-Bac-β1 that is assembled on a polyisoprene carrier prior to its transfer to protein (Young, 2002). The *C. jejuni* Pgl pathway and

Table 10.3 Separation strategies for LLO analysis

Separation strategy	Application
Reversed-phase HPLC	Garrett *et al.* (2007)
Normal-phase HPLC	Guan *et al.* (2010)
Porous graphite carbon	Reid *et al.* (2009)
PR-NH_2 (aminopropyl silica)	Crossman *et al.* (2005)
Capillary electrophoresis	Reid *et al.* (2008)

HPLC, high-performance liquid chromatography.

its potential application in glycoengineering is covered in Chapter 4. The *C. jejuni* Pgl pathway has been extensively characterized at the genetic and protein levels, which made it an ideal model system to develop microbial glycolipidomics strategies. The intense interest in this system is due in part to the fact that it can serve as a simple model system for the eukaryotic N-linked protein glycosylation pathway (Guerry and Szymanski, 2008; Nothaft and Szymanski, 2010).

While affinity-capture CE–MS is not as sensitive as some of the classic techniques described above, LLOs could be detected and characterized from a sample size containing approximately 10^{12}–10^{13} bacterial cells. While not as sensitive as traditional methods such as FACE, this sample size is amenable to the laboratory setting. This method allows for rapid enrichment of LLOs from a crude mixture of lipids using an immobilized lectin (soybean agglutinin) with specificity towards the GalNAc residues in the *C. jejuni* LLOs of interest. Owing to the low abundance LLOs associated with the Pgl pathway and instrument sensitivity, intermediates could only be detected in select isogenic mutants in the pathway. This is most probably due to the fact that in the presence of polypeptide acceptors, LLOs are used as they are formed. Despite the inability to detect LLOs in wild-type cells, analyses of isogenic mutants in the oligosaccharyltransferase (*pglB*) and membrane flippase (*pglK*) revealed an accumulation of undecaprenyl-pyrophosphate (Und-PP)-heptasaccharide (Table 10.4). Tandem MS (MS/MS) experiments in the negative-ion mode also provided further evidence for the assembly of the *C. jejuni N*-glycan on undecaprenyl phosphate (Und-P). The lack of accumulating LLOs, in isogenic mutants affecting the assembly of the oligosaccharide on Und-P, such as the galactosyltransferases PglH and PglJ was unexpected in light of previous reports of glycoprotein production in *C. jejuni* (Table 10.4) (Kelly, 2006). Further investigation of these results indicated the lack of accumulating truncated LLOs in *C. jejuni* was due to feedback inhibition (Reid *et al.*, 2008).

The application of affinity-capture mass spectrometry to the analysis of membrane-associated processes in carbohydrate biosynthesis pathways can provide a simple and rapid method to gather structural information for the oligosaccharides under study. Unlike conventional techniques, radioisotope labelling and chromatographic separation are not required for analysis (Lehrman, 2007). Additionally, by varying the lectin used for affinity capture, this method can be applied to a wide range of systems. Alternatively, an immobilized carbohydrate-specific antibody could be used instead of a lectin. Analysis of complex bacterial carbohydrate pathways such as those for pilin *O*-glycosylation and peptidoglycan biosynthesis may provide new insights into the synthesis and control of these virulence factors. Targeted metabolomic profiling using techniques such as affinity-capture CE–MS is a powerful tool

Table 10.4 Summary of results of *N*-glycan intermediate analyses in *C. jejuni*

Strain	Function	Glycoprotein phenotype[1]	LLOs
11168	–	Heptasaccharide	None
pglB	Oligosaccharyl transferase	Unmodified	Yes
pglD	Acetyltransferase	Minor heptasaccharide[2]	None
pglH	Processive GalNAc transferase	Unmodified	None
pglI	Glc transferase	Hexasaccharide	None
pglJ	Second GalNAc transferase	Unmodified	None
pglK	Flippase	Unmodified	Yes

[1]Kelly *et al.* (2006).

[2]Minor levels of N-glycosylation observed in this mutant.

GalNAc, *N*-acetylgalactosamine; Glc, glucose.

for bacterial glycobiology and glycoengineering. It provides a high-throughput method for assessing the role of genes in a biosynthetic pathway and allows for identification of potential regulation mechanisms by intermediates in the pathway. With the recent advances in glycoengineering in bacteria (Fisher *et al.*, 2011; Lizak *et al.*, 2011; Messner *et al.*, 2008), affinity-capture CE–MS also allows for a more comprehensive assessment of the efficiency and potential bottlenecks associated with producing potential designer glycoconjugates in bacteria.

High-performance liquid chromatography

High-performance liquid chromatography (HPLC) is a powerful tool for the separation of biomolecules. Traditionally, the analysis of mono- and polysaccharides employed high-pH anion exchange chromatography (HPAEC) (Andersen *et al.*, 1994; Bray and Clarke, 1992; Clarke *et al.*, 1995; Currie and Perry, 2006; Weitzhandler *et al.*, 1996). While a powerful and versatile method, it requires high salt and pH conditions that are not typically conducive for mating in hybrid analytical techniques such as LC–MS (Bruggink *et al.*, 2005; Dong *et al.*, 2010). There is a growing body of literature on the application of traditional reversed-phase (RP), normal phase (NP) and hydrophilic interaction liquid chromatography (HILIC) to the separation and analysis of LLOs and in particular polyisoprene LLOs (Garrett *et al.*, 2007; Guan *et al.*, 2010; Kelleher *et al.*, 2001; Maillet *et al.*, 2011; Piretti *et al.*, 1995). While many of these techniques were developed for the analysis of eukaryotic LLOs, they have the potential to be applicable to bacterial glycobiology as well.

RP–HPLC has been utilized for the analysis of polyisoprenes (Carlson *et al.*, 2000), polyisoprenyl phosphates (Barreteau *et al.*, 2009) and polyisoprenyl-phosphate mono- and oligosaccharides with varying degrees of success (Garrett *et al.*, 2007; Glover *et al.*, 2005; Trent *et al.*, 2001). A wide variety of stationary phases have been employed including C-18, C-8, and C-4. The challenge to effectively using RP–HPLC for the separation of LLOs is to find a stationary phase that adequately retains and separates the LLOs of interest. Owing to the amphipathic nature of these molecules, separation can be difficult. RP–HPLC and RP–HPLC–MS have been used most extensively in the separation and analysis of eukaryotic LLOs (Carlson *et al.*, 2000; Garrett *et al.*, 2007; Lehrman, 2007; Haeuptle *et al.*, 2010). In relation to bacterial glycolipidomics, RP–HPLC has been successfully employed in the separation and analysis of undecaprenol (Und) and Und-P in *E. coli* and *Staphylococcus aureus* (Barreteau *et al.*, 2009). Und and Und-P were extracted from isolated membranes using the classical Bligh–Dyer extraction. The isoprenoids were separated using a C-4 column and an isocratic organic solvent mobile phase with analyte detection at 210 nm. Under these conditions, the authors were able to effectively separate and quantify levels of Und, Und-P, and Und-PP (Barreteau *et al*, 2009). While this technique was capable of separating and quantifying levels of polyisoprenes from bacterial membranes, the lack of reported separation of bacterial LLOs underlines the difficulty in effectively separating these amphipathic molecules using RP–HPLC.

Recently, NP-HPLC has been used successfully to separate and analyse polyisoprenes and LLOs (Bamba *et al.*, 2004, 2005). The use of NP stationary phases can allow for the separation of compounds that are too hydrophobic or hydrophilic for traditional RP–HPLC and has the added advantage of being able to separate isomers (Wuhrer *et al.*, 2003). In NP chromatography, the stationary phase is more polar than the mobile phase. NP stationary phases come in two flavours, non-bonded and bonded. Silica is the most common and versatile of the non-bonded phases. The major drawback to non-bonded silica stationary phases arises when aqueous mobile phases are used, which can result in difficulty in achieving reproducible retention times. Of the bonded silica phases, amino-propyl silica (Pr-NH_2) has proven to be the most successful for the separation of LLOs (Hall *et al.*, 1997; Bodennec *et al.*, 2000, 2003; Kelleher *et al.*, 2001; Lehrman, 2007). The major drawback to Pr-NH_2 columns is analyte-mediated derivatization that results in inactivation of the stationary phase. The use of Pr-NH_2 stationary phases has met with a great deal of success in the analysis of eukaryotic LLOs associated with protein N-glycosylation (Kaluzny *et al.*, 1985; Kelleher *et al.*, 2001).

In terms of applications to microbial LLO analysis, NP-HPLC has been successfully employed for the analysis of lipid-mediated steps in S-layer oligosaccharide biosynthesis of the archaeon *Haloferax volcanii* (Guan *et al.*, 2010). For a detailed description of glycosylation in Archaea refer to Chapter 7. The use of a non-bonded silica stationary phase with a chloroform:methanol:water mobile phase allowed for the first observation that archaea sequentially assemble their *N*-glycan pentasaccharide on a dolichyl phosphate carrier. This method was able to show that *H. volcanii* contains a population of C_{55} and C_{60} dolichyl phosphates with two saturated isoprene units at the α and ω positions. This differs from both Bacteria and Eukarya, considering that Bacteria contain only unsaturated polyprenyl phosphate, while eukaryotic dolichols are unsaturated only at the α-position. Through the use of deletion mutants in the *H. volcanii* S-layer glycosylation pathway, the authors were able to detect and characterize the accumulated LLOs. Additionally, Guan and coworkers, were able to demonstrate the presence of dolichol-phosphate-hexose and showed that it is formed by the action of AlgD, providing evidence that Archaea utilize dolichol-phosphate linked sugar precursors for S-layer glycan synthesis (Guan *et al.*, 2010).

Porous graphite carbon

Porous graphite carbon (PGC) is a conductive crystalline material made up of smooth graphene sheets. Its unique structure and physico-chemical properties cause carbon-based adsorbents to exhibit a retention mechanism that is different from conventional reversed-phase packing materials (Colin *et al.*, 1982; Liang *et al.*, 2003; Pereira, 2008). The unique properties of carbon columns allows for the resolution of structurally similar compounds such as geometric isomers and diastereomers, (Pereira, 2008) as well as polar (Hanai, 2003; West *et al.*, 2010) and non-polar (Delobel *et al.*, 2006; Roy *et al.*, 2006; Gaudin *et al.*, 2007) compounds. The unique separation modes reported for PGC provide an intriguing stationary phase for the separation of polyisoprene-linked

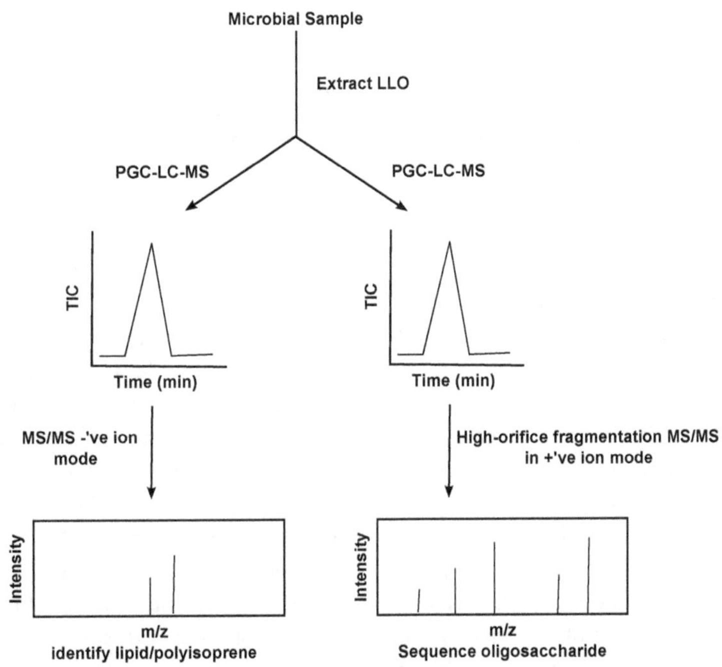

Figure 10.2 General workflow for the analysis of bacterial LLOs by PGC–LCMS.

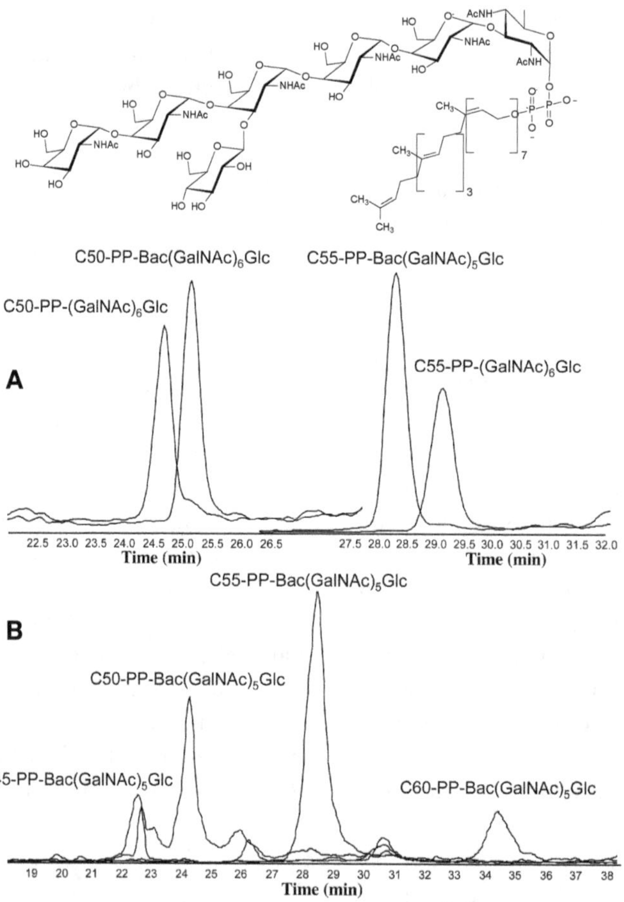

Figure 10.3 PGC–LC–MS analysis of *C. jejuni* LLOs. (A) Reconstructed ion chromatograms of LLOs isolated from a *C. jejuni* flippase mutant (*pglK*) which is unable to glycosylate proteins. LLOs are separated according to polyisoprene length (C_{50} or C_{55}) as well as minor structural changes in the oligosaccharide (HexNAc versus Bac at reducing end). (B) Reconstructed ion chromatograms of LLOs showing chromatographic separation by the size of the carrier polyisoprene. Note: Bac is 2,4-diacetamido-2,4,6-trideoxy-D-glucose.

oligosaccharides with the possibility of differentiating these based on either the carbohydrate portion or the lipid carrier, depending on solvent conditions. PGC has unique properties as a stationary phase in HPLC. PGC behaves as a strongly retentive alkyl-bonded silica gel for non-polar analytes; however, its retention and selectivity behaviour towards polar and structurally related compounds is very different (Pabst *et al.*, 2007; Pereira, 2008). The retention of analytes by graphite carbon from aqueous/organic eluents is determined by the balance of two factors: (1) hydrophobicity, which is primarily a solution effect that tends to drive analytes out of solution and (2) the interaction of polarizable or polarized groups in the analyte with the flat graphite surface. The strength of the interaction depends on both the molecular area of the analyte (shape) in contact with the graphite surface and upon the nature and type of functional groups at the point of interaction with the graphite (Pereira, 2008). Under the correct conditions, PGC has the potential to be able to separate LLOs with small differences which would be a tremendous advantage in glycolipidomics (Reid *et al.*, 2009). PGC–LC–MS provides a simple, straight-forward workflow that can provide both survey information on the composition of a sample, as well as detailed structural characterization of LLOs of interest (Fig. 10.2). When LLOs from the *C. jejuni* Pgl pathway (see Chapter 4 for background) were separated by PGC–LC–MS, they were resolved based on polyisoprene size and structural differences in the oligosaccharide portion (Fig. 10.3). This method demonstrated for the first time that lipid-linked carbohydrate synthesis in bacteria proceeds on polyisoprenes other than C_{50} or C_{55}. In *C. jejuni*, the Pgl *N*-glycan is synthesized on polyisoprenes ranging from C_{45} to C_{60}. A detailed examination of the extracted ion chromatograms from Fig. 10.3 revealed the presence of additional – albeit minor – peaks with masses corresponding to the same LLOs labelled in the figure: Presumably, cis/trans isomers of the polyisoprenes are present as well, and are also amenable to chromatographic separation on PGC (Reid *et al*, 2009). This is consistent with currently proposed mechanisms of retention on graphite stationary phases, since cis/trans isomerism would affect the general shape of the molecules, thereby affecting their interaction with the flat graphite rings (Pereira, 2008). Moreover, as shown in Table 10.5, hydroxylated forms of the polyisoprene carrier were observed in the *C. jejuni* flippase mutant *pglK*. These hydroxylated forms could be a step in the degradation of accumulated LLOs to allow for recycling of the polyisoprene for use in other pathways. In order for methods like PGC–LC–MS to be a practical alternative to FACE and OST assays in bacterial glycomics and glycolipidomics, it needs to be able to detect LLOs from a minimal sample size. The sparingly small size of biological sample required for analysis makes PGC–LC–MS (10^4 bacterial cells) a practical alternative to FACE for LLO analysis. Moreover, PGC–LC–MS allowed for structural characterization of the major LLO components (Reid *et al.*, 2009). In addition, to these features PGC–LC–MS has the added advantage of being a simple extraction procedure which is compatible with downstream analysis without extensive off-line purification or chemical derivatization prior to analysis.

Concluding remarks and future directions

Bacteria produce an array of glycan-based structures ranging from glycoproteins to CPS and LPS which are involved in diverse roles in the life cycle of the bacterium and in host–pathogen interactions. Bacteria produce huge combinatorial variations of glycan structures, which coupled to the lack of genetic data has hampered studies on the role of these glycoconjugates in pathogenesis. With the rapid technological improvements in MS analysis and front-end separation strategies, a new era in bacterial glycomics and glycolipidomics is emerging (Kay *et al.*, 2010). With success in the development of glycoengineering in yeast (Oh *et al.*, 2008; Wei *et al.*, 2008), attention has turned to the exploitation of bacterial glycosylation pathways for glycoengineering (Langdon *et al.*, 2009; Calo *et al.*, 2010; Fisher *et al.*, 2011; Ristl *et al.*, 2011). As microbial geneticists attempt to exploit these microbial pathways for commercial applications, there will be a tremendous need for analytical tools that can effectively characterize the intermediates in these glycoengineered

Table 10.5 Polyisoprenylpyrophosphate lipids observed in the *C. jejuni pglK* mutant as determined by MS/MS fragmentation of the corresponding LLOs

Observed LLO ion ($[M-2H]^{-2}$)	Theoretical mass of LLO (Da)[1]	Observed lipid ion ($[M-H]^-$)	Theoretical mass of lipid (Da)	Composition of LLO
1097.6	2196.1	771.5/789.5	790.5	C_{45}-PP-Bac(HexNAc)$_5$Hex
1131.5	2264.1	839.5/857.5	858.6	C_{50}-PP-Bac(HexNAc)$_5$Hex
1139.6	2280.1	855.5/873.5	874.6	OH-C_{50}-PP-Bac(HexNAc)$_5$Hex
1165.4	2332.2	907.5/925.5	925.6	C_{55}-PP-Bac(HexNAc)$_5$Hex
1173.6	2348.2	923.5/941.5	942.6	OH-C_{55}-PP-Bac(HexNAc)$_5$Hex
1199.8	2400.3	975.5/993.5	994.7	C_{60}-PP-Bac(HexNAc)$_5$Hex

[1]Bac, 2,4-diacetamido-2,4,6-trideoxy-D-glucose.

HexNAc, *N*-acetylhexosamine; Hex, hexose.

organisms to ensure fidelity of the designer glycan synthesis. Additionally, the development of analytical tools for the characterization of LLOs will help elucidate previously uncharacterized steps in bacterial glycan synthesis.

References

Aas, F.E., Vik, Å., Vedde, J., Koomey, M., and Egge-Jacobsen, W. (2007). *Neisseria gonorrhoeae* O-linked pilin glycosylation: functional analyses define both the biosynthetic pathway and glycan structure. Mol. Microbiol. *65*, 607–624.

Andersen, D.C., Goochee, C.F., Cooper, G., and Weitzhandler, M. (1994). Monosaccharide and oligosaccharide analysis of isoelectric focusing-separated and blotted granulocyte colony-stimulating factor glycoforms using high-pH anion-exchange chromatography with pulsed amperometric detection. Glycobiology *4*, 459–467.

Andreotti, A., Bonaduce, I., Colombini, M.P., Gautier, G., Modugno, F., and Ribechini, E. (2006). Combined GC/MS analytical procedure for the characterization of glycerolipid, waxy, resinous, and proteinaceous materials in a unique paint microsample. Anal. Chem. *78*, 4490–4500.

Ballou, C.E., and Dell, A. (1985). Polysaccharide–lipid interaction analyzed by fast-atom-bombardment mass spectrometry. Carbohydr. Res. *140*, 139–143.

Bamba, T., Fukusaki, E.-I., Nakazawa, Y., and Kobayashi, A. (2004). Rapid and high-resolution analysis of geometric polyprenol homologues by connected octadecylsilylated monolithic silica columns in high-performance liquid chromatography. J. Sep. Sci. *27*, 293–296.

Bamba, T., Fukusaki, E., Minakuchi, H., Nakazawa, Y., and Kobayashi, A. (2005). Separation of polyprenol and dolichol by monolithic silica capillary column chromatography. J. Lipid Res. *46*, 2295–2298.

Barreteau, H., Magnet, S., Ghachi, M.E., Touzé, T., Arthur, M., Mengin-Lecreulx, D., and Blanot, D. (2009). Quantitative high-performance liquid chromatography analysis of the pool levels of undecaprenyl phosphate and its derivatives in bacterial membranes. J. Chromatogr. B. *877*, 213–220.

Bino, R.J., Hall, R.D., Fiehn, O., Kopka, J., Saito, K., Draper, J., Nikolau, B.J., Mendes, P., Roessner-Tunali, U., Beale, M.H., *et al.* (2004). Potential of metabolomics as a functional genomics tool. Trends Plant Sci. *9*, 418–425.

Bird, S.S., Marur, V.R., Sniatynski, M.J., Greenberg, H.K., and Kristal, B.S. (2011). Lipidomics profiling by high-resolution LC–MS and high-energy collisional dissociation fragmentation: focus on characterization of mitochondrial cardiolipins and monolysocardiolipins. Anal. Chem. *83*, 940–949.

Bligh, E.G., and Dyer, W.J. (1959). A rapid method of total lipid extraction and purification. Biochem. Cell Biol. *37*, 911–917.

Bodennec, J., Koul, O., Aguado, I., Brichon, G., Zwingelstein, G., and Portoukalian, J. (2000). A procedure for fractionation of sphingolipid classes by solid-phase extraction on aminopropyl cartridges. J. Lipid Res. *41*, 1524–1531.

Bodennec, J., Pelled, D., and Futerman, A.H. (2003). Aminopropyl solid phase extraction and 2 D TLC of neutral glycosphingolipids and neutral lysoglycosphingolipids. J. Lipid Res. *44*, 218–226.

Bray, M.R., and Clarke, A.J. (1992). Action pattern of xylo-oligosaccharide hydrolysis by *Schizophyllum commune* xylanase A. Eur. J. Biochem. *204*, 191–196.

Bruggink, C., Wuhrer, M., Koeleman, C., Barreto, V., Lui, Y., Pohl, C., Ingendoh, A., Hokke, C., and Deelder, A. (2005). Oligosaccharide analysis by capillary-scale high-pH anion-exchange chromatography with on-line ion-trap mass spectrometry. J. Chromatogr. B Analyt. Technol. Biomed. Life Sci. *828*, 136–143.

Bundy, J., and Fenselau, C. (1999). Lectin-based affinity capture for MALDI–MS analysis of bacteria. Anal. Chem. *71*, 1460–1463.

Byss, M., Triska, J., and Elhottova, D. (2007). GC–MS–MS analysis of bacterial fatty acids in heavily creosote-contaminated soil samples. Anal. Bioanal. Chem. *387*, 1573–1577.

Calo, D., Eilam, Y., Lichtenstein, R.G., and Eichler, J. (2010). Towards glycoengineering in archaea: replacement of *Haloferax volcanii* AglD with homologous glycosyltransferases from other halophilic archaea. Appl. Environ. Microbiol. *76*, 5684–5692.

Carlson, T., Skorupinska-Tudek, K., Hertel, J., Chojnacki, T., Olsson, J.M., and Swiezewska, E. (2000). Single polyprenol and dolichol isolation by semipreparative high-performance liquid chromatography technique. J. Lipid Res. *41*, 1177–1180.

Clarke, V.A., Platt, N., and Butters, T.D. (1995). Cloning and expression of the β-N-Acetylglucosaminidase gene from *Streptococcus pneumoniae.* J. Biol. Chem. *270*, 8805–8814.

Colin, H., Guiochon, G., and Jandera, P. (1982). The solvent eluotropic strength on carbon adsorbents. Chromatographia *15*, 133–139.

Crossman, Z.M., Ineson, P., and Evershed, R.P. (2005). The use of 13C labelling of bacterial lipids in the characterisation of ambient methane-oxidising bacteria in soils. Org. Geochem. *36*, 769–778.

Currie, H.A., and Perry, C.C. (2006). Resolution of complex monosaccharide mixtures from plant cell wall isolates by high pH anion exchange chromatography. J. Chromatogr. A. *1128*, 90–96.

Delobel, A., Roy, S., Touboul, D., Gaudin, K., Germain, D.P., Baillet, A., Brion, F., Prognon, P., Chaminade, P., and Laprévote, O. (2006). Atmospheric pressure photoionization coupled to porous graphitic carbon liquid chromatography for the analysis of globotriaosylceramides. Application to Fabry disease. J. Mass Spectrom. *41*, 50–58.

Desai, P.T., Walsh, M.K., and Weimer, B.C. (2008). Solid phase capture of pathogenic bacteria using gangliosides and detection with real time PCR. Appl. Environ. Microbiol.*74*, 2254–2258.

Desiderio, C., Rossetti, D.V., Iavarone, F., Messana, I., and Castagnola, M. (2010). Capillary electrophoresis-mass spectrometry: Recent trends in clinical proteomics. J. Pharm. Biomed. Anal. *53,* 1161–1169.

Dong, M., Wu, M., Wang, F., Qin, H., Han, G., Dong, J., Wu, R.Ä., Ye, M., Liu, Z., and Zou, H. (2010). Coupling strong anion-exchange monolithic capillary with MALDI-TOF MS for sensitive detection of phosphopeptides in protein digest. Anal. Chem. *82,* 2907–2915.

Eliyahu, D. (2009). GC–MS for characterization and identification of ant semiochemicals. Cold Spring Harb. Protoc. *2009,* pdb prot5249.

Faridmoayer, A., Fentabil, M.A., Mills, D.C., Klassen, J.S., and Feldman, M.F. (2007). Functional characterization of bacterial oligosaccharyltransferases involved in O-linked protein glycosylation. J. Bacteriol. *189,* 8088–8098.

Faridmoayer, A., Fentabil, M.A., Haurat, M.F., Yi, W., Woodward, R., Wang, P.G., and Feldman, M.F. (2008). Extreme substrate promiscuity of the *Neisseria* oligosaccharyl transferase involved in protein O-glycosylation. J. Biol. Chem. *283,* 34596–34604.

Fisher, A.C., Haitjema, C.H., Guarino, C., Celik, E., Endicott, C.E., Reading, C.A., Merritt, J.H., Ptak, A.C., Zhang, S., and DeLisa, M.P. (2011). Production of secretory and extracellular N-linked glycoproteins in *Escherichia coli*. Appl. Environ. Microbiol. *77,* 871–881.

Fuchs, B., Suss, R., Nimptsch, A., and Schiller, J. (2009). MALDI-TOF–MS directly combined with TLC: A review of the current state. Chromatographia *69,* 95–105.

Gao, N., and Lehrman, M.A. (2006). Non-radioactive analysis of lipid-linked oligosaccharide compositions by fluorophore-assisted carbohydrate electrophoresis. In Methods in Enzymology, Minoru, F., ed. (New York, Academic Press), pp. 3–20.

Garrett, T.A., Guan, Z., and Raetz, C.R. (2007). Analysis of ubiquinones, dolichols, and dolichol diphosphate-oligosaccharides by liquid chromatography-electrospray ionization-mass spectrometry. Methods Enzymol. *432,* 117–143.

Gaudin, K., Hanai, T., Chaminade, P., and Baillet, A. (2007). Retention behaviour of polyunsaturated fatty acid methyl esters on porous graphitic carbon. J. Chromatogr. A. *1157,* 56–64.

Glover, K.J., Weerapana, E., and Imperiali, B. (2005). *In vitro* assembly of the undecaprenylpyrophosphate-linked heptasaccharide for prokaryotic N-linked glycosylation. Proc. Nat. Acad. Sci. U.S.A. *102,* 14255–14259.

Gopal, P.K., Sullivan, P.A., and Smart, J.B. (2001). Utilisation of galacto-oligosaccharides as selective substrates for growth by lactic acid bacteria including *Bifidobacterium lactis* DR10 and *Lactobacillus rhamnosus* DR20. Int. Dairy J. *11,* 19–25.

Guan, Z., Breazeale, S.D., and Raetz, C.R.H. (2005). Extraction and identification by mass spectrometry of undecaprenyl diphosphate-MurNAc-pentapeptide-GlcNAc from *Escherichia coli*. Anal. Biochem. *345,* 336–339.

Guan, Z., Naparstek, S., Kaminski, L., Konrad, Z., and Eichler, J. (2010). Distinct glycan-charged phosphodolichol carriers are required for the assembly of the pentasaccharide N-linked to the *Haloferax volcanii* S-layer glycoprotein. Mol. Microbiol. *78,* 1294–1303.

Guerry, P., and Szymanski, C.M. (2008). Campylobacter sugars sticking out. Trends Microbiol. *16,* 428–435.

Haeuptle, M.A., H,lsmeier, A.J., and Hennet, T. (2010). HPLC and mass spectrometry analysis of dolichol-phosphates at the cell culture scale. Anal. Biochem. *396,* 133–138.

Hall, C.W., McLachlan, K.R., Krag, S.S., and Robbins, A.R. (1997). Reduced utilization of Man5GlcNAc2-P–P-lipid in a Lec9 mutant of Chinese hamster ovary cells: Analysis of the steps in oligosaccharide-lipid assembly. J. Cell. Biochem. *67,* 201–215.

Hanai, T. (2003). Separation of polar compounds using carbon colums. *J.* Chromatogr. A *989,* 183–196.

Hermentin, P., Witzel, R., Doenges, R., Bauer, R., Haupt, H., Patel, T., Parekh, R.B., and Brazel, D. (1992). The mapping by high-pH anion-exchange chromatography with pulsed amperometric detection and capillary electrophoresis of the carbohydrate moieties of human plasma alpha 1-acid glycoprotein. Anal. Biochem. *206,* 419–429.

Kaluzny, M.A., Duncan, L.A., Merritt, M.V., and Epps, D.E. (1985). Rapid separation of lipid classes in high yield and purity using bonded phase columns. J. Lipid Res. *26,* 135–140.

Kay, E., Lesk, V.I., Tamaddoni-Nezhad, A., Hitchen, P.G., Dell, A., Sternberg, M.J., Muggleton, S., and Wren, B.W. (2010). Systems analysis of bacterial glycomes. Biochem. Soc. Trans. *38,* 1290–1293.

Kelleher, D.J., Karaoglu, D., and Gilmore, R. (2001). Large-scale isolation of dolichol-linked oligosaccharides with homogeneous oligosaccharide structures: determination of steady-state dolichol-linked oligosaccharide compositions. Glycobiology *11,* 321–333.

Kelly, J., Jarrell, H., Millar, L., Tessier, L., Fiori, L.M., Lau, P.C., Allan, B., and Szymanski, C.M. (2006). Biosynthesis of the N-linked glycan in *Campylobacter jejuni* and addition onto protein through block transfer. J. Bacteriol. *188,* 2427–2434.

Kohda, D., Yamada, M., Igura, M., Kamishikiryo, J., and Maenaka, K. (2007). New oligosaccharyltransferase assay method. Glycobiology *17,* 1175–1182.

Kuntz, C., Sonnenbichler, J., Sonnenbichler, I., Sumper, M., and Zeitler, R. (1997). Isolation and characterization of dolichol-linked oligosaccharides from *Haloferax volcanii*. Glycobiology *7,* 897–904.

Kussak, A., and Weintraub, A. (2002). Quadrupole ion-trap mass spectrometry to locate fatty acids on lipid A from Gram-negative bacteria. Anal. Biochem. *307,* 131–137.

Langdon, R.H., Cuccui, J., and Wren, B.W. (2009). N-linked glycosylation in bacteria: an unexpected application. Future Microbiol. *4,* 401–412.

Lehrman, M.A. (2007). Teaching dolichol-linked oligosaccharides more tricks with alternatives to metabolic radiolabelling. Glycobiology *17,* 75R-85R.

Li, J., and Richards, J.C. (2007). Application of capillary electrophoresis mass spectrometry to the characterization of bacterial lipopolysaccharides. Mass Spectrom. Rev. *26*, 35–50.

Li, Y.L., Su, X., Stahl, P.D., and Gross, M.L. (2007). Quantification of diacylglycerol molecular species in biological samples by electrospray ionization mass spectrometry after one-step derivatization. Anal. Chem. *79*, 1569–1574.

Liang, C., Dai, S., and Guiochon, G. (2003). A graphitized-carbon monolithic column. Anal. Chem. *75*, 4904–4912.

Lizak, C., Fan, Y.-Y., Weber, T.C., and Aebi, M. (2011). N-linked glycosylation of antibody fragments in *Escherichia coli*. Bioconj. Chem. ASAP. *22*, 488–496.

Lodowska, J., Mazurek, U., Kurkiewicz, S., Wilczok, A., Swierczek, G., Tam, I., Pytel, A., and Wilczok, T. (1999). GC–MS analysis of fatty acids in phospholipids from *Chlorella vulgaris* cultivated synchronously in the presence of propranolol. Acta. Pol. Pharm. *56*, 21–27.

Maillet, F., Poinsot, V., Andre, O., Puech-Pages, V., Haouy, A., Gueunier, M., Cromer, L., Giraudet, D., Formey, D., Niebel, A., *et al.* (2011). Fungal lipochitooligosaccharide symbiotic signals in arbuscular mycorrhiza. Nature *469*, 58–63.

Mashego, M.R., Wu, L., Van Dam, J.C., Ras, C., Vinke, J.L., Van Winden, W.A., Van Gulik, W.M., and Heijnen, J.J. (2004). MIRACLE: mass isotopomer ratio analysis of U-13C-labelled extracts. A new method for accurate quantification of changes in concentrations of intracellular metabolites. Biotechnol. Bioeng. *85*, 620–628.

Mashego, M.R., Rumbold, K., De Mey, M., Vandamme, E., Soetaert, W., and Heijnen, J.J. (2007). Microbial metabolomics: past, present and future methodologies. Biotechnol. Lett. *29*, 1–16.

Masson, L., and Holbein, B.E. (1985). Role of lipid intermediate(s) in the synthesis of serogroup B *Neisseria meningitidis* capsular polysaccharide. J. Bacteriol. *161*, 861–867.

Matyash, V., Liebisch, G., Kurzchalia, T.V., Shevchenko, A., and Schwudke, D. (2008). Lipid extraction by methyl-tert-butyl ether for high-throughput lipidomics. J. Lipid Res. *49*, 1137–1146.

Messner, P., Steiner, K., Zarschler, K., and Scheffer, C. (2008). S-layer nanoglycobiology of bacteria. Carbohydr. Res. *343*, 1934–1951.

Moini, M. (2002). Capillary electrophoresis mass spectrometry and its application to the analysis of biological mixtures. Anal. Bioanal. Chem. *373*, 466–480.

Murphy, R., Raetz, C.R., Reynolds, C.M., and Barkley, R.M. (2005). Mass spectrometry advances in lipidomics: collision-induced decomposition of Kdo2-lipid A. Prostaglandins Other Lipid Mediat. *77*, 131–140.

Nakanishi, H., Ogiso, H., and Taguchi, R. (2009). Qualitative and quantitative analyses of phospholipids by LC–MS for lipidomics. Meth. Mol. Biol. *579*, 287–313.

Nothaft, H., and Szymanski, C.M. (2010). Protein glycosylation in bacteria: sweeter than ever. Nat. Rev. Microb. *8*, 765–778.

Nygren, H., Seppanen-Laakso, T., Castillo, S., Hyotylainen, T., and Oresic, M. (2011). Liquid chromatography-mass spectrometry (LC–MS)-based lipidomics for studies of body fluids and tissues. Meth. Mol. Biol. *708*, 247–257.

Oh, D.B., Park, J.S., Kim, M.W., Cheon, S.A., Kim, E.J., Moon, H.Y., Kwon, O., Rhee, S.K., and Kang, H.A. (2008). Glycoengineering of the methylotrophic yeast *Hansenula polymorpha* for the production of glycoproteins with trimannosyl core N-glycan by blocking core oligosaccharide assembly. Biotechnol. J. *3*, 659–668.

Oursel, D., Loutelier-Bourhis, C., Orange, N., Chevalier, S., Norris, V., and Lange, C.M. (2007). Identification and relative quantification of fatty acids in Escherichia coli membranes by gas chromatography/mass spectrometry. Rapid Commun. Mass Spectrom. *21*, 3229–3233.

Pabst, M., Bondili, J.S., Stadlmann, J., Mach, L., and Altmann, F. (2007). Mass + retention time = structure: A strategy for the analysis of N-glycans by carbon LC-ESI–MS and its application to fibrin N-glycans. Anal. Chem. *79*, 5051–5057.

Paul, C., Lyle, E., Beveridge, T., Tapping, R., Kropinski, A., and Vinogradov, E. (2009). Characterization of the cell surface glycolipid from *Spirochaeta aurantia*. Glycoconj. J. *26*, 1097–1108.

Pereira, L. (2008). Porous graphitic carbon as a stationary phase in HPLC: Theory and applications. J. Liquid chromatograph. Rel. Technol. *31*, 1687–1731.

Piretti, M., Pagliuca, G., and Tarozzi, G. (1995). Simultaneous reversed-phase high-performance liquid chromatographic separation of non-polar isoprenoid lipids and their determination. J. Chromatogr. B Biomed. Appl. *674*, 177–185.

Reid, C.W., Stupak, J., Chen, M.M., Imperiali, B., Li, J., and Szymanski, C.M. (2008). Affinity-capture tandem mass spectrometric characterization of polyprenyl-linked oligosaccharides: tool to study protein N-glycosylation pathways. Anal. Chem. *80*, 5468–5475.

Reid, C.W., Stupak, J., Szymanski, C.M., and Li, J. (2009). Analysis of bacterial lipid-linked oligosaccharide intermediates using porous graphitic carbon liquid chromatography-electrospray ionization mass spectrometry: heterogeneity in the polyisoprenyl carrier revealed. Anal. Chem. *81*, 8472–8478.

Ristl, R., Steiner, K., Zarschler, K., Zayni, S., Messner, P., and Schaffer, C. (2011). The s-layer glycome-adding to the sugar coat of bacteria. Int. J. Microbiol. *16*, 127870.

Roberts, L.D., McCombie, G., Titman, C.M., and Griffin, J.L. (2008). A matter of fat: An introduction to lipidomic profiling methods. J. Chromatogr. B *871*, 174–181.

Roy, S., Delobel, A., Gaudin, K., Touboul, D., Germain, D.P., Baillet, A., Prognon, P., LaprÈvote, O., and Chaminade, P. (2006). Liquid chromatography on porous graphitic carbon with atmospheric pressure photoionization mass spectrometry and tandem mass

spectrometry for the analysis of glycosphingolipids. J. Chromatogr. A *1117,* 154–162.
Schiller, J., Suss, R., Fuchs, B., Muller, M., Zschornig, O., and Arnold, K. (2007). MALDI–TOF MS in lipidomics. Front. Biosci. *12,* 2568–2579.
Suntornsuk, L. (2010). Recent advances of capillary electrophoresis in pharmaceutical analysis. Anal. Bioanal. Chem. *398,* 29–52.
Szymanski, C.M., and Wren, B.W. (2005). Protein glycosylation in bacterial mucosal pathogens. Nat. Rev. Microbiol. 3, 225–237.
Takayama, K., Qureshi, N., Hyver, K., Honovich, J., Cotter, R.J., Mascagni, P., and Schneider, H. (1986). Characterization of a structural series of lipid A obtained from the lipopolysaccharides of *Neisseria gonorrhoeae.* Combined laser desorption and fast atom bombardment mass spectral analysis of high performance liquid chromatography-purified dimethyl derivatives. J. Biol. Chem. *261,* 10624–10631.
Trent, M.S., Ribeiro, A.A., Doerrler, W.T., Lin, S., Cotter, R.J., and Raetz, C.R. (2001). Accumulation of a polyisoprene-linked amino sugar in polymyxin-resistant *Salmonella typhimurium* and *Escherichia coli*: structural characterization and transfer to lipid A in the periplasm. J. Biol. Chem. *276,* 43132–43144.
Troy, F.A., Frerman, F.E., and Heath, E.C. (1971). The biosynthesis of capsular polysaccharide in *Aerobacter aerogenes.* J. Biol. Chem. *246,* 118–133.
Vieler, A., Wilhelm, C., Goss, R., S¸fl, R., and Schiller, J.r. (2007). The lipid composition of the unicellular green alga *Chlamydomonas reinhardtii* and the diatom *Cyclotella meneghiniana* investigated by MALDI–TOF MS and TLC. Chem. Phys. Lipids *150,* 143–155.
Wang, X., Ribeiro, A.A., Guan, Z., and Raetz, C.R. (2009). Identification of undecaprenyl phosphate-beta-D-galactosamine in *Francisella novicida* and its function in lipid A modification. Biochemistry *48,* 1162–1172.
Wei, Y., Li, C., Huang, W., Li, B., Strome, S., and Wang, L.X. (2008). Glycoengineering of human IgG1-Fc through combined yeast expression and *in vitro* chemoenzymatic glycosylation. Biochemistry *47,* 10294–10304.
Weitzhandler, M., Slingsby, R., Jagodzinski, J., Pohl, C., Narayanan, L., and Avdalovic, N. (1996). Eliminating monosaccharide peak tailing in high pH anion-exchange chromatography with pulsed amperometric detection. Anal. Biochem. *241,* 135–136.
West, C., Elfakir, C., and Lafosse, M. (2010). Porous graphitic carbon: A versatile stationary phase for liquid chromatography. J. Chromatogr. A. *1217,* 3201–3216.
Wuhrer, M., Koeleman, C.A.M., Deelder, A.M., and Hokke, C.H. (2003). Normal-phase nanoscale liquid chromatography, mass spectrometry of underivatized oligosaccharides at low-femtomole sensitivity. Anal. Chem. *76,* 833–838.
Young, D., and Buchanan, T. (1983). A serological test for leprosy with a glycolipid specific for *Mycobacterium leprae.* Science *221,* 1057–1059.
Young, N.M., Brisson, J.R., Kelly, J., Watson, D.C., Tessier, L., Lanthier, P.H., Jarrell, H.C., Cadotte, N., St. Michael, F., Aberg, E., *et al.* (2002). Structure of *N*-linked glycan present on multiple glycoproteins in the Gram-negative bacterium, *Campylobacter jejuni.* 277, 42530–42539.

Synthesis of 4-Acetamidohexoses in Bacteria: Structural Insights from the Bacillosamine and Nonulosonic Acid Pathways

11

Allan Matte, Ian C. Schoenhofen, Traian Sulea, Miroslaw Cygler and N. Martin Young

Abstract

Many hexose sugars in bacteria undergo a variety of modifications, including oxidation/reduction, amination and acetylation, as part of biosynthesis into their final biologically active forms. Enzymes that catalyse these reactions normally utilize nucleotide-linked sugar substrates, utilizing the nucleotide as an 'ancient handle' to bind and orient the sugar within the enzymes' active site. We and others have focused efforts on elucidating structure–function relationships for a subset of such biosynthetic enzymes, those associated with the synthesis of trideoxy-diacetamidohexoses, and the nonulsonate sugars subsequently derived from them. With structural information combined with site-directed mutagenesis, enzymatic analysis and molecular modelling, these studies have been essential to understanding the chemistry of how these enzymes bind their substrates and effect catalysis. Enzymes having different folds, such as *N*-acetyltransferases, can utilize different scaffolds to attain the same 4-acetamido-sugar product. In the case of dehydratases/epimerases and aminotransferases, the enzymes have a conserved structure, but utilize subtle differences within their active site to confer substrate binding and the nature of the final product. These studies depict the structural relationships between these enzymes, while at the same time high-lighting important differences that are beginning to reveal their function at the molecular level.

Introduction

Bacteria synthesize a far wider range of complex sugars that are much more diverse than carbohydrates found in eukaryotes. These sugars are incorporated into structures such as polysaccharides, glycoproteins and antibiotics, most of which are covered in other chapters of this book. Our focus here is on the biosynthetic pathways of one family of complex amino sugars, the trideoxy-diacetamidohexoses (TDHs), and their further elaboration into nonulosonic acids (NulOs). The latter are nine-carbon keto sugars; the best known member of this family is neuraminic acid, which occurs in both prokaryotes and eukaryotes (Chen and Varki, 2010). Specifically, we will describe structure/function studies of enzymes involved in the synthesis of two TDHs, 2,4-diacetamido-2,4,6-trideoxy-D-Man and 2,4-diacetamido-2,4,6-trideoxy-L-Alt, and their extension into the NulOs, legionaminic acid (Leg) and pseudaminic acid (Pse), respectively (Fig. 11.1).

In various *Campylobacter* and *Helicobacter* spp., Leg, Pse, as well as 2,4-diacetamido-2,4,6-trideoxy-D-Glc (*N,N*'-diacetylbacillosamine or BacdiNAc) are key components of *N*- and *O*-linked glycoproteins. BacdiNAc is the precursor to the mannose sugar above and is the linkage sugar in *N*-linked glycoproteins (Young *et al.*, 2002) while the NulOs occur as single *O*-linked sugars on flagella (Logan, 2006). All three of these sugars were originally identified as constituents of *O*-chains, for example the NulOs occur in lipopolysaccharides of *Pseudomonas aeruginosa* (Knirel *et al.*, 1984) and *Legionella pneumophila* (Zähringer *et*

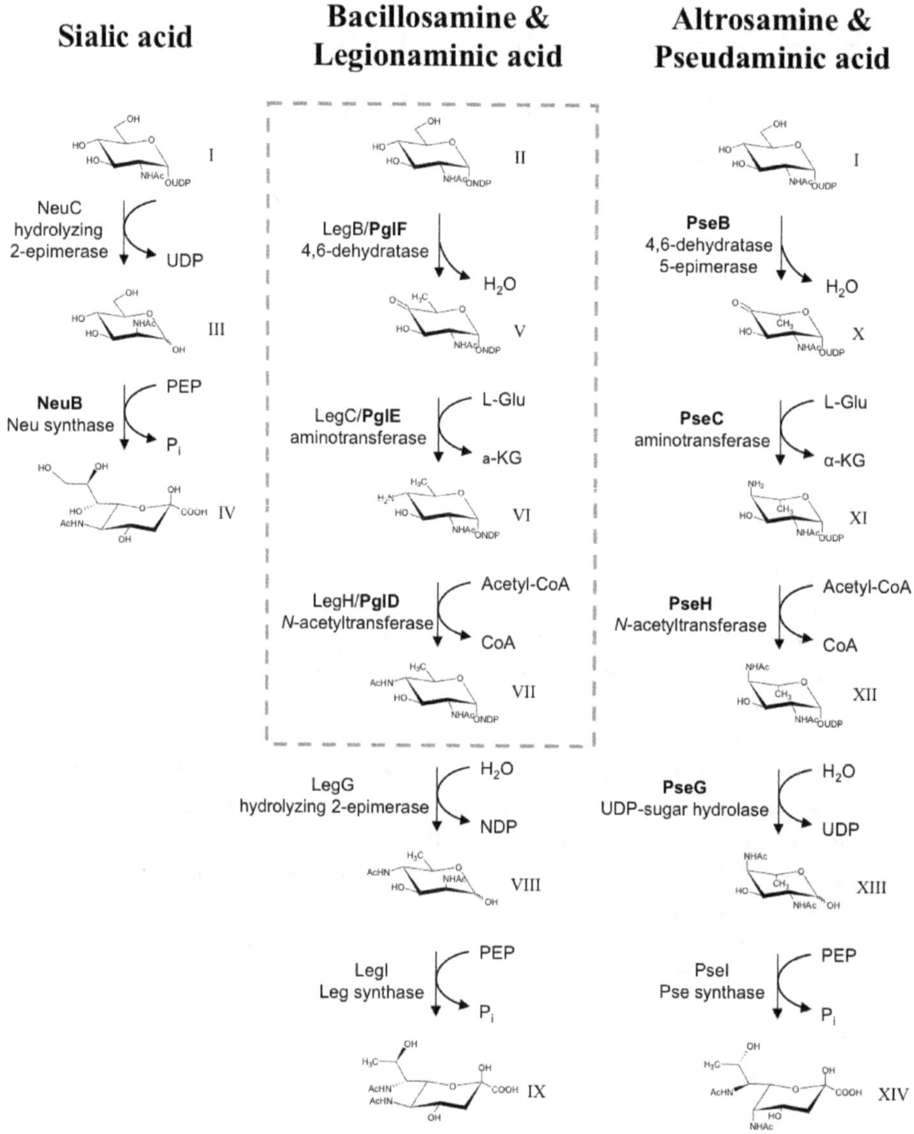

Figure 11.1 Bacterial biosynthetic pathways leading to the TDHs 2,4-diacetamido-2,4,6-trideoxy-D-mannose (VIII) or 2,4-diacetamido-2,4,6-trideoxy-L-altrose (XIII) and their extension into the nonulosonates legionaminic acid (IX; Leg5Ac7Ac) and pseudaminic acid (Pse5Ac7Ac), respectively. The bacterial biosynthetic pathway leading to sialic acid (Neu5Ac) is shown on the left as reference. The dashed box identifies the biosynthetic steps that are common between the general *N*-linked protein glycosylation pathway (Pgl), leading to NDP-BacdiNAc, and the legionaminic acid biosynthetic pathway (Leg). Note that the NDP utilized by Pgl enzymes is UDP, whereas the NDP utilized by Leg enzymes may be either UDP or GDP. The enzymes shown in bold are those for which atomic coordinates have been deposited in the Protein Data Bank (see Table 11.1). Except for the nonulosonates, the carbohydrate structures are all shown in $^4C_1$ chair conformations for comparison purposes. Stereochemical differences between Leg and Pse are introduced during the first, second and fourth biosynthetic steps. The carbohydrates are: (I) UDP-GlcNAc, (II) NDP-GlcNAc, (III) ManNAc, (IV) Neu5Ac or sialic acid, (V) NDP-2-acetamido-2,6-dideoxy-α-D-*xylo*-hexos-4-ulose, (VI) NDP-4-amino-4,6-dideoxy-α-D-GlcNAc, (VII) NDP-2,4-diacetamido-2,4,6-trideoxy-α-D-Glc or NDP-BacdiNAc, (VIII) 2,4-diacetamido-2,4,6-trideoxy-D-Man, (IX) Leg5Ac7Ac or legionaminic acid, (X) UDP-2-acetamido-2,6-dideoxy-β-L-*arabino*-hexos-4-ulose, (XI) UDP-4-amino-4,6-dideoxy-β-L-AltNAc, (XII) UDP-2,4-diacetamido-2,4,6-trideoxy-β-L-Alt, (XIII) 2,4-diacetamido-2,4,6-trideoxy-L-Alt, (XIV) Pse5Ac7Ac or pseudaminic acid.

al., 1995), from which their names are derived. Many variants of the basic structures of the two NulOs are known, arising chiefly from modifications of the two amino groups at C-5 and C-7 by addition of different acyl groups, with the most common *N*-acyl group being an acetyl moiety.

The basic synthetic pathways for the various NulOs and their TDH precursors use similar enzyme chemistries to carry out five synthetic steps from an NDP-GlcNAc precursor to a NulO (Fig. 11.1). For the enzymes whose structures will be described here, the pathway precursor is UDP-GlcNAc. In some bacteria, for example in *C. jejuni*, a pathway to Leg starting from GDP-GlcNAc has been found (Schoenhofen *et al.*, 2009). For four of the five synthetic reactions, the *C. jejuni* and *H. pylori* enzymes catalysing them show considerable sequence and structural similarities. They lead to different forms of NulOs because the individual enzymes give rise to stereochemically different products, by proceeding with inversion or retention of configuration at the key carbon atom. Theoretically, this could give rise to eight possible TDHs and 16 possible NulOs, but only four NulOs are known (Leg, 4-epiLeg, 8-epiLeg and Pse; Schoenhofen *et al.*, 2006a). The differences in the active sites of these homologous enzymes which give rise to the opposite configurations are of key interest to structural biologists. In the remaining *N*-acetylation step, where sequence and structural similarities are lacking, no change in configuration occurs. Unlike the other synthetic reactions, members of two distinct families of enzymes are utilized in this step, although both use acetyl-CoA as the donor. The families are the left-handed β-helix (LβH) and the GCN5-related (GNAT) *N*-acetyltransferases.

In the first two steps in the BacdiNAc pathway, carried out by a dehydratase, PglF, and an aminotransferase, PglE, the configurations at C-5 and C-4 are retained to form UDP-4-amino-4,6-dideoxy-α-D-GlcNAc. In contrast, in the pathway to the Pse precursor, the corresponding enzymes PseB (or FlaA1) and PseC invert the configurations of these carbon atoms, forming UDP-4-amino-4,6-dideoxy-β-L-AltNAc (Fig. 11.1). The new 4-amino group is then acetylated by PglD (an LβH enzyme) or PseH (a GNAT enzyme) respectively, to create UDP-BacdiNAc and UDP-2,4-diacetamido-2,4,6-trideoxy-β-L-Alt. The NulO pathways also require removal of the nucleotide from these nucleotide-sugar products, which proceeds with retention of configuration in the Pse pathway (PseG), but with inversion at C-2 in the Leg pathway (LegG or NeuC homologue), forming a ManNAc derivative analogously to the neuraminic acid pathway. This step and the next are carried out by enzymes which are homologues of those that synthesize the prototypic NulO, neuraminic acid. They were therefore termed NeuC and NeuB before it was recognized that their products were different NulOs. Similarly NeuA was used for the two CMP-NulO synthases, which make the substrates for the glycosyltransferases that incorporate the NulOs into glycoproteins or polysaccharides.

The NeuB, or NulO synthase catalysed reaction adds three more carbons to the free TDH, from a phosphoenolpyruvate donor. The stereochemistry in the Leg and Pse pathways is the same, but it can be inferred from the occurrence of 4-epiLeg that the opposite configuration can be formed (Schoenhofen *et al.*, 2006a). Structural data are available for the NeuB (Gunawan *et al.*, 2005) and NeuA (Mosimann *et al.*, 2001) enzymes from the Neu pathway, but not for their Leg and Pse homologues.

Several structures have now been determined, primarily for the dehydratase, aminotransferase and acetyltransferase enzymes involved in synthesis of Bac and Pse, as summarized in Table 11.1. In the following, we provide a detailed account of the current data available on structure–function relationships for these enzymes. Reviews of structure–function studies of nucleotide-linked sugar modifying enzymes (Field and Naismith, 2003) as well as a recent account summarizing current structural information on aminotransferases and *N*-acetyltransferases involved in the biosynthesis of unusual microbial sugars (Holden *et al.*, 2010) are available to the interested reader.

The dehydratase/epimerases

These enzymes belong to the short-chain dehydrogenase/reductase (SDR) superfamily (Kallberg *et al.*, 2010), specifically to the subgroup with sugar substrates described by Favia *et al.* (2008). Though

Table 11.1 Summary of structural data for enzymes involved in the biosynthesis of bacillosamine and nonulosonic acids. Atomic coordinates are available through the RCSB Protein Data Bank (PDB) at (http://www.rcsb.org/pdb/home/home.do)

Sugar	Enzyme (specific source)[1]	Function	PDB code	Ligand	Reference
Pse	PseB (Hp0840 or FlaA1)	C-4,6 dehydratase/C-5 epimerase	2GNA	NADP/UDP-Gal	Ishiyama *et al*. (2006)
			2GN4	NADPH/UDP-GlcNAc	
			2GN6	NADP/UDP-GlcNAc	
			2GN8	NADP/UDP	
			2GN9	NADP/UDP-Glc	
	PseC (Hp0366)	C-4 aminotransferase	2FN6	None	Schoenhofen *et al*. (2006c)
			2FNI	Pyridoxal-5′-phosphate	
			2FNU	UDP-4-amino-4,6-dideoxy-β-L-AltNAc	
	PseH (Hp0327)	*N*-acetyltransferase	N/A	–	Unpublished results
	PseG (Cj1312)	UDP–sugar hydrolase	3HBM	None	Rangarajan *et al*. (2009)
			3HBN	UDP	
Bac	PglF (Cj1120c)	C-4,6-dehydratase	N/A	–	Unpublished results
	PglE (Cj1121c)	C-4 aminotransferase	1O69	(2-Amino-4-formyl-5-hydroxy-6-methylpyridin-3-yl)methyl dihydrogen phosphate	Badger *et al*. (2005)
			1O61	Pyridoxal-5′-phosphate	
			1O62	None	
	PglD (Cj1123c)	*N*-acetyltransferase	3BFP	None	Olivier and Imperiali (2008); Rangarajan *et al*. (2008); unpublished results
			2VHE	Coenzyme A	
			3BSW	Citrate	

			3BSY	Acetyl-coenzyme A	
			3BSS	UDP-2-acetamido-4-amino-2,4,6-trideoxy-α-D-glucopyranose	
			2NPO	None	
Sialic acid	NeuC	Hydrolysing 2-epimerase	N/A	–	–
	NeuB	Sialic acid synthase	1XUU	Malate/Mn^{2+}	Gunawan *et al*. (2005); Liu *et al*. (2009); unpublished results
			1XUZ	5-Deoxy-5-{[(1S)–1-hydroxyethyl]amino}-D-glucitol/Mn^{2+}/phosphoenolpyruvate	
			2WQP	5-(acetylamino)–3,5-dideoxy-2-O-phosphono-D-*erythro*-L-manno-nononic acid/Mn^{2+}	
			3CM4	Malate	
			2ZDR	(4S)–2-methyl-2,4-pentanediol/Mg^{2+}	
	NeuA	CMP acylneuraminate synthetase	1EZI	None	Mosimann *et al*. (2001)
			1EYR	CDP	

their SDR domains are similar in sequence (30% identity, 53% similarity) the Bac pathway enzyme PglF also has an N-terminal membrane anchor domain, followed by a linker region, which do not occur in PseB. Although structural data for PglF have been collected and are being analysed (A. Matte *et al.*, unpublished work), this discussion is limited to PseB.

The *H. pylori* dehydratase, PseB

PseB (also known as FlaA1), which had originally been characterized as a C-4,6 dehydratase/C-4 reductase, was shown to utilize UDP-GlcNAc as substrate and NADPH as the cofactor (Creuzenet *et al.*, 2000; Creuzenet, 2004). Subsequent studies established that PseB is in fact a bifunctional C-4,6 dehydratase/C-5 epimerase (Ishiyama *et al.*, 2006; Schoenhofen *et al.*, 2006a; Morrison *et al.*, 2008). While initially believed to be a dimer (Creuzenet *et al.*, 2000), electron microscopic analysis of single particles as well as X-ray structure determination of PseB later revealed it to be a hexamer (Ishiyama *et al.*, 2006) (Fig. 11.2A). The PseB protomer is bilobal and adopts a structural variation of the classical Rossmann fold. Both the PseB sequence and structure are consistent with it being a member of the (SDR) superfamily, and appears to be the first structure of this family that also performs the C-5 epimerization step.

The PseB enzyme as crystallized contained bound $NADP^+$, indicating that this cofactor co-purifies with the enzyme. Structures of PseB were determined with $NADP^+$ and bound UDP-galactose (PDB code **2GNA**) or UDP-GlcNAc (PDB code **2GN6**) provide insights into substrate binding and catalysis. While UDP-galactose binds to the enzyme, it is not a substrate but instead possibly a competitive inhibitor (Creuzenet *et al.*, 2000; McNally *et al.*, 2008). Based on the crystal structure and the similarity to other SDR family members, Thr-131, Tyr-141 as well as Asp-132 and Lys-133, were implicated as potential catalytic residues (Ishiyama *et al.*, 2006) (Fig. 11.2B). Subsequently, mutations of Asp-126, Lys-127 and Tyr-135 in *C. jejuni* PseB, corresponding to Asp-132, Lys-133 and Tyr-141 of *H. pylori* PseB, were prepared and characterized, implicating these residues as also affecting catalysis (Morrison *et al.*, 2008).

The proposed catalytic mechanism for *H. pylori* PseB as determined from the *C. jejuni* enzyme (Morrison *et al.*, 2008) is similar to the chemistry proposed for dTDP-glucose-4,6-dehydratase, RlmB (Allard *et al.*, 2002) and is depicted in Fig. 11.2C. The overall reaction can be summarized in three steps. Initial oxidation to form the 4-keto group involves both Tyr-135 (Tyr-141 in *H. pylori* PseB) acting as a base and $NADP^+$ being transiently reduced by accepting the axial C-4 proton (Morrison *et al.*, 2008). A key role for Tyr-135 in the reaction is consistent with the large effect on reaction rate observed when this residue is mutated to Phe. The second step involves deprotonation at C-5, which could involve Lys-127 (Lys-133 in *H. pylori* PseB) acting as a base, given the location of this residue in the PseB crystal structure (Ishiyama *et al.*, 2006). A key role for Lys-127 is consistent with the much lower activity exhibited by the K127A mutant enzyme. This dehydration step is concomitant with the release of water and formation of an additional double bond at C-5, i.e. an enone intermediate. The third step involves hydride transfer to C-6 from NADPH and protonation at C-5 by a residue functioning as an acid. The chirality at the C-5 position will be determined by how the proton is added at the reduction step; if it is at the same face as during the dehydration, then the C-5 configuration will be retained, while if it occurs at the opposite face, it will be inverted. As pointed out by Morrison *et al.* (2008), given that PseB is an inverting dehydratase, it is most unlikely that abstraction and donation of the proton at C-5 are performed by the same residue. Further experiments will be required to determine if Asp-126 (Asp-132 in *H. pylori* PseB), or another residue, acts as an acid to facilitate the epimerase activity of PseB.

The aminotransferases

PglE and PseC perform pyridoxal-5′-phosphate (PLP) dependent amino group transfers using glutamate as the amino donor. While they differ in the configuration of both the substrate ketosugar and the amino group in the product, the complicated reaction chemistry they employ is expected to be very similar. The enzymes show moderate

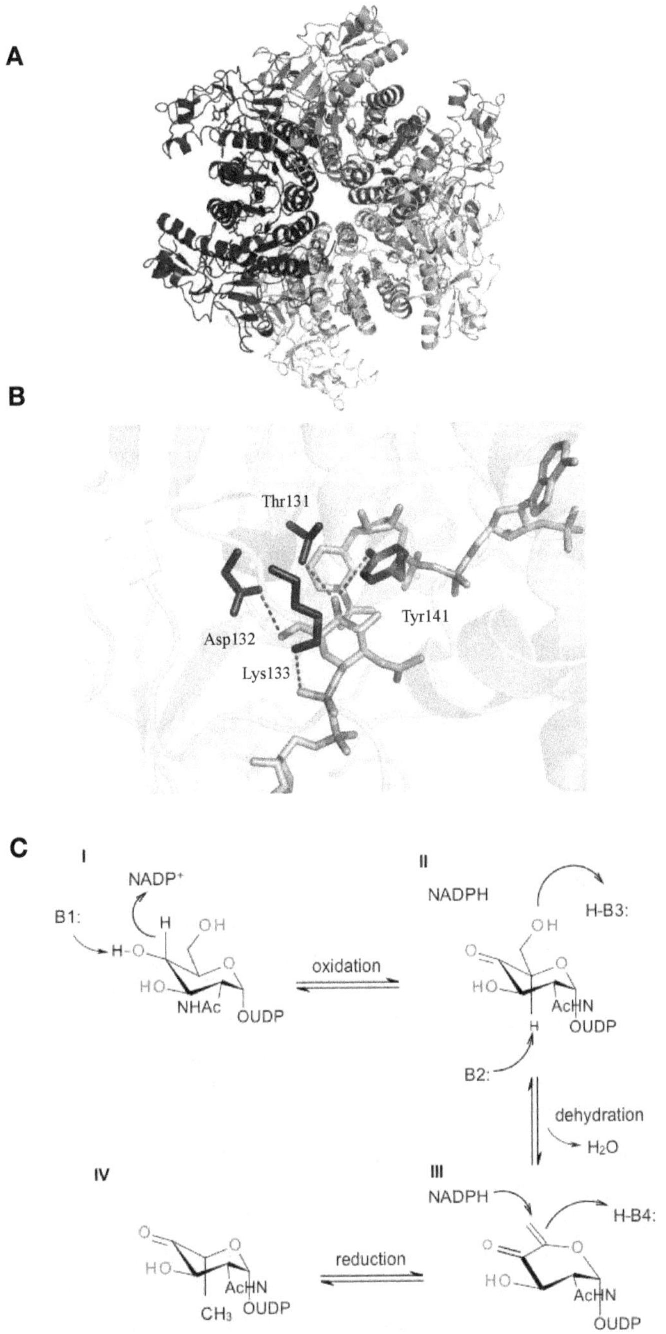

Figure 11.2 Crystal structure of the C-4,6 dehydratase/C-5 epimerase PseB (FlaA1) from the Pse biosynthetic pathway of *H. pylori* bound to $NADP^+$ and UDP-GlcNAc (**PDB 2GN6**). (A) Structure of the PseB hexamer, with each of the three dimers depicted (light grey/medium grey/dark grey) and the $NADP^+$ or UDP-GlcNAc ligands presented in stick representation. (B) Active site region, showing $NADP^+$, UDP-GlcNAc, and key active site residues (Thr-131, Asp-132, Lys-133, Tyr-141). These and subsequent figures were prepared using PyMol (www.pymol.org). (C) Proposed catalytic mechanism for PseB. The first step is oxidation to form the 4-keto group with reduction of $NADP^+$ by transfer of the proton from C-4 (Morrison *et al.*, 2008). The second step is de-protonation and dehydration at C-5, resulting in an additional double bond, forming an enone intermediate. The final step involves hydride transfer from NADPH to C-6 and protonation at C-5 by a residue functioning as an acid.

homology (24% identity and 47% similarity) and share the AAT family fold.

The *H. pylori* aminotransferase, PseC

PseC from *H. pylori* performs PLP-dependent amino transfer using glutamate as the amino donor to generate UDP-4-amino-4,6-dideoxy-β-L-AltNAc (Schoenhofen *et al.*, 2006a,b,c) from the UDP-4-ketosugar product of PseB. The corresponding enzyme from *C. jejuni*, designated Cj1294, has also been characterized, although the product was identified as a GalNAc instead of the AltNAc sugar (Obhi and Creuzenet, 2005). The crystal structure of PseC has been determined in apo form (PDB code **2FN6**), as an internal aldimine between Lys183^{NZ} and PLP (PDB code **2FN1**) as well as an external aldimine between PLP and the UDP-4-amino-4,6-dideoxy-β-L-AltNAc product (PDB code **2FNU**; Schoenhofen *et al.*, 2006c). The enzyme is a dimer, with each subunit comprising two α/β/α domains (Fig. 11.3A). Structurally, PseC adopts the PLP-dependent aminotransferase fold and is a member of the GABA-aminotransferase-like structural family as classified within the Structural Classification of Proteins (SCOP) database (Murzin *et al.*, 1995). Within this group, the crystal structures of *Caulobacter crescentus* GDP-perosamine synthase (Cook and Holden, 2008; PDB code **3BN1**; Cook *et al.*, 2008; PDB codes **3DR4**, **3DR7**), *Salmonella typhimurium* ArnB (Noland *et al.*, 2002; PDB codes **1MDO**, **1MDX**, **1MDZ**) and DesI from *Streptomyces venezuelae* (Burgie and Holden, 2007; PDB code **2PO3**) are especially noteworthy, as they represent other examples of sugar aminotransferases that transfer an amino moiety to the C-4 keto position. The respective substrates for these enzymes are GDP-4-keto-6-dideoxy-D-mannose, an intermediate in *O*-antigen biosynthesis, UDP-4-keto-4-deoxy-L-arabinose, an intermediate in the lipid A modification pathway and dTDP-4-keto-4,6-dideoxyglucose, an intermediate in the synthesis of desosamine, an unusual sugar found within macrolide antibiotics.

Structural comparison of the PseC internal or external aldimine shows there are no significant structural changes compared with the apo enzyme. Residues forming the active site region are depicted in Fig. 11.3B. The phosphate moiety of PLP is orientated via interactions with the side-chains of Thr-57 and Ser-178, as well as main-chain amides from Ala-56 and Thr-57. The α3 helix dipole also contributes to charge neutralization of the phosphate. The PLP pyridine ring is sandwiched by Phe-84 and Ser-156. Asp-154 of PseC is a key active site residue, as it maintains the pyridinium nitrogen atom in its protonated state, necessary for cofactor function. As expected, the K183R mutation resulted in a significant loss of activity *ca.* the wild-type enzyme, this lysine being responsible for formation of the Schiff base with PLP (Schoenhofen *et al.*, 2006c; Fig. 11.3C).

The crystal structure of the PseC-external aldimine complex provided key information on how the substrate binds to this enzyme. The PseC product, UDP-4-amino-4,6-dideoxy-β-L-AltNAc reacted with PLP and in the reverse reaction generated the external aldimine. The UDP–sugar density is somewhat more disordered as compared with the PLP, based on weaker electron density and somewhat higher average temperature factors. Interestingly, the UDP–sugar makes contacts with both subunits, showing how the PseC dimer is important for substrate recognition. Only the N-terminal domain interacts with the altrose sugar, with Phe84 stacking against one face, and His189 and Ile192 on the opposite face. Binding of the sugar by PseC predominantly involves van der Waals interactions, with only one hydrogen bond between the sugar O-3 atom and the side-chain of Tyr316. While the sugar has few direct interactions with the enzyme, the diphosphate and uracil moieties make many more direct contacts, suggestive of PseC specificity for UDP-linked substrates. The relative flexibility and few direct contacts involving the sugar may be important in catalysis, permitting conformational changes that may be important during the catalytic cycle.

The issue of how enzymes such as PseC and ArnB donate the amino group specifically to the axial position in the sugar, as compared with DesI and PglE, where the amino group is donated equatorially, has been partially addressed based on the crystal structure of DesI bound to the external aldimine containing dTDP-4-amino-4,6-dideoxyglucose (Burgie and Holden, 2007; PDB code

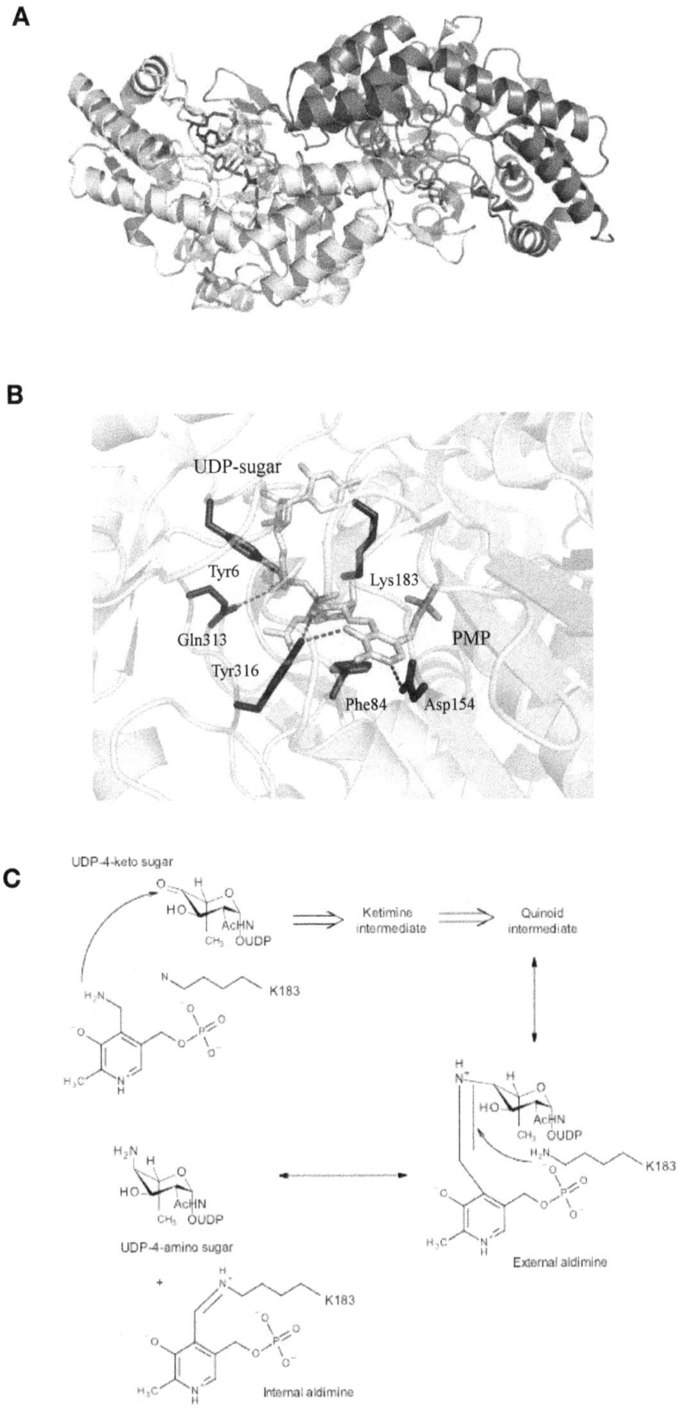

Figure 11.3 (A) Crystal structure of *H. pylori* PseC bound to its external aldimine between PLP and the UDP-4-amino-4,6-dideoxy-β-L-AltNAc product (**PDB 2FNU**) sowing the dimer organization. (B) Close-up of the PseC active site, showing the key residues Tyr6, Phe84, Asp154, Lys183, Gln313 and Tyr316. (C) Proposed mechanism in abbreviated format for PseC (Schoenhofen *et al.*, 2006c). Lys183 plays a key role via alternatively forming an internal aldimine (with PLP) and external aldimine (with UDP-sugar) during the catalytic cycle, converting the UDP-4-keto sugar to an UDP-4-amino sugar.

2PO3). The PLP-binding regions between PseC and DesI are very similar and both enzymes share a stacking interaction between a Phe residue and the hexose ring. However, the substrates bind to these enzymes in different conformations resulting from changes in the dihedral angles about the phosphoryl groups linking the nucleotide to the respective hexose. This results in an ~180° rotation of the hexose between PseC and DesI, accounting for the axial versus equatorial addition of the amino group (Burgie and Holden, 2007). Both PseC and ArnB, despite their relatedness in structure, belong to different sequence families, and have differences in the portions of their substrate binding clefts that would recognize the sugar. The His180 residue of PseC was found to be structurally equivalent to an Asn in PglE, although the H180N mutation in PseC did not affect either the activity or the enzyme product, indicating that this residue is not a specificity determinant. While comparing the external aldimine structures of PseC and DesI provides significant insight into the stereochemistry of amino transfer, understanding the contributions of the various active site residues towards sugar specificity will require further studies.

The *C. jejuni* aminotransferase, PglE

The corresponding enzyme in the Bac biosynthetic pathway, Cj1121c (PglE) has been characterized biochemically (Olivier *et al.*, 2006; Schoenhofen *et al.*, 2006a; Vijayakumar *et al.*, 2006). One brief structure report of PglE as one of many targets from the Joint Centre for Structural Genomics (JCSG) has been made (Badger *et al.*, 2005) and coordinates have been deposited with the Protein Data bank (PDB code **1O69**; with PLP–PDB code **1O61**; apo-form, PDB code **1O62**), classifying it as an aminotransferase. The UDP–sugar binding region within the active site adopts quite a different shape from that of PseC (Schoenhofen *et al.*, 2006c), although the exact nature of how the substrate binds has not been determined.

The acetyltransferases

While many crystal structures of bacterial acetyltransferases have been determined, relatively few of these are of enzymes involved in *N*- or *O*-acetylation of nucleotide-linked sugars. In the following we summarize structural and functional data for two such enzymes which transfer an acetyl group from acetyl-coenzyme A (AcCoA) to the 4-amino moiety of the sugar substrate. The first is from the Bac pathway, and it belongs to the LβH family of acyltransferases. The second is from the GNAT family and while its substrate, TDP-Fuc4N, differs considerably from the PseC product, this enzyme nevertheless is a good model for the Pse pathway *N*-acetyltransferase, PseH.

The *C. jejuni* acetyltransferase, PglD

Synthesis of UDP-*N,N'*-diacetylbacillosamine (GDP-BacdiNAc) from the corresponding UDP-4-amino sugar produced by PglE is carried out by the PglD *N*-acetyltransferase, Cj1123c (Olivier *et al.*, 2006). The identity of the resulting product generated by PglD was verified by NMR and ESI–MS analysis, proving that this enzyme catalyses acetyl transfer to the 4-amino position. PglD is an efficient acetyltransferase, showing a $k_{cat}/K_M = 1.18 \times 10^6$/min/mM (Olivier *et al.*, 2006). The PglD sequence contains an imperfect but readily detectable 'hexapeptide repeat' signature motif, defining each of the three β-strands for each rung of a β-helix (Choi *et al.*, 2008). Crystal structure determination of PglD independently by two laboratories confirmed the left-handed β-helix (LβH) fold for this enzyme (Olivier and Imperialli, 2008; Rangarajan *et al.*, 2008; Fig. 11.4A). This trimeric LβH fold is common to other sugar *N*- and *O*-acetyltransferases, as well as to various enzymes of microbial origin. However, PglD is unusual among LβH enzymes in that it possesses an N-terminal α/β domain and is classified as a distinct structural family within the SCOP database (Murzin *et al.*, 1995).

The first reported structures of *C. jejuni* PglD by Rangarajan *et al.* (2008) were for the apo-enzyme (PDB code **3BFP**) as well as the complex with coenzyme A (PDB code **2VHE**). Another crystal structure (PDB code **2NPO**) of *C. jejuni* PglD has been deposited but not otherwise described. Each PglD subunit consists of an N-terminal α/β domain (residues 2–72) and six turns of LβH (residues 75–187) (Fig. 11.4A). Structural comparison of the apo and CoA-bound forms of PglD showed that the C-terminal end of

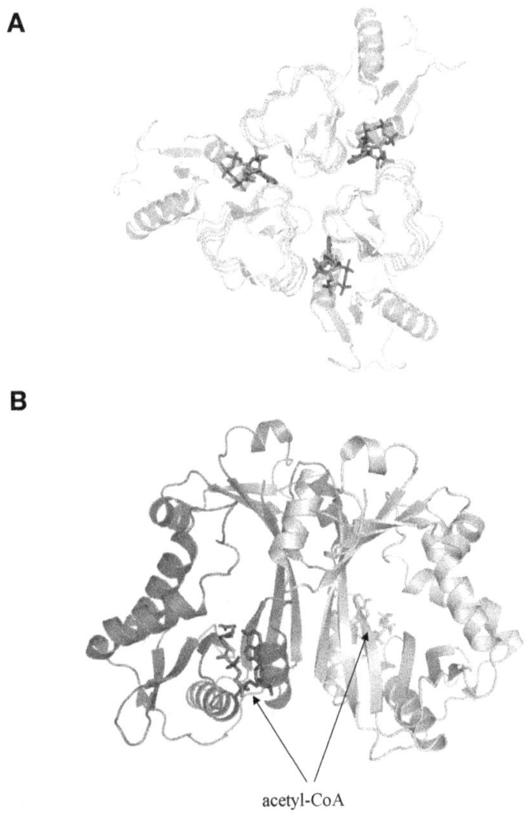

Figure 11.4 Crystal structures of two *N*-acetyltransferases representing two distinct protein folds that have been adopted in the biosynthesis of 4-acetamidohexoses. (A) Crystal structure of *C. jejuni* PglD bound to acetyl-CoA (**PDB 3BSY**), an *N*-acetyltransferase from the bacillosamine biosynthetic pathway that adopts the left-handed β-helix fold. (B) Crystal structure of *E. coli* WecD bound to acetyl-CoA (**PDB 2FT0**), an *N*-acetyltransferase that forms TDP-Fuc4NAc belonging to the GNAT-fold. The two subunits of the dimer are shaded light and dark. In both structures acetyl-CoA is depicted in stick representation.

the LβH, residues 189–195, adopt an 'extended' conformation in the apo-form, such that it crosses over and interacts with an adjacent subunit of the trimer, but folds to extend the β-helix when bound to CoA.

A sample of PglD that co-purified with CoA resulted in the co-crystal structure and was found to bind at the interface of a pair of LβH subunits, such that three CoA molecules bind to a PglD trimer. This result immediately suggested that formation of the PglD oligomer is critical for its function. The more polar phosphate-containing portions of CoA are seen to be solvent exposed, and the more apolar regions buried. The adenine ring forms several hydrophobic contacts with residues from both subunits, including Val-188, Met-195, Val-178′, Pro-191′ and Val-190′ (prime numbers indicate residues participating from another subunit). The mainly apolar pantethiene moiety forms contacts with Phe-152, Ile-155, Val-137, Ala-142′, Leu-160′ and Pro-161′. The only polar contacts with this portion of CoA are to the main-chain amides of Ile-155 and Gly-173. Efforts to co-crystallize PglD with the UDP–sugar substrate were unsuccessful, although its binding site could be inferred based on a shallow cavity located proximal to the S atom of CoA. Several residues were identified in this region that could participate in substrate binding or catalysis, including Asn-118, His-134, Glu-124′ and His-125′. To evaluate their potential contribution, each residue was mutated to alanine and the activity of PglD mutant enzymes measured. The largest effect on activity was observed for H125A, E124A and N118A mutant enzymes, and a lesser effect for H134A.

In the absence of an experimental structure, efforts were made to model PglD in complex with the tetrahedral intermediate, resulting from attack of the C-4 amino group of UDP-Bac4N on the carbonyl carbon of acetyl-CoA (Fig. 11.5B). Owing to the nature of the PglD molecular surface, the substrate was not predicted to bind in an extended conformation but rather adopt an 'L' shape with respect to CoA. The Bac4N sugar was also accommodated between two subunits like the CoA, but binding of the UDP fragment resided entirely in one monomer, specifically in the distinct N-terminal α/β domain. A key predicted interaction includes a H-bond between His-125′ and the C-4 amino group, suggestive of a role for His-125′ in activating the amino moiety by removing a proton. This would be consistent with the decrease in activity observed for the H125A mutant enzyme. The substrate 4-amino group is predicted to be stabilized via water–mediated interactions with His-15, Asn-118 and Glu-124′. The main chain amide of Gly-143′ is predicted to help stabilize the oxyanion. No specific residue could be identified that would protonate the S atom of CoA, thereby promoting breakdown of the intermediate and product release. Alternatively, the CoA-S^- product could be released from the active site owing to electrostatic repulsion between the S^- atom and Glu124.

Crystal structures of *C. jejuni* PglD have also been solved in the presence of citrate (PDB code **3BSW**), acetyl-CoA (PDB code **3BSY**) and the UDP-Bac4N substrate (PDB code **3BSS**) (Olivier and Imperialli, 2008). Importantly, the carbonyl O atom of the acetyl-CoA thioester forms H-bonds to Asn-118 and His-134. Trapping of crystals with bound UDP-Bac4N is critical, as it provides an experimental view of substrate binding. Interestingly, Met-40–Thr-45, corresponding to helix α2 within the α/β domain in the apo enzyme, unwinds to form a coiled structure in order to accommodate UDP–sugar binding entirely within the N-terminal α/β domain as predicted by computational docking. The important interaction of His125′ with the C-4 amino group of the substrate, as well a four orders of magnitude drop in k_{cat}/K_M with the H125A mutant enzyme, support its role as a general base activating the substrate. The protonated form of the 4-amino substrate is expected to have a pK_a of ~7.6, consistent with the need for a base to remove this proton. A superposition of the PglD crystal structures independently complexed with the UDP-Bac4N and acetyl-CoA substrates shows a suboptimal geometry for attack, with the nucleophilic C-4 amino group almost in plane with the acetyl thioester group, and at 3.6 Å from the electrophilic acetyl carbonyl C and 2.2 Å from the acetyl methyl C. This suggests some rearrangement of the reacting groups is needed prior to forming the tetrahedral intermediate adduct. Nevertheless, the overlay of these independent structures prompted a mechanism with roles for Asn-118 and His-134 in stabilizing the oxyanion intermediate, with His-125′ returned to the neutral state by donating a proton to Glu-124. This same residue is predicted to protonate the CoA thiolate anion (Olivier and Imperialli, 2008).

The *E. coli* acetyltransferase, WecD

The structure of WecD was determined both in apo form (PDB code **2FS5**) and bound to acetyl-CoA (PDB code **2FT0**) (Hung *et al.*, 2006). The enzyme forms dimers both in the crystal structure and in solution, with each monomer adopting the GNAT *N*-acetyltransferase fold (Fig. 11.4B). There are few characterized examples of nucleotide-sugar acetyltransferases that adopt the GNAT fold, one other being the GlcNAc6p *N*-acetyltransferase from *S. cerevisiae* which introduces NAc at C-2 of the sugar (Peneff *et al.*, 2001). Each subunit can be divided into two domains, an N-terminal domain (residues 3–69, 219–224) and a C-terminal domain (70–218), the latter domain consisting of a seven-stranded mixed β-sheet flanked by four α-helices constituting the GNAT fold.

Each monomer contains a tunnel, which is funnel-like on one face and narrow on the opposite face, and represents the acetyl-CoA binding site. The acetyl moiety binds nearest to the tunnel opening, with the remainder of the CoA extending away on the funnel-shaped side of the molecule, such that it is largely solvent-exposed.

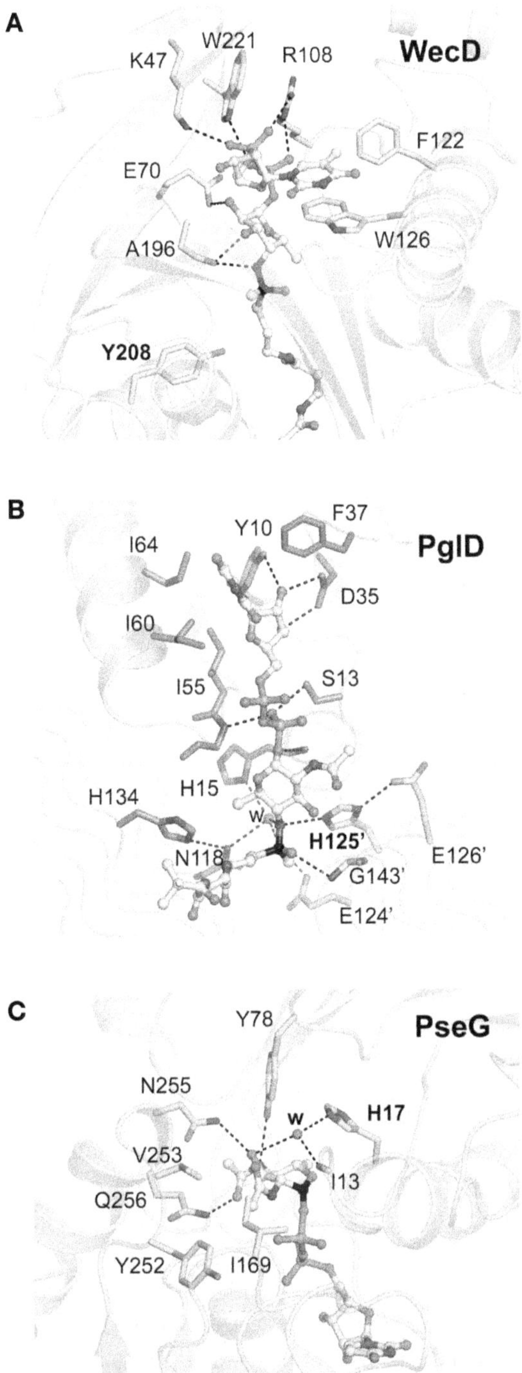

Figure 11.5 Molecular models of enzyme-bound substrates. (A) WecD acetyltransferase with bound TDP-Fuc4N-AcCoA tetrahedral intermediate. (B) PglD acetyltransferase with bound UDP-Bac4N-AcCoA tetrahedral intermediate. (C) PseG hydrolase with bound UDP-6-deoxy-AltdiNAc substrate. Proteins are rendered as translucent ribbons, with selected binding site residues shown as sticks and labelled. Bold labels indicate putative catalytic residues and water molecule. Substrates are shown in ball-and-stick representation. The C atoms coloured in black represent reactive centres. Hydrogen bonds are shown with dashed lines. For clarity, the CoA portion of the adduct is shown only partially in (A) and (B).

Importantly, Tyr-208 was identified within 3 Å of the acetyl-CoA S atom and may be involved in H-bonding. Most other polar contacts with CoA, with the exception of Asn-201, involve the main chain atoms Leu-168 and Gly-170. Apolar interactions with the CoA pantetheine moiety are mediated by Phe-104, Leu-168, Ala-169 and Ala-204. The CoA pyrophosphate is anchored by a salt bridge to Arg-207. There are few interactions between the adenine base and protein, as reflected by weaker density for this region of the cofactor.

No crystal structures of WecD with dTDP or TDP have so far been reported. In an attempt to gain insight into substrate binding, flexible molecular docking by Monte Carlo energy minimization (MCM) conformational sampling of the TDP-Fuc4N substrate as a tetrahedral intermediate adduct with acetyl-CoA was employed (Hung *et al.*, 2006). This model (Fig. 11.5A) positions the TDP-Fuc4N binding site just on the opposite side of the tunnel to the acetyl group of acetyl-CoA. The sugar is predicted to form polar and apolar contacts with Glu-70, Phe-104, Tyr-123 and Ala-196. The diphosphate is located proximal to Lys-47, Arg-108 and Trp-223, suitable for neutralizing the negative charge. The Phe-122, Trp-126 and Phe-135 side-chains potentially stack against the thymidine ring.

The Tyr-208 residue is probably important either in appropriately orienting the acetyl group for transfer, or donating a proton to the developing thiolate anion during the last step in catalysis, as implicated in structures and mutagenesis studies of related acetyltransferases. However, no residue in WecD could be identified that could function as a general base, abstracting a proton from the C-4 amino group of the substrate, with the closest candidate groups, from residues Glu68 and Glu70, being over 5 Å distant. A similar scenario has emerged in some, but not all, GNAT acetyltransferases, with a variety of possible mechanisms being proposed. Deprotonation of the C-4 amino group from the free charged state to generate the neutral nucleophile may be promoted by the hydrophobic environment as the substrate enters the binding site tunnel (Hung *et al.*, 2006). One of the most interesting possibilities for further deprotonation of this amino group from the tetrahedral intermediate, put forth by Holden and co-workers for the *N*-acetyltransferases QdtC (Thoden *et al.*, 2009) and WlbB (Thoden and Holden, 2010), enzymes with the LβH fold, is that the enzyme does not require a protein side-chain to act as a catalytic base, but instead the S atom of CoA could function to accept the proton. The extent to which this hypothesis will hold up to further experimental scrutiny for acetyltransferases having both the LβH and GNAT folds remains to be established.

Extension of TDHs to form nonulosonic acids

Nonulosonic acids are made from NDP-TDH species in two steps. In the first step, the nucleotide is removed to generate the free sugar, and this can occur with retention or inversion at C-2. For example, in the neuraminic acid pathway the best known Nulo, neuraminic acid, the UDP-GlcNAc substrate is transformed into ManNAc by the enzyme NeuC, a hydrolysing 2-epimerase. The HexNAc is then extended by addition of a three-carbon fragment from phosphoenolpyruvate (NeuB or homologue), at its reducing end. In order to form a substrate for the relevant glycosyltransferases, the nonulosonic acid is first activated by attachment of CMP from CTP (NeuA or homologue). The Leg/Pse pathway enzymes were named by analogy to the Neu ones before their true substrates were characterized, hence they were termed NeuC, NeuB and NeuA. Structural studies of the Pse and Leg enzymes have begun with the structure of the PseG enzyme; we also describe briefly here the related enzymes from the Neu pathway.

The *C. jejuni* UDP–sugar hydrolase, PseG

Metabolomics analysis by CE-ESMS (capillary electrophoresis-electrospray mass spectrometry) and NMR analysis of sugars from *C. jejuni* 81-176 sequential *pse* gene knockout strains implicated PseG as a putative UDP–sugar hydrolase (McNally *et al.*, 2006). Hydrolysis of the nucleotide-sugar occurs after acetylation at the C-4 amino group, catalysed by PseH, and prior to condensation of 2,4-diacetamido-2,4,6-trideoxy-L-altrose with PEP to form pseudaminic acid, catalysed by PseI.

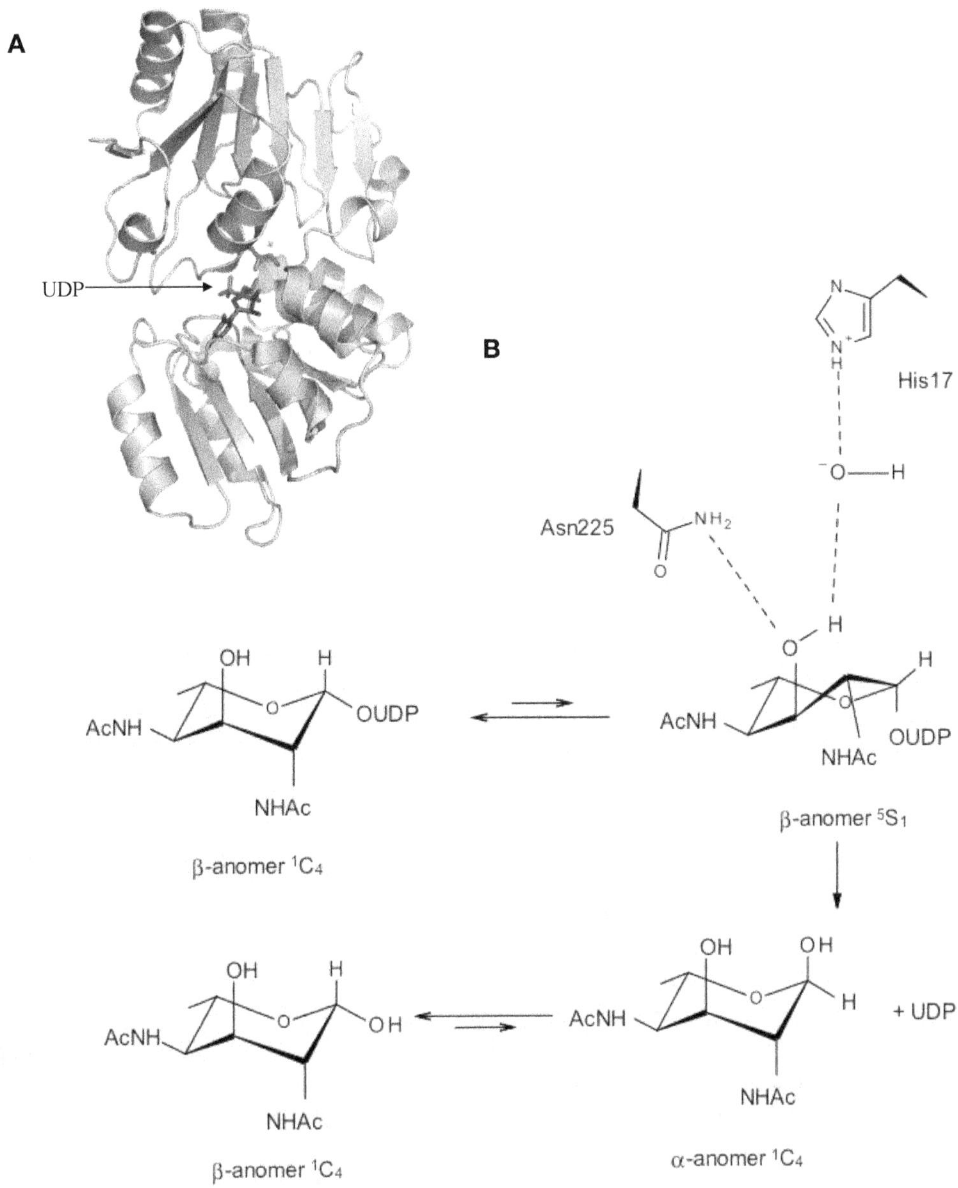

Figure 11.6 Crystal structure of *C. jejuni* PseG bound to UDP (**PDB 3HBN**), the nucleotide-sugar hydrolase of pseudaminic acid biosynthesis. (A) PseG adopts the GT-B fold and binds its substrate in a similar location between two α/β domains, as does the glycosyltransferase MurG. (B) Proposed catalytic mechanism for PseG (Rangarajan *et al.*, 2009). The key catalytic residue, His17, functions as a general base abstracting a proton from a water molecule and once activated attacks the C-1 atom of the sugar, resulting in nucleotide hydolysis. Modelling data implicate the sugar adopting a twist-boat conformation in forming the Michaelis complex with the enzyme.

Definitive assignment of *C. jejuni* PseG as the UDP–sugar hydrolase and its detailed biochemical characterization were carried out by Liu and Tanner (2006). The hydrolase activity of PseG was found to be metal-ion independent, to proceeds with inversion of configuration at C-1 and to use a

C–O bond cleavage mechanism (Liu and Tanner, 2006). Sequence comparisons showed that PseG is most related to MurG, an inverting metal-independent UDP–sugar glycosyltransferase involved in peptidoglycan synthesis that belongs to the glycosyltransferase B (GT-B) family (Hu *et al.*, 2003). Of three possible mechanistic scenarios envisaged by Liu and Tanner (2006), the available experimental evidence supported one in which a water molecule would attack the anomeric C-1 atom, displacing UDP and resulting in inversion of configuration at C-1.

To gain further insight into the structure and function of PseG and its possible relationship to MurG, its crystal structure was determined (Rangarajan *et al.*, 2009) in apo-form (PDB code **3HBM**) and bound to its UDP product (PDB code **3HBN**). PseG appears to be a monomer both in solution and in the crystal structure, and comprises two structurally related α/β domains, with a parallel β-sheet forming the core of each domain (Fig. 11.6A). The crystal structure of PseG verified its relatedness to MurG, with both enzymes adopting the GT-B fold. The domains are connected by a short α-helix, with the domains moving towards one another and partially closing upon binding UDP. A molecule of the UDP product binds at the domain interface. One face of the UDP-binding site is formed by the sequence motif DX_5GXGHX_2R making contacts with the β-phosphate. This sequence motif can be potentially used to identify functional homologs of PseG in other bacteria. Residues involved in UDP-binding, located mainly in the C-terminal domain, include Arg-143 and Glu-239 (to the 2′ and 3′ ribose hydroxyls), primarily main-chain amide interactions with the phosphoryl groups, with Phe-15, Cys-163, Ala-189, Ile-219 and Leu-236 forming a hydrophobic pocket for binding the uracil base.

As the crystal structure of the UDP–sugar substrate bound to either wild-type or mutant PseG enzymes could not be obtained, efforts were directed towards protein-ligand docking and molecular dynamics in order to model the enzyme–substrate complex. A glycerol molecule, arising from the cryo-protecting solution used to flash-cool crystals prior to X-ray data collection, was observed adjacent to the UDP-binding site of PseG, binding in a similar location as UDP-GlcNAc to MurG (Hu *et al.*, 2003). Different starting UDP–sugar conformations, corresponding to the $^1C_4$, $^4C_1$ chairs, or the $^5S_1$ twist-boat were subject to MCM conformational sampling for docking to PseG. In all cases the lowest-energy docking solution corresponded to the $^5S_1$ twist-boat conformation (Rangarajan *et al.*, 2009). Further hybrid QM/MM unconstrained molecular dynamics simulations are consistent with the $^5S_1$ twist-boat most favourably representing the sugar conformation in the PseG Michaelis complex (Fig. 11.5C). A structured water molecule (Wat110) is appropriately positioned (3.5 Å) for in-line attack at C-1, with inversion of configuration, as compared with $^1C_4$ or $^4C_1$ conformers which display less favourable geometry with respect to the attacking water. This $^5S_1$ twist-boat may also more closely resemble the PseG transition state, thereby reducing the activation energy barrier in catalysis.

Based on the modelling results, predicted interactions between PseG and sugar include Gly-16, His-17, Tyr-78, Ile-169, Ser-234, Tyr-252, Val-253, Asn-255, Gln-256 and the Arg-20–Asp-101 salt bridge. His-17 is a key residue, implicated as the general base that activates the water molecule for nucleophilic attack at C-1. Both of the H17F and H17L mutant enzymes showed greatly diminished activity, while apparently not significantly affecting the PseG structure, arguing for an important role in catalysis. The corresponding H17N mutant enzyme retained ~10% activity, suggesting that an Asn residue can partially fulfil the role of polarizing and activating Wat110. The N255A mutant enzyme showed ~1% enzyme activity, while Y78F was least affected. The predicted role of Asn255 is to stabilize the $^5S_1$ twist-boat sugar conformation, via interaction its 3-OH, and is consistent with the large decrease in activity observed for the N255A mutant enzyme. A postulated mechanism for PseG is presented in Fig. 11.6B.

PEP synthase/CMP-synthetase (NeuB/NeuA)

No crystal structures have so far been determined for either the PEP-condensing synthase or the nucleotide-activating CMP synthetase for either

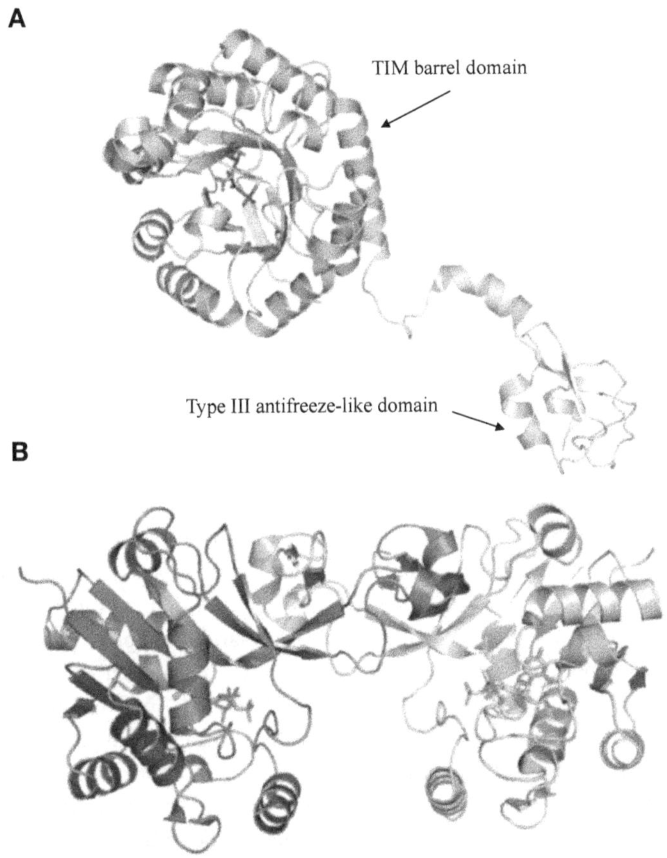

Figure 11.7 (A) Crystal structures of *N. meningitidis* NeuB sialic acid synthase bound to PEP, Mn^{2+} and ManNAc (**PDB 1XUZ**). Only one subunit of the dimer is displayed for clarity. (B) Structure of the NeuA synthetase bound to CDP (**PDB 1EYR**).

the Leg or Pse pathways. However, the corresponding, sequence-related enzymes have been structurally determined for the sialic acid biosynthetic pathway in *N. meningitidis*, corresponding to NeuB (Gunawan *et al.*, 2005) and NeuA (Moisimann *et al.*, 2001), respectively.

Crystal structures of NeuB bound to malate (PDB code **1XUU**) as well as to PEP and reduced ManNAc, having a hydroxyl moiety rather than aldehyde at C-1 (PDB code **1XUZ**) have been reported (Gunawan *et al.*, 2005). In the co-crystal structure the reduced ManNAc adopts an open chain form as opposed to a closed ring. The NeuA structure is a functional homodimer, with each subunit consisting of a N-terminal TIM barrel domain containing the active site, and a C-terminal domain possessing a structure similar to that of type III antifreeze proteins. The dimer adopts a domain-swapped arrangement, such that the antifreeze domain of one subunit interacts with the TIM barrel domain of the other subunit (Fig. 11.7A). This domain swapping is of functional importance, as a conserved residue in this domain, Arg314, participates in binding the ManNAc substrate. A Mn^{2+} ion was found coordinated to both His-215 and His-236, as well as to the O2P atom of PEP and O1 of ManANAc, implicating this metal in activating the carbonyl of the sugar substrate for catalysis. The expected catalytic mechanism is predicted to involve attack of C-3 of PEP at the C-1 carbonyl carbon of ManNAc, with formation of a tetrahedral intermediate during the catalytic cycle. The expected structure of the tetrahedral intermediate was subsequently used to synthesize a potent inhibitor of this enzyme, and the NeuB co-crystal structure was determined at 1.75 Å resolution (Liu *et al.*, 2009). The inhibitor showed a K_i(app) of 3.1 μM. This study provided further insight into the reaction catalysed by NeuB by substantiating the role of Mn^{2+} in electrophilic catalysis, as well as indicating that a metal-coordinated water molecule acts on the oxocarbonium intermediate during catalysis.

The crystal structures of NeuA in apo-form (PDB code **1EYR**), as well as bound to cytosine-5′-diphosphate, CDP (PDB code **1EZI**) have been reported (Mosimann *et al.*, 2001). Like NeuB, NeuA is a dimeric enzyme, although the fold of each subunit is completely different from that of NeuB. A ~190-residue globular N-terminal α/β domain is connected to an approximately 35 residue domain whose function appears to be in mediating dimer formation (Fig. 11.7B). Comparing the apo and CDP-bound forms of NeuA reveals ordering of the P-loop (residues 10–22), which plays a key role in catalysis. As would be expected, a putative Mg^{2+} binding site, involving Asp-209, Asp-211 and substrate phosphoryl groups, as well as an oxyanion hole, formed in part by the main chain amides of Ser-15 and Gly-17, suitable for interacting with the CTP-γ-phosphate, were observed in the structure. A pocket suitable for binding NeuAc was identified adjacent to the nucleotide-binding site and close to the dimerization domain. Modelling of NeuAc bound to this pocket identified a number of potential substrate-binding residues, although no specific residue acting as a general base to initiate nucleophilic attack of the NeuAc substrate at the CTP γ-phosphate could be identified.

Concluding remarks

Much has been learned in the past 10 years on the occurrence and biosynthesis of 4-acetamidohexose sugars, and from future bacterial genome sequencing it is anticipated that new family members will continue to be identified. There are obvious parallels in the chemistries utilized in the synthesis of Pse, Bac and Leg, with combinations of dehyratases to generate a C-4 keto group, conversion of the C-4 keto to a C-4 amino, and acetyl-CoA-dependent acetyl transfer to the C-4 amino by an *N*-acetyltransferase, yielding a 4-acetamidohexose, the theme of this chapter. While important advances have been made in structure–function relationships of the corresponding biosynthetic enzymes, much remains to be learned. We are in our infancy in terms of understanding how these pathways and enzymes evolved. The completely different structures and sequences of the respective *N*-acetyltransferases from the Pse and Bac pathways is a case in point, suggestive of completely independent origins for these sugar biosynthetic pathways. In addition, the subtleties in protein structure associated with substrate recognition and catalysis for sugars having similar structures remains an open question that is only slowly being unravelled. In the future, we can look forward to utilization of both the sugars and their respective biosynthetic enzymes as tools for the preparation of novel glycoconjugates with a variety of potential applications. For example, since recent evidence suggests that human sialidases will ineffectively hydrolyse Leg from glycoconjugates containing this sugar (Watson *et al.*, 2011), one particular utility may be to increase the serum half-life of current therapeutic glycoproteins and antibodies via incorporation of this novel sugar.

Acknowledgements

We wish to thank many of our colleagues who have contributed invaluably to the studies described in this review. X-ray data collection for the structural studies described in this review was performed at beamlines X8C, X12B, X25 and X29 of the National Synchrotron Light Source, Brookhaven National Laboratory. Financial support for this research has been provided by the National Research Council Canada, the NRC Genomics and Health Initiative (GHI) and the Canadian Institutes for Health Research (CIHR). This is NRCC publication No. 53143.

References

Allard, S.T.M., Beis, K., Giraud, M.-F., Hegeman, A.D., Whitfield, C., Graninger, M., Messner, P., Allen, A.G., Maskell, D.J., and Naismith, J.H. (2002). Toward a structural understanding of the dehydratase mechanism. Structure *10*, 81–92.

Badger, J., Sauder, J.M., Adams, J.M., Antonysamy, S., Bain, K., Bergseid, M.G., Buchanan, S.G., Buchanan, M.D., Batiyenko, Y., Christopher, J.A., *et al.* (2005).

Structural analysis of a set of proteins resulting from a bacterial genomics project. Proteins *60,* 787–796.

Burgie, E.S., and Holden, H.M. (2007). Molecular architecture of DesI: A key enzyme in the biosynthesis of desosamine. Biochemistry *46,* 8999–9006.

Chen, X., and Varki, A. (2010). Advances in the biology and chemistry of sialic acids. ACS Chem. Biol. *5,* 163–176.

Choi, J.H., Govaerts, C., May, B.C.H., and Cohen, F.E. (2008). Analysis of the sequence and structural features of the left-handed β-helical fold. Proteins *73,* 150–160.

Cook, P.D., and Holden, H.M. (2008). GDP-perosamine synthase: structural analysis and production of a novel trideoxysugar. Biochemistry *47,* 2833–2840.

Cook, P.D., Carney, A.E., and Holden, H.M. (2008). Accomodation of GDP-linked sugars in the active site of GDP-perosamine synthase. Biochemistry *47,* 10685–10693.

Creuzenet, C. (2004). Characterization of Cj*1293,* a new UDP-GlcNAc C6 dehydratase from *Campylobacter jejuni.* FEBS Lett. *559,* 136–140.

Creuzenet, C., Schur, M.J., Li, J., Wakarchuk, W.W., and Lam, J.S. (2000). FlaA1, a new bifunctional UDP-GlcNAc C_6 dehydratase/C_4 reductase from *Helicobacter pylori.* J. Biol. Chem. *275,* 34873–34880.

Favia, A.D., Nobeli, I., Glaser, F., and Thronton, J.M. (2008). Molecular docking for substrate identification: The short-chain dehydrogenases/reductases. J. Mol. Biol. *375,* 855–874.

Field, R.A., and Naismith, J.H. (2003). Structural and mechanistic basis of bacterial sugar nucleotide-modifying enzymes. Biochemistry *42,* 7637–7647.

Gunawan, J., Simard, D., Gilbert, M., Lovering, A.L., Wakarchuk, W.W., Tanner, M.E., and Strynadka, N.C.J. (2005). Structural and mechanistic analysis of sialic acid synthase NeuB from *Neisseria meningitidis* in complex with Mn^{2+}, phosphoenolpyruvate, and *N*-acetylmannosaminitol. J. Biol. Chem. *280,* 3555–3563.

Holden, H.M., Cook, P.D., and Thoden, J.B. (2010). Biosynthetic enzymes of unusual microbial sugars. Curr. Opin. Struct. Biol. *20,* 1–8.

Hu, Y., Chen, L., Ha, S., Gross, B., Falcone, B., Walker, D., Mokhtarzadeh, M., and Walker, S. (2003). Crystal structure of the MurG:UDP–GlcNAc complex reveals common structural principles of a superfamily of glycosyltransferases. Proc. Natl. Acad. Sci. U.S.A. *100,* 845–849.

Hung, M.-N., Rangarajan, E., Munger, C., Nadeau, G., Sulea, T., and Matte, A. (2006). Crystal structure of TDP-fucosamine acetyltransferase (WecD) from *Escherichia coli,* an enzyme required for enterobacterial common antigen synthesis. J. Bacteriol. *188,* 5606–5617.

Ishiyama, N., Creuzenet, C., Miller, W.L., Demendi, M., Anderson, E.M., Harauz, G., Lam, J.S., and Berghuis, A.M. (2006). Structural studies of FlaA1 from *Helicobacter pylori* reveal the mechanism for inverting 4,6-dehydratase activity. J. Biol. Chem. *281,* 24489–24495.

Kallberg, Y., Oppermann, U., and Persson, B. (2010). Classification of the short-chain dehydrogenase/reductase superfamily using hidden Markov models. FEBS J. *277,* 2375–2386.

Knirel, Y.A., Vinogradov, E.V., L'vov, V.L., Kocharova, N.A., Shashkov, A.S., Dmitriev, B.A., and Kochetkov, N.K. (1984). Sialic acids of a new type from the lipopolysaccharide of *Pseudomonas aeruginosa* and *Shigella boydii.* Carbohydr. Res. *133,* C5–C8.

Liu, F., and Tanner, M.E. (2006). PseG of pseudaminic acid biosynthesis. A UDP–sugar hydrolase as a masked glycosyltransferase. J. Biol. Chem. *281,* 20902–20909.

Liu, F., Lee, H.J., Strynadka, N., and Tanner, M.E. (2009). Inhibition of *Neisseria meningitides* sialic acid synthase by a tetrahedral intermediate analogue. Biochemistry *48,* 9194–9201.

Logan, S.M. (2006). Flagellar glycosylation – a new component of the motility repertoire? Microbiology *152,* 1249–1262.

McNally, D.J., Hui, J.P.M., Aubry, A.J., Mui, K.K.K., Guerry, P., Brisson, J.-R., Logan, S.M., and Soo, E.C. (2006). Functional characterization of the flagellar glycosylation locus in *Campylobacter jejuni* 81-176 using a focused metabolomics approach. J. Biol. Chem. *281,* 18489–18498.

McNally, D.J., Schoenhofen, I.C., Houliston, R.S., Khieu, N.H., Whitfield, D.M., Logan, S.M., Jarrell, H.C., and Brisson, J.R. (2008). CMP-pseudaminic acid is a natural potent inhibitor of PseB, the first enzyme of the pseudaminic acid pathway in *Campylobacter jejuni* and *Helicobacter pylori.* ChemMedChem 3, 55–59.

Morrison, J.P., Schoenhofen, I.C., and Tanner, M.E. (2008). Mechanistic studies on PseB of pseudaminic acid biosynthesis: A UDP-acetylglucosamine 5-inverting 4,6-dehydratase. Bioorg. Chem. *36,* 312–320.

Mosimann, S.C., Gilbert, M., Dombroswki, D., To, R., Wakarchuk, W., and Strynadka, N.C. (2001). Structure of a sialic acid-activating synthetase, CMP-acylneuraminate synthetase in the presence and absence of CDP. J. Biol. Chem. *276,* 8190–8196.

Murzin, A.G., Brenner, S.E., Hubbart, T., and Chothia, C. (1995). SCOP: a structural classification of proteins database for the investigation of sequences and structures. J. Mol. Biol. *247,* 536–540.

Noland, B.W., Newman, J.M., Hendle, J., Badger, J., Christopher, J.A., Tresser, J., Buchanan, M.D., Wright, T.A., Rutter, M.E., Sanderson, W.E., *et al.* (2002). Structural studies of *Salmonella typhimurium* ArnB (PmrH) aminotrasnferase: A 4-amino-4-deoxy-L-arabinose lipopolysaccharide-modifying enzyme. Structure *10,* 1569–1580.

Obhi, R.K., and Creuzenet, C. (2005). Biochemical characterization of the *Campylobacter jejuni* Cj1294, a novel UDP-4-keto-6-deoxy-GlcNAc aminotransferase that generates UDP-4-amino-4,6-dideoxy-GalNAc. J. Biol. Chem. *280,* 20902–20908.

Olivier, N.B., and Imperialli, B. (2008). Crystal structure and catalytic mechanism of PglD from *Campylobacter jejuni.* J. Biol. Chem. *283,* 27937–27946.

Olivier, N.B., Chen, M.M., Behr, J.R., and Imperiali, B. (2006). *In vitro* biosynthesis of

UDP-*N,N'*-diacetylbacillosamine by enzymes of the *Campylobacter jejuni* general protein glycosylation system. Biochemistry *45,* 13659–13669.

Peneff, C., Mengin-Lecreulx, D., and Bourne, Y. (2001). The crystal structures of apo and complexed *Saccharomyces cerevisiae* GNA1 shed light on the catalytic mechanism of an amino-sugar *N*-acetyltransferase. J. Biol. Chem. *276,* 16328–16334.

Rangarajan, E.S., Ruane, K.M., Sulea, T., Watson, D.C., Proteau, A., Leclerc, S., Cygler, M., Matte, A., and Young, N.M. (2008). Structure and active site residues of PglD, an *N*-acetyltransferase from the bacillosamine synthetic pathway required for *N*-glycan synthesis in *Campylobacter jejuni*. Biochemistry *47,* 1827–1836.

Rangarajan, E.S., Proteau, A., Cui, Q., Logan, S.M., Potetinova, Z., Whitfield, D., Purisima, E.O., Cygler, M., Matte, A., Sulea, T., *et al.* (2009). Structural and functional analysis of *Campylobacter jejuni* PseG. A UDP–sugar hydrolase from the pseudaminic acid biosynthetic pathway. J. Biol. Chem. *284,* 20989–21000.

Schoenhofen, I.C., McNally, D.J., Vinogradov, E., Whitfield, D., Young, N.M., Dick, S., Wakarchuk, W.W., Brisson, J.-R., and Logan, S.M. (2006a). Functional characterization of dehydratase/aminotrasferase pairs from *Helicobacter* and *Campylobacter*. Enzymes distinguishig the pseudaminic acid and bacillosamine biosynthetic pathways. J. Biol. Chem. *281,* 723–732.

Schoenhofen, I.C., McNally, D.J., Brisson, J.R., and Logan, S.M. (2006b). Elucidation of the CMP-pseudaminic acid pathway in *Helicobacter pylori*: synthesis from UDP-N-acetylglucosamine by a single enzymatic reaction. Glycobiology *16,* 8C-14C.

Schoenhofen, I.C., Lunin, V.V., Julien, J.-P., Li, Y., Ajamian, E., Matte, A., Cygler, M., Brisson, J.-R., Aubry, A., Logan, S.M., Bhatia, S., Wakarchuk, W.W., and Young, N.M. (2006c). Structural and functional characterization of PseC, an aminotransferase involved in the biosynthesis of pseudaminic acid, an essential flagellar modification in *Helicobacter pylori*. J. Biol. Chem. *281,* 8907–8916.

Schoenhofen, I.C., Vinogradov, E., Whitfield, D.M., Brisson, J.-R., and Logan, S.M. (2009). The CMP-legionaminic acid pathway in *Campylobacter*. Biosynthesis involving novel GDP-linked precursors. Glycobiology *19,* 715–725.

Thoden, J.B., and Holden, H.M. (2010). Molecular structure of WlbB, a bacterial N-acetyltransferase involved in the biosynthesis of 2,3-diacetamido-2,3-dideoxy-D-mannuronic acid. Biochemistry *49,* 4644–4653.

Thoden, J.B., Cook, P.D., Schäffer, C., Messner, P., and Holden, H.M. (2009). Structural and functional studies of QtdC: an N-acetyltransferase required for the biosynthesis of dTDP-3-acetamido-3,6-dideoxy-alpha-D-glucose. Biochemistry *48,* 2699–2709.

Vijayakumar, S., Merkx-Jacques, A., Ratnayake, D.B., Gryski, I., Obhi, R.K., Houle, S., Dozois, C.M., and Creuzenet, C. (2006). Cj1121c, a novel UDP-4-keto-6-deoxy-GlcNAc aminotransferase essential for protein glycosylation and virulence in *Campylobacter jejuni*. J. Biol. Chem. *281,* 27733–27743.

Young, N.M., Brisson, J.-R., Kelly, J., Watson, D.C., Tessier, L., Lanthier, P.H., Jarrell, H.C., Cadotte, N., St Michael, F., Aberg, E., *et al.* (2002). Structure of the N-linked glycan present on multiple glycoproteins in the gram-negative bacterium *Campylobacter jejuni*. J. Biol. Chem. *277,* 42530–42539.

Watson, D.C., Leclerc, S., Wakarchuk, W.W., and Young, N.M. (2011). Enzymatic synthesis and properties of glycoconjugates with legionaminic acid as a replacement for neuraminic acid. Glycobiology *21,* 99–108.

Zähringer, U., Knirel, Y.A., Lindner, B., Helbig, J.H., Sonesson, A., Marre, R., and Rietschel, E.T. (1995). The lipopolysaccharide of *Legionella pneumophila* serogroup 1 (strain Philadelphia 1): chemical structure and biological significance. Prog. Clin. Biol. Res. *392,* 113–139.

Application of Bacterial Glycosyltransferases in the Synthesis of Bioactive Glycans

12

Warren Wakarchuk

Abstract

It is now accepted that complex glycans play major roles in biology, such as the development of the embryo, the function of the immune system, microbial and viral pathogenesis and cellular communication, to name just a few. The many faceted roles that glycans play in biology makes them a challenge to understand on functional level, and the complexity of the structures themselves makes them daunting targets for chemical synthesis, which is required for examination of their binding interactions and for future development of carbohydrate based therapeutics. In order to facilitate the synthesis of complex glycans, we have been examining glycosyltransferases which make strategic linkages in biologically active glycans. Many of the mammalian enzymes have not been as easy to express as active recombinant proteins, and many have a more restricted acceptor specificity that limits their use for synthesis. Our focus has been on the use of bacterial enzymes from pathogens which make molecular mimics of host glycans, and which have been shown to be potent catalysts for carbohydrate synthesis. This chapter will provide a review on a variety of bacterial enzymes that we and others have enabled for *in vitro* synthetic carbohydrate chemistry, as well as some promising *in vivo* production strategies for bioactive carbohydrates.

Introduction

Cell surface carbohydrates are key molecules in molecular recognition events between cells. It is well established that carbohydrate-mediated molecular recognition is used extensively in biology for signal transduction in the immune system (Crocker *et al.*, 2007; Schauer, 2009; Sperandio *et al.*, 2009), during development (Haltiwanger and Lowe, 2004) as well as many other normal biological interactions. It is not surprising then that human pathogens have developed the ability to exploit these molecular recognition events to evade detection by the host defences, and to use host receptors to dampen the immune response (Carlin *et al.*, 2009) or to aid in colonization (Harvey *et al.*, 2000). One successful strategy of human bacterial pathogens is to mimic host cell surface glycan structures (molecular mimicry) in their capsular polysaccharide (CPS) or lipopolysaccharide (LPS). The structures being mimicked include the carbohydrate head groups of common glycolipids, e.g. lacto-*N*-neotetraose (Pavliak *et al.*, 1993); glycosaminoglycans (hyaluronic acid and unsulfated chondroitin; DeAngelis, 2002); Lewis blood group antigens (Monteiro *et al.*, 2000); and polysialic acid (Bhattacharjee *et al.*, 1975, 1976). The carbohydrate structural diversity in several pathogens with molecular mimics has been studied in detail in *Campylobacter jejuni* (Gilbert, 2008), *Helicobacter pylori* (Monteiro, 2001), *Neisseria meningitidis* (Zhu et *al.*, 2002), *N. gonorrhoeae* (Tong et *al.*, 2002), *Haemophilus influenzae* (Schweda et *al.*, 2007), various *Streptococcus* species both Group A (Maclennan, 1956; Stollerman and Dale, 2008) and Group B (Cieslewicz *et al.*, 2005), some isolates of *Escherichia coli* (Barry and Goebel, 1957; Vann *et al.*, 1981; Lidholt and Fjelstad, 1997; Steenbergen and Vimr, 2003), and certain species of *Pasteurella* (Adlam, 1987;

DeAngelis, 2002). These structures are detailed in Table 12.1.

Recently the crystallographic structures of the enzymes involved in the synthesis of bacterial molecular mimic structures were the subject of a review (Wakarchuk, 2011).

In order to exploit these bacterial glycosyltransferases as catalysts for synthesis of bioactive carbohydrates, it is important to understand structure–function details about these enzymes and how they differ from their mammalian counterparts. An excellent and very detailed review of glycosyltransferase structure and function was recently published by Lairson *et al.* (2008) so the comments here about basic enzymology of glycosyltransferases will be limited since the aforementioned review is very comprehensive. Carbohydrate active enzymes have been classified into families in the sequence database for carbohydrate active enzymes,CAZy (Cantarel *et al.*, 2009). At present this database contains 92 families of glycosyltransferases, and 118 families of glycoside hydrolases obtained from the analysis of over 1182 bacterial genomes (as well as ~200 other organisms). Many of the synthetically useful bacterial glycosyltransferase enzymes are found in enzyme families having only bacterial members (e.g. CAZy GT-42 and GT-52), while a few are found in very large families which have representation from all branches of the tree of life (e.g. CAZy GT-2 and GT-4). Glycosyltransferases catalyse the formation of glycosidic bonds from activated sugar donors and a variety of acceptor substrates. In bacteria, the activated sugar donors can be nucleoside sugar monophosphates (e.g. CMP-Neu5Ac), nucleoside sugar diphosphates (e.g. UDP-Gal, GDP-Man etc.), lipid sugar phosphates (e.g. bactoprenol phosphate mannose), lipid sugar pyrophosphates, (e.g. bactoprenol pyrophosphate oligosaccharides), or unsubstituted sugar phosphates (e.g. mannose-1-phosphate). All of the bacterial enzymes that have been employed for enzymatic synthesis discussed here use sugar-nucleotides as donor substrates, although it has been shown that some bacterial enzymes can use alternative synthetic donors (Lougheed *et al.*, 1999; Lairson *et al.*, 2007).

The reaction stereochemistry helps to further define the families of glycosyltransferases, and as in the case of the glycoside hydrolases, there are two possible stereochemical outcomes based on the donor sugar. These are respectively retention or inversion of the anomeric configuration of the transferred sugar. The schematic in Fig. 12.1 diagrams these two possible mechanisms. The inverting glycosyltransferases use a direct displacement S_N2-like mechanism involving a catalytic base. The base activates a nucleophile on the acceptor, by removing a proton and then the acceptor oxygen attacks the anomeric carbon of the donor substrate to form the glycosidic bond. The mechanism for the retaining glycosyltransferases was thought to be similar to the two-step mechanism originally postulated for retaining glycosidases by Koshland (1953), in which a covalent intermediate is first formed and then is hydrolysed to release the product with net retention of configuration. However, the retaining glycosyltransferase mechanism is still the subject of some debate in the literature, principally because no convincing evidence for a covalent intermediate has yet been documented. In fact, protein structures of retaining GT enzymes show that the active sites lack a suitable nucleophile of the type found in retaining glycosidases. The currently proposed mechanism for retaining glycosyltransferases postulates a short lived ion-pair, which could collapse to regenerate the starting material or could undergo front-side attack by an incoming nucleophile, leading to a product with retained stereochemistry. For a detailed discussion of glycosyltransferase mechanisms, the reader is referred to the review by Lairson *et al.* (2008).

This chapter will focus on the application of bacterial glycosyltransferases for the synthesis of biologically active glycoconjugates. It will show how effective the exploitation of these enzymes has been in area of chemoenzymatic synthesis of a wide range of molecules with many applications.

Synthesis of lacto-*N*-neotetraose and related structures

Human milk contains a large variety of reducing oligosaccharides which are considered to be important pre-biotics for the infant gut microflora, and as specific inhibitors of binding of pathogens

Table 12.1 Bacterial mimics of mammalian glycans

Name of host glycan	Glycan composition	Bacterial source	References
Lacto-*N*-neotetraose	{Neu5Ac-α-2,3}-Gal-β-1,4-GlcNAc-β-1,3-Gal-β-1,4-Glc-R	*Neisseria meningitidis, N. gonorrhoeae, Streptococcus pneumoniae, Group B Streptococci, Histophilus somnus*	Pavliak *et al.* (1993), Cieslewicz *et al.* (2005), St. Michael *et al.* (2005a)
Lacto-*N*-tetraose	{Neu5Ac-α–2,3}-Gal-β-1,3-GlcNAc-β-1,3-Gal-β-1,4-Glc-R	*Histophilus somnus*	St. Michael *et al.* (2005a)
Globotriose/Gb3	{Neu5Ac-α-2,6}-Gal-α-1,4-Gal-β-1,4-Glc-R	*Neisseria meningitidis, N. gonorrhoeae, Haemophilus influenzae*	Wakarchuk *et al.* (1998b), Schweda *et al.* (2007)
Globotetraose/Gb4	GalNAc-β-1,3-Gal-α-1,4-Gal-β-1,4-Glc-R	*Haemophilus influenzae, Campylobacter jejuni*	Schweda *et al.* (2007)
Forssman antigen	GalNAc-α-1,3-GalNAc-β-1,3-Gal-α-1,4-Gal-β-1,4-Glc-R	*Pasteurella multocida*	St. Michael *et al.* (2005b)
3′-Sialyllactose/GM3	Neu5Ac-α-2,3-Gal-β-1,4-Glc-R	*Haemophilus influenzae, Campylobacter jejuni*	Schweda *et al.* (2007), Gilbert (2008)
GM2	GalNAc-β-1,4-[Neu5Ac-α-2,3-]Gal-β-1,4-Glc-R	*Campylobacter jejuni*	Gilbert (2008)
GM1a	Gal-β-1,3-GalNAc-β-1,4-[Neu5Ac-α-2,3-]Gal-β-1,3-Hep-R	*Campylobacter jejuni*	Gilbert (2008)
GD1a	Neu5Ac-α-2,8-Neu5Ac-α-2,3-Gal-β-1,3-GalNAc-β-1,4-Gal-β-1,3-Hep-R	*Campylobacter jejuni*	Gilbert (2008)
GD3	Neu5Ac-α-2,8-Neu5Ac-α-2,3-Gal-β-1,4-Glc-R	*Campylobacter jejuni*	Gilbert (2008)
GT1a	Neu5Ac-α-2,8-Neu5Ac-α-2,3-Gal-β-1,3-GalNAc-β-1,4-[Neu5Ac-α-2,3-]Gal-β-1,3-Hep-R	*Campylobacter jejuni*	Gilbert (2008)
Lewis X	Gal-β-1,4-[Fuc-α-1,3]-GlcNAc-β-R	*Helicobacter pylori*	Monteiro (2001)
Lewis Y	Fuc-α-1,2-Gal-β-1,4-[Fuc-α-1,3]-GlcNAc-β-R	*Helicobacter pylori*	Monteiro (2001)
Lewis A	Gal-β-1,3-[Fuc-α-1,4]-GlcNAc-β-R	*Helicobacter pylori*	Monteiro (2001)
Lewis B	Fuc-α-1,2-Gal-β-1,3-[Fuc-α-1,4]-GlcNAc-β-R	*Helicobacter pylori*	Monteiro (2001)
Hyaluronic acid	$[$-GlcA-β-1,3-GlcNAc-β-1,4-GlcA-β-1,3-GlcNAc-β-1,4-$]_n$	*Streptococcus pyogenes, Pasteurella multocida*	Maclennan (1956), Rosner *et al.* (1992)
N-acetyl-Heparosan	$[$-GlcA-α-1,4-GlcNAc-β-1,4-GlcA-α-1,4-GlcNAc-β-1,4-$]_n$	*Pasteurella multocida, Escherichia coli K5*	Vann *et al.* (1981); DeAngelis *et al.* (2002)
Chondroitin	$[$-GlcA-β-1,3-GalNAc-β-1,4-GlcA-β-1,3-GalNAc-β-1,4-$]_n$	*Pasteurella multocida, Escherichia coli K4*[i]	Lidholt and Fjelstad (1997), DeAngelis *et al.* (2002)
Polysialic acid	$[$Neu5Ac-α-2,8-Neu5Ac-α-2,8$]_n$	*Neisseria meningitidis group B. Pasteurella haemolytica*	Bhattacharjee *et al.* (1975), Adlam (1987)

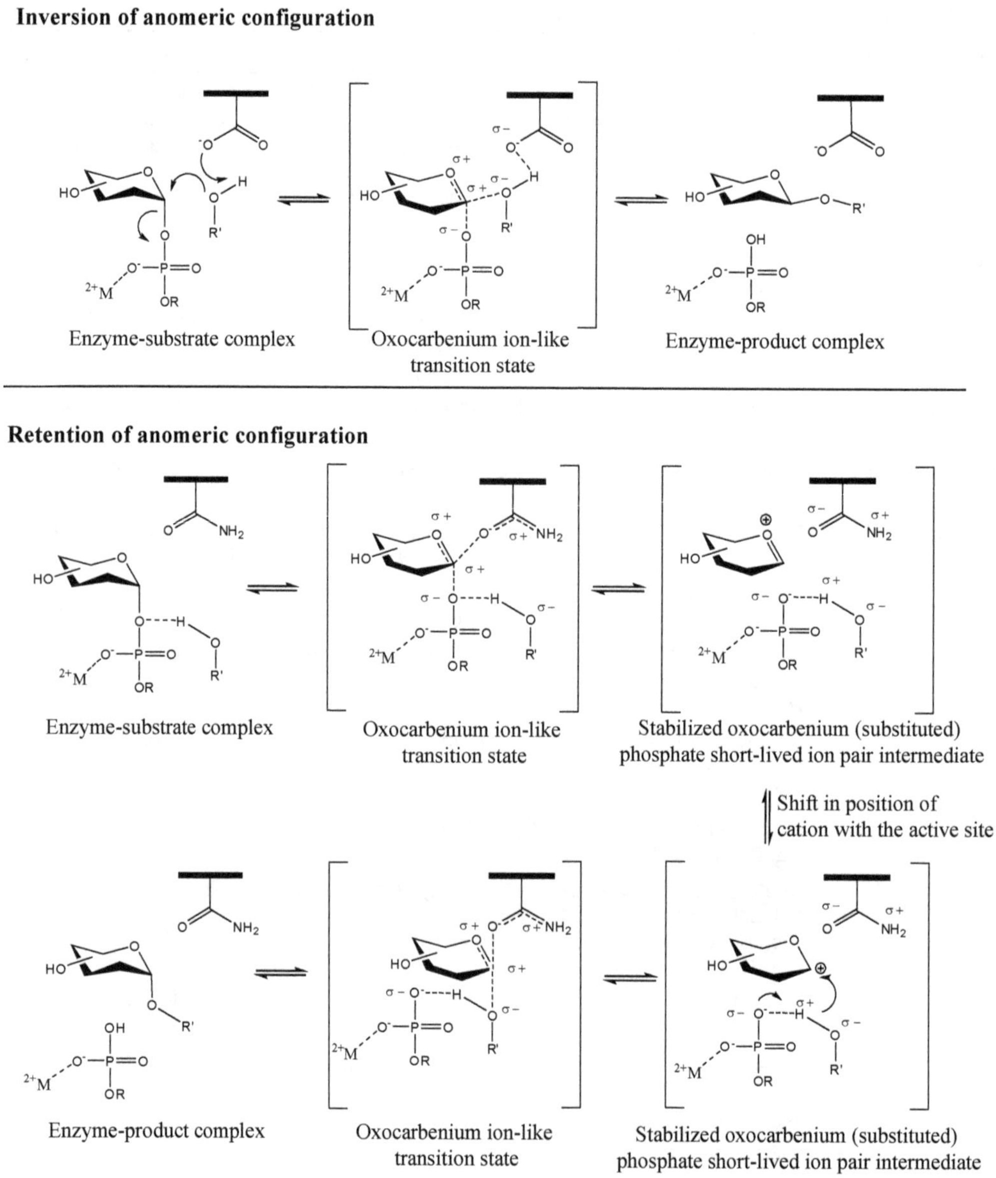

Figure 12.1 Schematic of glycosyltransferase reaction mechanisms. This figure is adapted from Lairson *et al.* (2008). The top panel shows the reaction path for the 'inverting' mechanism which uses a catalytic base to generate a nucleophile that attacks the acceptor sugar. The lower panel depicts a current model for the 'retaining' mechanism, which is thought to proceed through a short-lived ion pair, which allows a front side attack by an incoming nucleophile.

and their toxins (Kobata, 2010; Newburg *et al.*, 2005). The tetrasaccharide lacto-*N*-neotetraose (LNnT) is abundant in human milk and is mimicked by several successful human pathogens (Table 12.1). It is also the core structure for many other oligosaccharides found in milk, and as head groups of glycolipids. In order to assess the possibility of using these oligosaccharides as anti-infectives or probiotics, an efficient large-scale chemoenzymatic synthesis route needed to be devised.

The enzymes for the biosynthesis of the

LNnT and the α-2,3-sialic acid capped form were cloned from *N. meningitidis* and expressed as active recombinant proteins (Gilbert *et al.*, 1996; Wakarchuk *et al.*, 1996, 1998a; Blixt *et al.*, 1999). The key enzymes were LgtA, a β-1,3-*N*-acetylglucosaminyl transferase (CAZy GT-2), and LgtB, a β-1,4-galactosyltransferase (CAZy GT-25). The *E. coli*-produced enzymes were shown to be highly efficient catalysts for *in vitro* synthesis of LNnT derivatives (Yan *et al.*, 2000). The cost of sugar nucleotide donors for synthesis applications can be prohibitive. For example UDP-Gal costs around $200 for 100 mg from Calbiochem (four times the cost of UDP-Glc), the cost of CMP-Neu5Ac is around $7000 per gram when purchased at that scale from Nacalai USA, Inc. Both of these are generally used at 1.5 equivalents per reaction, making a 1 g scale synthesis of an oligosaccharide terminating in sialic acid very expensive. For reactions using UDP-Gal, the enzyme UDP-4′-Glc/Gal epimerase (GalE) can be used to generate UDP-Gal from the less expensive UDP-Glc during the reaction. In an effort to streamline the use of the LgtB glycosyltransferase a fusion protein was generated between the two enzymes, giving rise to GalE-LgtB (Blixt *et al.*, 2001) (Fig. 12.2). This fusion protein has been successfully applied to the synthesis of a variety of LacNAc containing oligosaccharides.

The LNnT oligosaccharide in human milk is often capped with a terminal α-2,3- (or a-2,6) linked sialic acid residue, and the bacterial mimic is capped with an α-2,3-linked sialic acid. There was good evidence that this pentasaccharide was a good anti-infective *in vitro* and in animal models for *Streptococcus pneumoniae* (Idanpaan-Heikkila *et al.*, 1997; Barthelson *et al.*, 1998). The sialyltransferase from *Neisseria* which caps this structure (Nst or Lst) was the first bacterial sialyltransferase to be expressed as recombinant protein (Gilbert *et al.*, 1996) and used for chemoenzymatic synthesis (Yan *et al.*, 2001). The Lst protein was not very soluble on its own so it was produced as a fusion protein with the *N. meningitidis* CMP-Neu5Ac synthetase which improved the overall solubility and madethe sugar nucleotide donor CMP-Neu5Ac from CTP and sialic acid (Gilbert *et al.*, 1998). This paper clearly showed that 100 gram scale synthesis of 3′-sialyllactose could be achieved with high yields, and subsequently this was scaled up to 15 kg (Dr Jim Paulson, The Scripps Research Institute, personal communication). Commercial-scale synthesis of LNnT with bacterial enzymes was achieved by NEOSE technologies using versions of these bacterial enzymes and the company undertook a clinical trial with the α-2,3-sialylated form of LNnT to examine treatment of otitis media in young children. This clinical trial was unfortunately not successful for a variety of possible reasons (Ukkonen *et al.*, 2000), but the technology for large-scale production of enzymes for *in vitro* enzymatic synthesis had

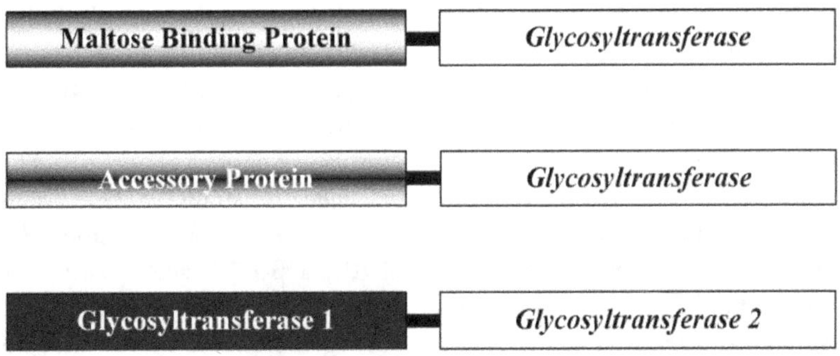

Figure 12.2 Protein engineering of glycosyltransferases to improve synthetic ability. Fusions with the *E. coli* Maltose binding protein (MalE or MPB) can increase the solubility of bacterial and other glycosyltransferases, while fusions with sugar nucleotide synthases/epimerases (CMP-Neu5Ac synthease, UDP-Glc(NAc)/Gal(NAc)-4′-epimerase) can improve the efficiency of the *in vitro* synthesis reaction by providing increased local concentrations of sugar nucleotide donors. Fusing two enzymes required for a reaction sequence can also increase the *in vitro* reaction efficiency.

clearly been demonstrated, and the utility of such oligosaccharides as anti-infectives or probiotics might still be a viable approach in the future (Kobata, 2010; Zivkovic *et al.*, 2010).

More recently homologues of LgtA and LgtB have been described from *H. pylori*; these are JHP1032 a β-1,3-*N*-acetylglucosaminyl transferase (CAZy GT-8) (Logan *et al.*, 2005), and HP0826 a β-1,4-galactosyltransferase (CAZy GT-25) (Logan *et al.*, 2000). *Helicobacter* strains do not make an LNnT mimic, but they do synthesize *N*-acetyllactosamine (LacNAc), which is then further elaborated into Lewis blood group antigens by various fucosyltransferases (Wang *et al.*, 2000). The *H. pylori* enzymes that make LacNAc have been applied to the synthesis of poly LacNAc structures in an effort to simulate an extra-cellular matrix for the growth of cultured mammalian cells (Sauerzapfe *et al.*, 2009). It should also be mentioned that the *Neisserial* enzymes LgtA/B can also be used to make soluble poly LacNAc structures at the gram scale (Vasiliu *et al.*, 2006). The HP0826 enzyme has been shown to be a versatile catalyst for the *in vitro* synthesis of novel LacNAc derivatives including a thio-linked version which is resistant to β-galactosidase (Namdjou *et al.*, 2008). The further application of these enzymes for the production of bioactive carbohydrates is certainly anticipated.

Globotriose (blood group P^k epitope), globotetraose and the Forsmann antigen

Another common carbohydrate structure from *Neisseria* and *Haemophilus influenzae* is the trisaccharide Gal-α-1,4-Gal-β-1,4-Glc-β-R (Table 12.1) which is the head group of the glycosphingolipid known as globotriose or Gb3, and a human P blood group antigen (Spitalnik and Spitalnik, 1995). In humans, these carbohydrate epitopes are the target for the binding subunit of some of the bacterial AB_5 toxins from *Shigella* and *E. coli* (Merritt and Hol, 1995). Work on the purified toxin binding subunit SLT-1b showed that the toxin could be neutralized by synthetic structures tailored to fit all five binding subunits (Kitov *et al.*, 2000), leading to the idea that an effective therapeutic for this toxin could be developed (Mulvey *et al.*, 2003). The critical enzyme for the preparative synthesis of the terminal Gal-α-1,4-Gal-β structure for binding SLT-1b was LgtC (CAZy GT-8) from *N. meningitidis* immunotype L1 (Persson *et al.*, 2001). This α-galactosyltransferase from *N. meningitidis* was engineered to over-express as a soluble protein by deletion of the C-terminal 20 amino acids (Wakarchuk *et al.*, 1998a), and was also produced as UDP-4′-Glc/Gal epimerase fusion for preparative work (Kitov *et al.*, 2003). The LgtC enzyme was also shown to be remarkably tolerant of chemically modified acceptors and donors (Lougheed *et al.*, 1999; Kitov *et al.*, 2003; Lairson *et al.*, 2007). Recently the use of such a chemoenzymatic synthetic compound *in vivo* has been enhanced by using the elegant approach of supramolecular templating to couple the toxin binding to the binding of a second ligand to a pentraxin serum protein (Solomon *et al.*, 2005). The pre-organization of binding provides high-avidity inhibitors for multimeric receptors. It is probable that similar approaches may be examined for the inhibition of other AB_5 toxins, or perhaps to other multivalent ligands such as viral particles.

The P^k blood group trisaccharide is the base structure for other glycolipid head groups some of which are well known tumour markers (Hakomori and Zhang, 1997). Globotetraose (GalNAc-β-1,3-Gal-α-1,4-Gal-β-1,4-Glc-β-R), globopentaose (Gal-β-1,3-GalNAc-β-1,3-Gal-α-1,4-Gal-β-1,4-Glc-β-R), Globo H (Fuc-α-1,2-Gal-β-1,3-GalNAc-β-1,3-Gal-α-1,4-Gal-β-1,4-Glc-β-R) and the Forssman antigen (GalNAc-α-1,3-GalNAc-β-1,3-Gal-α-1,4-Gal-β-1,4-Glc-β-R) are all P^k-based structures that have been synthesized chemoenzymically employing bacterial enzymes for at least part of the synthesis. Globotetraose is the base structure for the cancer antigen Globo H and is synthesized in *H. influenzae* strain Rd (Hood *et al.*, 2001) (Table 12.1). The key enzyme from *H. influenzae* was LgtD, a β-1,3-*N*-acetylgalactosaminyltransferase which used P^k as an acceptor, and which could use globotetraose as an acceptor to produce globopentaose (Randriantsoa *et al.*, 2007). For the synthesis of Globo H a terminal α-1,2-fucose residue is required. A gene encoding an α-1,2-fucosyl transferase was identified in the O-chain biosynthesis cluster of an *E. coli* strain of the 0128 serotype (Shao *et al.*, 2003), and the enzyme has been employed in the chemoenzymatic synthesis of Globo H (Wang *et al.*, 2008). Vaccine efforts with these oligosaccharides are ongoing (Astronomo and Burton, 2010) and despite recent advances in automated chemical synthesis of carbohydrates (Seeberger, 2008), chemoenzymatic synthesis will probably remain a viable option to produce these oligosaccharides. The rare Forssman antigen has been suggested as another cancer marker, but at present it is not considered a normal glycolipid modification in humans. Nevertheless as it is a rare glycolipid, access to preparative quantities will permit the examination of its binding partners. The enzymatic synthesis of this pentasaccharide was accomplished in multi-milligram scale using a combination of enzymes from *C. jejuni* and *P. multocida* (Houliston *et al.*, 2009). The enzymes used were the α-1,4-galactosyltransferase CgtD (CAZy GT-32), and a β-1,3-*N*-acetylgalactosaminyltransferase, CgtE (CAZy GT-2). The terminal α-1,3-*N*-acetylgalactosaminyltransferase (CAZy GT-4) PM1138 came from the genome strain of *P. multocida* PM70 (St. Michael *et al.*, 2005b). In this work the authors do caution that even though CgtE is an efficient catalyst at this scale, the LgtD protein described above from *H. influenzae* may be a better catalyst (Fig. 12.3).

Lewis antigens

The Lewis blood group antigens are another class of carbohydrates that have been identified as tumour markers (Kannagi, 2003) and as critical structures in the inflammatory response (Fukuda *et al.*, 1999). The sialyl-Lewis X structure, α-2,3-NeuAc-Gal-β-1,4-GlcNAc[α-1,3-Fuc]-R was

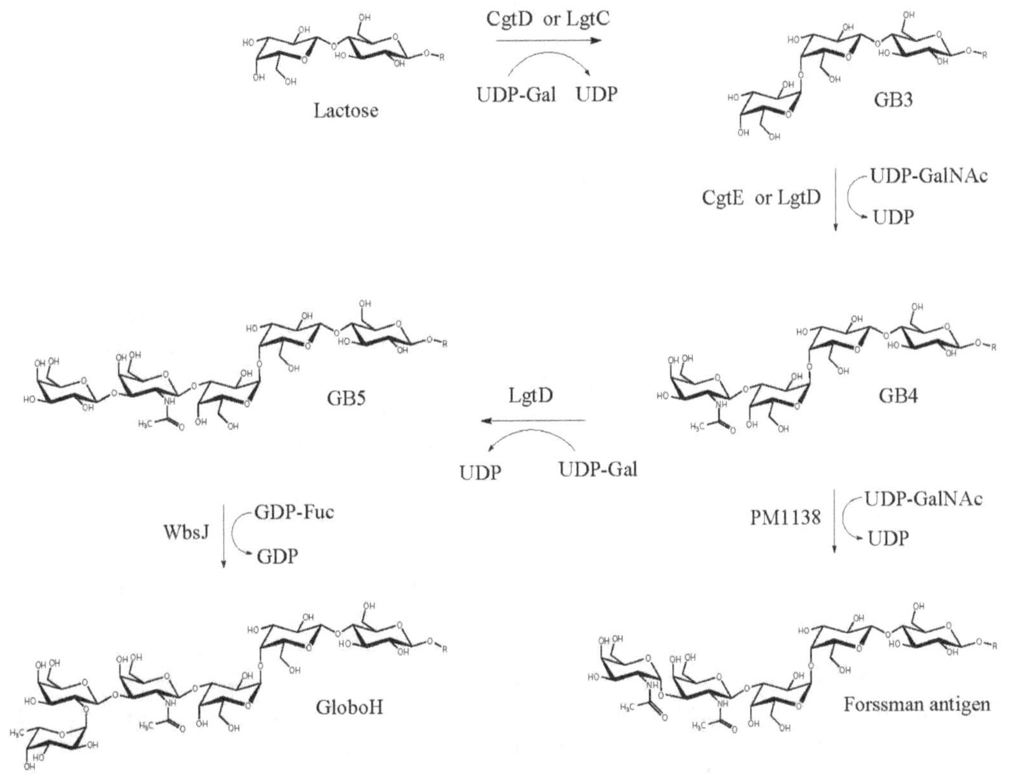

Figure 12.3 Synthetic routes to GloboH or Forssman antigen using bacterial glycosyltransferases. In this synthetic scheme there are alternative enzymes: CgtD or LgtC to make Gb3, CgtE or LgtD to make Gb4. With Gb4 as an intermediate there are two choices for further elongation: either to the Forssman antigen with Pm1138 or Gb5 with LgtD and then GloboH using WbsJ from *E. coli* O128.

identified as potential therapeutic molecule for modulation of inflammation based on its interaction with the cell surface lectins (selectins) involved in the leucocyte adhesion cascade (Sperandio *et al.*, 2009) and has been the subject of many synthetic efforts both chemical and chemoenzymatic (Pudelko *et al.*, 2010). The Lewis X and the type 1 linkage variant of Lewis A are both made in *H. pylori* and these organisms have provided useful fucosyltransferase enzymes for the synthesis of the Lewis X/Y/A structures (Ge *et al.*, 1997; Wang *et al.*, 1999; Rasko *et al.*, 2000) (Table 12.1). Recently the utility of the *H. pylori* α-1,3-fucosyltransferase was demonstrated when it was used to generate a library of modified Lewis X compounds (Wang *et al.*, 2009). Currently, it is not known if the *H. pylori* fucosyltransferases will be of value in the commercial-scale synthesis of fucosylated glycans, but their synthetic utility *in vitro* has clearly been demonstrated. Two tetrasaccharides carrying the trisaccharide Lewis X motif on a GlcNAc or a Gal residue have been produced on the gram scale by high-cell-density cultures of metabolically engineered *Escherichia coli* strains that overexpressed the *H. pylori futA* gene for α-3 fucosyltransferase and the *N. meningitidis lgtB* gene for β-4 galactosyltransferase (Dumon *et al.*, 2006) (see below for more details on the living factory approach to synthesis).

Gangliosides

The sialic acid containing glycosphingolipids are known as gangliosides and are ubiquitous in the mammalian cell membrane and are highly enriched in neural tissue. They provide a number of important functions for the cell including their role as co-receptors in signal transduction events, and stabilizing nerve fibres (Wennekes *et al.*, 2009). Gangliosides are also well known tumour markers, and have been the targets of therapeutic cancer vaccines (Fredman *et al.*, 2003; Labrada *et al.*, 2010). In addition to cancer vaccines, the ganglioside GM1a has shown promise as a treatment for Parkinson's disease (Schneider *et al.*, 2010). These types of glycolipids are mimicked in the lipo-oligosaccharide of many strains of *C. jejuni* (Gilbert, 2008), but have not yet been found in any other bacterial species. The biosynthetic enzymes for the ganglioside mimic structures from *C. jejuni* were identified in an operon encoding all of the glycosyltransferase genes as well as the sialic acid biosynthesis genes (Gilbert *et al.*, 2000). There are four enzymes from *C. jejuni* which are useful for the *in vitro* synthesis. They are: the α-2,3-sialyltransferase to produce GM3 glycan, Cst-II; the β-1,4-*N*-acetylgalactosaminyltransferase to produce a GM2/GD2 glycan, CgtA; and the β-1,3-galactosyltransferase, CgtB, to produce the GM1a/GD1a/GT1a structures. The sialyltransferase Cst-II has multiple alleles one of which permits the enzyme to operate as an α-2,8-sialyltransferase for the production of di- and tri-sialylated gangliosides, GD3/GD2/GT1a (Gilbert *et al.*, 2002). In addition the mono-functional α-2,3-sialyltransferase enzyme Cst-I was also identified from a strain of *C. jejuni* that also expressed Cst-II, and while its *in vivo* function is not known, it is a very potent catalyst for the synthesis of a variety of α-2,3-sialosides (Chiu *et al.*, 2007). These enzymes have been employed in the gram scale *in vitro* synthesis of many soluble ganglioside glycans (Blixt *et al.*, 2005). These ganglioside analogues have many applications, such as the use of soluble GM1a analogues to inhibit the binding of cholera toxin and *E. coli* heat-labile toxin to authentic GM1a (Sisu *et al.*, 2009). These compounds may find an application as a means to remove anti-ganglioside antibodies from patients with Guillain–Barré syndrome (Townson *et al.*, 2007). Other efforts at using these oligosaccharides on surfaces for diagnostics and immunoadsorbants are continuing (Pukin *et al.*, 2008). One cautionary note about the use of bacterial sialyltransferases for synthesis is that they can all catalyse the hydrolysis (back) reaction to various degrees, especially in the absence of acceptor (Gilbert and Wakarchuk, unpublished data; Cheng *et al.*, 2008). This is not a concern as long as the reactions are monitored, and stopped before hydrolysis becomes the dominant reaction.

Glycosaminoglycans

Many of the enzymes mentioned so far that have been employed for chemoenzymatic synthesis are involved in the biosynthesis of the outer core structure of short chain lipo-oligosaccharides from Gram-negative bacteria like *Neisseria, Haemophilus* and *Campylobacter*. Another source of unique enzymes for the synthesis of molecular mimic structures are genetic loci for capsular polysaccharides, especially those from *P. multocida* strains which individually produces hyaluronic acid, and un-sulfated chondroitin, and unepimerized *N*-acetyl-heparosan (DeAngelis, 2002). Un-sulfated chondroitin, and unepimerized heparosan can also be found in *E. coli* K4 (Lidholt and Fjelstad, 1997) and *E. coli* K5, respectively (Vann *et al.*, 1981) (Table 12.1). Glycosaminoglycans are currently employed as therapeutics for a wide variety of indications, including anti-coagulants, joint viscoelastic supplementation and eye surgery, to name a few (DeAngelis, 2008; Laremore *et al.*, 2009). In 2008 there were a number of deaths that occurred in patients being treated with animal-derived heparin that turned out to be contaminated with over sulfated chondrotin sulphate (Blossom *et al.*, 2008). This kind of contamination could be completely avoided with material derived from the enzymatic or chemical synthesis of these polymers.

The chain length of theses polymers is also a critical factor as the length has been shown to affect the biological activity dramatically (DeAngelis, 2008; Mousa, 2010; Wolny *et al.*, 2010). The ability to control the size of polymers made by enzymatic synthesis could help the development of compounds for specific applications, with decreased potential for side effects when these carbohydrates are used as drugs. Recently, the *P. multocida* heparin synthase was demonstrated to be valuable for the *in vitro* chemoenzymatic synthesis along with various mammalian heparin modifying enzymes (Liu *et al.*, 2010). These heparin sulphate oligosaccharides of four or five disaccharide repeats were as potent as the commercially available chemically synthesized heparin pentamer. The enzymes from *P. multocida* have been used for controlled *in vitro* synthesis of hyaluronan polymers of various molecular weights, which should provide the opportunity to further evaluate the biological activity of these carbohydrates (DeAngelis, 2008).

Production of oligosaccharides by recombinant bacterial strains – living factories

Many of the uses of oligosaccharides for therapy demand large quantities, and for this reason alternatives have been investigated to the expensive *in vitro* enzymatic methods which require isolated enzymes and substantial quantities of sugar nucleotides. One early alternative approach was called bacterial coupling and this involved growing two bacterial strains separately, one producing sugar nucleotide precursors and one producing the transferase(s). When extracts of these strains were mixed with an acceptor like lactose, high levels of globotriose oligosaccharide production could be achieved (Koizumi *et al.*, 1998). Another more elegant approach is called the living factory and was pioneered by the group of Eric Samain at Centre de Recherches sur les Macromolécules Végétales (CNRS), in Grenoble, France. Their approach was to engineer an *E. coli* strain to contain multiple bacterial glycosyltransferases for a particular biosynthetic pathway, and have the strain build various oligosaccharides on a simple lactose building block (Fig. 12.4). The *E. coli* strain is *lacZ* deficient and so will not degrade lactose, but still imports it via the lactose transporter (LacY) into the cells where it is a substrate for a variety of these transferases. The strains also carry a *nanA* knock out which prevents degradation of exogenously added sialic acid by the sialic acid aldolase. This approach has been used very successfully for many of the oligosaccharides mentioned in the previous section, and has been scaled up using fermentation technology, making this a very attractive approach to obtain the natural oligosaccharide portions of neutral glycolipids, gangliosides, and many of the human milk oligosaccharides (Bettler *et al.*, 1999; Priem *et al.*, 2002; Antoine *et al.*, 2003; Randriantsoa *et al.*, 2007; Fierfort and Samain, 2008; Yavuz *et al.*, 2008). One potential problem with the living factory approach is that some glycosyltransferases can use more than one donor (e.g. some GalNAc transferases can use UDP-Gal at significant rates) which would give mixed products that are hard to separate. Recently, a company called Elictyl has begun marketing these types of oligosaccharides, making them widely accessible to researchers in the field (http://www.elicityl.fr).

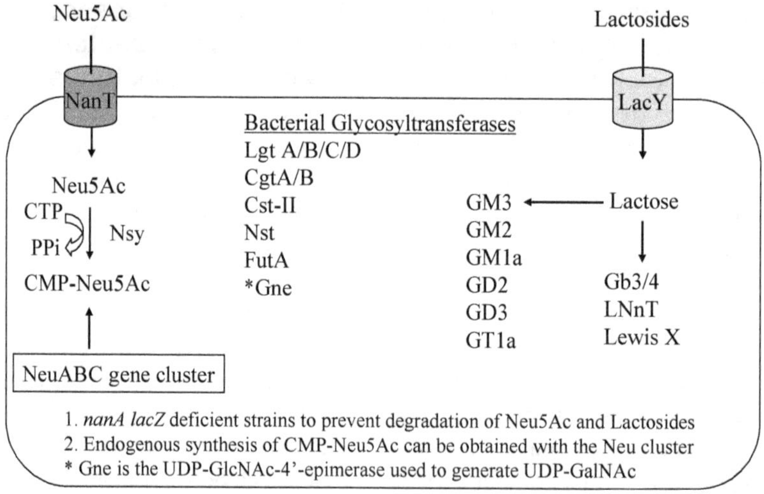

Figure 12.4 Schematic of the living factory approach to oligosaccharide synthesis. Given the cost of sugar nucleotide synthesis and the possible problems of incomplete *in vitro* synthesis, workers at the CNRS in Grenoble, France, have developed an elegant *in vivo* synthetic strategy for the production of a variety of oligosaccharides using enzymes *from Neisseria, Campylobacter, Haemophilus, and Helicobacter*. The figure depicts an composite *E. coli* cell which has all of the machinery for the living factory. Oligosaccharides are produced by feeding a *lacZ nanA* deficient strain of *E. coli* the simple lactose and sialic acid building blocks (imported by LacY and NanT) and having the cell do all the work of elaborating the oligosaccharide when the correct combination of glycosyltransferases and accessory enzymes like epimerases is present. The requirement to feed sialic acid can be overcome by adding the genes for sialic acid synthesis (NeuA-C). The cell can then do all the synthetic work of elaborating the oligosaccharide.

One application of this technology to the synthesis of gangliosides has been to couple the *in vivo* derived oligosaccharide to a sphingosine lipid, using a glycosynthase enzyme (Vaughan *et al.*, 2006; Hancock *et al.*, 2009), which can be done on a suitably large scale. Glycosynthases are engineered hydrolase enzymes which do not catalyse hydrolysis but catalyse synthesis when the oligosaccharide being transferred has a suitable leaving group. This approach circumvents the need to perform multiple purifications during the synthesis of a complex ganglioside. Glycosynthases have been developed for many synthetic reactions and are very useful additions to enzymatic carbohydrate synthesis, but this topic is beyond the scope of this review and so the reader is directed to reviews on this topic (Perugino *et al.*, 2004; Shaikh and Withers, 2008).

Another *in vivo* synthesis strategy has been employed to produce 'receptor-mimic probiotics', which are harmless *E. coli* strains displaying the carbohydrate receptors for a number of bacterial AB_5 toxins elaborated by enteric pathogens (Paton *et al.*, 2006, 2010). These *E. coli* strains were produced by expressing the genes for biosynthesis of: Gb3, Gb4, GM1a and GM2, in a mutant lacking a normal outer core LPS structure. The genes used to produce these toxin-receptor mimics are the same ones described above from *Neisseria* and *Campylobacter*, that were used for *in vitro* chemoenzymatic synthesis. While these designer organisms have been proven in lab tests to protect animals from lethal doses of toxins, they have not been developed further for human clinical use out of perceived fear of using genetically modified bacteria for treatments and that some kind of immune response might be generated to normal human glycan structures as a result of these modified surface glycans (Paton *et al.*, 2010).

Production of oligosaccharides libraries for use in glycan arrays and other applications

The determination of ligands involved in carbohydrate–protein interactions has been greatly facilitated by the use of fabricated glycan arrays on various surfaces (Paulson *et al.*, 2006; Taylor and Drickamer, 2009). These arrays allow screening of a glycan binding protein against many possible binding partners in a single experiment, after which specific glycans can be studied in more detail. One such array has been fabricated by the Consortium for Functional Glycomics, and in Dec. 2010 had a library of 511 glycans on it. In the context of the fabrication, the glycans needed to be tagged at the reducing end so that they could be covalently attached to the surfaces. For this reason core oligosaccharides needed to be synthesized with functionalized spacers and then elaborated into various complex glycans. Chemo-enzymatic synthesis was used to produce many of the glycans on the array, and the versatile bacterial glycosyltransferases, some of which have been described in this chapter, were used to generate key linkages for some of these compounds (Blixt and Razi, 2006). Bacterial glycosyltransferases were used in the production of approximately 200 of the 511 glycans on the array (Dr Jim Paulson, The Scripps Research Institute, personal communication).

Other glycan libraries have also been made using bacterial glycosyltransferases, especially the GT-80 sialyltransferases from *Pasteurella* and *Photobacterium* (Yu *et al.*, 2005; Chokhawala *et al.*, 2008; Cao *et al.*, 2009). The large variety of compounds that have been made with modified acceptors and donors demonstrated the synthetic versatility of these bacterial sialyltransferases. This substrate promiscuity has in general been observed in many bacterial glycosyltransferases, so the application of these enzymes in the continued development of compounds for probing glycobiology seems very probable.

Concluding remarks

In the last 15 years, bacterial glycosyltransferases have been employed in the synthesis of a wide variety of carbohydrates and glycoconjugates based on human glycoconjugate structures. Their ease of production in *E. coli* and often relaxed acceptor specificity has contributed to the exploitation of these enzymes. Recently, improvements to the activity of two of the enzymes described here, Cst-II and CgtB, have been made using directed evolution with a fluorescent cell sorting technique (Aharoni *et al.*, 2006; Yang *et al.*, 2010). The application of directed evolution will permit the development of improved catalysts for new synthetic applications based on specific acceptors or modified donors. In addition to the improvement through genetic manipulation, we now have access to a large number of uncharacterized glycosyltransferases by virtue of the explosive growth of genomic sequence determination. The current CAZy database holds several thousand enzymes in families where multiple activities have been documented (Cantarel *et al.*, 2009), and mining these families for new enzyme activities could reveal useful synthetic catalysts, if we can devise ways of screening for these activities. The development of screens for glycosyltransferase enzyme activities is more difficult than many other enzymes, since the formation of new products must be detected usually with binding proteins (lectins, or antibodies) rather than the release of products as has been applied to glycosidases. However, it may be possible to use the same methods developed for the directed evolution to screen libraries of unknown glycosyltransferases for specific activities. While chemical synthesis of carbohydrates has become quite efficient and automation is now being applied (Seeberger, 2008) there will remain a need to use enzymes for the foreseeable future and the application of molecular biology will undoubtedly ensure development of new catalysts for glycoconjugate synthesis will continue to be a fruitful research endeavour.

Acknowledgements

I would like to thank Dr N. Martin Young, Dr Jamie Rich and Dr Michel Gilbert for help proofreading the manuscript.

References

Adlam, C., Knight, J. M., Mugridge, A., Williams, J. M., and Lindon, J. C. (1987). Production of colominic acid by *Pasteurella haemolytica* serotype A2 organisms FEBS Microbiol. Lett. *42*, 23–25.

Aharoni, A., Thieme, K., Chiu, C.P., Buchini, S., Lairson, L.L., Chen, H., Strynadka, N.C., Wakarchuk, W.W., and Withers, S.G. (2006). High-throughput screening methodology for the directed evolution of glycosyltransferases. Nat. Methods 3, 609–614.

Antoine, T., Priem, B., Heyraud, A., Greffe, L., Gilbert, M., Wakarchuk, W.W., Lam, J.S., and Samain,

E. (2003). Large-scale *in vivo* synthesis of the carbohydrate moieties of gangliosides GM1 and GM2 by metabolically engineered *Escherichia coli*. Chembiochemistry *4*, 406–412.

Astronomo, R.D., and Burton, D.R. (2010). Carbohydrate vaccines: developing sweet solutions to sticky situations? Nat. Rev. Drug Discov. *9*, 308–324.

Barry, G.T., and Goebel, W.F. (1957). Colominic acid, a substance of bacterial origin related to sialic acid. Nature *179*, 206.

Barthelson, R., Mobasseri, A., Zopf, D., and Simon, P. (1998). Adherence of *Streptococcus pneumoniae* to respiratory epithelial cells is inhibited by sialylated oligosaccharides. Infect. Immun. *66*, 1439–1444.

Bettler, E., Samain, E., Chazalet, V., Bosso, C., Heyraud, A., Joziasse, D.H., Wakarchuk, W.W., Imberty, A., and Geremia, A.R. (1999). The living factory: *in vivo* production of N-acetyllactosamine containing carbohydrates in *E. coli*. Glycoconj. J. *16*, 205–212.

Bhattacharjee, A.K., Jennings, H.J., Kenny, C.P., Martin, A., and Smith, I.C. (1975). Structural determination of the sialic acid polysaccharide antigens of *Neisseria meningitidis* serogroups B and C with carbon 13 nuclear magnetic resonance. J. Biol. Chem. *250*, 1926–1932.

Bhattacharjee, A.K., Jennings, H.J., Kenny, C.P., Martin, A., and Smith, I.C. (1976). Structural determination of the polysaccharide antigens of *Neisseria meningitidis* serogroups, Y. W-*135*, and BO1. Can. J. Biochem. *54*, 1–8.

Blixt, O., and Razi, N. (2006). Chemoenzymatic synthesis of glycan libraries. In Methods in Enzymology, Fukuda, M., ed. (Academic Press), pp. 137–153.

Blixt, O., van, D.I., Norberg, T., and van den Eijnden, D.H. (1999). High-level expression of the *Neisseria meningitidis lgtA* gene in *Escherichia coli* and characterization of the encoded N-acetylglucosaminyltransferase as a useful catalyst in the synthesis of GlcNAc beta 1-->3Gal and GalNAc beta 1-->3Gal linkages. Glycobiology *9*, 1061–1071.

Blixt, O., Brown, J., Schur, M.J., Wakarchuk, W., and Paulson, J.C. (2001). Efficient preparation of natural and synthetic galactosides with a recombinant beta-1,4-galactosyltransferase-/UDP-4′-gal epimerase fusion protein. J. Org. Chem. *66*, 2442–2448.

Blixt, O., Vasiliu, D., Allin, K., Jacobsen, N., Warnock, D., Razi, N., Paulson, J.C., Bernatchez, S., Gilbert, M., and Wakarchuk, W. (2005). Chemoenzymatic synthesis of 2-azidoethyl-ganglio-oligosaccharides GD3, GT3, GM2, GD2, GT2, GM1, and GD1a. Carbohydr. Res. *340*, 1963–1972.

Blossom, D.B., Kallen, A.J., Patel, P.R., Elward, A., Robinson, L., Gao, G., Langer, R., Perkins, K.M., Jaeger, J.L., Kurkjian, K.M., *et al.* (2008). Outbreak of adverse reactions associated with contaminated heparin. New Engl. J. Med. *359*, 2674–2684.

Cantarel, B.L., Coutinho, P.M., Rancurel, C., Bernard, T., Lombard, V., and Henrissat, B. (2009). The Carbohydrate-Active EnZymes database (CAZy): an expert resource for Glycogenomics. Nucleic Acids Res. *37*, D233–D238.

Cao, H., Muthana, S., Li, Y., Cheng, J., and Chen, X. (2009). Parallel chemoenzymatic synthesis of sialosides containing a C5-diversified sialic acid. Bioorg. Med. Chem. Lett. *19*, 5869–5871.

Carlin, A.F., Uchiyama, S., Chang, Y.-C., Lewis, A.L., Nizet, V., and Varki, A. (2009). Molecular mimicry of host sialylated glycans allows a bacterial pathogen to engage neutrophil Siglec-9 and dampen the innate immune response. Blood *113*, 3333–3336.

Cheng, J., Yu, H., Lau, K., Huang, S., Chokhawala, H.A., Li, Y., Tiwari, V.K., and Chen, X. (2008). Multifunctionality of *Campylobacter jejuni* sialyltransferase CstII: Characterization of GD3/GT3 oligosaccharide synthase, GD3 oligosaccharide sialidase, and trans-sialidase activities. Glycobiology *18*, 686–697.

Chiu, C.P., Lairson, L.L., Gilbert, M., Wakarchuk, W.W., Withers, S.G., and Strynadka, N.C. (2007). Structural analysis of the alpha-2,3-sialyltransferase Cst-I from *Campylobacter jejuni* in apo and substrate-analogue bound forms. Biochemistry *46*, 7196–7204.

Chokhawala, H.A., Huang, S., Lau, K., Yu, H., Cheng, J., Thon, V., Hurtado-Ziola, N., Guerrero, J.A., Varki, A., and Chen, X. (2008). Combinatorial Chemoenzymatic Synthesis and High-Throughput Screening of Sialosides. ACS Chem. Biol. *3*, 567–576.

Cieslewicz, M.J., Chaffin, D., Glusman, G., Kasper, D., Madan, A., Rodrigues, S., Fahey, J., Wessels, M.R., and Rubens, C.E. (2005). Structural and genetic diversity of group B *Streptococcus* capsular polysaccharides. Infect. Immun. *73*, 3096–3103.

Crocker, P.R., Paulson, J.C., and Varki, A. (2007). Siglecs and their roles in the immune system. Nat. Rev. Immunol. *7*, 255–266.

DeAngelis, P.L. (2002). Microbial glycosaminoglycan glycosyltransferases. Glycobiology *12*, 9R-16.

DeAngelis, P.L. (2008). Monodisperse hyaluronan polymers: synthesis and potential applications. Curr. Pharm. Biotechnol. *9*, 246–248.

DeAngelis, P.L., Gunay, N.S., Toida, T., Mao, W.-j., and Linhardt, R.J. (2002). Identification of the capsular polysaccharides of Type D and F *Pasteurella multocida* as unmodified heparin and chondroitin, respectively. Carbohydr. Res. *337*, 1547–1552.

Dumon, C., Bosso, C., Utille, J.P., Heyraud, A., and Samain, E. (2006). Production of Lewis x tetrasaccharides by metabolically engineered *Escherichia coli*. Chembiochemistry *7*, 359–365.

Fierfort, N., and Samain, E. (2008). Genetic engineering of *Escherichia coli* for the economical production of sialylated oligosaccharides. J. Biotechnol. *134*, 261–265.

Fredman, P., Hedberg, K., and Brezicka, T. (2003). Gangliosides as therapeutic targets for cancer. BioDrugs *17*, 155–167.

Fukuda, M., Hiraoka, N., and Yeh, J.C. (1999). C-Type lectins and sialyl lewis X oligosaccharides: versatile roles in cell–cell interaction. J. Cell Biol. *147*, 467–470.

Ge, Z., Chan, N.W.C., Palcic, M.M., and Taylor, D.E. (1997). Cloning and heterologous expression of an α1,3-Fucosyltransferase gene from the gastric pathogen *Helicobacter pylori*. J. Biol. Chem. *272*, 21357–21363.

Gilbert, M., Watson, D.C., Cunningham, A.M., Jennings, M.P., Young, N.M., and Wakarchuk, W.W. (1996). Cloning of the lipooligosaccharide alpha-2,3-sialyltransferase from the bacterial pathogens *Neisseria meningitidis* and *Neisseria gonorrhoeae*. J. Biol. Chem. *271*, 28271–28276.

Gilbert, M., Bayer, R., Cunningham, A.M., DeFrees, S., Gao, Y., Watson, D.C., Young, N.M., and Wakarchuk, W.W. (1998). The synthesis of sialylated oligosaccharides using a CMP-Neu5Ac synthetase/sialyltransferase fusion. Nat. Biotechnol. *16*, 769–772.

Gilbert, M., Brisson, J.R., Karwaski, M.F., Michniewicz, J., Cunningham, A.M., Wu, Y., Young, N.M., and Wakarchuk, W.W. (2000). Biosynthesis of ganglioside mimics in *Campylobacter jejuni* OH4384. Identification of the glycosyltransferase genes, enzymatic synthesis of model compounds, and characterization of nanomole amounts by 600-mhz (1)h and (13)c NMR analysis. J. Biol. Chem. *275*, 3896–3906.

Gilbert, M., Karwaski, M.F., Bernatchez, S., Young, N.M., Taboada, E., Michniewicz, J., Cunningham, A.M., and Wakarchuk, W.W. (2002). The genetic bases for the variation in the lipo-oligosaccharide of the mucosal pathogen, *Campylobacter jejuni*. Biosynthesis of sialylated ganglioside mimics in the core oligosaccharide. J. Biol. Chem. *277*, 327–337.

Gilbert, M., Parker, C.T., and Moran, A.P. (2008). *Campylobacter jejuni* lipooligosaccharides: structure and function. In Campylobacter 3rd edn, Irving Nachamkin, C.M.S., and Blaser, M.J. eds. (Washington DC, ASM Press), pp. 483–504.

Hakomori, S.-I., and Zhang, Y. (1997). Glycosphingolipid antigens and cancer therapy. Chem. Biol. *4*, 97–104.

Haltiwanger, R.S., and Lowe, J.B. (2004). Role of glycosylation in development. Annu. Rev. Biochem. *73*, 491–537.

Hancock, S.M., Rich, J.R., Caines, M.E.C., Strynadka, N.C.J., and Withers, S.G. (2009). Designer enzymes for glycosphingolipid synthesis by directed evolution. Nat. Chem. Biol. *5*, 508–514.

Harvey, H.A., Porat, N., Campbell, C.A., Jennings, M., Gibson, B.W., Phillips, N.J., Apicella, M.A., and Blake, M.S. (2000). Gonococcal lipooligosaccharide is a ligand for the asialoglycoprotein receptor on human sperm. Mol. Microbiol. *36*, 1059–1070.

Hood, D.W., Cox, A.D., Wakarchuk, W.W., Schur, M., Schweda, E.K., Walsh, S.L., Deadman, M.E., Martin, A., Moxon, E.R., and Richards, J.C. (2001). Genetic basis for expression of the major globotetraose-containing lipopolysaccharide from *H. influenzae* strain Rd (RM118). Glycobiology *11*, 957–967.

Houliston, R.S., Bernatchez, S., Karwaski, M.F., Mandrell, R.E., Jarrell, H.C., Wakarchuk, W.W., and Gilbert, M. (2009). Complete chemoenzymatic synthesis of the Forssman antigen using novel glycosyltransferases identified in *Campylobacter jejuni* and *Pasteurella multocida*. Glycobiology *19*, 153–159.

Idanpaan-Heikkila, I., Simon, P.M., Zopf, D., Vullo, T., Cahill, P., Sokol, K., and Tuomanen, E. (1997). Oligosaccharides interfere with the establishment and progression of experimental pneumococcal pneumonia. J. Infect. Dis. *176*, 704–712.

Kannagi, R. (2003). Molecular mechanism for cancer-associated induction of sialyl Lewis X and sialyl Lewis A expression – The Warburg effect revisited. Glycoconjug. J. *20*, 353–364.

Kitov, P.I., Sadowska, J.M., Mulvey, G., Armstrong, G.D., Ling, H., Pannu, N.S., Read, R.J., and Bundle, D.R. (2000). Shiga-like toxins are neutralized by tailored multivalent carbohydrate ligands. Nature *403*, 669–672.

Kitov, P.I., Paszkiewicz, E., Wakarchuk, W.W., and Bundle, D.R. (2003). Preparative-scale chemoenzymatic synthesis of large carbohydrate assemblies using alpha (1-->4)-galactosyltransferase/UDP-4′-Gal-epimerase fusion protein. Methods Enzymol. *362*, 86–105.

Kobata, A. (2010). Structures and application of oligosaccharides in human milk. Proc. Jpn. Acad. Ser. B Phys. Biol. Sci. *86*, 731–747.

Koizumi, S., Endo, T., Tabata, K., and Ozaki, A. (1998). Large-scale production of UDP-galactose and globotriose by coupling metabolically engineered bacteria. Nat. Biotechnol. *16*, 847–850.

Koshland, D.E. (1953). Sterechemistry and the mechanism of enzymatic reactions. Biol. Rev. Camb. Phil. Soc. *28*, 416–436.

Labrada, M., Clavell, M., Bebelagua, Y., León, J.d., Alonso, D.F., Gabri, M.R., Veloso, R.C., Vérez, V., and Fernández, L.E. (2010). Direct validation of NGcGM3 ganglioside as a new target for cancer immunotherapy. Exp. Opin. Biol. Ther. *10*, 153–162.

Lairson, L.L., Wakarchuk, W.W., and Withers, S.G. (2007). Alternative donor substrates for inverting and retaining glycosyltransferases. Chem. Commun. (Camb.), 365–367.

Lairson, L.L., Henrissat, B., Davies, G.J., and Withers, S.G. (2008). Glycosyltransferases: structures, functions, and mechanisms. Annu. Rev. Biochem. *77*, 521–555.

Laremore, T.N., Zhang, F., Dordick, J.S., Liu, J., and Linhardt, R.J. (2009). Recent progress and applications in glycosaminoglycan and heparin research. Curr. Opin. Chem. Biol. *13*, 633–640.

Lidholt, K., and Fjelstad, M. (1997). Biosynthesis of the *Escherichia coli* K4 capsule polysaccharide. J. Biol. Chem. *272*, 2682–2687.

Liu, R., Xu, Y., Chen, M., Weïwer, M., Zhou, X., Bridges, A.S., DeAngelis, P.L., Zhang, Q., Linhardt, R.J., and Liu, J. (2010). Chemoenzymatic design of heparan sulfate oligosaccharides. J. Biol. Chem. *285*, 34240–34249.

Logan, S.M., Conlan, J.W., Monteiro, M.A., Wakarchuk, W.W., and Altman, E. (2000). Functional genomics of *Helicobacter pylori*: identification of a beta-1,4 galactosyltransferase and generation of mutants with altered lipopolysaccharide. Mol. Microbiol. *35*, 1156–1167.

Logan, S.M., Altman, E., Mykytczuk, O., Brisson, J.R., Chandan, V., Schur, M.J., St Michael, F., Masson, A., Leclerc, S., Hiratsuka, K., *et al.* (2005). Novel biosynthetic functions of lipopolysaccharide rfaJ homologs from *Helicobacter pylori*. Glycobiology *15*, 721–733.

Lougheed, B., Ly, H.D., Wakarchuk, W.W., and Withers, S.G. (1999). Glycosyl fluorides can function as substrates for nucleotide phosphosugar-dependent glycosyltransferases. J. Biol. Chem. *274*, 37717–37722.

Maclennan, A.P. (1956). The Production of Capsules, Hyaluronic Acid and Hyaluronidase by Group A and Group C *Streptococci*. J. Gen. Microbiol. *14*, 134–142.

Merritt, E.A., and Hol, W.G.J. (1995). AB5 toxins. Curr. Opin. Struct. Biol. *5*, 165–171.

Monteiro, M.A. (2001). *Helicobacter pylori*: a wolf in sheep's clothing: the glycotype families of *Helicobacter pylori* lipopolysaccharides expressing histo-blood groups: structure, biosynthesis, and role in pathogenesis. Adv. Carbohydr. Chem. Biochem. *57*, 99–158.

Monteiro, M.A., Appelmelk, B.J., Rasko, D.A., Moran, A.P., Hynes, S.O., MacLean, L.L., Chan, K.H., Michael, F.S., Logan, S.M., O'Rourke, J., *et al.* (2000). Lipopolysaccharide structures of *Helicobacter pylori* genomic strains 26695 and J99, mouse model *H. pylori*Sydney strain, *H. pylori* P466 carrying sialyl Lewis, X., and *H. pylori* UA915 expressing Lewis B classification of *H. pylori* lipopolysaccharides into glycotype families. Eur. J. Biochem. *267*, 305–320.

Mousa, S.A. (2010). Heparin and low-molecular weight heparins in thrombosis and beyond. Methods Mol. Biol. *663*, 109–132.

Mulvey, G.L., Marcato, P., Kitov, P.I., Sadowska, J., Bundle, D.R., and Armstrong, G.D. (2003). Assessment in mice of the therapeutic potential of tailored, multivalent Shiga toxin carbohydrate ligands. J. Infect. Dis. *187*, 640–649.

Namdjou, D.J., Chen, H.M., Vinogradov, E., Brochu, D., Withers, S.G., and Wakarchuk, W.W. (2008). A beta-1,4-galactosyltransferase from *Helicobacter pylori* is an efficient and versatile biocatalyst displaying a novel activity for thioglycoside synthesis. Chembiochemistry *9*, 1632–1640.

Newburg, D.S., Ruiz-Palacios, G.M., and Morrow, A.L. (2005). Human milk glycans protect infants against enteric pathogens. Annu Rev Nutr *25*, 37–58.

Paton, A.W., Morona, R., and Paton, J.C. (2006). Designer probiotics for prevention of enteric infections. Nat. Rev. Microbiol. *4*, 193–200.

Paton, A.W., Morona, R., and Paton, J.C. (2010). Receptor-mimic probiotics: potential therapeutics for bacterial toxin-mediated enteric diseases. Exp. Rev. Gastroenterol. Hepatol. *4*, 253–255.

Paulson, J.C., Blixt, O., and Collins, B.E. (2006). Sweet spots in functional glycomics. Nat. Chem. Biol. *2*, 238–248.

Pavliak, V., Brisson, J.R., Michon, F., Uhrin, D., and Jennings, H.J. (1993). Structure of the sialylated L3 lipopolysaccharide of *Neisseria meningitidis*. J. Biol. Chem. *268*, 14146–14152.

Persson, K., Ly, H.D., Dieckelmann, M., Wakarchuk, W.W., Withers, S.G., and Strynadka, N.C. (2001). Crystal structure of the retaining galactosyltransferase LgtC from *Neisseria meningitidis* in complex with donor and acceptor sugar analogs. Nat. Struct. Biol. *8*, 166–175.

Perugino, G., Trincone, A., Rossi, M., and Moracci, M. (2004). Oligosaccharide synthesis by glycosynthases. Trends Biotechnol. *22*, 31–37.

Priem, B., Gilbert, M., Wakarchuk, W.W., Heyraud, A., and Samain, E. (2002). A new fermentation process allows large-scale production of human milk oligosaccharides by metabolically engineered bacteria. Glycobiology *12*, 235–240.

Pudelko, M., Bull, J., and Kunz, H. (2010). Chemical and chemoenzymatic synthesis of glycopeptide selectin ligands containing sialyl Lewis X structures. Chembiochemistry *11*, 904–930.

Pukin, A.V., Weijers, C.A., van Lagen, B., Wechselberger, R., Sun, B., Gilbert, M., Karwaski, M.F., Florack, D.E., Jacobs, B.C., Tio-Gillen, A.P., *et al.* (2008). GM3, GM2 and GM1 mimics designed for biosensing: chemoenzymatic synthesis, target affinities and 900 MHz NMR analysis. Carbohydr. Res. *343*, 636–650.

Randriantsoa, M., Drouillard, S., Breton, C., and Samain, E. (2007). Synthesis of globopentaose using a novel beta-1,3-galactosyltransferase activity of the *Haemophilus influenzae* beta-1,3-N-acetylgalactosaminyltransferase LgtD. FEBS Lett. *581*, 2652–2656.

Rasko, D.A., Wang, G., Palcic, M.M., and Taylor, D.E. (2000). Cloning and Characterization of the α(1,3/4) Fucosyltransferase of *Helicobacter pylori*. J. Biol. Chem. *275*, 4988–4994.

Rosner, H., Grimmecke, H.-D., Knirel, Y.A., and Shashkov, A.S. (1992). Hyaluronic acid and a (1-->4)-[beta]-D-xylan, extracellular polysaccharides of *Pasteurella multocida* (Carter type A) strain 880. Carbohydr. Res. *223*, 329–333.

Sauerzapfe, B., Krenek, K., Schmiedel, J., Wakarchuk, W.W., Pelantova, H., Kren, V., and Elling, L. (2009). Chemo-enzymatic synthesis of poly N-acetyllactosamine (poly LacNAc) structures and their characterization for CGL2-galectin-mediated binding of ECM glycoproteins to biomaterial surfaces. Glycoconj. J. *26*, 141–159.

Schauer, R. (2009). Sialic acids as regulators of molecular and cellular interactions. Curr. Opin. Struct. Biol. *19*, 507–514.

Schneider, J.S., Sendek, S., Daskalakis, C., and Cambi, F. (2010). GM1 ganglioside in Parkinson's disease: Results of a five year open study. Journal of the Neurol. Sci. *292*, 45–51.

Schweda, E.K.H., Richards, J.C., Hood, D.W., and Moxon, E.R. (2007). Expression and structural diversity of the lipopolysaccharide of *Haemophilus influenzae*: Implication in virulence. Int. J. Med. Microbiol. *297*, 297–306.

Seeberger, P.H. (2008). Automated carbohydrate synthesis as platform to address fundamental aspects of glycobiology – current status and future challenges. Carbohydr. Res. *343*, 1889–1896.

Shaikh, F.A., and Withers, S.G. (2008). Teaching old enzymes new tricks: engineering and evolution of glycosidases and glycosyl transferases for improved glycoside synthesis. Biochem. Cell. Biol. *86*, 169–177.

Shao, J., Li, M., Jia, Q., Lu, Y., and Wang, P.G. (2003). Sequence of Escherichia coli O128 antigen biosynthesis cluster and functional identification of an alpha-1,2-fucosyltransferase. FEBS Lett. *553,* 99–103.

Sisu, C., Baron, A.J., Branderhorst, H.M., Connell, S.D., Weijers, C.A., de Vries, R., Hayes, E.D., Pukin, A.V., Gilbert, M., Pieters, R.J., *et al.* (2009). The influence of ligand valency on aggregation mechanisms for inhibiting bacterial toxins. Chembiochemistry *10,* 329–337.

Solomon, D., Kitov, P.I., Paszkiewicz, E., Grant, G.A., Sadowska, J.M., and Bundle, D.R. (2005). Heterobifunctional multivalent inhibitor-adaptor mediates specific aggregation between Shiga toxin and a pentraxin. Org. Lett. *7,* 4369–4372.

Sperandio, M., Gleissner, C.A., and Ley, K. (2009). Glycosylation in immune cell trafficking. Immunol. Rev. *230,* 97–113.

Spitalnik, P.F., and Spitalnik, S.L. (1995). The P blood group system: Biochemical, serological, and clinical aspects. Trans. Med. Rev. *9,* 110–122.

St. Michael, F., Li, J., Howard, M.D., Duncan, A.J., Inzana, T.J., and Cox, A.D. (2005a). Structural analysis of the oligosaccharide of *Histophilus somni* (*Haemophilus somnus*) strain 2336 and identification of several lipooligosaccharide biosynthesis gene homologues. Carbohydr. Res. *340,* 665–672.

St. Michael, F., Vinogradov, E., Li, J., and Cox, A.D. (2005b). Structural analysis of the lipopolysaccharide from *Pasteurella multocida* genome strain Pm70 and identification of the putative lipopolysaccharide glycosyltransferases. Glycobiology *15,* 323–333.

Steenbergen, S.M., and Vimr, E.R. (2003). Functional relationships of the sialyltransferases involved in expression of the polysialic acid capsules of *Escherichia coli* K1 and K92 and Neisseria meningitidis groups B or C. J. Biol. Chem. *278,* 15349–15359.

Stollerman, G.H., and Dale, J.B. (2008). The importance of the Group A *Streptococcus* capsule in the pathogenesis of human infections: a historical perspective. Clin. Infect. Dis. *46,* 1038–1045.

Taylor, M.E., and Drickamer, K. (2009). Structural insights into what glycan arrays tell us about how glycan-binding proteins interact with their ligands. Glycobiology *19,* 1155–1162.

Tong, Y., Arking, D., Ye, S., Reinhold, B., Reinhold, V., and Stein, D.C. (2002). *Neisseria gonorrhoeae* strain PID2 simultaneously expresses six chemically related lipooligosaccharide structures. Glycobiology *12,* 523–533.

Townson, K., Boffey, J., Nicholl, D., Veitch, J., Bundle, D., Zhang, P., Samain, E., Antoine, T., Bernardi, A., Arosio, D., *et al.* (2007). Solid phase immunoadsorption for therapeutic and analytical studies on neuropathy-associated anti-GM1 antibodies. Glycobiology *17,* 294–303.

Ukkonen, P., Varis, K., Jernfors, M., Herva, E., Jokinen, J., Ruokokoski, E., Zopf, D., and Kilpi, T. (2000). Treatment of acute otitis media with an antiadhesive oligosaccharide: a randomised, double-blind, placebo-controlled trial. Lancet *356,* 1398–1402.

Vann, W.F., Schmidt, M.A., Jann, B., and Jann, K. (1981). The structure of the capsular polysaccharide (K5 antigen) of urinary-tract-infective *Escherichia coli* *010*,K5:H4. Eur. J. Biochem. *116,* 359–364.

Vasiliu, D., Razi, N., Zhang, Y., Jacobsen, N., Allin, K., Liu, X., Hoffmann, J., Bohorov, O., and Blixt, O. (2006). Large-scale chemoenzymatic synthesis of blood group and tumor-associated poly N-acetyllactosamine antigens. Carbohydr. Res. *341,* 1447–1457.

Vaughan, M.D., Johnson, K., DeFrees, S., Tang, X., Warren, R.A.J., and Withers, S.G. (2006). Glycosynthase-mediated synthesis of glycosphingolipids. J. Am. Chem. Soc. *128,* 6300–6301.

Wakarchuk, W. (2011). Glycosyltransferases involved in molecular mimicry. In Carbohydrate-Modifying Biocatalysts, Grunwald, P., ed. (London, UK, Pan Stanford Publishing Pte Ltd), pp. x–xx. In press.

Wakarchuk, W.W., Cunningham, A., Watson, D.C., and Young, N.M. (1998a). Role of paired basic residues in the expression of active recombinant galactosyltransferases from the bacterial pathogen *Neisseria meningitidis.* Protein Eng. *11,* 295–302.

Wakarchuk, W.W., Gilbert, M., Martin, A., Wu, Y., Brisson, J.R., Thibault, P., and Richards, J.C. (1998b). Structure of an alpha-2,6-sialylated lipooligosaccharide from *Neisseria meningitidis* immunotype L1. Eur. J. Biochem. *254,* 626–633.

Wakarchuk, W., Martin, A., Jennings, M.P., Moxon, E.R., and Richards, J.C. (1996). Functional relationships of the genetic locus encoding the glycosyltransferase enzymes involved in expression of the lacto-N-neotetraose terminal lipopolysaccharide structure in *Neisseria meningitidis.* J. Biol. Chem. *271,* 19166–19173.

Wang, G., Boulton, P.G., Chan, N.W.C., Palcic, M.M., and Taylor, D.E. (1999). Novel *Helicobacter pylori* alpha-1,2-fucosyltransferase, a key enzyme in the synthesis of Lewis antigens. Microbiology *145,* 3245–3253.

Wang, G., Ge, Z., Rasko, D.A., and Taylor, D.E. (2000). Lewis antigens in *Helicobacter pylori*: biosynthesis and phase variation. Mol. Microbiol. *36,* 1187–1196.

Wang, W., Hu, T., Frantom, P.A., Zheng, T., Gerwe, B., del Amo, D.S., Garret, S., Seidel, R.D., and Wu, P. (2009). Chemoenzymatic synthesis of GDP-L-fucose and the Lewis X glycan derivatives. Proc. Natl. Acad. Sci. U.S.A. *106,* 16096–16101.

Wang, Z., Gilbert, M., Eguchi, H., Yu, H., Cheng, J., Muthana, S., Zhou, L., Wang, P.G., Chen, X., and Huang, X. (2008). Chemoenzymatic syntheses of tumor-associated carbohydrate antigen Globo-H and stage-specific embryonic antigen 4. Adv. Synth. Catal. *350,* 1717–1728.

Wennekes, T., van den Berg, Richard, J.B.H.N., Boot, Rolf, G., van der Marel, Gijsbert, A., Overkleeft, Herman, S., and Aerts, Johannes, M.F.G. (2009). Glycosphingolipids – nature, function, and pharmacological modulation. Angewandte Chemie Int. Ed. *48,* 8848–8869.

Wolny, P.M., Banerji, S., Gounou, C., Brisson, A.R., Day, A.J., Jackson, D.G., and Richter, R.P. (2010). Analysis of CD44–hyaluronan interactions in an artificial membrane system. J. Biol. Chem. *285,* 30170–30180.

Yan, F., Wakarchuk, W.W., Gilbert, M., Richards, J.C., and Whitfield, D.M. (2000). Polymer-supported and chemoenzymatic synthesis of the *Neisseria meningitidis* pentasaccharide: a methodological comparison. Carbohydr. Res. *328,* 3–16.

Yan, F., Gilbert, M., Wakarchuk, W.W., Brisson, J.R., and Whitfield, D.M. (2001). Chemoenzymatic iterative synthesis of difficult linkages of oligosaccharides on soluble polymeric supports. Org. Lett. *3,* 3265–3268.

Yang, G., Rich, J.R., Gilbert, M., Wakarchuk, W.W., Feng, Y., and Withers, S.G. (2010). Fluorescence activated cell sorting as a general ultra-high-throughput screening method for directed evolution of glycosyltransferases. J. Am. Chem. Soc. *132,* 10570–10577.

Yavuz, E., Drouillard, S., Samain, E., Roberts, I., and Priem, B. (2008). Glucuronylation in *Escherichia coli* for the bacterial synthesis of the carbohydrate moiety of nonsulfated HNK-1. Glycobiology *18,* 152–157.

Yu, H., Chokhawala, H., Karpel, R., Wu, B., Zhang, J., Zhang, Y., Jia, Q., and Chen, X. (2005). A multifunctional *Pasteurella multocida* sialyltransferase: a powerful tool for the synthesis of sialoside libraries. J. Am. Chem. Soc. *127,* 17618–17619.

Zhu, P., Klutch, M.J., Bash, M.C., Tsang, R.S., Ng, L.K., and Tsai, C.M. (2002). Genetic diversity of three lgt loci for biosynthesis of lipooligosaccharide (LOS) in *Neisseria* species. Microbiology *148,* 1833–1844.

Zivkovic, A.M., German, J.B., Lebrilla, C.B., and Mills, D.A. (2010). Human milk glycobiome and its impact on the infant gastrointestinal microbiota. Proc. Natl. Acad. Sci. *108(1),* 4653–4658.

13 Metabolic Labelling of Bacterial Glycans with Chemical Reporters

Danielle H. Dube

Abstract

Though long believed to be absent from bacteria, glycoproteins are now known to be synthesized in a number of bacterial species. Traditional methods to study glycoproteins have revealed fascinating glycan structures that are exclusively found in bacteria and are frequently linked to pathogenesis. In recent years, these methods have been augmented by a complementary approach, termed metabolic oligosaccharide engineering (MOE), to facilitate large-scale systematic studies of the entire complement of glycan structures in bacteria, referred to as bacterial glycomics. In MOE, bacterial glycans are metabolically labelled with unique chemical functionalities, called chemical reporters. Labelling bacterial glycans in this manner facilitates glycoprotein detection and enrichment. In addition to enabling glycoprotein profiling, the labelled glycans can undergo selective covalent bond formation, thereby permitting further applications. For example, labelled glycans are poised to disrupt the bacterial surface coat, target bacterial cells with toxins, trap glycan-based host–pathogen interactions, and image dynamic changes in glycosylation. This chapter focuses on MOE methodology, its application to the study of bacterial glycoproteins, and its future role in treating infectious disease.

Introduction

Owing to the absence of subcellular organelles, bacteria were long believed to lack the ability to synthesize glycoproteins. Emerging evidence, though, has shown unequivocally the existence of glycoproteins in several bacterial species and the presence of bacterial glycosylation enzymes that synthesize these structures (Benz and Schmidt, 2002; Power and Jennings, 2003; Szymanski and Wren, 2005; Fletcher *et al.*, 2009). Moreover, mounting evidence demonstrates that certain bacterial pathogens synthesize glycoproteins with distinctive, non-human glycans that are linked to virulence (Schmidt *et al.*, 2003; Dube *et al.*, 2010). Therefore, continued efforts to identify bacterial glycoproteins and elucidate their function have the potential to reveal mechanisms of pathogenesis and unveil novel targets for therapeutic intervention.

Although bacterial glycoproteins are frequently linked to pathogenesis and provide avenues for targeted intervention, most bacterial glycomes remain poorly understood. With the exception of a small number of well-characterized examples (Szymanski *et al.*, 1999; Linton *et al.*, 2002; Schirm *et al.*, 2003; Kelly *et al.*, 2006; Schoenhofen *et al.*, 2006), fundamental questions in bacterial glycobiology, such as the identity of glycan-modified proteins, the structure of the modifying glycan, the machinery responsible for synthesizing these structures, and finally the biological function of these glycoproteins, remain largely unanswered. To address these questions, bacterial glycoproteins must be inventoried.

A suite of strategies is being utilized to discover novel bacterial glycoproteins and elucidate their functional roles (Balonova *et al.*, 2009; Dube *et al.*, 2010). Traditional methods rely on carbohydrate-binding proteins (lectins) or periodic acid/hydrazide chemistry to label and detect glycoproteins (Doig *et al.*, 1996; Linton *et al.*, 2002; Hsu *et al.*, 2006). The former method takes advantage of a lectin's affinity for particular glycan structures, whereas the latter approach non-specifically exploits the ubiquity of *cis*-diols in glycans. These approaches have proven utility in bacterial glycomics, as they have been used to identify the majority of bacterial glycoproteins characterized to date. A complementary approach termed metabolic oligosaccharide engineering (MOE) has recently augmented traditional approaches to studying bacterial glycoproteins. In MOE, endogenous carbohydrate biosynthetic pathways are utilized to metabolically label bacterial glycans with unique chemical motifs, called chemical reporters (Keppler *et al.*, 2001; Dube and Bertozzi, 2003; Prescher and Bertozzi, 2006). Once labelled, the bacterial glycans can be detected, enriched, and identified, thus enabling glycomics efforts (Laughlin *et al.*, 2006; Laughlin and Bertozzi, 2007). In addition to facilitating bacterial glycomics efforts, MOE permits further applications such as disrupting bacterial surface glycans, targeting bacterial cells based on distinctive glycan expression, identifying bacterial glycan–protein interactions, and imaging bacterial glycan dynamics (Dube and Bertozzi, 2003). Below, I provide an overview of MOE methodology, its application to the study of bacterial glycoproteins, and its potential impact on the development of glycosylation-based therapeutic strategies.

Metabolic oligosaccharide engineering methodology

MOE is a versatile method that introduces subtle structural modifications into monosaccharide residues within cellular glycans (Keppler *et al.*, 2001; Dube and Bertozzi, 2003). MOE was pioneered in mammalian cells by Reutter and colleagues, who demonstrated that the cell's biosynthetic machinery can be employed to install unnatural monosaccharides into glycans (Kayser *et al.*, 1992; Keppler *et al.*, 1995, 2001). Briefly, synthetically derived unnatural monosaccharides fed to cells or animals intercept permissive carbohydrate biosynthetic pathways, are transformed into activated sugars, and are ultimately appended to cellular glycans. The ability to engineer cell surface glycosylation in this manner provides a means of labelling glycoconjugates and modulating carbohydrate-mediated interactions. As an extension of this method, Bertozzi and coworkers demonstrated that monosaccharides containing small, unique functional groups known as chemical reporters can be processed by mammalian cells and incorporated into their glycans (Fig. 13.1) (Mahal *et al.*, 1997; Saxon and Bertozzi, 2000). Glycans containing chemical reporters can then be tagged with exquisitely selective reactive partners (Fig. 13.1) (Mahal *et al.*, 1997; Saxon and Bertozzi, 2000). This two-step process enables the covalent attachment of myriad biological probes, such as epitope tags to facilitate glycoprotein enrichment and identification, bulky groups to perturb biological interactions, therapeutic compounds for targeted cell destruction, cross-linkers for identification of glycan interactions, and fluorophores to permit glycan visualization.

To be applicable to the study of bacterial glycans, MOE requires that an unnatural monosaccharide be taken up by the cell and processed by the cell's metabolic pathways. Unnatural monosaccharides are typically accessed by chemical or chemoenzymatic syntheses (Goon *et al.*, 2003b; Koenigs *et al.*, 2009; Liu *et al.*, 2009). Transporters embedded in the bacterial membrane enable the uptake of some charged monosaccharides (Schilling *et al.*, 2001; Goon *et al.*, 2003b). In the absence of the appropriate transporter, however, temporary masking of hydrophilic moieties on the sugar with hydrophobic groups maximizes the efficiency of unnatural monosaccharide uptake by the cell (Sarkar *et al.*, 1995; Luchansky *et al.*, 2004); hydrophobic acyl groups (e.g. acetyl, butyl) increase transport across the lipophilic membrane approximately 900-fold and are removed by non-specific esterases once the sugar is in the cell (Jones *et al.*, 2004). In one reported example, metabolic incorporation of an unnatural sugar into bacterial glycoproteins was only detectable if bacteria were supplemented with a sugar that had its hydroxyl groups masked by acetyl groups (Koenigs *et al.*, 2009).

To maximize the likelihood that the cell's metabolic pathways will process the unnatural

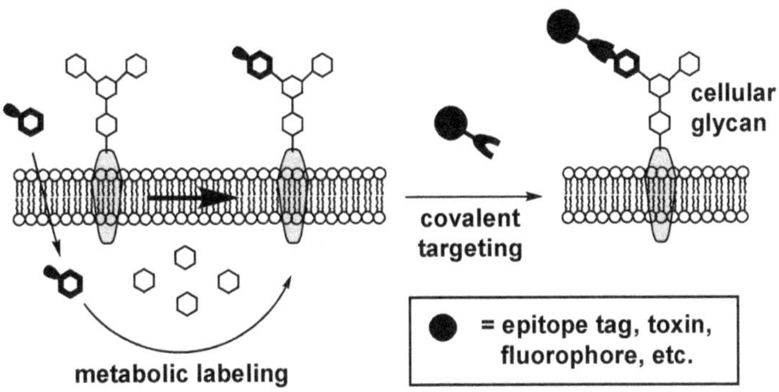

Figure 13.1 Metabolic oligosaccharide engineering (MOE). Cellular glycans are metabolically labelled with an unnatural sugar that contains a chemical reporter (). In a second step, the chemical reporter is tagged with a selective reactive partner ( ) for the attachment of myriad biological probes (●).

monosaccharide, the sugar should be a close analogue of an endogenous monosaccharide. The modification must introduce so minimal a structural perturbation that the unnatural sugar remains a substrate of carbohydrate processing enzymes. The modifications permitted depend upon the substrate flexibility of the biosynthetic enzymes in a particular pathway, which can be experimentally determined. Tolerated modifications in bacteria range from the addition of a few extra atoms, such as the azide ($-N_3$) chemical reporter (Liu *et al.*, 2009), to the addition of a long seven-carbon heptanoyl group (Goon *et al.*, 2003b). If a bottleneck step is identified, supplementing cells with substrate analogues that intercept the pathway downstream of bottleneck enzymes can be used to overcome modification constraints.

MOE to install chemical reporters in bacterial glycans has additional restrictions. The chemical reporter itself must meet the requirements of bioorthogonality: it must be stable in water; it must be unreactive with endogenous biological functional groups; it must be non-toxic; and it must have a unique mode of reactivity that enables selective chemical elaboration with a reactive partner that is bioorthogonal (Prescher and Bertozzi, 2005). Several chemical reporters and bioorthogonal chemistries have been developed to meet these requirements (Table 13.1). For example, the ketone is a chemical reporter that has been synthetically introduced into sugar analogues and reacts with amino-oxy or hydrazide probes (Mahal *et al.*, 1997; Yarema *et al.*, 1998). Similarly, the alkyne is elaborated by copper-catalysed azide-alkyne cycloaddition (click chemistry) (Rostovtsev *et al.*, 2002; Speers *et al.*, 2003). Finally, the azide is the most popular chemical reporter owing to its small size and selective reactivity with myriad partners, including (1) phosphine probes via Staudinger ligation (Saxon and Bertozzi, 2000), (2) alkyne probes via copper-catalysed click chemistry, (3) cyclooctyne probes via copper-free click chemistry (Agard *et al.*, 2004; Baskin *et al.*, 2007), and (4) cyclopropenone probes via photo-triggered click chemistry (Poloukhtine *et al.*, 2009). All of these chemistries are compatible with live cells (Saxon and Bertozzi, 2000; Agard *et al.*, 2004; Baskin *et al.*, 2007; Hong *et al.*, 2010). The Staudinger ligation and copper-free click chemistry even proceed in living animals (e.g. mice, zebrafish, and nematodes), where these chemistries tag azide-labelled glycans on cells (Prescher *et al.*, 2004; Dube *et al.*, 2006; Laughlin *et al.*, 2008; Laughlin and Bertozzi, 2009b; Chang *et al.*, 2010). Ketone-containing and azide-containing variants of several sugars are commercially available (supplied by Thermo Scientific, Invitrogen and Sigma), as are corresponding reactive partners.

In addition to these bioorthogonal reporters, several groups have reported the metabolic incorporation of sugar analogues containing chemical cross-linkers that facilitate unequivocal identification of glycan-protein binding interactions (Han *et al.*, 2005; Tanaka and Kohler, 2008; Bond *et al.*, 2010; Yu *et al.*, 2010). Metabolically incorporated photo-cross-linking reporters include diazirine and aryl azide groups (Table 13.1). Once irradiated, these groups form reactive species that crosslink to nearby biological interaction partners.

Metabolic labelling of bacterial glycans: progress and future directions

MOE has found wide-ranging utility in eukaryotic cells and in animals, where it has been used to profile glycoproteins, disrupt surface glycans, target cells based on glycan expression, identify glycan–protein interactions, and image glycan dynamics.

Table 13.1 Chemical reporters and their reactions used in MOE. R = glycan. R' = biological probe except where noted.

chemical reporter	reactive partner	covalent product
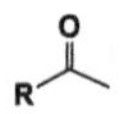 ketone	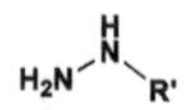 hydrazide	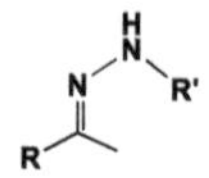
	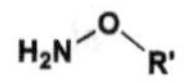 amino-oxy	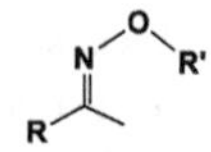
R–N_3 azide	Staudinger ligation 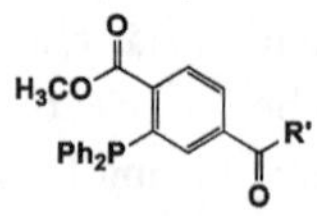	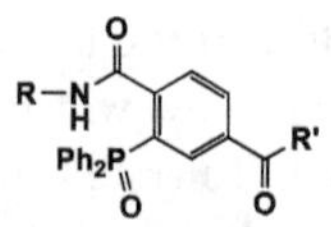
	Click chemistry R'—≡ , Cu(I), ligand	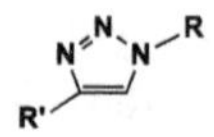
	Cu-free click chemistry 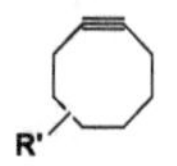	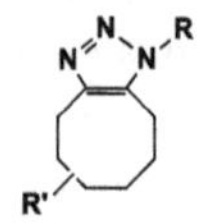
	Phototriggered click chemistry 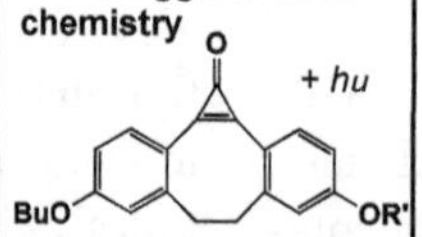	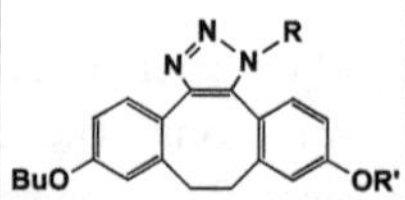
R—≡ alkyne	Click chemistry R'—N_3 , Cu(I), ligand	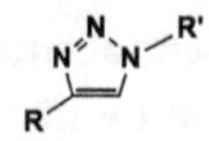
 diazirine	Photocrosslinking H—R' + *hu* R' = interaction partner	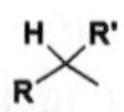
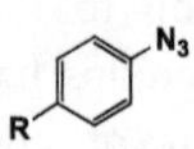 aryl azide	Photocrosslinking H—R' + *hu* R' = interaction partner	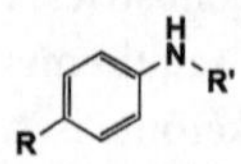

Below I describe the progress that has been made towards extending these MOE applications to bacterial systems and future directions of the field.

Profiling bacterial glycoproteins

MOE provides a means for proteomic analysis of glycoproteins and for rapidly identifying glycoprotein fingerprints associated with particular cell types – information that could prompt the development of glycosylation-based therapeutics and diagnostics. For this application, glycans are first metabolically labelled with a chemical reporter and then covalently modified with an epitope tag (Fig. 13.2A). For example, azide-containing analogues of ManNAc, *N*-acetylglucosamine (GlcNAc), *N*-acetylgalactosamine (GalNAc), and fucose were employed to profile eukaryotic glycoproteins modified with sialic acid, O-GlcNAc, mucin-type O-linked glycans, and fucose, respectively (Saxon and Bertozzi, 2000; Hang *et al.*, 2003; Vocadlo *et al.*, 2003; Sprung *et al.*, 2005; Rabuka *et al.*, 2006; Sawa *et al.*, 2006). Following cell lysis, labelled glycans were covalently elaborated with reactive partners containing epitope tags, such as biotin and the FLAG peptide, to enable detection by Western blot analysis and enrichment by affinity-column chromatography. Western blot analyses indicated the presence of glycosylated proteins and revealed glycoprotein fingerprints, while enrichment enabled analytical characterization of modified proteins and their modifying glycans (Fig. 13.2A). MOE-based glycoprotein profiling efforts have been carried out successfully in cancer cells, zebrafish, and mice (Prescher *et al.*, 2004; Sprung *et al.*, 2005; Dube *et al.*, 2006; Nandi *et al.*, 2006; Laughlin *et al.*, 2008), and recently have been extended to bacterial systems (Koenigs *et al.*, 2009).

Similar to work done previously with MOE in eukaryotic systems, my laboratory demonstrated that MOE can be used to profile bacterial glycoproteins (Koenigs *et al.*, 2009). Intrigued by the known links between bacterial glycosylation and virulence, yet relative paucity of information about bacterial glycoproteins as a whole, we sought to address fundamental questions in bacterial glycobiology using MOE. Given that *N*-acetylglucosamine (GlcNAc) is a common precursor of bacterial monosaccharides, we hypothesized that the azide-containing sugar *N*-azidoacetylglucosamine (GlcNAz), an analogue of GlcNAc, could metabolically label a

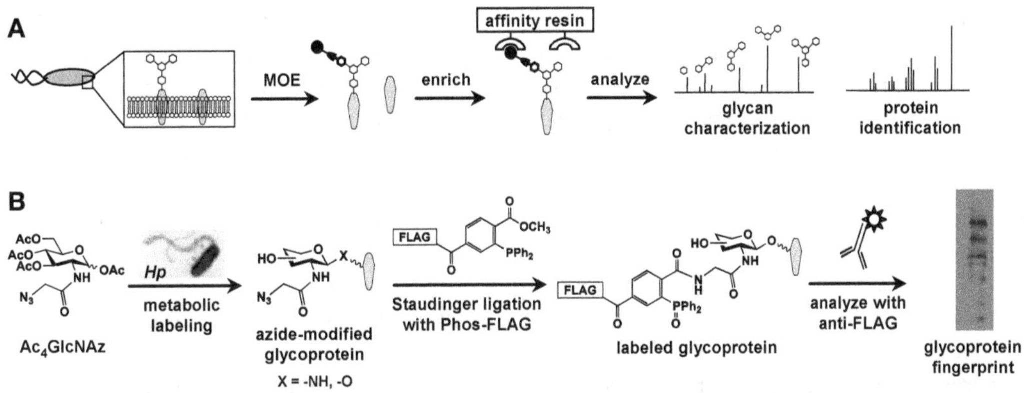

Figure 13.2 Metabolic profiling of bacterial glycoproteins. (A) General strategy for enriching and identifying bacterial glycoproteins using MOE. First, bacterial glycans are metabolically labelled with a chemical reporter and elaborated with a reactive partner containing an epitope tag (●). Next, glycoproteins are enriched from other cellular proteins by virtue of an affinity resin that binds exclusively to the epitope tag. Finally, enriched glycoproteins are analysed to characterize bacterial glycans and identify glycosylated proteins. (B) As an example of metabolic profiling of bacterial glycoproteins, Koenigs *et al.* demonstrated that *Helicobacter pylori*'s (*Hp*'s) glycoproteins can be labelled using the azide-containing sugar $Ac_4GlcNAz$, can be subsequently reacted via Staudinger ligation with Phos-FLAG, and can then be detected by Western blot analysis. A large number of uncharacterized N- and O-linked glycoproteins were revealed.

broad range of bacterial glycoproteins. To test this hypothesis, we utilized the membrane permeable peracetylated GlcNAz (Ac_4GlcNAz) in metabolic labelling experiments. As a test organism, we selected *H. pylori*, a bacterium with two previously characterized glycoproteins with a known link to virulence. In our experiments, *H. pylori* were supplemented with Ac_4GlcNAz and later lysed. The harvested lysate was probed for the presence of azide-labelled proteins using the Staudinger ligation with a phosphine probe comprising a FLAG peptide (Phos-FLAG) (Fig. 13.2B). *H. pylori* supplemented with Ac_4GlcNAz displayed a large number of azide-labelled proteins (Fig. 13.2B), suggesting that *H. pylori* synthesizes far more than the two previously characterized flagellin glycoproteins (Koenigs *et al.*, 2009). Treatment of *H. pylori*'s azide-labelled proteins with enzymes that remove asparagine-linked (N-linked) and serine/threonine-linked (O-linked) glycans led to a diminishment of azide-dependent signal. These data provide strong evidence that *H. pylori* synthesizes a suite of both N-linked and O-linked glycoproteins. Efforts are currently under way in my laboratory to enrich *H. pylori*'s glycoproteins using methodology developed to enrich azide-labelled proteins from mammalian cells (Laughlin and Bertozzi, 2007). Once enriched, these glycoproteins and their glycans will be identified and characterized. Follow-up experiments will be aimed at identifying *H. pylori*'s N-linked and O-linked general glycosylation systems, determining the role of *H. pylori*'s glycans in pathogenesis, and targeting *H. pylori* based on its unique glycans. This work will reveal further mechanisms of *H. pylori*'s pathogenesis and unveil novel targets for therapeutic intervention. Ultimately, the information gained by MOE could lead to the development of new strategies to eradicate *H. pylori* infection.

Future experiments in this area should focus on establishing the generality of MOE for labelling bacterial glycans, as MOE has the potential to have a substantial impact in the area of bacterial glycomics. Discovery of the full complement of bacterial glycoproteins will answer basic questions in the field of bacterial glycobiology and begin to unravel the evolutionary significance of these underexplored biomolecules. Moreover, given the link between bacterial glycoproteins and virulence (Benz and Schmidt, 2002; Schmidt *et al.*, 2003; Dube *et al.*, 2010), these analyses are likely to reveal novel drug targets. Finally, analyses of glycoprotein profiles will reveal fingerprints that are unique to particular pathogens, which could lead to the development of new diagnostics.

Disrupting bacterial surface glycans

Carbohydrate binding interactions mediate cellular binding events that are crucial in microbial infection (Reid *et al.*, 2010), development (Haltiwanger and Lowe, 2004), and the immune response (Rosen, 2004). For example, the influenza virus binds to sialic acid residues as a means to invade host cells, polysialic acid (PSA) modulates the adhesion of neural cells, and fucosylated glycans are required for neuronal development (Johnson and Pekas, 1964; Acheson *et al.*, 1991; Kalovidouris *et al.*, 2005; Murrey *et al.*, 2006). The ability to modulate these binding events provides an opportunity to prevent microbial infection, control development, and tune the immune response. MOE provides a means to disrupt glycan structures; when cells process an unnatural sugar, glycan structures are modified or truncated (Fig. 13.3A), often causing altered glycan function (Dube and Bertozzi, 2003). For example, Reutter and coworkers demonstrated that metabolically incorporated *N*-acyl homologues of sialic acid disrupt influenza virus infection of mammalian cells, presumably by inhibiting viral binding to the modified cell surface sialic acid (Keppler *et al.*, 1995). Moreover, Mahal, Charter and others found that an *N*-acyl homologue of sialic acid inhibits PSA formation and therefore alters neuronal development (Mahal *et al.*, 2001; Buttner *et al.*, 2002; Charter *et al.*, 2002). Additionally, Hsieh-Wilson and colleagues reported that metabolic incorporation of the chain-terminating sugar 2-deoxy-galactose into neuronal glycans inhibited addition of fucose at the 2-hydroxyl position of galactose, ultimately leading to under-fucosylated neurons with stunted neurite outgrowth and delayed synapse formation (Murrey *et al.*, 2006). Finally, there are a number of reports describing unnatural sialic acids interfering with lectin binding at the cell surface (Keppler *et al.*, 1995;

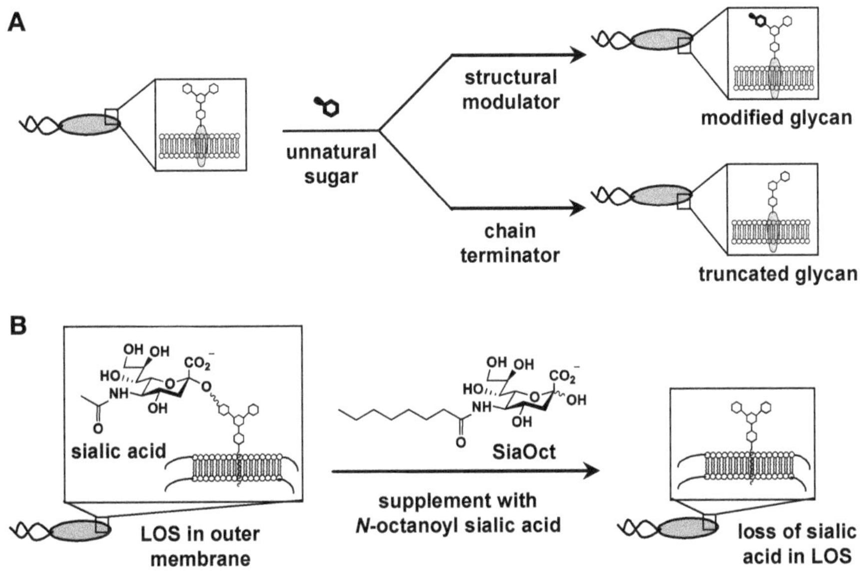

Figure 13.3 Disrupting bacterial surface glycans. (A) General strategy for disrupting bacterial glycans using MOE. First, bacterial cells are supplemented with an unnatural sugar. The unnatural sugar can either act as a structural modulator and be incorporated into modified glycans (top) or act as a chain terminator and lead to truncated glycans (bottom). In each case, the structure of the glycan is disrupted; this disruption can potentially alter biological function. (B) In an example of this application, Goon *et al.* demonstrated that supplementing *Haemophilus ducreyi* with *N*-octanoyl sialic acid (SiaOct) leads to loss of sialic acid in *H. ducreyi*'s lipo-oligosaccharide (LOS).

Herrmann *et al.*, 1997; Oetke *et al.*, 2002; Luchansky and Bertozzi, 2004). In each of these cases, modulating glycan structure altered glycan function.

Similar to reports that MOE can disrupt eukaryotic glycans, there is also direct evidence that MOE can alter bacterial glycan structures. In a set of metabolic feeding experiments, Bertozzi, Gibson and coworkers demonstrated that the extent of *Haemophilus ducreyi*'s lipo-oligosaccharide (LOS) sialylation can be diminished by supplementing *H. ducreyi* with elongated, *N*-acyl homologues of sialic acid (Goon *et al.*, 2003b). Remarkably, feeding *H. ducreyi* the eight-carbon analogue *N*-octanoyl sialic acid leads to complete loss of sialic acid in *H. ducreyi*'s LOS (Fig. 13.3B). In contrast, shorter analogues, such as *N*-propanoyl sialic acid, are incorporated into *H. ducreyi*'s LOS in place of sialic acid residues. This example demonstrates that, depending on the structure, unnatural sugar analogues can metabolically inhibit bacterial glycosylation enzymes and lead to truncated glycans, or they can be processed by the cell and incorporated into modified glycan structures. By disrupting *H. ducreyi*'s LOS structure in this manner, the role of LOS sialylation in *H. ducreyi*'s adherence to host cells, immunogenicity, and virulence can be determined.

The observation that MOE can modulate bacterial glycan structures should stimulate an explosion of interest in this area. Given the importance of bacterial glycans in pathogenesis, disrupting bacterial glycan structures using MOE has the potential to modulate the virulence of pathogens and interfere with microbial infections. In addition, altering the structures of bacterial glycans using MOE provides an approach to decipher the biological functions and importance of these structures. Future efforts will reveal a variety of unnatural monosaccharides that prevent the formation of bacterial glycans linked to virulence, monosaccharides that ultimately have the potential to treat infectious disease.

Targeting bacterial cells based on distinctive glycan expression

Glycosylation changes with the physiological state of the cell, and specific glycans are found preferentially or exclusively on certain cells (Dube and Bertozzi, 2005). Therefore, one might use the expression of a distinctive glycan structure as the basis for selectively targeting a population of invasive cells, such as bacterial pathogens, for destruction. MOE provides a unique opportunity to achieve this goal, as it has been employed to modify a subset of glycans with chemical reporters, ultimately enabling covalent targeting of modified cells with toxic compounds (Fig. 13.4A). Bertozzi and coworkers utilized MOE to metabolically replace sialic acid residues on cancer cells with a ketone-containing analogue of sialic acid, thus covering cancer cells with ketones (Mahal *et al.*, 1997; Yarema *et al.*, 1998). In a kinetic study using biotin hydrazide as a model drug, Nauman *et al.* discovered that the number of molecules covalently delivered to ketone-covered cells is on par with the number delivered by antibody-based targeting methods (Nauman and Bertozzi, 2001). Towards a similar goal, Azoulay *et al.* developed phosphine-based prodrugs that

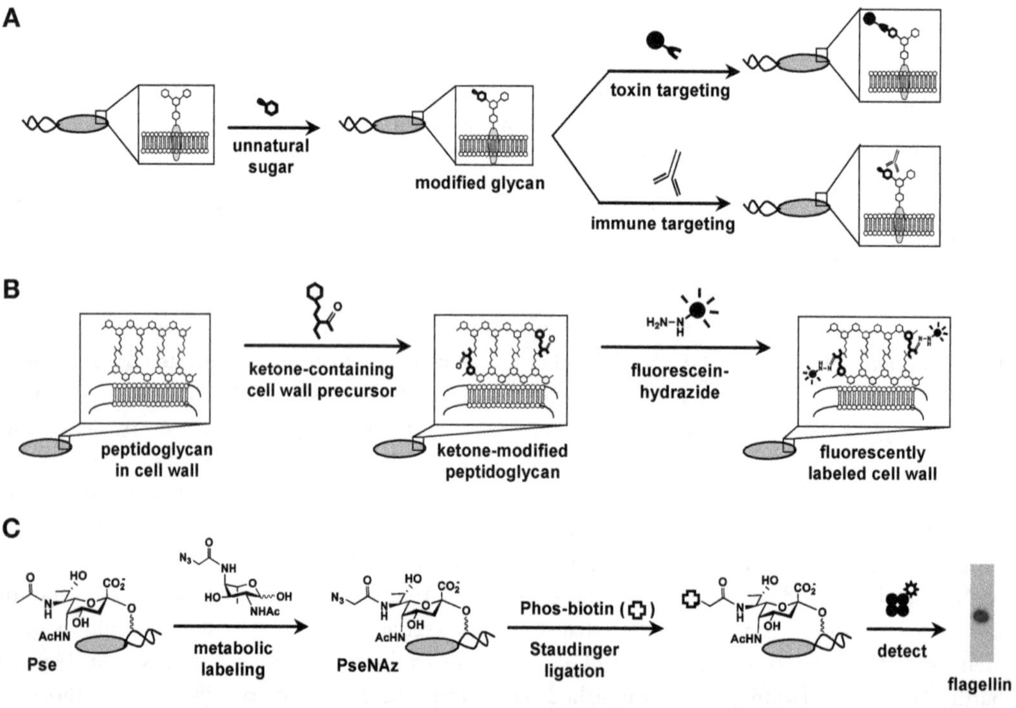

Figure 13.4 Selective targeting of bacterial cells based on distinctive glycans. (A) General strategy for using MOE to target bacterial cells based on distinctive glycan expression. First, cells are treated with an unnatural sugar that is exclusively incorporated into a bacterial glycan. Modified glycans are then targeted (top) covalently with a reactive partner containing a toxin or (bottom) by the host's immune system, ultimately causing bacterial cell death. (B) As an example of metabolic labelling of a glycan exclusively found on bacteria, Sadamoto *et al.* metabolically labelled a glycan found in the bacterial cell wall by supplementing *Lactobacillus* sp. with a ketone-containing cell wall precursor. Metabolically incorporated ketones were then reacted with fluorescein-hydrazide and visualized by fluorescence microscopy. (C) In another example of selective bacterial glycan labelling, Liu *et al.* were able to metabolically replace *C. jejuni*'s flagellin monosaccharide pseudaminic acid (Pse) with azidopseudaminic acid (PseNAz) by supplementing cells with 6-deoxy-AltNAc4NAz (shown), an azide-containing analogue of a dedicated Pse precursor. PseNAz residues were reacted with Phos-biotin via Staudinger ligation and detected with streptavidin. Western blot analysis revealed that only flagellin glycoproteins were modified with azides.

liberate their payload upon Staudinger ligation with azides, offering an opportunity to selectively damage cells covered with azide-labelled glycans (Azoulay *et al.*, 2006). These studies suggest that covalent delivery of drugs to cells based on glycan presentation may be a viable therapeutic treatment.

As an alternative targeting modality, MOE has been employed to stimulate an immune response against cells based on the display of an unnatural sugar (Fig. 13.4A). When conjugated to an immunostimulant, an unnatural monosaccharide causes a potent immune response. Stimulated immune cells then recognize cells bearing the unnatural monosaccharide and target them for destruction. Jennings and colleagues pioneered this approach to target cancer cells in mice containing PSA with *N*-propanoyl sialic acid residues (Liu *et al.*, 2000). Lemieux *et al.* demonstrated that an immune response can also be raised against *N*-levulinoyl sialic acid (SiaLev), leading to complement-mediated lysis of SiaLev covered cells in culture (Lemieux and Bertozzi, 2001). In more recent work, Guo and coworkers established that immunotherapy based on the incorporation of *N*-phenylacetyl sialic acid analogues into the tumour-associated carbohydrate antigens sialyl Tn and GM3 leads to potent and selective cancer cell destruction (Chefalo *et al.*, 2006; Wu and Guo, 2006). These results suggest that immune targeting of cells bearing unnatural glycans is feasible. A similar strategy should enable selective targeting of bacterial cells.

To be applicable to destroying bacterial cells, these targeting modalities require selective metabolic labelling of a glycan that is exclusively found on bacteria and is absent from humans. Two studies have established that it is possible to label glycans exclusively found on bacterial cells with chemical reporters, thus offering the opportunity to selectively target bacteria based on their unique glycans. The first report of *selective* metabolic labelling of bacterial glycans with chemical reporters employed a ketone-modified glycopeptide analogue to modify the bacterial cell wall (Sadamoto *et al.*, 2002). Nishimura and coworkers metabolically introduced ketones into the cell wall of Gram-positive bacteria by supplementing the bacteria with a ketone-analogue of the cell wall precursor UDP-*N*-acetylmuramic acid pentapeptide (Fig. 13.4B). The authors then visualized ketones incorporated into the cell wall using a fluorescein hydrazide probe (Fig. 13.4B). This work is particularly exciting because the ketone analogue is a substrate of bacterial but not of human cells and should therefore enable selective labelling of bacteria with ketones. Because ketone-hydrazide chemistry is robust even in the context of a living animal (Rideout *et al.*, 1990; Rideout, 1994), bacteria labelled with ketones are poised for selective covalent targeting with toxins within a host. Alternatively, the ketone-modified cell wall precursor could be used as a vaccine to elicit an immune response against bacteria.

In a second example of applying MOE for selective bacterial glycan modification, Logan, Tanner and colleagues exclusively labelled *C. jejuni*'s pseudaminic acid residues on flagellin proteins with azides (Liu *et al.*, 2009). Pseudaminic acid is present on only a small number of pathogenic bacteria, and neither pseudaminic acid nor any enzyme (or analogue) in its biosynthetic pathway is found in mammalian cells (Schirm *et al.*, 2003; Schoenhofen *et al.*, 2006). Therefore, pseudaminic acid biosynthesis can be exploited to incorporate azides selectively onto pathogenic bacterial cells. Tanner *et al.* established that *C. jejuni*'s pseudaminic acid biosynthetic pathway is permissive of azide-containing sugars (Liu *et al.*, 2009). Specifically, the azide-containing sugar 6-deoxy-AltNAc-4-NAz, an analogue of pseudaminic acid's dedicated metabolic precursor 6-deoxy-AltdiNAc, can be processed by *C. jejuni*'s pseudaminic acid biosynthetic pathway and metabolically replace pseudaminic acid with azido-pseudaminic acid on *C. jejuni* cell surfaces (Fig. 13.4C). Further, their work revealed that *C. jejuni*'s azide-modified flagellin glycoproteins can be tagged via Staudinger ligation with phosphine-biotin (Fig. 13.4C). This work sets the stage for using the Staudinger ligation, a reaction that reflects the distribution of cell-surface azides even within a living animal (Dube *et al.*, 2006), to selectively deliver phosphine-based reactive partners containing toxins (Azoulay *et al.*, 2006) to azide-covered bacteria within a host. Moreover, this report lays the groundwork for stimulating an immune response against *C.*

jejuni based on expression of azido-pseudaminic acid.

The preponderance of distinctive glycans on bacterial cells suggests a number of exciting prospects (Dube *et al.*, 2010). Future efforts should first focus on assessing the viability of covalently targeting bacteria based on their unique glycans or prompting an immune response based on these structures. Once the viability of these approaches has been established, a plethora of unusual glycan structures exclusively found on pathogenic bacteria could be targeted. If these approaches prove successful, they have the potential to establish a new paradigm for antibiotic and bacterial vaccine design.

Identifying bacterial glycan–protein interactions

Glycan–protein interactions are crucial for mediating cellular events ranging from bacterial adhesion to fetal implantation (Genbacev *et al.*, 2003; Reid *et al.*, 2010). Given the critical role of glycan–protein interactions in mediating cellular binding events, delineating these interactions will unveil novel drug targets. MOE provides a means to unambiguously identify the binding partners of glycans. Briefly, eukaryotic glycoproteins and glycolipids have been metabolically labelled with monosaccharides containing photo-activatable cross-linkers, such as aryl azides or diazirines (Fig. 13.5) (Luchansky *et al.*, 2004; Han *et al.*, 2005; Bond and Kohler, 2007; Tanaka and Kohler, 2008). After exposing cultured cells to light, glycoproteins and glycolipids of interest were purified with their binding partners covalently attached (Fig. 13.5). Mass spectrometry analysis facilitated the identification of candidate ligands, and immunochemical analysis confirmed results. Paulson and coworkers pioneered this approach using an aryl azide-bearing sialic acid analogue to identify ligands of the glycoprotein CD22 (Han *et al.*, 2005). More recently, Kohler and colleagues extended this technology using diazirine-bearing sialic acid analogues to identify ligands of CD22 and the glycolipid GM1 (Bond *et al.*, 2010; Yu *et al.*, 2010).

Similar to previous work performed in eukaryotes, the use of metabolically incorporated cross-linkers will enable the definitive identification of bacterial glycan–protein interactions. This application of MOE in bacteria will provide insight into host–pathogen interactions. For example, a cross-linker incorporated into *C. jejuni*'s pseudaminic acid residues would facilitate the

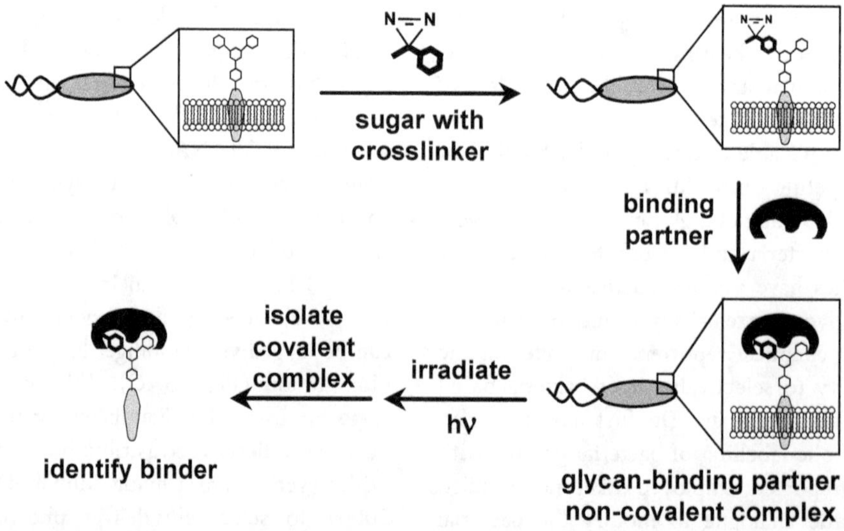

Figure 13.5 General strategy for identifying bacterial glycan–protein interactions. First, cells are treated with an unnatural sugar that contains a photo-activable cross-linker (diazirine shown). Next, cells are incubated with potential binding partners and irradiated, resulting in covalent attachment of a binding partner to the glycan. The complex is then purified to enable identification of the binding partner.

discovery of this abundant flagellin monosaccharide's binding partners. Given pseudaminic acid's link to *C. jejuni*'s virulence (Goon *et al.*, 2003a), such studies are likely to identify other virulence factors and yield insight into how *C. jejuni* adheres to host cells. A similar approach could be utilized to establish the role of other unusual glycan structures in bacterial function and pathogenesis. Ultimately, this information will guide the development of antibacterials that inactivate key glycan–protein interactions. Efforts should focus on identifying photo-activable metabolic substrates containing cross-linkers that can intercept bacterial carbohydrate biosynthetic pathways.

Imaging bacterial glycan dynamics

Imaging the glycome has the potential to enhance our understanding of glycan-mediated processes and to afford novel diagnostics for disease. MOE is the only method currently available to image glycans *in vivo* (Laughlin and Bertozzi, 2009a). For this application, the incorporation of chemical reporters into glycans permits ligation of detectable fluorescent probes (Fig. 13.6). The Bertozzi and Wong groups have developed a number of phosphine and alkyne-based probes for imaging azide-containing glycans (Laughlin and Bertozzi, 2009a). Their designs include a number of 'smart' probes that become fluorescent only after reaction with the azide target, leading to enhanced resolution relative to probes that are always fluorescent (Lemieux *et al.*, 2003; Sivakumar *et al.*, 2004; Sawa *et al.*, 2006; Hsu *et al.*, 2007; Hangauer and Bertozzi, 2008). Most recently, Cohen *et al.* reported a phosphine-based probe for real-time bioluminescence imaging of glycans on live cells (Cohen *et al.*, 2010). MOE has enabled glycan imaging in cell-based systems, in mice, in zebrafish, and in *C. elegans* (Prescher *et al.*, 2004; Baskin *et al.*, 2007; Laughlin *et al.*, 2008; Laughlin and Bertozzi, 2009b; Chang *et al.*, 2010). Similar to work conducted in these systems, MOE will enable real-time imaging of bacterial glycan dynamics. For example, phosphine-based probes could be used to detect and diagnose particular bacterial infections, and these probes could monitor bacterial glycan dynamics in a host as infection progresses. Ultimately, imaging bacterial glycan dynamics will unveil physiologically relevant information about the initiation, spread, and treatment of bacterial disease.

Conclusion

Bacterial glycomics is an exciting area that promises to yield insights into bacterial pathogenesis and to unveil novel drug targets. Metabolic oligosaccharide engineering can complement existing strategies to discover bacterial glycoproteins and glycolipids. Moreover, MOE provides a versatile means to study and disrupt bacterial glycans. Although metabolic labelling of bacterial glycans with chemical reporters is in its infancy, there are ample precedents in mammalian systems to suggest the impact MOE can have in bacterial systems. Future efforts will focus on identifying metabolic substrates to label bacterial sugars, and utilizing installed chemical reporters to profile bacterial glycoproteins, disrupt bacterial glycans, target bacterial cells based on their distinctive glycans, identify bacterial glycan–protein interactions, and image bacterial glycan dynamics. Given the versatility of this approach, there will no doubt be other creative applications of MOE to study bacterial glycosylation. Work in this area will lead to a deeper understanding of the role of glycans

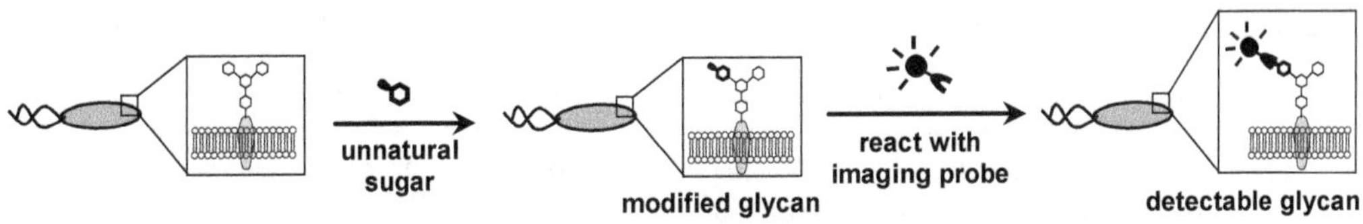

Figure 13.6 General strategy for imaging bacterial glycan dynamics. First, cells are treated with an unnatural sugar that contains a chemical reporter. Next, cells are incubated with a reactive partner containing an imaging probe. Detectable glycans are then visualized.

in host–pathogen interactions, information that will spur the development of glycosylation-based therapeutics.

Acknowledgements

I wish to thank Jennifer Prescher, Kanokwan Champasa, Bo Wang, and Max Nowicki for helpful comments on this chapter. This work was funded in part by a Camille and Henry Dreyfus Foundation Faculty Startup Award and research grants from Research Corporation (grant number: CCSA 7852) and the National Institute of General Medical Sciences (grant number: R15GM093867). The content is solely the responsibility of the author and does not necessarily represent the official views of the National Institute of General Medical Sciences or the National Institutes of Health.

References

Acheson, A., Sunshine, J.L., and Rutishauser, U. (1991). NCAM polysialic acid can regulate both cell–cell and cell–substrate interactions. J. Cell Biol. *114,* 143–153.

Agard, N.J., Prescher, J.A., and Bertozzi, C.R. (2004). A strain-promoted [3+2] azide-alkyne cycloaddition for covalent modification of blomolecules in living systems. J. Am. Chem. Soc. *126,* 15046–15047.

Azoulay, M., Tuffin, G., Sallem, W., and Florent, J.-C. (2006). A new drug-release method using the Staudinger ligation. Bioorg. Med. Chem. Lett. *16,* 3147–3149.

Balonova, L., Hernychova, L., and Bilkova, Z. (2009). Bioanalytical tools for the discovery of eukaryotic glycoproteins applied to the analysis of bacterial glycoproteins. Exp. Rev. Proteomics *6,* 75–85.

Baskin, J.M., Prescher, J.A., Laughlin, S.T., Agard, N.J., Chang, P.V., Miller, I.A., Lo, A., Codelli, J.A., and Bertozzit, C.R. (2007). Copper-free click chemistry for dynamic *in vivo* imaging. Proc. Natl. Acad. Sci. U.S.A. *104,* 16793–16797.

Benz, I., and Schmidt, M.A. (2002). Never say never again: protein glycosylation in pathogenic bacteria. Mol. Microbiol. *45,* 267 – 276.

Bond, M.R., and Kohler, J.J. (2007). Chemical methods for glycoprotein discovery. Curr. Opin. Chem. Biol. *11,* 52–58.

Bond, M.R., Whitman, C.M., and Kohler, J.J. (2010). Metabolically incorporated photocrosslinking sialic acid covalently captures a ganglioside–protein complex. Mol. BioSyst. *6,* 1796–1799.

Buttner, B., Kannicht, C., Schmidt, C., Loster, K., Reutter, W., Lee, H.Y., Nohring, S., and Horstkorte, R. (2002). Biochemical engineering of cell surface sialic acids stimulates axonal growth. J. Neurosci. *22,* 8869–8875.

Chang, P.V., Prescher, J.A., Sletten, E.M., Baskin, J.M., Miller, I.A., Agard, N.J., Lo, A., and Bertozzi, C.R. (2010). Copper-free click chemistry in living animals. Proc. Natl. Acad. Sci. U.S.A. *107,* 1821–1826.

Charter, N.W., Mahal, L.K., Koshland, D.E., Jr., and Bertozzi, C.R. (2002). Differential effects of unnatural sialic acids on the polysialylation of the neural cell adhesion molecule and neuronal behavior. J. Biol. Chem. *277,* 9255–9261.

Chefalo, P., Pan, Y., Nagy, N., Guo, Z., and Harding, C.V. (2006). Efficient metabolic engineering of GM3 on tumor cells by *N*-phenylacetyl-D-mannosamine. Biochemistry *45,* 3733–3739.

Cohen, A.S., Dubikovskaya, E.A., Rush, J.S., and Bertozzi, C.R. (2010). Real-time bioluminescence imaging of glycans on live cells. J. Am. Chem. Soc. *132,* 8563–8565.

Doig, P., Kinsella, N., Guerry, P., and Trust, T.J. (1996). Characterization of a post-translational modification of Campylobacter flagellin: identification of a sero-specific glycosyl moiety. Mol. Microbiol. *19,* 379–387.

Dube, D.H., and Bertozzi, C.R. (2003). Metabolic oligosaccharide engineering as a tool for glycobiology. Curr. Opin. Chem. Biol. *7,* 616–625.

Dube, D.H., and Bertozzi, C.R. (2005). Glycans in cancer and inflammation--potential for therapeutics and diagnostics. Nat. Rev. Drug Discov. *4,* 477–488.

Dube, D.H., Champasa, K., and Wang, B. (2011). Chemical tools to discover and target bacterial glycoproteins. Chem. Commun. *47,* 87–101.

Dube, D.H., Prescher, J.A., Quang, C.N., and Bertozzi, C.R. (2006). Probing mucin-type O-linked glycosylation in living animals. Proc. Natl. Acad. Sci. U.S.A. *103,* 4819–4824.

Fletcher, C.M., Coyne, M.J., Villa, O.F., Chatzidaki-Livanis, M., and Comstock, L.E. (2009). A general O-glycosylation system important to the physiology of a major human intestinal symbiont. Cell *137,* 321–221.

Genbacev, O.D., Prakobphol, A., Foulk, R.A., Krtolica, A.R., Ilic, D., Singer, M.S., Yang, Z.-Q., Kiessling, L.L., Rosen, S.D., and Fisher, S.J. (2003). Trophoblast L-selectin-mediated adhesion at the maternal–fetal interface. Science *299,* 405–408.

Goon, S., Kelly, J.F., Logan, S.M., Ewing, C.P., and Guerry, P. (2003a). Pseudaminic acid, the major modification on *Campylobacter* flagellin, is synthesized via the Cj1293 gene. Mol. Microbiol. *50,* 659–671.

Goon, S., Schilling, B., Tullius, M.V., Gibson, B.W., and Bertozzi, C.R. (2003b). Metabolic incorporation of unnatural sialic acids into *Haemophilus ducreyi* lipooligosaccharides. Proc. Natl. Acad. Sci. U.S.A. *100,* 3089–3094.

Haltiwanger, R.S., and Lowe, J.B. (2004). Role of glycosylation in development. Annu. Rev. Biochem. *73,* 491–537.

Han, S., Collins, B.E., Bengtson, P., and Paulson, J.C. (2005). Homomultimeric complexes of CD22 in B cells revealed by protein–glycan cross-linking. Nat. Chem. Biol. *1,* 93–97.

Hang, H.C., Yu, C., Kato, D.L., and Bertozzi, C.R. (2003). A metabolic labelling approach toward proteomic analysis of mucin-type O-linked glycosylation. Proc. Natl. Acad. Sci. U.S.A. *100,* 14846–14851.

Hangauer, M.J., and Bertozzi, C.R. (2008). A FRET-based fluorogenic phosphine for live-cell imaging with the Staudinger ligation. Angew. Chem. Int. Ed. *47,* 2394–2397.

Herrmann, M., von der Lieth, C.W., Stehling, P., Reutter, W., and Pawlita, M. (1997). Consequences of a subtle sialic acid modification on the murine polyomavirus receptor. J. Virol. *71,* 5922–5931.

Hong, V., Steinmetz, N.F., Manchester, M., and Finn, M.G. (2010). Labelling live cells by copper-catalysed alkyne azide click chemistry. Bioconj. Chem. *21,* 1912–1916.

Hsu, K.L., Pilobello, K.T., and Mahal, L.K. (2006). Analyzing the dynamic bacterial glycome with a lectin microarray approach. Nat. Chem. Biol. *2,* 153–157.

Hsu, T.-L., Hanson, S.R., Kishikawa, K., Wang, S.-K., Sawa, M., and Wong, C.-H. (2007). Alkynyl sugar analogs for the labelling and visualization of glycoconjugates in cells. Proc. Natl. Acad. Sci. U.S.A. *104,* 2614–2619.

Johnson, C.A., and Pekas, D.J.W., R. J. (1964). Neuraminidases and influenza virus infection in embryonated eggs. Science *143,* 1051–1052.

Jones, M.B., Teng, H., Rhee, J.K., Lahar, N., Baskaran, G., and Yarema, K.J. (2004). Characterization of the cellular uptake and metabolic conversion of acetylated *N*-acetylmannosamine (ManNAc) analogues to sialic acids. Biotech. Bioeng. *85,* 394–405.

Kalovidouris, S.A., Gama, C.I., Lee, L.W., and Hsieh-Wilson, L.C. (2005). A role for fucose alpha-1,2 galactose carbohydrates in neuronal growth. J. Am. Chem. Soc. *127,* 1340–1341.

Kayser, H., Zeitler, R., Kannicht, C., Grunow, D., Nuck, R., and Reutter, W. (1992). Biosynthesis of a nonphysiological sialic-acid in different rat organs, using *N*-propanoyl-D-hexosamines as precursors. J. Biol. Chem. *267,* 16934–16938.

Kelly, J., Jarrell, H., Millar, L., Tessier, L., Fiori, L.M., Lau, P.C., Allan, B., and Szymanski, C.M. (2006). Biosynthesis of the N-linked glycan in *Campylobacter jejuni* and addition onto protein through block transfer. J. Bacteriol. *188,* 2427–2434.

Keppler, O.T., Stehling, P., Herrmann, M., Kayser, H., Grunow, D., Reutter, W., and Pawlita, M. (1995). Biosynthetic modulation of sialic acid-dependent virus–receptor interactions of two primate polyoma viruses. J. Biol. Chem. *270,* 1308–1314.

Keppler, O.T., Horstkorte, R., Pawlita, M., Schmidts, C., and Reutter, W. (2001). Biochemical engineering of the *N*-acyl side-chain of sialic acid: biological implications. Glycobiology *11,* 11R-18R.

Koenigs, M.B., Richardson, E.A., and Dube, D.H. (2009). Metabolic profiling of *Helicobacter pylori* glycosylation. Mol. Biosyst. *5,* 909–912.

Laughlin, S.T., and Bertozzi, C.R. (2007). Metabolic labelling of glycans with azido sugars and subsequent glycan-profiling and visualization via Staudinger ligation. Nat. Prot. *2,* 2930 – 2944.

Laughlin, S.T., and Bertozzi, C.R. (2009a). Imaging the glycome. Proc. Natl. Acad. Sci. U.S.A. *106,* 12–17.

Laughlin, S.T., and Bertozzi, C.R. (2009b). *In vivo* imaging of *Caenorhabditis elegans* glycans. ACS Chem. Biol. *4,* 1068–1072.

Laughlin, S.T., Agard, N.J., Baskin, J.M., Carrico, I.S., Chang, P.V., Ganguli, A.S., Hangauer, M.J., Lo, A., Prescher, J.A., and Bertozzi, C.R. (2006). Metabolic labelling of glycans with azido sugars for visualization and glycoproteomics. Methods Enzymol. *415,* 230–250.

Laughlin, S.T., Baskin, J.M., Amacher, S.L., and Bertozzi, C.R. (2008). *In vivo* imaging of membrane-associated glycans in developing zebrafish. Science *320,* 664–667.

Lemieux, G.A., and Bertozzi, C.R. (2001). Modulating cell surface immunoreactivity by metabolic induction of unnatural carbohydrate antigens. Chem. Biol. *8,* 265–275.

Lemieux, G.A., De Graffenried, C.L., and Bertozzi, C.R. (2003). A fluorogenic dye activated by the staudinger ligation. J. Am. Chem. Soc. *125,* 4708–4709.

Linton, D., Allan, E., Karlyshev, A.V., Cronshaw, A.D., and Wren, B.W. (2002). Identification of *N*-acetylgalactosamine-containing glycoproteins PEB3 and CgpA in *Campylobacter jejuni.* Mol. Microbiol. *43,* 497–508.

Liu, F., Aubry, A.J., Schoenhofen, I.C., Logan, S.M., and Tanner, M.E. (2009). The engineering of bacteria bearing azido-pseudaminic acid-modified flagella. ChemBioChem *10,* 1317–1320.

Liu, T.M., Guo, Z.W., Yang, Q.L., Sad, S., and Jennings, H.J. (2000). Biochemical engineering of surface alpha 2–8 polysialic acid for immunotargeting tumor cells. J. Biol. Chem. *275,* 32832–32836.

Luchansky, S.J., and Bertozzi, C.R. (2004). Azido sialic acids can modulate cell–surface interactions. ChemBioChem *5,* 1706–1709.

Luchansky, S.J., Goon, S., and Bertozzi, C.R. (2004). Expanding the diversity of unnatural cell-surface sialic acids. ChemBioChem *5,* 371–374.

Mahal, L.K., Charter, N.W., Angata, K., Fukuda, M., Koshland, D.E., Jr., and Bertozzi, C.R. (2001). A small-molecule modulator of poly alpha 2,8-sialic acid expression on cultured neurons and tumor cells. Science *294,* 380–381.

Mahal, L.K., Yarema, K.J., and Bertozzi, C.R. (1997). Engineering chemical reactivity on cell surfaces through oligosaccharide biosynthesis. Science *276,* 1125–1128.

Murrey, H.E., Gama, C.I., Kalovidouris, S.A., Luo, W.I., Driggers, E.M., Porton, B., and Hsieh-Wilson, L.C. (2006). Protein fucosylation regulates synapsin Ia/ Ib expression and neuronal morphology in primary hippocampal neurons. Proc. Natl. Acad. Sci. U.S.A. *103,* 21–26.

Nandi, A., Sprung, R., Barma, D.K., Zhao, Y., Kim, S.C., and Falck, J.R. (2006). Global identification of O-GlcNAc-modified proteins. Anal. Chem. *78,* 452–458.

Nauman, D.A., and Bertozzi, C.R. (2001). Kinetic parameters for small-molecule drug delivery by covalent cell surface targeting. Biochim. Biophys. Acta *1568,* 147–154.

Oetke, C., Brossmer, R., Mantey, L.R., Hinderlich, S., Isecke, R., Reutter, W., Keppler, O.T., and Pawlita, M. (2002). Versatile biosynthetic engineering of sialic acid in living cells using synthetic sialic acid analogues. J. Biol. Chem. *277*, 6688–6695.

Poloukhtine, A.A., Mbua, N.E., Wolfert, M.A., Boons, G.J., and Popik, V.V. (2009). Selective labelling of living cells by a photo-triggered click reaction. J. Am. Chem. Soc. *131*, 15769–15776.

Power, P.M., and Jennings, M.P. (2003). The genetics of glycosylation in Gram-negative bacteria. FEMS Microbiol. Lett. *218*, 211–222.

Prescher, J.A., and Bertozzi, C.R. (2005). Chemistry in living systems. Nat. Chem. Biol. *1*, 13–21.

Prescher, J.A., and Bertozzi, C.R. (2006). Chemical technologies for probing glycans. Cell *126*, 851–854.

Prescher, J.A., Dube, D.H., and Bertozzi, C.R. (2004). Chemical remodelling of cell surfaces in living animals. Nature *430*, 873–877.

Rabuka, D., Hubbard, S.C., Laughlin, S.T., Argade, S.P., and Bertozzi, C.R. (2006). A chemical reporter strategy to probe glycoprotein fucosylation. J. Am. Chem. Soc. *128*, 12078–12079.

Reid, C.W., Fulton, K.M., and Twine, S.M. (2010). Never take candy from a stranger: the role of the bacterial glycome in host–pathogen interactions. Future Microbiol. *5*, 267–288.

Rideout, D. (1994). Self-assembling drugs: a new approach to biochemical modulation in cancer chemotherapy. Cancer Invest. *12*, 189–202.

Rideout, D., Calogeropoulou, T., Jaworski, J., and McCarthy, M. (1990). Synergism through direct covalent bonding between agents: a strategy for rational design of chemotherapeutic combinations. Biopolymers *29*, 247–262.

Rosen, S.D. (2004). Ligands for L-selectin: homing, inflammation, and beyond. Annu. Rev. Immunol. *22*, 129–156.

Rostovtsev, V.V., Green, L.G., Fokin, V.V., and Sharpless, K.B. (2002). A stepwise Huisgen cycloaddition process: Copper(I)-catalysed regioselective 'ligation' of azides and terminal alkynes. Angew. Chem. Int. Ed. *41*, 2596–2599.

Sadamoto, R., Niikura, K., Sears, P.S., Liu, H.T., Wong, C.H., Suksomcheep, A., Tomita, F., Monde, K., and Nishimura, S.I. (2002). Cell-wall engineering of living bacteria. J. Am. Chem. Soc. *124*, 9018–9019.

Sarkar, A.K., Fritz, T.A., Taylor, W.H., and Esko, J.D. (1995). Disaccharide uptake and priming in animal cells: inhibition of sialyl Lewis X by acetylated Gal beta 1→4GlcNAc beta-O-naphthalenemethanol. Proc. Natl. Acad. Sci. U.S.A. *92*, 3323–3327.

Sawa, M., Hsu, T.-L., Itoh, T., Sugiyama, M., Hanson, S.R., Vogt, P.K., and Wong, C.-H. (2006). Glycoproteomic probes for fluorescent imaging of fucosylated glycans *in vivo*. Proc. Natl. Acad. Sci. U.S.A. *103*, 12371–12376.

Saxon, E., and Bertozzi, C.R. (2000). Cell surface engineering by a modified Staudinger reaction. Science *287*, 2007–2010.

Schilling, B., Goon, S., Samuels, N.M., Gaucher, S.P., Leary, J.A., Bertozzi, C.R., and Gibson, B.W. (2001). Biosynthesis of sialylated lipooligosaccharides in *Haemophilis ducreyi* is dependent on exogenous sialic acid and not mannosamine. Incorporation studies using N-acylmannosamine analogues, *N*-glycolylneuraminic acid, and C-13-labelled N-acetylneuraminic acid. Biochemistry *40*, 12666–12677.

Schirm, M., Soo, E.C., Aubry, A.J., Austin, J., Thibault, P., and Logan, S.M. (2003). Structural, genetic and functional characterization of the flagellin glycosylation process in *Helicobacter pylori*. Mol. Microbiol. *48*, 1579–1592.

Schmidt, M.A., Riley, L.W., and Benz, I. (2003). Sweet new world: glycoproteins in bacterial pathogens. Trends Microbiol. *11*, 554–561.

Schoenhofen, I.C., McNally, D.J., Brisson, J.-R., and Logan, S.M. (2006). Elucidation of the CMP-pseudaminic acid pathway in *Helicobacter pylori*: synthesis from UDP-N-acetylglucosamine by a single enzymatic reaction. Glycobiology *16*, 8C-14.

Sivakumar, K., Xie, F., Cash, B.M., Long, S., Barnhill, H.N., and Wang, Q. (2004). A fluorogenic 1,3-dipolar cycloaddition reaction of 3-azidocoumarins and acetylenes. Org. Lett. *6*, 4603–4606.

Speers, A.E., Adam, G.C., and Cravatt, B.F. (2003). Activity-based protein profiling *in vivo* using a copper(I)-catalysed azide-alkyne [3+2] cyloaddition. J. Am. Chem. Soc. *123*, 4686–4687.

Sprung, R., Nandi, A., Chen, Y., Kim, S.C., Barma, D., Falck, J.R., and Zhao, Y.M. (2005). Tagging-via-substrate strategy for probing O-GlcNAc modified proteins. J. Prot. Res. *4*, 950–957.

Szymanski, C.M., and Wren, B.W. (2005). Protein glycosylation in bacterial mucosal pathogens. Nat. Rev. Microbiol. *3*, 225–237.

Szymanski, C.M., Yao, R., Ewing, C.P., Trust, T.J., and Guerry, P. (1999). Evidence for a system of general protein glycosylation in *Campylobacter jejuni*. Mol. Microbiol. *32*, 1022–1030.

Tanaka, Y., and Kohler, J.J. (2008). Photoactivatable crosslinking sugars for capturing glycoprotein Interactions. J. Am. Chem. Soc. *130*, 3278–3279.

Vocadlo, D.J., Hang, H.C., Kim, E.-J., Hanover, J.A., and Bertozzi, C.R. (2003). A chemical approach for identifying O-GlcNAc-modified proteins in cells. Proc. Natl. Acad. Sci. U.S.A. *100*, 9116–9121.

Wu, J., and Guo, Z. (2006). Improving the antigenicity of sTn antigen by modification of its sialic acid residue for development of glycoconjugate cancer vaccines. Bioconj. Chem. *17*, 1537–1544.

Yarema, K.J., Mahal, L.K., Bruehl, R.E., Rodriguez, E.C., and Bertozzi, C.R. (1998). Metabolic delivery of ketone groups to sialic acid residues – Application to cell surface glycoform engineering. J. Biol. Chem. *273*, 31168–31179.

Yu, S.-H., Bond, M.R., Whitman, C.M., Kohler, J.J., and Minoru, F. (2010). Metabolic labelling of glycoconjugates with photocrosslinking sugars. In Methods in Enzymology (Academic Press), pp. 541–562.

Exploiting the Bacterial Surface: The Successful Application of Glycoconjugate Vaccines

14

Robert A. Pon

Abstract

The simple act of covalently attaching poorly immunogenic carbohydrate residues to proteins renders the glycan more immunogenic leading to the generation of glycan-specific antibodies. This essential concept, founded over a century ago, is the basis for the field of anti-bacterial glycoconjugate vaccines. Glycoconjugate vaccines based on the glycans found in the capsular polysaccharides of invasive bacteria are now widely recognized as a highly safe and effective means to protect our most vulnerable population (namely young children) from the high rates of morbidity and mortality associated with invasive disease. This chapter will provide an overview of the immunobiology of glycoconjugate vaccines, general considerations for their design and synthesis, along with lessons we have learned from their use – all aimed at setting the tone for the next generation of glycoconjugate vaccines that the current era of glycomics will bring.

Introduction

Arguably one of the greatest medical achievements of the past 50 years has been the development and successful implementation of glycoconjugate vaccines (GCVs), being based almost exclusively on the chemical coupling of bacterial capsular polysaccharides (CPS) to protein carriers. These vaccines, beginning with the *Haemophilus influenzae* type b (Hib) conjugate vaccine (Schneerson *et al.*, 1980) have significantly reduced the mortality and morbidity associated with numerous invasive bacterial diseases, and have revamped the arsenal of vaccines used in routine infant immunization programs in developed countries. The success of these vaccines has relied upon the identification and isolation of virulent carbohydrate structures present within the envelope of pathogenic bacteria. The range and diversity of glycans found in bacteria attest to the importance of these structures for their survival; from glycoproteins and glycolipids within the cytoplasm and its surrounding envelope, to secreted products that alter the bacterium's local microenvironment. They are key substituents of *O*- and *N*-linked glycoproteins such as those found on pilins, flagellins, and S-layers; cell wall components of peptidoglycan, arabinogalactans, lipo- and teichoic acids, lipoarabinomannans; and substituents of outer leaflet structures such as lipopolysaccharides (LPS), lipooligosaccharides (LOS), capsular and exo-polysaccharides (Holst *et al.*, 2009). As these structures play important roles in cell integrity, barrier formation, adhesion, invasion, and notably immune system evasion, targeting these virulence factors with novel GCVs is an effective course of action.

In light of the advances being made in bacterial glycomics (much of which is detailed in the preceding chapters), the stage is now set for the development of new generation GCVs that span not only the sphere of pathogenic bacteria but will extend to diseases caused by viruses, fungi, parasites and aberrant eukaryotic cancers. These new generation GCVs will rely not only on current technology, but will also take advantage of our increasing knowledge of carbohydrate immunology and our ability to manipulate both *in vivo* and *in vitro* microbial and eukaryotic glycan synthesis to design and produce novel vaccine candidate glycoproteins or glycolipids. The focus of this chapter will be on what we have learned from the development of bacterial polysaccharide (PS)-based GCVs, how they have impacted invasive bacterial disease, challenges that remain, and how this knowledge will serve the development of new generation GCVs.

Bacterial polysaccharide vaccines – the first generation

The foundation for current bacterial GCVs dates back to the early 1900s with the recognition that secreted carbohydrate substances from *Streptococcus pneumoniae* (Dochez and Avery, 1917), later identified as their type-specific CPSs, were immunogenic in both mice (Schiemann and Casper, 1927) and humans (Tillett and Francis, Jr., 1929). Complete validation for their use as a prophylactic human vaccine came from the seminal study of MacLeod *et al.* (1945) who demonstrated a multivalent CPS vaccine (four pneumococcal serotypes) was 90% effective in preventing serotype specific adult pneumonia. Despite a lull in the development of subsequent CPS vaccines, the growing realization that antibiotics were not the panacea for all bacterial disease led to the resurgence and commercialization of several CPS vaccines against invasive disease caused by *S. pneumoniae*, Hib, *Neisseria meningitidis*, and *Salmonella enterica* serovar Typhi (Jennings and Pon, 1996). These PS vaccines have made a significant impact on invasive bacterial disease in healthy adults and adolescents, but they have proven to be deficient for populations most at risk including infants, the elderly, and the immunocompromised as outlined in the following section. Indeed, many individuals are unable to mount a CPS immune response regardless of their age due in part to CPS structures that share homology to host structures such as the CPSs of *N. meningitidis* serogroup B (NmB) and Group B *Streptococcus* (GBS) (Jennings and Pon, 1996). In general, protection afforded by these first generation vaccines was due to circulating levels of bactericidal or opsonophagocytic antibody (Ab) but as these levels wane, re-administration of the CPS vaccine often induced a state of hyporesponsiveness (Poolman and Borrow, 2011) as immunological memory is not associated with CPS vaccines. These significant deficiencies of plain CPS vaccines have precipitated the modern era of GCVs that are currently in use.

Glycoconjugate immunobiology

Adaptive immunity in the context of bacterial infections can be broadly broken down into cell-mediated or humoral immune responses. Cell-mediated immunity is critical for combating intracellular pathogens generally through the production of cytotoxic T cells that are highly adept at seeking out and killing bacteria hidden within host cells. Humoral immunity on the other hand, functions to protect against extracellular bacteria through the production of specific Abs that serve to target and clear these pathogens. Without doubt, the single most important adaptive measure to protect against invasive disease caused by encapsulated bacteria is the ability to mount a timely, specific, and effective Ab response to cell surface components (Robbins *et al.*, 1999). Circulating Ab levels specific to outer-membrane PSs such as CPS and LPS is the correlate of protection most often cited for the protective efficacy of PS-based vaccines, whereby bacteria decorated with specific Ab and complement molecules results in bacteriolysis (Anderson *et al.*, 1972) or clearance through the actions of host phagocytes (Licciardi *et al.*, 2011). Repetitive carbohydrate antigens such as those found in CPS and LPS are prototypical thymus-independent type II antigens (TI-2) that result in B-cell stimulation and Ab production without the direct involvement of T-helper cells (Stein, 1992). Failure of T-cell help is due to the poor ability of oligosaccharides to associate with major histocompatibility complex (MHC) class II molecules on antigen presenting cells required for recognition by T-cell receptors (TcRs). By definition, these PSs, composed of multiple oligosaccharide repeat units, are multivalent with respect to the number of B-cell epitopes that they contain; generally on the order of 4–6 sugar residues (Kabat *et al.*, 1988). A key physicochemical requirement of TI-2 antigens is that they contain a minimum of 20 epitopes per polymer spatially separated over 100 Å of cell surface for effective B-cell activation (Dintzis *et al.*, 1976; Robbins *et al.*, 1999). The high valency of the PS effectively cross-links multiple surface IgD/IgM (the B-cell receptor or BcR) and together with costimulation through complement receptor type 2 triggering (Griffioen *et al.*, 1993; Jeurissen *et al.*, 2004), engages the B cell to proliferate and differentiate into Ab secreting, PS specific plasma cells (Fig. 14.1A). The lack of cognate T-cell help has several important ramifications on the resultant immune response to PS immunogens. Typically, the ability to respond to PSs does not occur until late in ontogeny (age 1–3 years) and not maximally until the age of 6 years (Robbins

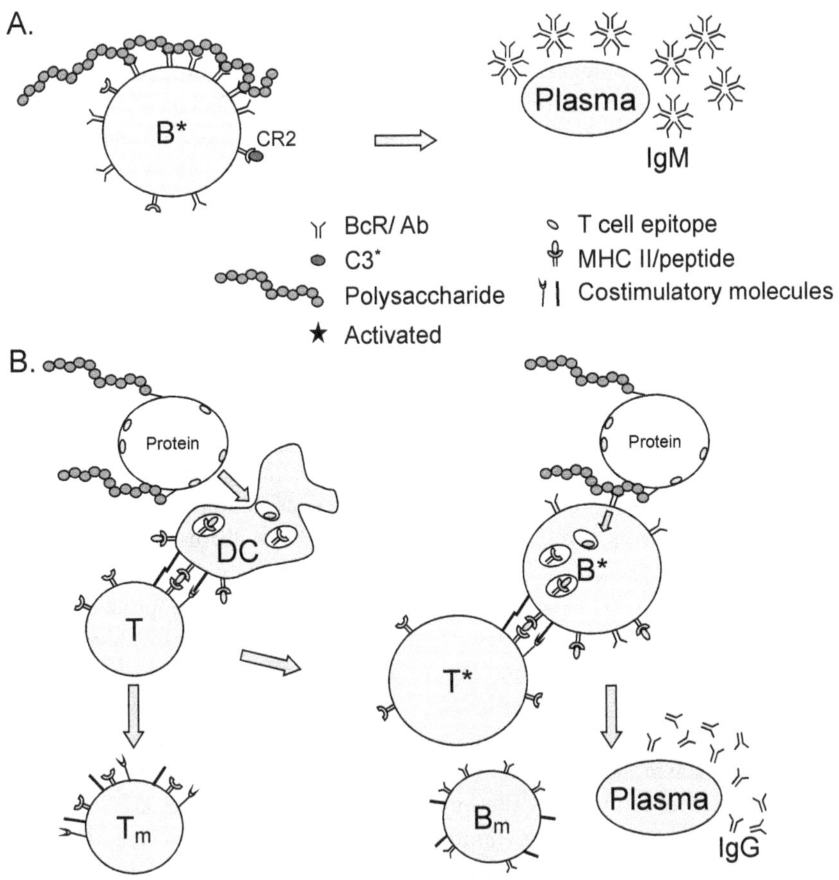

Figure 14.1 Immune response to polysaccharide and glycoconjugate vaccines. (A) Polysaccharides, by virtue of their iterated oligosaccharide repeat structures, are able to cross-link multiple B-cell receptors (surface Ig) to signal B-cell activation. Secondary costimulatory signals are thought to be mediated via complement receptor 2 binding of activated C3 split products. (B) Glycoconjugates are processed by dendritic cells where the protein is cleaved into peptides in endosomal compartments and bound to MHC class II molecules. Polysaccharides may be depolymerized via oxidative degradation but bind poorly to MHC class II. Carrier peptide/MHC II complexes are displayed on the DC surface where specific T-cell receptor recognition occurs (signal 1). Costimulation of T cells occurs through CD28–CD80/86 and CD40–CD40L interactions along with direction through cytokine release. B cells, by virtue of their specific B-cell receptor, capture glycoconjugate antigen and process it as described above to display carrier peptide/MHC class II molecules on their surface. Carrier specific T cells are capable of recognizing these complexes to deliver T-cell help through cognate interactions and focal cytokine release leading to efficient activation of B cells, their differentiation into affinity matured B cells and antibody secreting plasma cells. During this process, carrier specific memory T cells and polysaccharide specific memory B cells are generated.

et al., 1999) owing to the immaturity of splenic marginal zone B cells, the key cell type for TI-2 antigens (Guinamard *et al.*, 2000; Jeurissen *et al.*, 2004). Polysaccharides induce low affinity, non-class switched IgM Abs arising from a relatively restricted pool of B cell clonotypes (Scott *et al.*, 1992). Limited BcR heterogeneity is associated with a lack of affinity maturation – a mechanism to produce higher quality Ab (Scott *et al.*, 1992). Importantly, TI-2 antigens fail to induce anamnestic (booster) memory B-cell responses upon re-exposure to the immunizing antigen and in many instances this induces a hyporesponsive state (Poolman and Borrow, 2011). The quality and magnitude of a PS-induced Ab response can however, be influenced by non-specific T-cell help by production of cytokines and non-antigen driven cellular interactions (e.g. CD40-CD40L signalling) with PS activated B cells to help promote inefficient class switching to other Ab isotypes. Despite many shortcomings, PS vaccines can be quite efficacious in healthy adolescents and adults by producing relatively long-lived and protective circulating Ab.

Alternatively, the immune response to proteins, classified as thymus-dependent (TD) antigens, occurs early in life as a result of the exquisite interplay between professional antigen presenting dendritic cells (DCs), T cells and B cells; the process being characterized by cognate interactions and directed chemical signalling (Fig. 14.1B). The covalent coupling of TI-2 PS antigens to proteins imparts the TD nature of the protein onto the PS and this is due to the T cell response that is elicited by the protein component. The end result is the production of high frequency, affinity matured, isotype-switched Ab and the generation of long lived memory B cells. Mechanistically, PS glycoconjugates are internalized via pinocytosis by DCs where the protein is cleaved into peptide fragments in endosomal compartments through the actions of proteases. Whether PS is similarly depolymerized is not well known but the lack of sugar hydrolases in these compartments suggests other mechanisms would be responsible such as oxidative depolymerization (Duan *et al.*, 2008). Protein peptide fragments associate with MHC II molecules and are displayed on DCs in conjunction with secondary costimulatory ligands necessary for T-cell receptor recognition, activation, and differentiation into T helper cells that notably, are specific to the carrier protein and not to the PS elements of the glycoconjugate. During this sensitization of carrier specific T cells, a subset of cells will become long lived memory T cells that are capable of responding to the same carrier protein peptide with faster kinetics and less stringent requirements for costimulation (Crotty and Ahmed, 2004). Concomitant to this process, B cells residing in the spleen and lymph nodes that display surface Ig receptors specific for PS epitopes, capture and internalize glycoconjugates via receptor mediated endocytosis. Similar degradation of the glycoconjugate occurs with peptide fragments in association with MHC class II now displayed on those B cells that also possess BcRs that recognize specific PS. Carrier protein specific T cells are now capable of providing cognate T-cell help through specific recognition and binding of peptide/MHC II motifs, and together with upregulated B-cell adhesion, focal delivery of cytokines, and costimulatory signals, leads to the production of high frequency germinal centre derived plasma cells secreting isotype switched PS-specific Abs and the generation of long lived memory B cells (McHeyzer-Williams and McHeyzer-Williams, 2005). Although glycoconjugates are considered TD antigens akin to proteins, there are some important distinctions. Whereas proteins induce Ab of increasing affinity through somatic mutation within germinal centres, there is much less evidence that the same occurs with glycoconjugate induced B cells (Hougs *et al.*, 1999; Insel *et al.*, 1992) and is more akin to what is observed with TI-2 plain PS vaccines. GCVs, by virtue of induced T cell help, have helped to overcome two major shortcomings of PS vaccines. Because T-dependent help is functional shortly after birth, PS specific B cells gain the capacity to respond at this early age with the production of protective Ab. Furthermore, T-cell help also allows for the generation of immunological memory as demonstrated by the booster response that occurs upon re-challenge with the corresponding plain PS in GCV primed infants (Richmond *et al.*, 2001). The role that T-cell help plays in the maintenance of this immunological memory has come to the fore recently since B cell memory responses to the actual bacteria are not as robust as expected and vaccine failures have been observed (Trotter *et al.*, 2004). The key element is that GCVs give rise to a T-cell memory response specific to the carrier protein component of the vaccine but these same T cells are unable to function when they encounter the organism in question because there is no correlation between the proteins of the bacterium and the commonly used proteins for GCVs.

T-cell-mediated immunity to glycoconjugates

Most GCV studies have focused on how to use or enhance the beneficial effects of glycan-protein conjugation. T-cell involvement was restricted to protein elements because oligosaccharides associate poorly with MHC class II molecules. While this generality holds true with few exceptions (Deck *et al.*, 1995), there has been a paradigm shift with T cells now capable of recognizing glycoconjugates in the form of glycolipids when associated with MHC-like CD1 molecules (Sieling *et al.*, 1995). The burden of work in this area has focused on T-cell recognition of mycobacterial glyco-/lipid and although still at the investigational stage, these mycobacterial glycolipids represent a potential new category of GCV. T cell reactivities have been observed towards mycobacterial glucose monomycolate, phosphatidylinositol mannosides, lipoarabino- and lipomannans, and sulfoglycolipids (Cohen *et al.*, 2009). Antigen presenting cells such as DCs sample endocytic compartments where selected glycolipids are complexed with different isoforms of CD1 (CD1a-c and CD1d) and shuttled to the cell surface for display to T cells. In an analogous manner to peptide presentation, glycolipid/CD1 complexes are recognized by TcR where the fine specificity is due to the type and configuration of the polar sugar residues. Glycolipid reactive T cells are capable of releasing proinflammatory Th1 or anti-inflammatory Th2 cytokines depending on the microenvironment and are capable cytotoxic T cells able to efficiently kill intracellular mycobacteria (Felio *et al.*, 2009a). While it has not yet been directly established that CD1 restricted mycobacteria specific T cells are protective in mycobacterial disease, there is a great deal of promise that glycoconjugates such as these will form the basis of novel vaccines in which strong cell mediated immunity would be required (Felio *et al.*, 2009b).

Glycoconjugate vaccines – design considerations

Although the foundation for PS-based GCVs was aided tremendously by their plain CPS vaccine forerunners, there are several pre-clinical, empirically driven parameters that must be considered in the design of novel GCVs. The difficulty lies in how these parameters influence protective animal immune responses and their translation to human use, as it is unethical and cost prohibitive to hold comparative human trials for vaccine optimization. Analysis of vaccine construction coupled with extensive post-licensure surveillance studies have provided some important feedback to aid ongoing vaccine design. The following discussion is intended to familiarize the reader with these parameters, however for a more detailed understanding, there are several comprehensive reviews on the subject (Jennings and Sood, 1994; Jennings and Pon, 1996; Pon and Jennings, 2009; Zou and Jennings, 2009). An important additional consideration that often dictates the success or failure of new GCVs is the regulatory framework that must be adhered to in their design (Jones, 2008).

Carbohydrate antigen

The selection of candidate glycan is the single most important vaccine consideration and one that relies on a thorough understanding of the bacteria-host pathogenesis process, the challenges with obtaining or producing sufficient quantities of well characterized and pure materials, the correlates of protection from natural infection, the desired immunological response, and the potential pathogenic effects from host glycan similarities and/or inappropriate immune responses. Bacterial glycans can be traditionally isolated from culture sources, followed by enzymatic, physical, or chemical depolymerization to yield fragments or oligosaccharides that are often adequate as immunogens. Advances over the past decade in one pot and solid support oligosaccharide synthesis together with chemo-enzymatic approaches have brought synthetic oligosaccharides into the realm of vaccine feasibility whereas in the past, it was inconceivable that synthetic strategies could be scaled up while remaining cost effective (Boltje *et al.*, 2009; Hecht *et al.*, 2009). There still remain significant challenges with a synthetic approach but it offers some distinct advantages over naturally obtained glycans. For instance, synthetic glycans are chemically defined with a lack of heterogeneity that is often problematic with natural source glycan, they can reach high levels of purity, and they possess the ability to incorporate novel linkage strategies, which taken together, bode well for future synthetic GCV applications. In terms of soliciting Ab-mediated protection against extracellular bacteria, the key requirement is accessibility of antigen and one of the main reasons why CPS-based conjugates have been so successful. Capsular PSs with few exceptions are non-toxic, and the source of the most exposed and conserved protective epitopes on the surface of most bacteria- indeed invasive bacterial disease is rarely associated with unencapsulated organisms. LPS associated O-specific polysaccharides (O-SP) represent another class of important bacterial envelope antigens, with epidemiological studies correlating O-SP Ab levels with protection after natural infection (Robbins *et al.*, 1992). O-SPs however are subject to a tremendous amount of heterogeneity displaying many unique glycan structures within a bacterial species (Knirel, 2009), including host-related structures (Yuki *et al.*, 1993), so careful consideration must be given and priority placed on those serogroups/types most associated with significant morbidity and mortality and for which no other course of treatment is available. More conserved oligosaccharide structures, which theoretically would provide broader species coverage, can be found in core LPS (smooth) or lipooligosaccharide (LOS – rough) structures, in lipo-/teichoic acids and in cell wall associated glucans, however as vaccine targets, they suffer from restricted accessibility particularly if the organism has an associated capsule as is the case with most invasive bacteria.

Bacteria often display glycans with structural similarity to host tissues and selection of these targets requires careful consideration. These glycans are highly desired targets however, because they can encompass many bacterial sub-strains and are often not subject to variation. The human immune system has a complex system to protect against deleterious immune responses to host tissue otherwise known as autoimmunity. One of the fundamental means by which this is accomplished is through clonal deletion of T and B cells reactive to host antigens or through active tolerance mechanisms (anergy) to these same substrates. Breaking tolerance to host-mimicked glycans, such as what is thought to occur with both *Campylobacter jejuni* and *Helicobacter pylori* may contribute to disease. Expression of host ganglioside-like or Lewis antigen-like LOS structures in these organisms is believed to contribute to, if not mediate, some forms of Guillain-Barré syndrome or gastric cancers respectively (Moran, 2009). For this reason, as prophylactic GCV candidates, incorporation of host glycan structures will be subject to intense scrutiny for safety related concerns. GCVs directed towards tumour glycans are the exception to this regulatory hurdle since potent autoimmune anti-tumour responses are the desired outcome. GCVs for cancer therapy is one of the hottest areas of vaccine research and there are many reviews on the subject (Livingston and Ragupathi, 1997; Cipolla *et al.*, 2008; Astronomo *et al.*, 2010). Carbohydrate antigens relative to proteins are as a class less immunogenic owing to factors such as structural flexibility and receptor binding thermodynamics (Bundle and Young, 1992). For the reasons mentioned above, host-related glycans contribute to even poorer immune responses even in the face of strategies such as coupling to immunogenic proteins. A prime example is the response afforded by a NmB CPS-tetanus toxoid (TT) conjugate vaccine. As the CPS of NmB contains repeating units of α(2–8)-linked neuraminic acid or polysialic acid (PSA) that is also a key developmental antigen in mammalian neurogenesis (Troy, 1992), the immune response to the NmB–TT conjugate is remarkably poor despite the use of strong adjuvants (Jennings and Lugowski, 1981). A unique approach was employed to circumvent the poor immunogenicity of PSA that relied on chemical modification of the CPS to make it appear more foreign to the immune system (Fig. 14.2). Native *N*-acetyl groups of the repeating neuraminic acid residues were replaced with *N*-propionyl groups

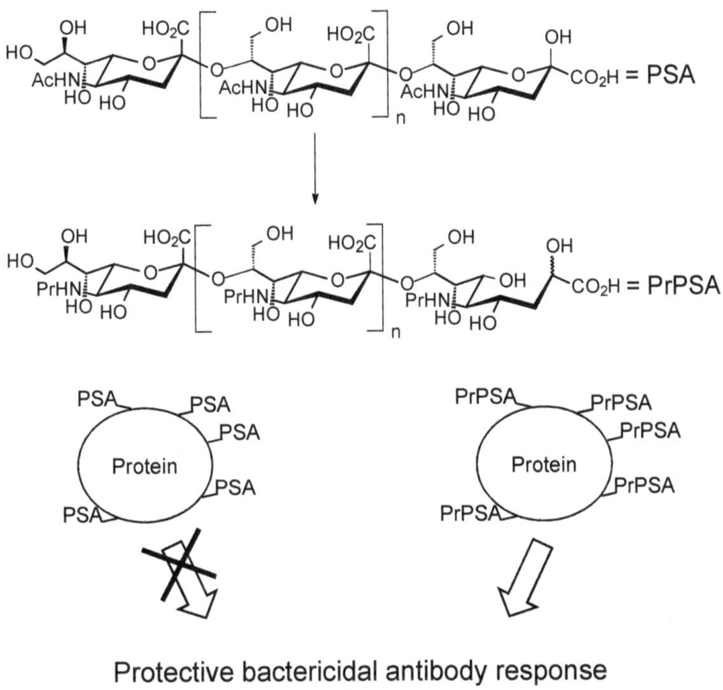

Figure 14.2 Chemical modification of polysaccharide structure to overcome poor immunogenicity. Poly *N*-acetyl neuraminic acid or PSA structures are found on mammalian nerve tissue and on the surfaces of bacteria such as *N. meningitidis*. Chemical manipulation to introduce *N*-propionyl groups to replace *N*-acetyl groups followed by conjugation to protein leads to the production of high titre protective bactericidal responses in contrast to no response with PSA- protein conjugates.

that differed only by an additional methylene. This small semi-synthetic structural change converted poorly immunogenic PSA conjugates into strongly immunogenic *N*-propionyl PSA conjugate vaccines. This vaccine was able to effectively raise bactericidal Ab in mice and protect them in NmB challenge experiments (Ashton *et al.*, 1989), and has proven to be safe with no observed vaccine-related complications in limited human trials (Krug *et al.*, 2004). These results validate chemical modification of host-resembling glycans as a tool to safely circumvent poor immunogenicity in vaccine design. Regardless, despite significant efforts to advance this vaccine, there still remains a great deal of apprehension to fully assess the NPrPSA-TT vaccine in advanced human trials due to the perceived possibility of autoimmune complications.

Studies investigating the NmB CPS phenomenon have revealed another important element regarding glycan selection in GCVs and that is the nature of the protective epitope. Several PS epitopes have now been shown to possess extended conformations necessary for productive Ab binding including the aforementioned CPS from NmB (Evans *et al.*, 1995), GBS (Jennings, 1998), and *S. pneumoniae* type 14 (Laferriere *et al.*, 1997). Glycoconjugate vaccine design needs to accommodate the ability to form these extended conformations. This may be due to host deletion of autoreactive B-cell clones specific to smaller non-conformational glycan present on tissues leaving only conformationally dependent B-cell clones left to recognize bacteria. The flip side of a protective Ab response that is conformationally restricted is the potential that deleterious autoimmune responses may be minimized, however this awaits further study.

Lastly, a key parameter in glycoconjugate design is the size and loading of the glycan fragment attached to the carrier protein, with the latter parameter more a reflection of the coupling chemistry (see below). As previously mentioned, there may be length dependencies that are critical for overcoming immune tolerance, however saccharide size in general is an important consideration. Glycan candidates range from native high molecular weight CPS (>200 kD) to short oligosaccharides, however to date, there are no approved vaccines based on short oligosaccharide fragments with the exception of a totally synthetic Hib conjugate vaccine with an average of eight repeating units of ribose ribitol phosphate (Verez-Bencomo *et al.*, 2004). An important distinction to make is between the minimal saccharide epitope required to induce an anti-bacterial response and the saccharide size that optimally induces in humans a protective, high titre, long-lasting memory response in the population most at risk (e.g. infants). The former question can be answered relatively easily using synthetic or depolymerized glycans in animal studies; however, the latter more relevant question can only be addressed through resource demanding clinical studies. Studies with synthetic overlapping structures of oligosaccharides found in type-specific pneumococcal CPSs indicated that immunogenicity and functional response with some serotypes increased progressively with an increasing number of oligosaccharide repeat units incorporated in the GCV, whereas a di-repeat unit was sufficient for some serotypes and the response did not markedly change with higher valency (Snippe *et al.*, 2008). Similarly, progressively better protective responses were obtained by increasing the number of ribose ribitol phosphate repeat units in Hib GCVs (Peeters *et al.*, 1992) and the biological pentasaccharide repeat unit of *Shigella flexneri* O-antigen (Mulard and Phalipon, 2008). In one of the few clinical assessments of chain length dependencies, although ~9 repeat units of ribose ribitol phosphate of Hib was shown to elicit optimal animal immune responses, it was determined that both Hib-$(PRP)_8$ and Hib–$(PRP)_{20}$ conjugate vaccines were equally adept at raising high titre Ab but only the latter conjugate was able to effectively prime infants for a booster vaccination (Anderson *et al.*, 1987). This study underlines the necessity for clinical evaluation of optimal saccharide sizing as the same relationship could not be established in animal studies. On the opposite end of the scale, the use of high molecular weight PSs for conjugation is similarly disadvantageous for two main reasons (1) loading onto proteins may be restricted owing to steric interference or inaccessibility of active groups required for coupling and (2) GCVs incorporating high molecular weight PS lose the advantage of the carrier protein and behave more like TI-2 antigens. This was elegantly demonstrated through a series of different length α(1–6)-linked dextran protein conjugates where smaller dextran fragments consistently produced stronger T-dependent immune responses relative to the longer fragments (Seppala and Makela, 1989). A similar conclusion was reached comparing three different sizes of GBS type III CPS (7, 14.5, and 27 kD) where the 14.5 kD conjugate provided the best protection when challenged with live group B streptococci (Wessels *et al.*, 1987). Thus, overall, the optimal saccharide size to be used in glycoconjugate design is largely chosen empirically based on animal studies and any refinement in humans will most likely only occur through vaccine failures.

Glycoconjugate chemistry

Methods developed to couple glycan to carrier protein or protein fragments are abundantly cited in the literature, however these protocols must conform to a number of conditions that are practically imposed by regulatory bodies and vaccine manufacturers and few, if any, fulfil all of the desired requirements. The basic requirements for the chemistry involved in GCV synthesis is the use of an efficient method of coupling, preferably performed in an aqueous environment, resulting in stable covalent linkages that are immunologically acceptable, preservation of the structural integrity of both the glycan and protein after conjugation, and assurance that residual coupling by-products left on the glycan or protein are non-toxic and not immunologically expressed. The resultant glycoconjugate should be defined, well characterizable, and optimized in terms of glycan loading and size, free from residual uncoupled glycan or protein, water soluble, and possess reproducible chemistries and immune performance. Complete details describing the most common coupling strategies that satisfy many of the above constraints can be found in several excellent reviews on the subject (Jennings and Sood, 1994; Jennings and Pon, 1996; Pozsgay and Kubler-Kielb, 2008; Zou and Jennings, 2009).

Coupling technology can be specific, selective, or random multi-point and is largely dictated by the physico-chemical properties of the glycan and to a lesser degree the carrier protein (Fig. 14.3). Specialized applications can also introduce a degree of order through scaffolding to evaluate multivalent clustering effects (Fig. 14.3C) as a means of inducing improved immune responses and/or an Ab response that selectively recognizes clustered bacterial antigen such as LOS within lipid rafts (Wang *et al.*, 2008). Large native PS by necessity requires random activation creating internal active sites for conjugation to protein as in the use of cyanogen bromide acting on sugar hydroxyl groups to form active cyanate esters (Zou and Jennings, 2009). These resultant glycoconjugates are ill–defined cross-linked products that often have poor solubilities and are difficult to characterize in terms of the epitope size between activation sites which pose problems for reproducibility in the manufacturing process (Fig. 14.3B). When a specific repetitive functionality is targeted in large molecular weight PS, such as the side chain moiety of sialic acid (*N*-acetylneuraminic acid) residues in GBS CPSs by using controlled amounts of activating agent (Kasper *et al.*, 1996), a more selective conjugate can be synthesized with better controllable parameters. Specific coupling approaches are more attainable using smaller PS fragments or ideally with synthetic substrates where attachment arms are built into the synthesis (Pozsgay and Kubler-Kielb, 2008). In terms of PS fragments, specific end residue activation leads to very defined, non-cross-linked conjugates that are more amenable to optimization strategies such as influence of glycan chain length or loading and thus are the preferred method for industry (Fig. 14.3A). The disadvantage of specific end unit activation is that PS fragments cannot be too large otherwise contact between activated terminal sugar and protein acceptor group is unfavoured but fortunately, this is rarely a problem due to the unwanted increased TI-2 nature of a large PS glycoconjugate as discussed above. Given the range of configurations presented above, although end unit activation is the preferred choice, licensed or late-stage GCVs have incorporated all variants and have proven to be effective in raising protective immune responses.

The specific chemistry and strategy involved in forming glycan and protein covalent bonds is

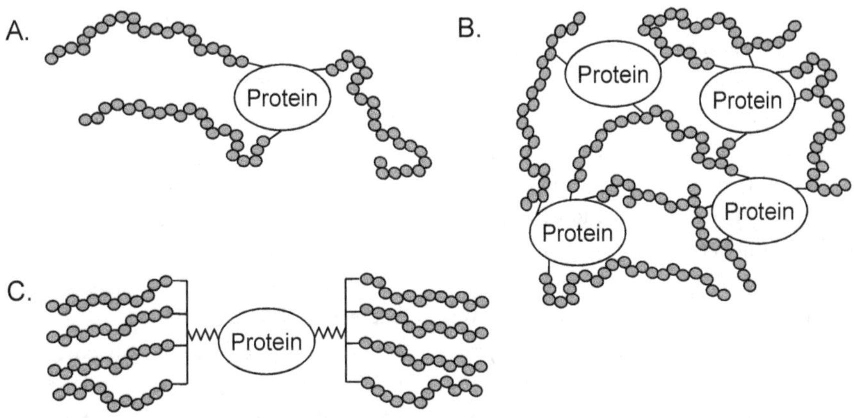

Figure 14.3 Schematic representation of different glycoconjugate designs. (A) Glycans activated through a single end terminus with or without the use of extension spacers produce non-cross-linked defined glycoconjugates. (B) Large native polysaccharides randomly activated via internal saccharide residues results in cross-linked product that can be challenging to define and reproduce. (C) Oligosaccharides can be introduced onto proteins using scaffolding to cluster glycan epitopes. This type of vaccine will give rise to antibodies that recognize clustered antigen such as those found on tumour cells.

virtually limitless, however the above-mentioned constraints for practicable glycoconjugate synthesis have reduced the options available. In general, amino, thio, hydroxyl, carbonyl, and carboxyl functionalities of both glycan and protein are employed (extensively reviewed in Jennings and Sood, 1994). The degree of activation plays an important role in the efficiency of the coupling, which is paramount especially if the glycan source is limiting, as well as contributing to both the physical and immunological properties of the final conjugate. As discussed above, multi-point activation leads to cross-linked products and over-activation can be detrimental by disrupting glycan epitopes as well as generating ill-defined insoluble product. Careful control is thus required and is governed largely by the choice of the activation route employed. Optimization of glycan loading, usually a function of the degree and reactivity of the activated groups on protein and glycan, can strongly influence the resultant GCV performance. For example, in a study examining the length and loading of *Shigella dysenteriae* O-SP (Pozsgay *et al.*, 1999), it was found that conjugates with low- or high-density dodecasaccharide antigen were suboptimal when compared with a conjugate with moderate incorporations (~10 chains/protein molecule). It was rationalized that higher incorporations mask important T-cell epitopes of the protein whereas low incorporations did not satisfactorily cluster Ig receptor binding on B cells. This same phenomenon has been observed in several other studies (Anderson *et al.*, 1989; Kabanova *et al.*, 2010).

Spacer elements are often employed in the design of GCVs and can be a natural extension of the activated and/or ring-opened attachment sugar or a variety of synthetic substrates employed to aid in the conjugation process. The main purpose for the use of spacers is to provide for greater flexibility in conjugate design and increase reactivity and control in forming protein-glycan covalent linkages. In addition, spacers also extend glycan residues away from the protein backbone where biological interactions with immune B cells may be hindered. A major requirement apart from the necessary reactive functionalities for conjugation is that the spacer should be immunologically silent. Numerous studies employing maleimide-based linkers have demonstrated that the majority of induced Abs are directed to the linkage region and only poor responses are raised against the glycan portion (Buskas *et al.*, 2004; Cox *et al.*, 2010). Similar findings were found using a squarate-based linker with GM3 glycoconjugates (R.A. Pon, unpublished). Linkers containing constrained cyclic structures are more prone to immunodominance versus flexible alkyl spacers. An example illustrating that immunodominance is not intrinsic entirely to the linker structure is given by the synthetic Hib-TT vaccine that employs maleimide esters and leads to protective anti-Hib Abs without undue spacer interference (Verez-Bencomo *et al.*, 2004). Spacer immunodominance in part, is a function of the immunogenicity of the glycan with greater spacer interference occurring with poorly immunogenic carbohydrate structures.

Finally, the conjugation method chosen must be mild enough to prevent glycan or protein structural changes. A case in point would be GCVs of NmB or GBS CPSs to protein where sialic residues govern the extended conformations of these PSs required to induce protective Ab responses. Conjugation conditions in a slightly acidic milieu leads to facile loss of sialic acids and consequent loss of PS conformation. Conjugation conditions must therefore be selected carefully based on the labile substituents present on the glycan of interest in order to preserve these potentially critical residues in the final product.

Carrier protein

The selection of an appropriate carrier protein is currently a hotly debated subject in the field of GCVs due to the increased use of relatively few carrier proteins and the possibility that excessive use will compromise the effectiveness of the antibacterial responses (in these cases those against bacterial CPSs). In theory, many proteins could function as a carrier but all currently approved GCVs employ proteins derived solely from bacterial sources; with the notable exception of the highly immunogenic keyhole limpet haemocyanin (KLH) that is derived from marine limpets and whose glycoconjugates are widely being evaluated in early phase cancer trials (Livingston, 1995). The current slate of carrier proteins in use includes the toxoids from tetanus (TT), diphtheria (DT), and a non-toxic mutant form of diphtheria, CRM_{197}, an outer membrane protein complex (OMP) of *N. meningitidis*, and more recently a non-typeable *H. influenzae* derived protein D, and a recombinant exotoxin of *Pseudomonas aeruginosa* (Table 14.1). The initial use of DT and TT was based on the fact that they were readily accessible, highly immunogenic, and efficacious vaccines in their own right, and were effective and licensed in infants. However one drawback of these toxoids was the very nature of the toxoiding process that eliminated many accessible lysine residues often required as the linkage substrate in glycan conjugation. CRM_{197} does not suffer this drawback since its point mutation renders the toxin non-toxic alleviating the requirement for the toxoiding process. As a result, CRM_{197} is the most prevalently used carrier protein since all lysine amino side chains are available for derivatization

Table 14.1 List of licensed and late clinical phase glycoconjugate vaccines

Target	Vaccine	Protein	Manufacturer
***Haemophilus influenzae* type b**	HibTiter®	CRM_{197}	Pfizer
	ActHib®	TT	Sanofi
	PedvaxHib®	OMP	Merck
	Hiberix®	TT	GSK
	Synthetic Quimi-Hib®	TT	Heber Biotech
Neisseria meningitidis			
Serogroup A	MenAfrica	TT	MVP
Serogroup C	Menjugate®	CRM_{197}	Novartis
	Meningitec®	CRM_{197}	Pfizer
	NeisVac-C®	TT	Baxter
A,C, Y, W-135	Menactra®	DT	Sanofi
A,C, Y, W-135	Menveo®	CRM_{197}	Novartis
Streptococcus pneumoniae			
4, 6B, 9V, 14, 18C, 19F, 23F	Prevnar-7®	CRM_{197}	Pfizer
Above+ 1, 3, 5, 6A, 7F, 19A	Prevnar-13®	CRM_{197}	Pfizer
1, 4, 5, 6B, 7F, 9V, 14, 18C, 19F, 23F	Synflorix®	NTHibPD, DT, TT	GSK
***Salmonella enterica* serovar Typhi**	Peda Typh®	TT	Bio-Med
	Vi-rEPA	r-EPA	1
***Group B streptococcus*[2]**			
Type Ia, Ib, II, III, V	–	TT	1
***Shigella flexneri* 2a**	–	rEPA	1
Shigella sonnei	–	CRM_{197}	1
***Staphylococcus aureus* serotype 5,8**	StaphVax®	rEPA	1

[1]Glycoconjugate vaccines that have been evaluated in late phase clinical trials – see www.fda.gov/BiologicsBloodVaccines/Vaccines/ApprovedProducts/default.htm for details.

[2]As individual vaccines not as a combination vaccine.

CRM_{197}, diphtheria toxin mutant; TT, tetanus toxoid; DT, diphtheria toxoid; NTHibPD, non-typeable *H. influenzae* protein D; r-EPA, recombinant exoprotein from *Pseudomonas aeruginosa;* GSK, GlaxoSmithKline; MVP, Meningitis Vaccine Project/PATH-WHO.

and the form of the protein is constant unlike batch to batch variation from the toxoiding process. OMP carrier proteins demonstrate unique immunological properties resulting in high initial Ab titres upon primary immunization with relatively poor booster effects. This effect is due to the presence of a mitogenic protein within the OMP complex which acts as a polyclonal stimulator (Liu *et al.*, 1992) and has been exploited in certain unique applications. For instance, Native American infants are highly susceptible to early Hib disease and a single Hib-OMP vaccination affords strong protective titres without the delay normally incurred with boosting regimes required by the other Hib GCVs (Harrison *et al.*, 1994). However, despite these kinetics, the complexity and variability of OMP vesicle preparations argues against their widespread use.

Given the few approved carrier proteins, the increase in serotype numbers included within multivalent vaccines [e.g. the recently approved 13-valent pneumococcal (Pn) CPS- CRM_{197} conjugate vaccine (Prevnar-13®)], and the growth in combination vaccines such as the proposed Globorix™ (DT, TT, pertussis, hepatitis B, and Hib and NmA and NmC GCVs), there is considerable concern that carrier protein overload will compromise vaccine effectiveness and limit new candidates. Principal among the concerns for vaccine interference is carrier induced epitope suppression (CIES) where a B-cell response to a glycan hapten is dwarfed by the B cell response that occurs to the carrier protein (Dagan *et al.*, 2010). Clearly there is a need for an expanded pool of new carrier proteins to limit CIES but also for carrier proteins that provide added value to a given GCV. For instance, the recently licensed decavalent Pn GCV (Synflorix™) is composed of eight different serotypes of Pn coupled to non-typeable *H. influenzae* derived protein D (+1 TT and 1 DT conjugate). The added advantage of the protein D carrier is that it is also active against *H. influenzae*-mediated acute otitis media (a middle ear infection), and a major health-related cost burden in industrialized countries (De Wals *et al.*, 2009). Taking this one step further, preliminary human trials have been conducted on a NmC GCV using a meningococcal dehydrogenase protein (P64k) as a carrier (Perez *et al.*, 2006). The conjugate was safe and raised bactericidal Ab but importantly, because the protein is derived from the same organism as its CPS, functional T-cell memory as described earlier should be available. The benefit from functional T-cell memory is a theoretical enhancement of anti-bacterial immunity but this awaits further human studies.

Adjuvants

Carbohydrate antigens are intrinsically poor immunogens relative to their protein cousins. As mentioned, conjugation to suitable proteins increases their immunogenicity significantly mainly by invoking key cell types such as DCs and T helper cells. Fundamental understanding of how pathogens signal danger with tailoring of the ensuing immune response has led to a surge in the field of immunomodulation and adjuvants for vaccine design. The only recognized adjuvant for use with GCVs is alum, a mixture of aluminium salts that favours Ab production. But as we now better understand the immune process that occurs with GCV administration, immunomodulatory agents can be selected to augment or deviate the ensuing immune response to the GCV such that it is better able to afford protection. For instance, serum bactericidal Ab- that is Ab possessing isotypes capable of fixing complement- is a requisite for clearing NmB. Using a chemically modified group B vaccine with alum as a basic adjuvant induces primarily non-complement fixing IgG1 Ab and consequently poor bactericidal activity in mice (Jennings, 1997). The IgG1 response is characteristic of a T helper type 2 response. However, if the vaccine is administered with a T helper type 1 promoting adjuvant such as monophosphoryl lipid A (a TLR4 agonist of DCs), the ensuing Ab response is skewed to the production of complement fixing IgG2a and consequently affords protection against live NmB challenge (Pon *et al.*, 1997). Similar manipulations using a host of immunomodulatory agents such as bacterial CpGs, cytokines, saponins, lipopeptides or through targeting DCs through C-type lectins, and the use of liposomal or microparticle vehicles represent future growth areas in glycoconjugate design (reviewed in Guy, 2009).

Glycoconjugate vaccines: lessons learned from the clinic

A list of the current licensed GCVs or those that are in late clinical trials are given in Table 14.1. A quick survey reveals that the majority of the approved vaccines is directed towards invasive bacterial disease, targeting CPS or O-SP of respiratory or enteric bacteria. While it is beyond the scope of this review, the clinical history and impact of each of these vaccines, particularly in the most vulnerable population namely infants, is a valuable complement to this overview and is highly recommended reading (Oosterhuis-Kafeja *et al.*, 2007; Pace and Pollard, 2007; Pon and Jennings, 2009). In all cases, the introduction of routine GCV vaccination in infants and children has significantly decreased the morbidity, and mortality associated with invasive disease. The first of such vaccines to the market was directed towards Hib, the leading cause of childhood meningitis at the time. Collectively, the vaccines against Hib possessed an overall efficacy rate of 84% from data obtained in 8 randomized trials with over 365,000 children, reaching 100% efficacy in some trials (Obonyo and Lau, 2006). In the USA, prevalence rates of Hib disease dropped >99% after the introduction of routine vaccination (Anon, 2002) with similar spectacular declines recorded in other Western countries (Chandran *et al.*, 2005). The introduction of the NmC CPS conjugate vaccines in England, Spain, and Canada in 2000, largely implemented to curb the steep rise in NmC disease (26 to 34% over four years) owing to the presence of a hypervirulent NmC clone, were equally as effective with 1 year efficacies between 92–97% (Balmer *et al.*, 2002; De Wals *et al.*, 2004; Larrauri *et al.*, 2005). *S. pneumoniae* (Pn) is a leading cause of bacteraemia and meningitis in young children, the elderly, and the immunocompromised and is a major contributor to childhood middle ear infections (otitis media). The first Pn CPS-based conjugate vaccine incorporated the seven serotypes of Pn most commonly encountered in North American children (PnC7) and has since been replaced by a similar vaccine with six additional serotypes (PnC13). In an efficacy trial, PnC7 resulted in >90% of children developing Ab to all serotypes with invasive Pn disease caused by vaccine serotypes reduced by 97.4% and 89% when all serotypes were accounted for (Black *et al.*, 2000). PnC7 has also had a modest positive impact on acute otitis media, a localized mucosal surface disease in children, with protection ranging from 57–67% for vaccine serotypes (Fletcher and Fritzell, 2007). And finally, typhoid fever is a major cause of morbidity and mortality with an estimated 22 million cases annually, predominantly in South Asia, Africa and South America (Crump *et al.*, 2004). The introduction of a *S. enterica* serovar Typhi CPS conjugate vaccine (Vi-rEPA), incorporating a recombinant exoprotein of *P. aeruginosa* as a carrier protein, to Vietnamese children with endemic *S. typhi* disease resulted in 91.1% protection over 27 months (Lin *et al.*, 2001) and 89% over 46 months (Mai *et al.*, 2003). It is clear from this sampling that GCVs are extraordinarily safe, robust, and effective. However, post-license surveillance studies have also identified some important lessons to consider with new generation GCVs.

Design

There are few clinical studies specifically designed to evaluate the impact of vaccine design factors and often this data must be extracted from post trial data. For example, it appears as though the DT carrier protein (not the mutant CRM_{197}) may be inferior to other carrier proteins. An early Hib-DT vaccine was found to be less immunogenic and ineffective in infants (Anon, 1991) relative to CRM_{197}, OMP, or TT versions and has since been removed from the market. Similarly, a combination *N. meningitidis* serogroups A,C,Y, and W-135- DT vaccine was also found to be deficient in young children (Rennels *et al.*, 2004) resulting in licensing in adolescents over the age of 11 years (since been revised to children 2+ years). Furthermore, for as yet unknown reasons, there has been an association of this vaccine with Guillain-Barré syndrome (Anon, 2009). Finally, an 11-valent Pn GCV including six serotypes conjugated to DT underwent clinical trials (Lucero *et al.*, 2004) but has not been carried forward. It is suspected that important T-cell epitopes may have been lost through the toxoiding process, rendering DT less immunogenic.

Of direct relevance to synthetic GCVs where glycan is by necessity relatively small, two independent human trials with Hib (Daum *et al.*, 1997) and Pn (Anderson *et al.*, 1987) oligosaccharides have established that longer fragments are superior to shorter glycans, consistent with most animal studies. Furthermore, a back to back comparison of three different NmC GCVs established one vaccine to be superior in infants. Notably, one of the differences between the vaccines was the use of NmC PS as opposed to short-chain oligosaccharides in the two suboptimal constructs (Richmond *et al.*, 2001). These findings however, do not preclude the use and development of oligosaccharide-based glycoconjugates since not enough clinical information is available to properly assess the merits/pitfalls of synthetic GCVs.

Bacterial carriage effects

The routine use of infant conjugate vaccines to protect against invasive disease has had an unexpected beneficial spin-off effect with the observation that mass Hib vaccination reduced Hib disease in the unvaccinated population (known as herd immunity), and this correlated with a decreased nasopharyngeal carriage of Hib, thereby reducing transmission from child to child (Takala *et al.*, 1991). The implication of this finding is that systemic immunization is able to clear organisms at the mucosal level (Robbins *et al.*, 1999). Mucosal reduction of bacteria upon routine GCV immunization has now been verified for NmC (Maiden and Stuart, 2002), Pn (Dagan *et al.*, 2003) and is suspected to be the mechanism of action for protection against enteric bacteria with the use of *Shigella flexneri, Shigella sonnei,* and *Shigella dysenteriae* O-SP GCVs (Cohen *et al.*, 1997; Pozsgay *et al.*, 1999; Passwell *et al.*, 2003). A striking example of this effect was found with the PnC7 conjugate vaccine where herd immunity was responsible not only for protecting non-immunized children but also for decreased carriage of Hib and a 32% serotype-related reduction in Pn disease in adults and the elderly (Anon, 2005). Given this significant mode of protection from glycoconjugate vaccination, carriage and herd immunity are important correlates of vaccine efficacy.

Immune correlates of vaccine efficacy

The effort to develop an NmC conjugate vaccine in England in the most expedient time frame possible led for the first time to licensure of new vaccines without proceeding through lengthy phase III efficacy trials. Licensure was based on pre-clinical establishment of immune correlates of protection including serum bactericidal Ab levels, immune memory induction, and Ab avidity increases/anamnestic responses. Initial optimism in the UK over the high level of vaccine effectiveness measured in the first year following vaccine implementation fell when follow up studies showed a rapid decline in bactericidal Ab levels. Preclinical studies had suggested however, that long term protection would be afforded by B-cell memory induction which was still operational in immunized children despite their decline in Ab levels (Borrow *et al.*, 2003). The follow up studies however were able to correlate the decline in bactericidal Ab levels with vaccine failures that were occurring in children who were still able to mount efficient memory responses. The current theory to explain why vaccine failures occurred in these children with intact long term memory is that there is a window of susceptibility present between infection and when a sufficiently strong memory B-cell response can be mounted to produce protective Ab. This delay may be qualitatively related to the fact that T-cell memory was not present when NmC was encountered. This result has precipitated a change in the immunization regime by providing a booster dose of NmC conjugate in the second year to raise serum bactericidal Ab to protective levels. The development of appropriate immune correlates of efficacy is now a high priority with all GCVs in order to effectively track protective efficacy. A single parameter such as specific Ab level is not sufficient and functional outcomes should also be included such as bactericidal or opsonophagocytic Ab levels, memory induction, bacterial carriage, and increased Ab quality (Licciardi *et al.*, 2011).

Serotype replacement

H. influenzae-related disease is restricted to one type whereas *N. meningitidis*, Group B *Streptococcus, S. pneumoniae,* and candidate enteric bacteria all possess multiple serotypes/serogroups. Highlighting this diversity is *S. pneumoniae* with over 90 different serotypes (Pon and Jennings, 2009). Serotypes of Pn associated with disease are highly variable in different populations, age groups, and geographical locations. The initial PnC7 conjugate vaccine included seven serotypes to cover the majority of infant disease in North America and has proven to be extremely effective in protecting against serotype associated disease. However follow up studies have revealed that carriage of vaccine serotypes is suppressed in vaccinated children, but this has been offset by the appearance of replacement serotypes such as 3, 6A, 7F, 19A, and 35B (Hanage *et al.*, 2007; Pichichero and Casey, 2007). Although overall Pn disease is not yet on the rise, the proportion of Pn disease is changing dramatically as exemplified by 40% of Pn disease being caused by serotype 19A in a 2005 study versus virtually no contribution a decade ago (Messina *et al.*, 2007). The appearance of these new serotypes within the population has prompted the development and licensing of the PnC13 GCV yet one has to wonder whether the same phenomenon will probably repeat itself. A similar fear stems from the effective control that NmC vaccination has had on group C meningococcal disease and the possibility that the endemic hypervirulent ST-11 clone will acquire the group B capsular PS through transformation, resulting in loss of vaccine control (Alcala *et al.*, 2002). It is for these reasons that close post-license surveillance is required to assess the effect of vaccine pressures.

Vaccine interference

The influx of new childhood vaccines together with their combinations, and the expansion of multivalent vaccines such as the tetravalent meningococcal and the 13-valent pneumococcal vaccines, all using shared carrier proteins, has precipitated a concern that vaccine interference from these combinations will compromise the effectiveness of the individual vaccines. The initial warning sign occurred with a resurgence of Hib disease in children (McVernon *et al.*, 2003) vaccinated with the Hib conjugate vaccine in combination with acellular but not whole cell pertussis vaccine (Pichichero *et al.*, 1997). This interference phenomenon was also observed with an 11-valent PnC vaccine composed of seven serotypes attached to TT and 4 attached to DT, administered in combination with acellular pertussis vaccine (Dagan *et al.*, 2004). In this cohort, the children responded poorly to the TT conjugates and abnormally to the DT conjugates upon boosting. This was originally attributed to carrier induced epitope suppression described previously. However it appears now to be related to adjuvancy that occurs with the whole but not with the acellular pertussis vaccine (Dagan *et al.*, 2010). Not all interference is negative. There are several reports that document increased response to Hib-TT GCVs when administered with the NmC-TT GCV (Tejedor *et al.*, 2006). The mechanisms of vaccine interference are complex and not yet well understood making predictions difficult but this does reinforce the value of strong post surveillance monitoring of new vaccines or combinations.

Summary

There is no doubt that the development of PS-based GCVs against invasive bacteria is one of the most significant advances in modern medicine, founded upon basic principles of immunology, carbohydrate and protein chemistry in tandem with empirical studies to tease out optimal conditions for their development. It has been based on a strong knowledge of bacterial virulence, the ability to isolate and define primarily CPS structures, the chemical principles to modify and attach them to proteins, and an understanding of how these glycans interact with the immune system to afford protection. This knowledge has resulted in numerous highly safe and effective vaccines, primarily for infants who are most susceptible to invasive bacterial disease, where in some instances, GCVs have led to the virtual eradication of the target disease. We have over the course of this vaccine development period learned important lessons for best practices in GCV design and how these vaccines behave in the immunized population. We are now poised to apply this collective knowledge to new generation GCVs not only directed towards bacteria but towards viruses, fungi, parasites, and cancers. There have been exponential advances over the past decade in the field of microbial and mammalian glycomics owing to a much better understanding of the genetics behind glycan expression and assembly, the molecular interactions and biological outcomes of glycan–glycan receptors, and the interaction of microbial glycans with the human immune system; all of which would not be possible without the tremendous progress made in high throughput glycan analysis. We are only now starting to discover new carbohydrate structures such as *N*- and *O*-linked bacterial glycoproteins and how these may relate to bacterial virulence and be possible vaccine targets. The genetic pathways underlying the formation of these new glycan structures are even being exploited to aid in the production of new glycoconjugates for novel vaccine use, an example of which is the engineering of innocuous bacteria with oligosaccharyltransferases and custom glycan gene cassettes to potentially derive living bacterial vectors expressing custom GCVs (Hug and Feldman, 2011). Our efforts to this point have focused on humoral responses to extracellular surface structures without applying immunological focusing. As we add in our new knowledge for immunomodulating glycan or glycoconjugate responses, we will see improved activities from current GCV strategies and develop previously untapped areas such as glycan-based cellular immunity towards intracytoplasmic pathogens such as *Mycobacterium tuberculosis*. As we unravel the language of the glycome, new therapeutic strategies of which GCVs is but one, will inevitably arise with the promise of further medical breakthroughs the likes of which we have not yet seen.

References

Alcala, B., Arreaza, L., Salcedo, C., Uria, M.J., De La Fuente, L., and Vazquez, J.A. (2002). Capsule switching among C:2b:P1.2,5 meningococcal epidemic strains after mass immunization campaign, Spain. Emerg. Infect. Dis. *8*, 1512–1514.

Anderson, P., Peter, G., Johnston, R.-B.J., Wetterlow, L.H., and Smith, D.H. (1972). Immunization of humans with polyribophosphate, the capsular antigen of Hemophilus influenzae, type b. J. Clin. Invest. *51*, 39–44.

Anderson, P., Pichichero, M., Edwards, K., Porch, C.R., and Insel, R. (1987). Priming and induction of *Haemophilus influenzae* type b capsular antibodies in early infancy by Dpo20, an oligosaccharide–protein conjugate vaccine. J. Pediatr. *111*, 644–650.

Anderson, P.W., Pichichero, M.E., Stein, E.C., Porcelli, S., Betts, R.F.C.D.M., Korones, D., Insel, R.A., Zahradnik, J.M., and Eby, R. (1989). Effect of oligosaccharide chain length, exposed terminal group, and hapten loading on the antibody response of human adults and infants to vaccines consisting of Haemophilus influenzae type b capsular antigen unterminally coupled to the diphtheria protein CRM197. J. Immunol. *142*, 2464–2468.

Anon (1991). Haemophilus b conjugate vaccines for prevention of *Haemophilus influenzae* type b disease among infants and children two months of age and older. Recommendations of the immunization practices advisory committee (ACIP). MMWR Recomm. Rep. *40*, 1–7.

Anon (2002). Progress toward elimination of *Haemophilus influenzae* type b invasive disease among infants and children – United States, 1998–2000. MMWR Morb. Mortal. Wkly. Rep. *51*, 234–237.

Anon (2005). Direct and indirect effects of routine vaccination of children with 7-valent pneumococcal conjugate vaccine on incidence of invasive pneumococcal disease – United States, 1998–2003. MMWR Morb. Mortal. Wkly. Rep. *54*, 893–897.

Anon (2009). Updated recommendation from the Advisory Committee on Immunization Practices (ACIP) for revaccination of persons at prolonged increased risk for meningococcal disease. MMWR Morb. Mortal. Wkly. Rep. *58*, 1042–1043.

Ashton, F.E., Ryan, J.A., Michon, F., and Jennings, H.J. (1989). Protective efficacy of mouse serum to the *N*-propionyl derivative of meningococcal group B polysaccharide. Microb. Pathog. *6*, 455–458.

Astronomo, R.D., and Burton, D.R. (2010). Carbohydrate vaccines: developing sweet solutions to sticky situations? Nat. Rev. Drug Discov. *9*, 308–324.

Balmer, P., Borrow, R., and Miller, E. (2002). Impact of meningococcal C conjugate vaccine in the UK. J. Med. Microbiol. *51*, 717–722.

Black, S., Shinefield, H., Fireman, B., Lewis, E., Ray, P., Hansen, J.R., Elvin, L., Ensor, K.M., Hackell, J., Siber, G., *et al.* (2000). Efficacy, safety and immunogenicity of heptavalent pneumococcal conjugate vaccine in children. Northern California Kaiser Permanente Vaccine Study Center Group. Pediatr. Infect. Dis. J. *19*, 187–195.

Boltje, T.J., Buskas, T., and Boons, G.J. (2009). Opportunities and challenges in synthetic oligosaccharide and glycoconjugate research. Nat. Chem. *1*, 611–622.

Borrow, R., Goldblatt, D., Finn, A., Southern, J., Ashton, L., Andrews, N., Lal, G., Riley, C., Rahim, R., Cartwright, K., *et al.* (2003). Immunogenicity of, and immunologic memory to, a reduced primary schedule of meningococcal C-tetanus toxoid conjugate vaccine in infants in the United Kingdom. Infect. Immun. *71*, 5549–5555.

Bundle, D.R., and Young, N.M. (1992). Carbohydrate–protein interactions in antibodies and lectins. Curr. Opin. Struc. Biol. 2, 666–673.

Buskas, T., Li, Y., and Boons, G.J. (2004). The Immunogenicity of the tumor-associated antigen Lewis Y may be suppressed by a bifunctional cross-linker required for coupling to a carrier protein. Chem. Eur. J. *10*, 3517–3524.

Chandran, A., Watt, J.P., and Santosham, M. (2005). Prevention of *Haemophilus influenzae* type b disease: past success and future challenges. Expert Rev. Vaccines. *4*, 819–827.

Cipolla, L., Peri, F., and Airoldi, C. (2008). Glycoconjugates in cancer therapy. Anti-Cancer Agent Me. *8*, 92–121.

Cohen, D., Ashkenazi, S., Green, M.S., Gdalevich, M., Robin, G., Slepon, R., Yavzori, M., Orr, N., Block, C., Ashkenazi, I., *et al.* (1997). Double-blind vaccine-controlled randomised efficacy trial of an investigational Shigella sonnei conjugate vaccine in young adults. Lancet *349*, 155–159.

Cohen, N.R., Garg, S., and Brenner, M.B. (2009). Antigen presentation by CD1: Lipids, T Cells, and NKT cells in microbial immunity. In Advances in Immunology, Frederick, W.A., ed. (Oxford, Academic Press), pp. 1–94.

Cox, A., St.Michael, F., Neelamegan, D., Lacelle, S., Cairns, C., Giuliani, M., Biolchi, A., Hoe, J., Moxon, E., and Richards, J. (2010). Investigating the candidacy of LPS-based glycoconjugates to prevent invasive meningococcal disease: immunology of glycoconjugates with high carbohydrate loading. Glycoconjugate J. *27*, 643–648.

Crotty, S. and Ahmed, R. (2004). Immunological memory in humans. Semin. Immunol *16*, 197–203.

Crump, J.A., Luby, S.P., and Mintz, E.D. (2004). The global burden of typhoid fever. Bull. World Health Organ. *82*, 346–353.

Dagan, R., Givon-Lavi, N., Zamir, O., and Fraser, D. (2003). Effect of a nonavalent conjugate vaccine on carriage of antibiotic-resistant Streptococcus pneumoniae in day-care centers. Pediatr. Infect. Dis. J. *22*, 532–540.

Dagan, R., Goldblatt, D., Maleckar, J.R., Yaich, M., and Eskola, J. (2004). Reduction of antibody response to an 11-valent pneumococcal vaccine coadministered with a vaccine containing acellular pertussis components. Infect. Immun. *72*, 5383–5391.

Dagan, R., Poolman, J., and Siegrist, C.A. (2010). Glycoconjugate vaccines and immune interference: A review. Vaccine *28*, 5513–5523.

Daum, R.S., Hogerman, D., Rennels, M.B., Bewley, K., Malinoski, F., Rothstein, E., Reisinger, K., Block, S., Keyserling, H., and Steinhoff, M. (1997). Infant immunization with pneumococcal CRM197 vaccines: effect of saccharide size on immunogenicity and interactions with simultaneously administered vaccines. J. Infect. Dis. *176*, 445–455.

De Wals, P., Deceuninck, G., Boulianne, N., and De Serres, G. (2004). Effectiveness of a mass immunization campaign using serogroup C meningococcal conjugate vaccine. JAMA *292*, 2491–2494.

De Wals, P., Black, S., Borrow, R., and Pearce, D. (2009). Modeling the impact of a new vaccine on pneumococcal and nontypable Haemophilus influenzae diseases: A new simulation model. Clin. Ther. *31*, 2152–2169.

Deck, B., Elofsson, M., Kihlberg, J., and Unanue, E.R. (1995). Specificity of glycopeptide-specific T cells. J. Immunol. *155*, 1074–1078.

Dintzis, H.M., Dintzis, R.Z., and Vogelstein, B. (1976). Molecular determinants of immunogenicity: the immunon model of immune response. Proc. Natl. Acad. Sci. U.S.A. *73*, 3671–3675.

Dochez, A.R., and Avery, O.T. (1917). The elaboration of specific soluble substance by Pneumococcus during growth. J. Exp. Med. *26*, 477–493.

Duan, J., Avci, F.Y., and Kasper, D.L. (2008). Microbial carbohydrate depolymerization by antigen-presenting cells: Deamination prior to presentation by the MHCII pathway. Proc. Natl. Acad. Sci. U.S.A. *105*, 5183–5188.

Evans, S.V., Sigurskjold, B.W., Jennings, H.J., Brisson, J.R., To, R., Altman, E., Frosch, M., Weisgerber, C., and Kratzin, H. (1995). Evidence for the extended helical nature of polysaccharide epitopes. The 2.8 A resolution structure and thermodynamics of ligand binding of an antigen binding fragment specific for alpha-(2–8)-polysialic acid. Biochemistry *34*, 6737–6744.

Felio, K., Nguyen, H., Dascher, C.C., Choi, H.J., Li, S., Zimmer, M.I., Colmone, A., Moody, D.B., Brenner, M.B., and Wang, C.R. (2009a). CD1-restricted adaptive immune responses to Mycobacteria in human group 1 CD1 transgenic mice. J. Exp. Med. *206*, 2497–2509.

Felio, K., Nguyen, H., Dascher, C.C., Choi, H.J., Li, S., Zimmer, M.I., Colmone, A., Moody, D.B., Brenner, M.B., and Wang, C.R. (2009b). CD1-restricted adaptive immune responses to Mycobacteria in human group 1 CD1 transgenic mice. J. Exp. Med. *206*, 2497–2509.

Fletcher, M.A., and Fritzell, B. (2007). Brief review of the clinical effectiveness of PREVENAR against otitis media. Vaccine *25*, 2507–2512.

Griffioen, A.W., Franklin, S.W., Zegers, B.J.M., and Rijkers, G.T. (1993). Expression and Functional Characteristics of the Complement Receptor Type 2 on Adult and Neonatal B Lymphocytes. Clin. Immunol. Immunop. *69*, 1–8.

Guinamard, R., Okigaki, M., Schlessinger, J., and Ravetch, J.V. (2000). Absence of marginal zone B cells in Pyk-2-deficient mice defines their role in the humoral response. Nat. Immunol. *1*, 31–36.

Guy, B. (2009). Adjuvants for protein-and carbohydrates-based vaccines. In Carbohydrate-based Vaccines and Immunotherapies, Guo, Z.W., and Boons, G.J., eds. (New Jersey, John Wiley and Sons), pp. 89–115.

Hanage, W.P., Huang, S.S., Lipsitch, M., Bishop, C.J., Godoy, D., Pelton, S.I., Goldstein, R., Huot, H., and Finkelstein, J.A. (2007). Diversity and antibiotic resistance among nonvaccine serotypes of Streptococcus pneumoniae carriage isolates in the post-heptavalent conjugate vaccine era. J. Infect. Dis. *195*, 347–352.

Harrison, L.H., Tajkowski, C., Croll, J., Reid, R., Hu, D., Brenneman, G., Weatherholtz, R.C., and Santosham, M. (1994). Postlicensure effectiveness of the Haemophilus influenzae type b polysaccharide-*Neisseria meningitidis* outer-membrane protein complex conjugate vaccine among Navajo children. J. Pediatr. *125*, 571–576.

Hecht, M.L., Stallforth, P., Silva, D.V.N., Adibekian, A., and Seeberger, P.H. (2009). Recent advances in carbohydrate-based vaccines. Curr. Opin. Chem. Biol. *13*, 354–359.

Holst, O., Moran, A.P., and Brennan, P.J. (2009). Overview of the glycosylated components of the bacterial cell envelope. In Microbial Glycobiology: Structures, Relevance, and Applications, Moran, A.P., ed. (London, Academic Press), pp. 3–13.

Hougs, L., Juul, L., Ditzel, H.J., Heilmann, C., Svejgaard, A., and Barington, T. (1999). The first dose of a *Haemophilus influenzae* type b conjugate vaccine reactivates memory B cells: evidence for extensive clonal selection, intraclonal affinity maturation, and multiple isotype switches to IgA2. J. Immunol. *162*, 224–237.

Hug, I., and Feldman, M.F. (2011). Analogies and homologies in lipopolysaccharide and glycoprotein biosynthesis in bacteria. Glycobiology *21*, 138–151.

Insel, R.A., Adderson, E.E., and Carroll, W.L. (1992). The repertoire of human antibody to the *Haemophilus influenzae* type b capsular polysaccharide. Int. Rev. Immunol. *9*, 25–43.

Jennings, H.J. (1997). *N*-Propionylated group B meningococcal polysaccharide glycoconjugate vaccine against group B meningococcal meningitis. Int. J. Inf. Dis. *1*, 158–164.

Jennings, H.J. (1998). Polysaccharide vaccines against disease caused by Haemophilus influenzae, group B *Streptococcus*, and *Salmonella typhi*. Carbohyd. Eur. *21*, 17–23.

Jennings, H.J., and Lugowski, C. (1981). Immunochemistry of groups A, B, and C meningococcal polysaccharide-tetanus toxoid conjugates. J. Immunol. *127*, 1011–1018.

Jennings, H.J., and Pon, R.A. (1996). Polysaccharides and glycoconjugates as human vaccines. In Polysaccharides in Medicinal Applications, Dimitriu, S., ed. (New York, Marcel Dekker), pp. 473–479.

Jennings, H.J., and Sood, R.K. (1994). Synthetic glycoconjugates as human vaccines. In Neoglycoconjugates. Preparation and Applications, Lee, Y.C., and Lee, R.T., eds. (London, Academic Press), pp. 325–371.

Jeurissen, A., Ceuppens, J.L., and Bossuyt, X. (2004). T lymphocyte dependence of the antibody response to 'T lymphocyte independent type 2' antigens. Immunology *111*, 1–7.

Jones, C. (2008). The regulatory framework for glycoconjugate vaccines. In Carbohydrate-based Vaccines, Roy, R., ed. (Washington, Oxford University Press), pp. 21–35.

Kabanova, A., Margarit, I., Berti, F., Romano, M.R., Grandi, G., Bensi, G., Chiarot, E., Proietti, D., Swennen, E., Cappelletti, E., *et al.* (2010). Evaluation of a Group A Streptococcus synthetic oligosaccharide as vaccine candidate. Vaccine *29*, 104–114.

Kabat, E.A., Liao, J., Osserman, E.F., Gamian, A., Michon, F., and Jennings, H.J. (1988). The epitope associated with the binding of the capsular polysaccharide of the group B meningococcus and of *Escherichia coli* K1 to a human monoclonal macroglobulin, IgMNOV. J. Exp. Med. *168*, 699–711.

Kasper, D.L., Paoletti, L.C., Wessels, M.R., Guttormsen, H.K., Carey, V.J., Jennings, H.J., and Baker, C.J. (1996). Immune response to type III group B streptococcal polysaccharide-tetanus toxoid conjugate vaccine. J. Clin. Invest. *98*, 2308–2314.

Knirel, Y.A. (2009). O-Specific polysaccharides of Gram-negative bacteria. In Microbial Glycobiology: Structures, Relevance, and Applications, Moran, A.P., ed. (London, Academic Press), pp. 57–73.

Krug, L.M., Ragupathi, G., Ng, K.K., Hood, C., Jennings, H.J., Guo, Z.W., Kris, M.G., Miller, V., Pizzo, B., Tyson, L., *et al.* (2004). Vaccination of small cell lung cancer patients with polysialic acid or *N*-propionylated polysialic acid conjugated to keyhole limpet hemocyanin. Clin. Cancer Res. *10*, 916–923.

Laferriere, C.A., Sood, R.K., deMuys, J.M., Michon, F., and Jennings, H.J. (1997). The synthesis of Streptococcus pneumoniae polysaccharide-tetanus toxoid conjugates and the effect of chain length on immunogenicity. Vaccine *15*, 179–186.

Larrauri, A., Cano, R., Garcia, M., and Mateo, S. (2005). Impact and effectiveness of meningococcal C conjugate vaccine following its introduction in Spain. Vaccine 23, 4097–4100.

Licciardi, P.V., Balloch, A., Russell, F.M., Nahm, M.H., Mulholland, K., and Tang, M.L.K. (2011). Pneumococcal glycoconjugate vaccines produce antibody responses that strongly correlate with function. Nat. Rev. Drug Discov. *10*, 393.

Lin, F.Y., Ho, V.A., Khiem, H.B., Trach, D.D., Bay, P.V., Thanh, T.C., Kossaczka, Z., Bryla, D.A., Shiloach, J., Robbins, J.B., *et al.* (2001). The efficacy of a Salmonella typhi Vi conjugate vaccine in two-to-five-year-old children. N. Engl. J. Med. *344*, 1263–1269.

Liu, M.A., Friedman, A., Oliff, A.I., Tai, J., Martinez, D., Deck, R.R., Shieh, J.T., Jenkins, T.D., Donnelly, J.J., and Hawe, L.A. (1992). A vaccine carrier derived from Neisseria meningitidis with mitogenic activity for lymphocytes. Proc. Natl. Acad. Sci. U.S.A. *89*, 4633–4637.

Livingston, P.O. (1995). Approaches to augmenting the immunogenicity of melanoma gangliosides: from whole melanoma cells to ganglioside–KLH conjugate vaccines. Immunol. Rev. *145*, 147–166.

Livingston, P.O., and Ragupathi, G. (1997). Carbohydrate vaccines that induce antibodies against cancer. 2. Previous experience and future plans. Cancer Immunol. Immunother. *45*, 10–19.

Lucero, M.G., Puumalainen, T., Ugpo, J.M., Williams, G., Kayhty, H., and Nohynek, H. (2004). Similar antibody concentrations in Filipino infants at age 9 months, after 1 or 3 doses of an adjuvanted, 11-valent pneumococcal diphtheria/tetanus-conjugated vaccine: a randomized controlled trial. J. Infect. Dis. *189*, 2077–2084.

McHeyzer-Williams, L.J. and McHeyzer-Williams, M.G. (2005). Antigen-specific memory B cell development. Annu. Rev. Immunol 23, 487–513.

MacLeod, C.M., Hodges, R.G., Heidelberger, M., and Bernhard, W.G. (1945). Prevention of pneumococcal pneumonia by immunization with specific capsular polysaccharides. J. Exp. Med. *82*, 445–465.

McVernon, J., Andrews, N., Slack, M. P. E., and Ramsay, M. E. Risk of vaccine failure after *Haemophilus influenzae* type b (Hib) combination vaccines with acellular pertussis. Lancet *361*, 1521–1523.

Mai, N.L., Phan, V.B., Vo, A.H., Tran, C.T., Lin, F.Y., Bryla, D.A., Chu, C., Schiloach, J., Robbins, J.B., Schneerson, R., *et al.* (2003). Persistent efficacy of Vi conjugate vaccine against typhoid fever in young children. N. Engl. J. Med. *349*, 1390–1391.

Maiden, M.C., and Stuart, J.M. (2002). Carriage of serogroup C meningococci 1 year after meningococcal C conjugate polysaccharide vaccination. Lancet *359*, 1829–1831.

Messina, A.F., Katz-Gaynor, K., Barton, T., Ahmad, N., Ghaffar, F., Rasko, D., and McCracken, G.H.J. (2007). Impact of the pneumococcal conjugate vaccine on serotype distribution and antimicrobial resistance of invasive Streptococcus pneumoniae isolates in Dallas, TX, children from 1999 through 2005. Pediatr. Infect. Dis. J. *26*, 461–467.

Moran, A.P. (2009). Molecular mimicry of host glycosylated structures by bacteria. In Microbial Glycobiology: Structures, Relevance, and Applications, Moran, A.P., ed. (London, Academic Press), pp. 847–870.

Mulard, L.A., and Phalipon, A. (2008). From epitope characterization to the design of semi-synthetic glycoconjugate vaccines against *Shigella flexneri* 2a infection. In Carbohydrate-based Vaccines, Roy, R., ed. (Washington, Oxford University Press), pp. 105–136.

Obonyo, C.O., and Lau, J. (2006). Efficacy of *Haemophilus influenzae* type b vaccination of children: a meta-analysis. Eur. J. Clin. Microbiol. Infect. Dis. *25*, 90–97.

Oosterhuis-Kafeja, F., Beutels, P., and Van Damme, P. (2007). Immunogenicity, efficacy, safety and effectiveness of pneumococcal conjugate vaccines (1998–2006). Vaccine *25*, 2194–2212.

Pace, D., and Pollard, A.J. (2007). Meningococcal A, C, Y and W-135 polysaccharide–protein conjugate vaccines. Arch. Dis. Child 92, 909–915.

Passwell, J.H., Ashkenazi, S., Harlev, E., Miron, D., Ramon, R., Farzam, N., Lerner-Geva, L., Levi, Y., Chu, C., Shiloach, J., *et al.*, and The Israel Shigella Study Group. (2003). Safety and immunogenicity of Shigella sonnei-CRM9 and Shigella flexneri type 2a–rEPA conjugate vaccines in one- to four-year-old children. Pediatr. Infect. Dis. J. *22*, 701–706.

Peeters, C.C., Evenberg, D., Hoogerhout, P., Käyhty, H., Saarinen, L., van Boeckel, C.A., van der Marel, G.A., van Boom, J.H., and Poolman, J.T. (1992). Synthetic trimer and tetramer of 3-beta-D-ribose-(1–1)-D-ribitol-5-phosphate conjugated to protein induce antibody responses to Haemophilus influenzae type b capsular polysaccharide in mice and monkeys. Infect. Immun. *60*, 1826–1833.

Perez, A.E., Dickinson, F.O., Banderas, F., Serrano, T., Llanes, R., Guzmín, D., Diaz, P., Alvarez, A., Guirola, M.A., *et al.* (2006). Safety and preliminary immunogenicity of MenC/P64k, a meningococcal serogroup C conjugate vaccine with a new recombinant carrier. FEMS Immunol. Med. Mic. *46*, 386–392.

Pichichero, M.E., and Casey, J.R. (2007). Emergence of a multiresistant serotype 19A pneumococcal strain not included in the 7-valent conjugate vaccine as an otopathogen in children. JAMA *298*, 1772–1778.

Pichichero, M.E., Latiolais, T., Bernstein, D.I., Hosbach, P., Christian, E., Vidor, E., Meschievitz, C., and Daum, R.S. (1997). Vaccine antigen interactions after a combination diphtheria-tetanus toxoid-acellular pertussis/purified capsular polysaccharide of *Haemophilus influenzae* type b-tetanus toxoid vaccine in two-, four- and six-month-old infants. Pediatr. Infect. Dis. J. *16*, 863–870.

Pon, R.A., and Jennings, H.J. (2009). Carbohydrate-based antibacterial vaccines. In Carbohydrate-based Vaccines and Immunotherapies, Guo, Z.W., and Boons, G.J., eds. (New Jersey, John Wiley and Sons), pp. 117–166.

Pon, R.A., Lussier, M., Yang, Q.L., and Jennings, H.J. (1997). *N*-Propionylated group B meningococcal polysaccharide mimics a unique bactericidal capsular epitope in group B *Neisseria meningitidis*. J. Exp. Med. *185*, 1929–1938.

Poolman, J., and Borrow, R. (2011). Hyporesponsiveness and its clinical implications after vaccination with polysaccharide or glycoconjugate vaccines. Expert Rev. Vaccines. *10*, 307–22.

Pozsgay, V., and Kubler-Kielb, J. (2008). Conjugation methods towards synthetic vaccines. In Carbohydrate-based Vaccines, Roy, R., ed. (Washington, Oxford University Press), pp. 36–70.

Pozsgay, V., Chu, C., Pannell, L., Wolfe, J., Robbins, J.B., and Schneerson, R. (1999). Protein conjugates of synthetic saccharides elicit higher levels of serum IgG lipopolysaccharide antibodies in mice than do those of the O-specific polysaccharide from *Shigella dysenteriae* type 1. Proc. Natl. Acad. Sci. U.S.A. *96*, 5194–5197.

Rennels, M., King, J.J., Ryall, R., Papa, T., and Froeschle, J. (2004). Dosage escalation, safety and immunogenicity study of four dosages of a tetravalent meninogococcal polysaccharide diphtheria toxoid conjugate vaccine in infants. Pediatr. Infect. Dis. J. 23, 429–435.

Richmond, P., Borrow, R., Goldblatt, D., Findlow, J., Martin, S., Morris, R., Cartwright, K., and Miller, E. (2001). Ability of 3 different meningococcal C conjugate vaccines to induce immunologic memory after a single dose in UK toddlers. J. Infect. Dis. *183*, 160–163.

Robbins, J.B., Chu, C., and Schneerson, R. (1992). Hypothesis for vaccine development: protective immunity to enteric diseases caused by nontyphoidal Salmonellae and Shigellae may be conferred by serum IgG antibodies to the O-Specific polysaccharide of their lipopolysaccharides. Clin. Infect. Dis. *15*, 346–361.

Robbins, J.B., Schneerson, R., Szu, S.C., and Pozsgay, V. (1999). Bacterial polysaccharide–protein conjugate vaccines. Pure Appl. Chem. *71*, 745–754.

Schiemann, O., and Casper, W. (1927). Sind die spezifisch pracipitablen Substanzen der 3 Pneumokokkentypen Haptene? Med. Microbiol. Immunol. *108*, 220–257.

Schneerson, R., Barrera, O., Sutton, A., and Robbins, J.B. (1980). Preparation, characterization, and immunogenicity of Haemophilus influenzae type b polysaccharide–protein conjugates. J. Exp. Med. *152*, 361–376.

Scott, M.G., Zachau, H.G., and Nahm, M.H. (1992). The human antibody V region repertoire to the type B capsular polysaccharide of Haemophilus influenzae. Int. Rev. Immunol. 9, 45–55.

Seppala, I., and Makela, O. (1989). Antigenicity of dextran–protein conjugates in mice. Effect of molecular weight of the carbohydrate and comparison of two modes of coupling. J. Immunol. *143*, 1259–1264.

Sieling, P.A., Chatterjee, D., Porcelli, S.A., Prigozy, T.I., Mazzaccaro, R.J., Soriano, T., Bloom, B.R., Brenner, M.B., Kronenberg, M., Brennan, P.J., *et al.* (1995). CD1-restricted T cell recognition of microbial lipoglycan antigens. Science *269*, 227–230.

Snippe, H., Jansen, W.T., and Kamerling, J.P. (2008). Immunology of experimental synthetic carbohydrate–protein conjugate vaccines against Streptococcus pneumoniae serotypes. In Carbohydrate-based Vaccines, Roy, R., ed. (Washington, Oxford University Press), pp. 85–104.

Stein, K.E. (1992). Thymus-independent and thymus-dependent responses to polysaccharide antigens. J. Infect. Dis. *165 (Suppl 1)*, S49-S52.

Takala, A.K., Eskola, J., Leinonen, M., Kayhty, H., Nissinen, A., Pekkanen, E., and Makela, P.H. (1991). Reduction of oropharyngeal carriage of Haemophilus influenzae type b (Hib) in children immunized with an Hib conjugate vaccine. J Infect. Dis. *164*, 982–986.

Tejedor, J.C., Moro, M., Ruiz-Contreras, J., Castro, J., Gomez-Campdera, J.A., Navarro, M.L., Merino, J.M., Martin-Ancel, A., Roca, J., Garcia-del-Rio, M., *et al.* (2006). Immunogenicity and reactogenicity of primary immunization with a hexavalent diphtheria-tetanus-acellular pertussis-hepatitis B-inactivated polio-Haemophilus influenzae type B vaccine coadministered with two doses of a meningococcal C-tetanus toxoid conjugate vaccine. Pediatr. Infect. Dis. J. *25*, 713–720.

Tillett, W.S., and Francis, T., Jr. (1929). Cutaneous reactions to the polysaccharides and proteins of pneumococcus in lobar pneumonia. J. Exp. Med. *50*, 687–701.

Trotter, C.L., Andrews, N.J., Kaczmarski, E.B., Miller, E., and Ramsay, M.E. (2004). Effectiveness of meningococcal serogroup C conjugate vaccine 4 years after introduction. Lancet *364*, 365–367.

Troy, F.A. (1992). Polysialylation: from bacteria to brains. Glycobiology *2*, 5–23.

Verez-Bencomo, V., Fernandez-Santana, V., Hardy, E., Toledo, M.E., Rodriguez, M.C., Heynngnezz, L., Rodriguez, A., Baly, A., Herrera, L., Izquierdo, M., *et al.* (2004). A synthetic conjugate polysaccharide vaccine against *Haemophilus influenzae* type b. Science *305*, 522–525.

Wang, S.K., Liang, P.H., Astronomo, R.D., Hsu, T.L., Hsieh, S.L., Burton, D.R., and Wong, C.H. (2008). Targeting the carbohydrates on HIV-1: Interaction of oligomannose dendrons with human monoclonal antibody 2G12 and DC-SIGN. Proc. Natl. Acad. Sci. U.S.A. *105*, 3690–3695.

Wessels, M.R., Munoz, A., and Kasper, D.L. (1987). A model of high-affinity antibody binding to type III group B Streptococcus capsular polysaccharide. Proc. Natl. Acad. Sci. U.S.A. *84*, 9170–9174.

Yuki, N., Taki, T., Inagaki, F., Kasama, T., Takahashi, M., Saito, K., Handa, S., and Miyatake, T. (1993). A bacterium lipopolysaccharide that elicits Guillain-Barré syndrome has a GM1 ganglioside-like structure. J. Exp. Med. *178*, 1771–1775.

Zou, W., and Jennings, H.J. (2009). Preparation of glycoconjugate vaccines. In Carbohydrate-based vaccines and immunotherapies, Guo Z.W., and Boons, G.J., eds. (New Jersey, John Wiley and Sons), pp. 55–88.

Index

A

B

C

D

E

P

www.ingramcontent.com/pod-product-compliance
Lightning Source LLC
LaVergne TN
LVHW062120200926
848484LV00007B/1727
* 9 7 8 1 9 0 4 4 5 5 9 5 0 *